An Introduction to Modern Differential Equations

Henry Ricardo

Medgar Evers College of the City University of New York

Houghton Mifflin Company

Boston   New York

Editor in Chief: *Jack Shira*
Sponsoring Editor: *Lauren Schultz*
Editorial Associate: *Marika Hoe*
Associate Project Editor: *Cecilia Molinari*
Editorial Assistant: *Kristin Penta*
Senior Production/Design Coordinator: *Jodi O'Rourke*
Senior Manufacturing Coordinator: *Jane Spelman*
Marketing Manager: *Ben Rivera*

Printed in the U.S.A.

Library of Congress Control Number: 00-133863

ISBN Number: 0-618-042393

123456789 – DOW – 05   04   03   02

For Catherine, my *sine qua non,*
and for Cathy, Christine, Henry and Marta,
and Tomás Agustín

CONTENTS

v

7 Systems of Nonlinear Differential Equations   **313**

PREFACE

Philosophy

For more than a decade there has been a tangible movement to reform the way we teach certain mathematical subjects. It started with calculus and has expanded to include courses earlier and later in the typical mathematics sequence. Instruction in ordinary differential equations has experienced such an evolution, in terms of both content and pedagogy. What once may have been regarded as a "collection of special 'methods,'"[1] has gradually evolved to provide more valuable experiences for the student—experiences that one prominent mathematician/author has called *conceptualization, exploration*, and *higher-level problem solving*.[2] It is this spirit that has influenced the creation of this book.

The text presents a solid yet highly accessible introduction to differential equations, developing the concepts from a dynamical systems perspective and employing technology to treat topics graphically, numerically, and analytically. It is designed to be appropriate for a wide variety of students and to serve as a natural successor to any modern calculus sequence.

In particular, the book acknowledges that most differential equations cannot be solved in closed form and makes extensive use of qualitative and numerical methods to analyze solutions. To accommodate this shift in emphasis, some traditional material has been de-emphasized or omitted. The text includes discussions of several significant mathematical models, although there is no systematic attempt to teach the art of modeling.[3] Similarly, the text introduces only the minimal amount of linear algebra necessary for an analysis of systems.

This book is intended to be the text for the one-semester ordinary differential equations course that is typically offered at the sophomore-junior level, but with some differences. The prerequisite for the course is two semesters of calculus. No prior knowledge of multivariable calculus and linear algebra is needed, because essential concepts from these topics are developed within the text itself. This book is aimed primarily at students majoring in mathematics, the natural sciences, and engineering. However, students in economics, business, and the social sciences who have the necessary background should also benefit.

Use of Technology

This text assumes that the reader has access to a computer algebra system (CAS) or perhaps some specialized software that will enable him or her to construct the required graphs (solution curves, phase portraits, etc.) and numerical

---

1. S. L. Ross, *Ordinary Differential Equations, 3rd ed.* (New York: Wiley, 1984): 25.
2. W. E. Boyce, "New Directions in Elementary Differential Equations," *College Mathematics Journal*, (November 1994): 364.
3. See D. A. Sánchez, "Review of Ordinary Differential Equations Texts," *American Mathematical Monthly* 105 (1998), second paragraph of p. 382.

approximations. For example, a spreadsheet program can be used effectively to implement Euler's method of approximating solutions. Although I use *Maple*® in my own course, no specific software or hardware platform is assumed for this book. To a large extent, even a graphing calculator will suffice.

Pedagogical Features and Writing Style

This book is truly meant to be *read* by the students. The style is accessible without excessive mathematical formality and extraneous material, although it does provide a solid foundation upon which individual teachers (with the aid of the accompanying *Instructor's Guide*) can build according to their taste and the students' needs. Every chapter has an informal *Introduction* that sets the tone and motivates the material to come. I have tried to motivate the introduction of new concepts in various ways, including references to earlier, more elementary mathematics courses taken by the student. Each chapter concludes with a narrative *Summary* reminding the reader of the important concepts in the chapter. Within sections there are figures and tables to help students visualize or summarize concepts. There are many worked-out examples and exercises taken from biology, chemistry, and economics, as well as from traditional pure mathematics, physics, and engineering. In the text itself, I lead the student through qualitative and numerical analyses of problems that would have been difficult to handle before the ubiquitous presence of graphing calculators and computers. The exercises that appear at the end of each content section range from the routine to the challenging, the latter problems often requiring some exploration and/or theoretical justification ("proof"). Some exercises introduce students to supplementary (often traditional) concepts. I have provided answers to odd-numbered problems at the back of the book, with more detailed solutions to these problems in the separate *Student Solutions Manual*. Every chapter has at least one project following the *Summary*.

I have written the book the way I teach the course, using a colloquial and interactive style. The student is frequently urged to "Think about this," "Check this," or "Make sure you understand." In general there are no proofs of theorems except for those mathematical statements that can be justified by a sequence of fairly obvious calculations/algebraic manipulations. In fact, there is no formal labeling of facts as theorems, although key results are italicized within the text or boxed off. Also, brief historical remarks related to a particular concept or result are placed throughout the text without obstructing the flow. This is not a mathematical treatise but a friendly, informative, modern introduction to tools needed by students in many disciplines. I have enjoyed teaching such a course, and I believe my students have benefited from the experience. I sincerely hope that the user of this book also gains some insight into the modern theory and applications of differential equations.

Key Content Features

Chapters 1–3 introduce *basic concepts* and focus on the analytical, graphical, and numerical aspects of first-order differential equations. In later chapters, these aspects (including the *Superposition Principle*) are generalized in natural ways to higher-order equations and systems of equations. Chapter 1 contains an informal section on the role of *technology* in the study of differential equations.

Chapter 4 starts with methods of solving important *second-order homogeneous and nonhomogeneous linear equations with constant coefficients* and discusses applications to electrical circuits and spring-mass problems. The high point of the chapter is the demonstration that any higher-order differential equation is equivalent to a *system* of first-order equations. The student is introduced to the *qualitative* analysis of systems (*phase portraits*), the *existence and uniqueness* of solutions of systems, and the extension of *numerical methods* for first-order equations to systems of first-order equations. Among the examples treated in this chapter are two forms of a *predator-prey* system (one linear, one nonlinear), an *arms race* illustration, and several *spring-mass* systems (including one showing *resonance*).

Chapter 5 begins with a brief introduction to the *matrix algebra* concepts necessary for the systematic exposition of *two-dimensional systems of autonomous linear equations* to follow. (This treatment is supplemented by Appendix B.) The importance of linearity is emphasized and the *Superposition Principle* is discussed again. The *stability* of such systems is completely characterized by means of the *eigenvalues* of the matrix of coefficients. *Spring-mass* systems are discussed in terms of their eigenvalues. There is also a brief introduction to the complexities of *nonhomogeneous* systems. Finally, via 3×3 and 4×4 examples, the student is shown how the ideas previously developed can be extended to *nth-order equations* and their equivalent systems.

Chapter 6 treats the *Laplace transform* and its applications to the solution of differential equations and systems of differential equations. This is perhaps the most traditional topic in the book; it is included because of its usefulness in many applied areas. In particular, it enables students to deal with *nonhomogeneous* linear equations and systems more easily and to handle *discontinuous driving forces*. The Laplace transform is applied to electrical circuit problems, the deflection of beams (a boundary-value problem), and spring-mass systems. However, in the spirit of the rest of the book, Section 6.6 shows the applicability of the Laplace transform to a *qualitative* analysis of linear differential equations.

Chapter 7 treats systems of *nonlinear* equations in a systematic way. The *stability* of nonlinear systems is analyzed. The important notion of a *linear approximation* to a nonlinear equation or system is developed, including the use of a qualitative result that we owe to Poincaré and Liapunov. Some important examples of nonlinear systems are treated in detail, including the *Lotka-Volterra equations*, the *undamped pendulum*, and the *van der Pol oscillator*. *Limit cycles* are discussed.

Appendices A–C present important prerequisite/corequisite material from *calculus* (*single-variable and multivariable*), *vector/matrix algebra*, and *complex numbers*, respectively. **Appendix D** supplements the text by introducing the *series solutions of ordinary differential equations*.

Supplements

- *Instructor's Guide with Solutions*   Includes solutions to all exercises in the text, chapter-by-chapter comments, *Maple* hints (and references to other software as well), additional examples and problems, and an extensive bibliography. This Guide is available for free to instructors who adopt the text.

- *Student Solutions Manual* Provides complete solutions to every odd-numbered exercise in the text.
- *SMARTHINKING*™ *Live On-line Tutoring* Houghton Mifflin has partnered with SMARTHINKING to provide an easy-to-use and effective on-line tutoring service. A dynamic **Whiteboard** and **Graphing Calculator function** enables students and e-structors to collaborate easily. SMARTHINKING offers three levels of service:

  - **Text-specific Tutoring** provides real-time, one-on-one instruction with a specially qualified e-structor.
  - **Questions Any Time** allows students to submit questions to the tutor outside the scheduled hours and receive a reply within 24 hours.
  - **Independent Study Resources** connect students with around-the-clock access to additional educational services, including interactive web sites, diagnostic tests and Frequently Asked Questions posed to SMARTHINKING e-structors.

- *A text-specific web site* Contains links to ordinary differential equations web sites, as well as some Maple labs, and other useful material. Visit **http://math.college.hmco.com** and follow the links to this textbook.

Acknowledgments

The approach and content of this book have been influenced primarily by three sources: (1) The Boston University Differential Equations Project; (2) The Consortium for Ordinary Differential Equations Experiments (**C•ODE•E);** and (3) The **Special Issue on Differential Equations**: *College Mathematics Journal,* Vol. 25, No. 5, (November 1994). I have also been heartened by David Sánchez's valuable review of recent texts in ordinary differential equations[4] and inspired by the recent MAA *Notes* volume, *Revolutions in Differential Equations: Exploring ODEs with Modern Technology*, which I had the privilege to review before publication.

I have found that it does, indeed, take a village to write a mathematics text. I have enjoyed the cooperation and candor of several classes of Medgar Evers College students who learned from this text while it was still under development. I single out Tamara Battle, Hibourahima Camara, Lenston Elliott, and Ayanna Moses as representatives of these patient students. I gratefully acknowledge the helpful comments of my colleague Tatyana Flesher on an early version of this text. I thank my chairperson, Darius Movasseghi, for his encouragement and for his support in such crucial areas as course scheduling and ensuring the availability of technology. I am grateful to my colleague Mahendra Kawatra for his continuing encouragement and support.

At Houghton Mifflin, I wish to thank Jack Shira, who was the first to express confidence in the philosophy and style of this book and has continued to support

---

4. Sánchez, loc. cit.

the project in many ways. I appreciate Paul Murphy's contributions while he was my editor. I am grateful for the professionalism, patience, and sense of humor shown by Marika Hoe and Cecilia Molinari in guiding me through the stages of writing, editing, and book production. I appreciate the successful efforts of Beverly Fusfield of Techsetters, Inc. and art editor George McLean in turning my many (often complex) figures into professional works of art. William Hoston did a splendid job as my accuracy reviewer. I benefited greatly from the comments and suggestions of my reviewers: Bill Goldbloom Bloch, Wheaton College; Beth Bradley, University of Louisville; Robert Bradshaw, Ohlone College; Martin Brown, Jefferson Community College; Dean Burbank, Gulf Coast Community College; Thomas W. Cairns, University of Tulsa; Benito Chen-Charpentier, University of Wyoming; Mark Farris, Midwestern State University; John H. Jaroma, Gettysburg College; Matthias Kawski, Arizona State University; Kevin Kreider, University of Akron; P. Gavin LaRose, Nebraska Wesleyan University; Michael A. McDonald, Occidental College; Douglas B. Meade, University of South Carolina; Roger Pinkham, Stevens Institute; Lila F. Roberts, Georgia Southern University; Bhagat Singh, University of Wisconsin-Manitowoc; Ann Sitomer, Portland Community College; Allan Struthers, Michigan Technological University; Ted J. Suffridge, University of Kentucky; Hossein T. Tehrani, University of Nevada; Luis Valdez-Sanchez, University of Texas at El Paso; and David Voss, Western Illinois University. I welcome any questions, additional comments, and suggestions for improvement. Please contact me via email at **henry@mec.cuny.edu**.

Above all, I am grateful to my wife, Catherine, for her love, steadfast encouragement, support, and patience—during the writing of this book and at all other times. I also thank her for her active assistance in proofreading and critiquing the manuscript throughout all its stages.

Henry Ricardo

1 | Introduction to Differential Equations

1.0 INTRODUCTION

What do the following situations have in common?

- An arms race between nations
- Tracking of the rate at which HIV-positive patients come to exhibit AIDS
- The dynamics of supply and demand in an economy
- The interaction between two or more species of animals on an island

The answer is that each of these areas of investigation can be modeled with differential equations. This means that the essential features of these problems can be represented using one or several differential equations, and the solutions of the mathematical problems provide insights into the future behavior of the systems being studied.

This book deals with *change*, with *flux*, with *flow*, and, in particular, with the *rate* at which change takes place. Every living thing changes. The tides fluctuate over the course of a day. Countries increase and diminish their stockpiles of weapons. The price of oil rises and falls. The proper framework of this course is **dynamics**—the study of systems that evolve over time.

The origin of dynamics (originally an area of physics) and of differential equations lies in the earliest work by the English scientist and mathematician Sir Isaac Newton (1642–1727) and the German philosopher and mathematician Gottfried Wilhelm Leibniz (1646–1716) in developing the new science of calculus in the seventeenth century. Newton in particular was concerned with determining the laws governing motion, whether of an apple falling from a tree or of the planets moving in their orbits. He was concerned with *rates of change*. However, you mustn't think that the subject of differential equations is all about physics. The same types of equations and the same kind of analysis of dynamical systems can be used to model and understand situations in biology, economics, military strategy, and chemistry, for example. Applications of this sort will be found throughout this book.

1

In the next section, we will introduce the language of differential equations and discuss some applications.

1.1   BASIC TERMINOLOGY

ORDINARY AND PARTIAL DIFFERENTIAL EQUATIONS

Ordinary Differential Equations

In general, an **ordinary differential equation** (**ODE**) is an equation that involves an unknown function of a single variable, its independent variable, and one or more of its derivatives.

EXAMPLE 1.1.1   An Ordinary Differential Equation

Here's a typical elementary ODE, with some of its components indicated:

$$\text{unknown function, } y \downarrow$$
$$3\frac{dy}{dt} = y$$
$$\text{independent variable, } t \uparrow$$

This equation describes an unknown function of t that is equal to three times its own derivative. Expressed another way, the differential equation describes a function whose rate of change is proportional to its size (value) at any given time, with constant of proportionality one-third.   ◆

In many dynamical applications, the independent variable is time, denoted by t, and we may denote the function's derivative using Newton's dot notation,[1] as in the equation $\ddot{x} + 3t\dot{x} + 2x = \sin(\omega t)$. You should be able to recognize a differential equation no matter what letters are used for the independent and dependent variables and no matter what derivative notation is employed. The context will determine what the various letters mean, and it's the *form* of the equation that should be recognized. For example, you should be able to see that the two ordinary differential equations

$$(A)\ \frac{d^2u}{dt^2} - 3\frac{du}{dt} + 7u = 0 \quad \text{and} \quad (B)\ \frac{d^2y}{dx^2} = 3\frac{dy}{dx} - 7y$$

are the same—that is, they are describing the same mathematical or physical behavior. In equation (A) the unknown function u depends on t, whereas in equation (B) the function y is a function of the independent variable x, but both equations describe the same relationship that involves the unknown function, its derivatives, and the independent variable. Each equation is describing a function whose second derivative equals three times its first derivative minus seven times itself.

---

1. In this notation, $\dot{x} = dx/dt$ and $\ddot{x} = d^2x/dt^2$.

The Leibniz notation for a derivative, $\dfrac{d(\ )}{d(\ )}$, is helpful because the independent variable (the fundamental quantity whose change is causing other changes) appears in the denominator, the dependent variable in the numerator. The three equations

$$\frac{dy}{dx} + 2xy = e^{-x^2}$$

$$x''(t) - 5x'(t) + 6x(t) = 0$$

$$\frac{dx}{dt} = \frac{3t^2 + 4t + 2}{2(x-1)}$$

leave no doubt about the relationship between independent and dependent variables. But in an equation such as $(w')^2 + 2t^3w' - 4t^2w = 0$, we must *infer* that the unknown function w is really $w(t)$, a function of the independent variable t.

The Order of an Ordinary Differential Equation

One way to classify differential equations is by their **order.** We say that an ordinary differential equation is of **order n,** or is an **nth-order equation,** if the highest derivative of the unknown function in the equation is the nth derivative. The equations

$$\frac{dy}{dx} + 2xy = e^{-x^2}$$

$$(w')^2 + 2t^3w' - 4t^2w = 0$$

$$\frac{dx}{dt} = \frac{3t^2 + 4t + 2}{2(x-1)}$$

are all *first-order* differential equations because the highest derivative in each equation is the first derivative. The equations

$$x''(t) - 5x'(t) + 6x(t) = 0$$

and

$$\ddot{x} + 3t\,\dot{x} + 2x = \sin(\omega t)$$

are second-order equations, and $e^{-x}y^{(5)} + (\sin x)y''' = 5e^x$ is of order 5.

A General Form for an Ordinary Differential Equation

If y is the unknown function with a single independent variable x, we can express an nth-order differential equation in a concise mathematical form as the relation

$$F(x, y, y', y'', y''', \ldots, y^{(n-1)}, y^{(n)}) = 0$$

or often as

$$y^{(n)} = G(x, y, y', y'', y''', \ldots, y^{(n-1)})$$

where $y^{(k)}$ denotes the kth derivative of y.

The next example shows what this form looks like in practice.

EXAMPLE 1.1.2   General Form for a Second-Order ODE

If y is an unknown function of x, then the second-order ordinary differential equation $2\dfrac{d^2y}{dx^2} + e^x\dfrac{dy}{dx} = y + \sin x$ can be written as

$$2\frac{d^2y}{dx^2} + e^x\frac{dy}{dx} - y - \sin x = 0$$

or as

$$\underbrace{2y'' + e^xy' - y - \sin x}_{F(x,\,y,\,y',\,y'')} = 0$$

Note that F denotes a mathematical expression involving the independent variable x, the unknown function y, and the first and second derivatives of y.

Alternatively, we could use ordinary algebra to solve the original differential equation for its highest derivative and write the equation as

$$y'' = \underbrace{\tfrac{1}{2}\sin x + \tfrac{1}{2}y - \tfrac{1}{2}e^xy'}_{G(x,\,y,\,y')}$$

◆

Partial Differential Equations

If we are dealing with functions of *several* variables and the derivatives involved are *partial* derivatives, then we have a **partial differential equation (PDE)**. (See Appendix A.7 if you are not familiar with partial derivatives.) For example, the partial differential equation $\dfrac{\partial^2 u}{\partial x^2} - \dfrac{1}{c^2}\dfrac{\partial^2 u}{\partial t^2} = 0$, which is called the *wave equation*, is of fundamental importance in many areas of physics and engineering. In this equation we are assuming that $u = u(x, t)$, a function of the two variables x and t. However, in this text, when we use the term *differential equation*, we'll mean an *ordinary* differential equation. Often we'll just write *equation*, if the context makes it clear that an ordinary differential equation is intended.

Linear and Nonlinear Ordinary Differential Equations

Another important way to categorize differential equations is in terms of whether they are linear or nonlinear. If y is a function of x, then the general form of a **linear ordinary differential equation** of order n is

$$a_n(x)y^{(n)} + a_{n-1}(x)y^{(n-1)} + \cdots + a_2(x)y'' + a_1(x)y' + a_0(x)y = f(x) \qquad (1.1.1)$$

where $a_n(x), a_{n-1}(x), \ldots, a_1(x), a_0(x)$, and $f(x)$ are functions of x. What is important here is that each coefficient function $a_i(x)$ depends on the independent variable x alone and doesn't have the dependent variable y or any of its derivatives in it. In particular, equation (1.1.1) involves no products or quotients of y and/or its derivatives.

EXAMPLE 1.1.3 **A Second-Order Linear Equation**

The equation $x'' + 3tx' + 2x = \sin(\omega t)$, where ω is a constant, is linear. We can see the form of this equation as follows:

$$\overbrace{1}^{a_2(t)} \cdot x'' + \overbrace{3t}^{a_1(t)} \cdot x' + \overbrace{2}^{a_0(t)} \cdot x = \overbrace{\sin(\omega t)}^{f(t)}$$

The coefficients of the various derivatives of the unknown function x are functions (sometimes constant) of the independent variable t alone. ◆

The next example shows that not all first-order equations are linear.

EXAMPLE 1.1.4 **A First-Order Nonlinear Equation**
(an HIV Infection Model)

The equation $\dfrac{dT}{dt} = s + rT\left(1 - \dfrac{T}{T_{\max}}\right) - \mu T$ models the growth and death of T cells, an important component of the immune system.[2] Here $T(t)$ is the number of T cells present at time t. If we rewrite the equation by removing parentheses, we get $\dfrac{dT}{dt} = s + rT - \left(\dfrac{r}{T_{\max}}\right)T^2 - \mu T$, and we see that there is a term involving the square of the unknown function. Therefore, the equation is not linear. ◆

In general, there are more systematic ways to analyze linear equations than to analyze nonlinear equations, and we'll see some of these methods in Chapters 2, 5, and 6. However, nonlinear equations are important and appear throughout this book. In particular, Chapter 7 is devoted to their analysis.

SYSTEMS OF ORDINARY DIFFERENTIAL EQUATIONS

In earlier mathematics courses, you have seen that sometimes you have to deal with *systems* of algebraic equations, such as

$$3x - 4y = -2$$
$$-5x + 2y = 7$$

Similarly, in working with differential equations you may find yourself confronting **systems of differential equations,** such as

$$\frac{dx}{dt} = -3x + y$$
$$\frac{dy}{dt} = x - 3y$$

___

2. E. K. Yeargers, R. W. Shonkwiler, and J. V. Herod, *An Introduction to the Mathematics of Biology: With Computer Algebra Models* (Boston: Birkhäuser, 1996): 341.

or

$$\dot{x} = -sx + sy$$
$$\dot{y} = -xz + rx - y$$
$$\dot{z} = xy - bz$$

where b, r, and s are constants. (Recall that $\dot{x} = \dfrac{dx}{dt}$, $\dot{y} = \dfrac{dy}{dt}$, and $\dot{z} = \dfrac{dz}{dt}$.) The last system arose in a famous study of meteorological conditions.

Note that each of these systems of differential equations has a different number of equations and that each equation in the first system is *linear,* whereas the last two equations in the second system are *nonlinear* because they contain products—xz in the second equation and xy in the third—of some of the unknown functions. Naturally, we'll call a system in which all equations are linear a **linear system,** and we'll refer to a system with at least one nonlinear equation as a **nonlinear system.** In Chapters 4, 5, 6, and 7, we'll see how systems of differential equations arise and learn how to analyze them. For now, just try to understand the *idea* of a system of differential equations.

EXERCISES 1.1

In Exercises 1–10, (a) identify the independent variable *and the* dependent variable *of each equation; (b) give the* order *of each differential equation; and (c) state whether the equation is* linear *or* nonlinear. *If your answer to (c) is nonlinear, explain why this is true.*

1. $y' = y - x^2$

2. $xy' = 2y$

3. $x'' + 5x = e^{-x}$

4. $(y')^2 + x = 3y$

5. $xy'(xy' + y) = 2y^2$

6. $\dfrac{d^2r}{dt^2} = 3\dfrac{dr}{dt} + \sin t$

7. $y^{(4)} + xy''' + e^x = 0$

8. $x^{(7)} + t^2x^{(5)} = xe^t$

9. $e^{y'} + 3xy = 0$

10. $t^2R''' - 4tR'' + R' + 3R = e^t$

11. For what value(s) of the constant a is the differential equation

$$\frac{d^2x}{dt^2} + (a^2 - a)x\frac{dx}{dt} = te^{(a-1)x}$$

a linear equation?

12. Classify each of the following systems as linear or nonlinear:

a. $\dfrac{dy}{dt} = x - 4xy$

$\dfrac{dx}{dt} = -3x + y$

b. $Q' = tQ - 3t^2R$

$R' = 3Q + 5R$

c. $\dot{x} = x - xy + z$
$\dot{y} = -2x + y - yz$
$\dot{z} = 3x - y + z$

d. $\dot{x} = 2x - ty + t^2z$
$\dot{y} = -2tx + y - z$
$\dot{z} = 3x - t^3y + z$

1.2   SOLUTIONS OF DIFFERENTIAL EQUATIONS

BASIC NOTIONS

In past mathematics courses, whenever you've encountered an equation you were probably asked to *solve* it, or find a *solution*. Simply put, a **solution** of a differential equation is a function that *satisfies* the equation: When you substitute this function into the differential equation, you get a true mathematical statement— an *identity*. Even before we begin learning formal solution methods in Chapter 2, we can *guess* the solutions of some simple differential equations. The next example shows how to guess intelligently.

EXAMPLE 1.2.1   Guessing and Verifying a Solution to an ODE

The first-order linear differential equation $\dfrac{dB}{dt} = kB$, where k is a given positive constant, is a simple model of a bank balance, $B(t)$, under compound interest t years after the initial investment. The rate of change of B at any instant is proportional to the *size* of B at that instant, with k as the constant of proportionality. This equation expresses the fact that the larger the bank balance at any time t, the faster it will grow.

You can guess what kind of function describes $B(t)$ if you think about the elementary functions you know and their derivatives. What kind of function has a derivative that is a constant multiple of itself? You should be able to see why $B(t)$ must be an *exponential* function of the form ae^{kt}, where a is any constant. By substituting $B(t) = ae^{kt}$ into the original differential equation, you can verify that you have guessed correctly. The left-hand side of the equation becomes $\dfrac{d(ae^{kt})}{dt}$, which equals kae^{kt}, and the right-hand side of the equation is $k(ae^{kt})$. The left-hand side equals the right-hand side, giving us an identity.

Anticipating an idea we'll discuss later in this section, we can let $t = 0$ in our solution function to conclude that $B(0) = ae^{k(0)} = a$—that is, the constant a must equal the initial investment. Finally, we can express the solution as $B(t) = B(0)e^{kt}$. ◆

More formally, a **solution** of the differential equation

$$F(x, y, y', y'', y''', \ldots, y^{(n-1)}, y^{(n)}) = 0, \text{ or } y^{(n)} = G(x, y, y', y'', y''', \ldots, y^{(n-1)})$$

on an interval (a, b) is a real-valued function $y = y(x)$ such that all the necessary derivatives of $y(x)$ exist on the interval and $y(x)$ satisfies the equation for every value of x in the interval. **Solving** a differential equation means finding all possible solutions of a given equation.

Note that we say "a" solution rather than "the" solution. A differential equation, if it has a solution at all, usually has more than one solution. Also, we should pay attention to the interval on which the solution may be defined. Later in this section and in Section 2.7, we will discuss in more detail the question of the existence and uniqueness of solutions. For now, let's just learn to recognize when a function is a solution of a differential equation, as in the next example.

EXAMPLE 1.2.2   **Verifying a Solution of a Second-Order Equation**
Suppose that someone claims that $x(t) = 5e^{3t} - 7e^{2t}$ is a solution of the second-order linear equation $x'' - 5x' + 6x = 0$ on the whole real line—that is, for all values of t in the interval $(-\infty, \infty)$. You can prove that this claim is correct by calculating $x'(t) = 15e^{3t} - 14e^{2t}$ and $x''(t) = 45e^{3t} - 28e^{2t}$ and then substituting these expressions into the original equation:

$$x''(t) - 5x'(t) + 6x(t)$$

$$= \overbrace{(45e^{3t} - 28e^{2t})}^{x''(t)} - 5\overbrace{(15e^{3t} - 14e^{2t})}^{x'(t)} + 6\overbrace{(5e^{3t} - 7e^{2t})}^{x(t)}$$
$$= 45e^{3t} - 28e^{2t} - 75e^{3t} + 70e^{2t} + 30e^{3t} - 42e^{2t}$$
$$= -30e^{3t} + 42e^{2t} + 30e^{3t} - 42e^{2t} = 0$$

Because $x(t) = 5e^{3t} - 7e^{2t}$ satisfies the original equation, we see that $x(t)$ is a solution. But this is not the only solution of the given differential equation. For example, $x_2(t) = -\pi e^{3t} + \dfrac{2}{3}e^{2t}$ is also a solution. (*Check this.*) We'll discuss this kind of situation in more detail a little later.   ◆

Implicit Solutions

Think back to the concept of *implicit functions* in calculus. The idea here is that sometimes functions are not defined cleanly (explicitly) by a formula in which the dependent variable (on one side) is expressed in terms of the independent variable and some constants (on the other side), as in the solution $x = x(t) = 5e^{3t} - 7e^{2t}$ of Example 1.2.2. For instance, you may be given the *relation* $x^2 + y^2 = 5$, which can be written in the form $G(x, y) = 0$, where $G(x, y) = x^2 + y^2 - 5$. The graph of this relation is a circle of radius $\sqrt{5}$ centered at the origin, and this graph does not represent a function. (*Why?*) However, this relation *does* define two functions *implicitly*: $y_1(x) = \sqrt{5 - x^2}$ and $y_2(x) = -\sqrt{5 - x^2}$, both having domains $[-5, 5]$. More advanced courses in analysis discuss when a relation actually defines one or more implicit functions. For now, just remember that even if you can't untangle a relation to get an explicit formula for a function, you can use implicit differentiation to find derivatives of any functions that may be buried in the relation.

In trying to solve differential equations, often we can't find an explicit solution and must be content with a solution defined implicitly.

EXAMPLE 1.2.3   **Verifying an Implicit Solution**
We want to show that any function y that satisfies the relation $G(x, y) = x^2 + y^2 - 5 = 0$ is a solution of the differential equation $\dfrac{dy}{dx} = -\dfrac{x}{y}$.

First we differentiate the relation implicitly, treating y as $y(x)$, an implicitly defined function of the independent variable x:

$$(1) \quad \frac{d}{dx}G(x, y) = \frac{d}{dx}(x^2 + y^2 - 5) = \frac{d}{dx}(0) = 0$$

$$(2) \quad 2x + 2y\overbrace{\frac{dy}{dx}}^{\text{Chain Rule}} - \frac{d}{dx}(5) = 0$$

$$(3) \quad 2x + 2y\frac{dy}{dx} = 0$$

Now we solve equation (3) for $\frac{dy}{dx}$, getting $\frac{dy}{dx} = \frac{-2x}{2y} = -\frac{x}{y}$ and proving that any function defined implicitly by the relation above is a solution of our differential equation. ◆

FAMILIES OF SOLUTIONS I

Next we want to discuss how many solutions a differential equation could have. For example, the equation $(y')^2 + 1 = 0$ has *no* real-valued solution (*think about this*), whereas the equation $|y'| + |y| = 0$ has exactly one solution, the function $y \equiv 0$. (*Why?*) As we saw in Example 1.2.2, the differential equation $x'' - 5x' + 6x = 0$ has at least two solutions.

The situation gets more complicated, as the next example shows.

EXAMPLE 1.2.4   An Infinite Family of Solutions
Suppose two students, Lenston and Jennifer, look at the simple first-order differential equation $\frac{dy}{dx} = f(x) = x^2 - 2x + 7$. A solution of this equation is a function of x whose first derivative equals $x^2 - 2x + 7$. Lenston thinks the solution is $\frac{x^3}{3} - x^2 + 7x$, and Jennifer thinks the solution is $\frac{x^3}{3} - x^2 + 7x - 10$. Both answers seem to be correct.

Solving this problem is simply a matter of integrating both sides of the equation:

$$y = \int dy = \int \frac{dy}{dx}\,dx = \int x^2 - 2x + 7\,dx$$

Because we are using an *indefinite* integral, there is always a constant of integration that we mustn't forget. The solution to our problem is actually an *infinite family of solutions*, $y(x) = \frac{x^3}{3} - x^2 + 7x + C$, where C is any real constant. Every concrete value of C gives us another member of the family. We have just solved our first differential equation in this course without guessing! Every time we performed an indefinite integration (found an antiderivative) in calculus class, we were solving a simple differential equation.   ◆

When describing the set of solutions of a first-order differential equation such as the one in the previous example, we usually refer to it as a **one-parameter family of solutions.** The *parameter* is the constant C. Each definite value of C gives us what is called a **particular solution** of the differential equation. In the last example, Lenston and Jennifer produced particular solutions, one with $C = 0$ and the other with $C = -10$. A particular solution is sometimes called an **integral** of the equation, and its graph is called an **integral curve** or a **solution curve.** Figure 1.1 shows three of the integral curves of the equation $\dfrac{dy}{dx} = x^2 - 2x + 7$, where $C = 15, 0$, and -10 (from top to bottom).

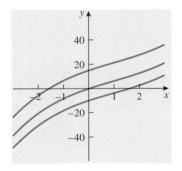

Figure 1.1

Integral curves of $\dfrac{dy}{dx} = x^2 - 2x + 7$ with
parameters $15, 0$, and -10

The curve passing through the origin is Lenston's particular solution; the solution curve passing through the point $(0, -10)$ is Jennifer's.

Initial-Value Problems (IVPs)

Now suppose that we want to solve a first-order differential equation for y, a function of the independent variable t, and we specify that one of its integral curves must pass through a particular point (t_0, y_0) in the plane. We are imposing the condition $y(t_0) = y_0$, which is called an **initial condition,** and the problem is then called an **initial-value problem** (**IVP**). Note that we are trying to pin down a particular solution this way. We find this solution by choosing a specific value of the constant of integration (the parameter).

Next we shall see how to solve a simple initial-value problem.

EXAMPLE 1.2.5 A First-Order Initial-Value Problem

Suppose that an object is moving along the x-axis in such a way that its instantaneous velocity at time t is given by $v(t) = 12 - t^2$. First we will find the position x of the object measured from the origin at any time $t > 0$.

Because the velocity function is the derivative of the position function, we can set up the first-order differential equation $\dfrac{dx}{dt} = 12 - t^2$ to describe our problem.

Simple integration of both sides yields

$$x(t) = \int dx = \int \frac{dx}{dt}\, dt = \int 12 - t^2\, dt = 12t - \frac{t^3}{3} + C$$

This last result tells us that the position of the object at an arbitrary time $t > 0$ can be described by any member of the one-parameter family $12t - \dfrac{t^3}{3} + C$, which is not a very satisfactory conclusion. But if we have some additional information, we can find a concrete value for C and end the uncertainty.

Suppose we know, for example, that the object is located at $x = -5$ when $t = 1$. Then we can use this *initial condition* to get

$$-5 = x(1) = 12(1) - \frac{1^3}{3} + C, \text{ or } -5 = \frac{35}{3} + C$$

This last equation implies that $C = \dfrac{-50}{3}$, so the position of the object at time t is given by the particular function $x(t) = 12t - \dfrac{t^3}{3} - \dfrac{50}{3} = 12t - \dfrac{(t^3 + 50)}{3}$.

We selected the initial condition $x(1) = -5$ randomly. Any other choice $x(t_0) = x_0$ would have led to a definite value for C and a particular solution of our problem. ◆

An Integral Form of an IVP Solution

If a first-order equation can be written in the form $y' = f(x)$—that is, if the right-hand side is a continuous (or piecewise continuous) function of the independent variable alone—then we can always express the solution to the IVP $y' = f(x), y(x_0) = y_0$ on an interval (a, b) as

$$y(x) = \int_{x_0}^{x} f(t)\, dt + y_0 \tag{1.2.1}$$

for x in (a, b). Note that we use the x value of the initial condition as the lower limit of integration and the y value of the initial condition as a particular constant of integration. We use t as a *dummy variable*. Given equation (1.2.1), the *Fundamental Theorem of Calculus (FTC)* (Appendix A.4) implies that $y' = f(x)$, and we see that $y(x_0) = \displaystyle\int_{x_0}^{x_0} f(t)\,dt + y_0 = 0 + y_0 = y_0$, which is what we want. This way of handling certain types of IVPs is common in physics and engineering texts. In Example 1.2.4, the solution of the equation with $y(-1) = 2$, for example, is

$$y(x) = \int_{-1}^{x} t^2 - 2t + 7\,dt + 2$$

$$= \left(\frac{t^3}{3} - t^2 + 7t\right)\Big|_{t=x} - \left(\frac{t^3}{3} - t^2 + 7t\right)\Big|_{t=-1} + 2$$

$$= \left(\frac{x^3}{3} - x^2 + 7x\right) - \left(\frac{-25}{3}\right) + 2 = \frac{x^3}{3} - x^2 + 7x + \frac{31}{3}$$

You should also solve this problem the way we did in Example 1.2.5—that is, without using a definite integral formula.

FAMILIES OF SOLUTIONS II

Although we have seen examples of first-order equations that have no solution or only one solution, in general we should expect a first-order differential equation to have an infinite set of solutions, described by a single parameter.

Extending our previous discussion, we state that an *n*th-order differential equation may have an **n-parameter family of solutions,** involving n arbitrary constants $C_1, C_2, C_3, \ldots, C_n$ (the parameters). For example, a solution of a second-order equation $y'' = g(t, y, y')$ may have *two* arbitrary constants. By prescribing the **initial conditions** $y(t_0) = y_0$ *and* $y'(t_0) = y_1$, we can determine specific values for these two constants and obtain a *particular* solution. Note that we use the same value, t_0, of the independent variable for each condition.

The next example shows how to deal with a second-order IVP.

EXAMPLE 1.2.6   A Second-Order IVP

We will show in Section 4.1 that any solution of the second-order linear equation $y'' + y = 0$ has the form $y(t) = A\cos t + B\sin t$ for arbitrary constants A and B. (You should verify that any function having the form indicated in the last sentence is a solution of the differential equation.) If a solution of this equation represents the *position* of a moving object relative to some fixed location, then the derivative of the solution represents the *velocity* of the particle at time t. If we specify, for example, the initial conditions $y(0) = 1$ and $y'(0) = 0$, we are saying that we want the position of the particle when we begin our study to be 1 unit in a positive direction from the fixed location and we want the velocity to be 0. In other words, our particle starts out at rest 1 unit (in a positive direction) from the fixed location.

We can use these initial conditions to find a particular solution of the original differential equation:

1. $y(0) = 1$ implies that $1 = y(0) = A\cos(0) + B\sin(0) = A$.
2. $y'(0) = 0$ implies that $0 = y'(0) = -A\sin(0) + B\cos(0) = B$.

Combining the results of (1) and (2), we find the particular solution $y(t) = \cos t$.   ◆

In general, if we're looking for the particular solution of the nth-degree equation $F(t, y, y', y'', y''', \ldots, y^{(n-1)}, y^{(n)}) = 0$ such that $y(t_0) = y_0, y'(t_0) = y_1, y''(t_0) = y_2, \ldots,$ and $y^{(n-1)}(t_0) = y_{n-1}$, where $y_0, y_1, y_2, \ldots, y_{n-1}$ are arbitrary real constants, we say that we are trying to solve an initial-value problem (IVP). (Later we will consider IVPs for *systems* of differential equations.) Right now we can't be sure if and when we can solve such a problem, but we will discuss the question of the *existence* and *uniqueness* of solutions in Chapters 2 and 4.

Boundary-Value Problems

For second- and higher-order differential equations, we can also determine a particular solution by specifying what are called **boundary conditions.** The idea here is to give conditions that must be satisfied by the solution function and/or its derivatives at *two different points* of the domain of the solution. The points chosen depend on the nature of the problem we are trying to solve and on the data we are given about the problem. For example, if you are analyzing the stresses on a steel girder of length L whose ends are imbedded in concrete, you may want to find $y(x)$, the bend or "give" at a point x units from one end if a load is placed somewhere on the beam (Figure 1.2). Note that the domain of y is $[0, L]$. In this problem it is natural to specify $y(0) = 0$ and $y(L) = 0$, reasonable values at the endpoints, or *boundaries,* of the solution interval. Graphically, we are requiring the solution y to pass through the points $(0, 0)$ and $(L, 0)$. (See Problem 25 in Exercises 1.2 for an applied problem of this type.)

If we try to find a particular solution of an equation (or system) that has boundary conditions, we say that we're trying to solve a **boundary-value problem** (**BVP**). The next example shows that, just as in the case of an initial-value problem, without further analysis we can't be sure whether there are solutions of a particular BVP or whether any solution we find is unique. In general, BVPs are harder to solve than IVPs. Although BVPs will appear in this book from time to time, we'll focus most of our attention on initial-value problems.

As the next example shows, some boundary-value problems have no solution, others have one solution, and some have (infinitely) many solutions.

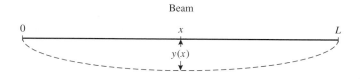

Figure 1.2
A solution $y(x)$ satisfying the boundary conditions $y(0) = 0$ and $y(L) = 0$

EXAMPLE 1.2.7   A BVP Can Have Many, One, or No Solutions

We'll use the second-order differential equation from Example 1.2.6, $y'' + y = 0$, which has the two-parameter family of solutions $y(t) = c_1 \cos t + c_2 \sin t$.

Now let's see what happens if we impose the boundary conditions $y(0) = 1$, $y(\pi) = 1$. The first condition implies that $1 = y(0) = c_1 \cos(0) + c_2 \sin(0) = c_1$, and the second condition tells us that $1 = y(\pi) = c_1 \cos(\pi) + c_2 \sin(\pi) = -c_1$. Because we can't have c_1 equaling 1 and -1 at the same time, this contradiction says that the boundary-value problem has *no solution*.

On the other hand, the boundary conditions $y(0) = 1$, $y(2\pi) = 1$ lead to a different conclusion. If we use the first condition, we get $1 = y(0) = c_1 \cos(0) + c_2 \sin(0) = c_1$. The second condition yields the result $1 = y(2\pi) = c_1 \cos(2\pi) + c_2 \sin(2\pi) = c_1$. The fact that we can't pin down the value of c_2 tells us that *any* value is all right. In other words, the BVP has *infinitely many solutions* of the form $y(t) = \cos t + c_2 \sin t$.

Finally, if we demand that $y(0) = 1$ and $y(\pi/4) = 1$, we find that $1 = y(0) = c_1 \cos(0) + c_2 \sin(0) = c_1$ and

$$1 = y\left(\frac{\pi}{4}\right) = c_1 \cos(\pi/4) + c_2 \sin\left(\frac{\pi}{4}\right) = c_1\left(\frac{\sqrt{2}}{2}\right) + c_2\left(\frac{\sqrt{2}}{2}\right)$$
$$= \frac{\sqrt{2}}{2} + c_2\left(\frac{\sqrt{2}}{2}\right)$$

which implies that $c_2 = \sqrt{2} - 1$. Therefore, this BVP has the *unique solution* $y(t) = \cos t + (\sqrt{2} - 1)\sin t$.   ◆

You should realize that for a general nth-order equation (or for a system of equations), there are many possible ways to specify boundary conditions, not always at the endpoints of solution intervals. The idea is to have a number of conditions that will enable us to solve for (specify) the appropriate number of arbitrary constants.

The following example shows how the use of boundary conditions can provide the solution of an interesting problem.

EXAMPLE 1.2.8   A Practical BVP

In the "Automobiles" section of the Sunday *New York Times* (August 10, 1997), we are told that a 1998 Olds Intrigue GL will go from 0 to 60 mph in 8 seconds. Assuming constant acceleration, we ask how far the car travels before it reaches 60 mph.

If $s(t)$ denotes the position of the car after t seconds, then we must calculate $s(8) - s(0)$, the total distance covered by the car in the 8-second interval. We know that the acceleration can be described as $a(t) = \dfrac{d^2 s}{dt^2}$, which in this problem equals some constant C, and we know that $s(0) = s'(0) = 0$—that is, our initial position is considered 0, and the velocity when we first put our foot on the gas pedal is also 0. The last bit of information we have is that $s'(8)$, the velocity at the end of 8 seconds, is 60 mph. Thus we have a second-order differential equation $\frac{d^2s}{dt^2} = C$ and some boundary conditions, and we must solve for the unknown function $s(t)$.

Now the basic rules of integral calculus tell us that when we find the anti-derivative of each side of the differential equation in the last paragraph, we get

$$\int \frac{d^2s}{dt^2} dt = \int C\,dt = Ct + C_1$$

where C_1 is a constant of integration. But $\int \frac{d^2s}{dt^2} dt = \frac{ds}{dt}$, so $\frac{ds}{dt} = Ct + C_1$. Integrating each side of this last equation gives us

$$s(t) = \frac{Ct^2}{2} + C_1 t + C_2$$

Thus we have an expression for $s(t)$, but it contains three arbitrary constants. Now we can use the initial condition $s(0) = 0$ to write

$$0 = s(0) = \frac{C(0)^2}{2} + C_1(0) + C_2$$

which boils down to $0 = C_2$, so we can say

$$s(t) = \frac{Ct^2}{2} + C_1 t$$

Because $s'(0) = 0$, we can see that $0 = s'(0) = (Ct + C_1)|_{t=0} = C_1$, and thus $s(t) = \frac{Ct^2}{2}$.

We still have one unknown constant, C, in our formula, but we know that at the end of 8 seconds, the velocity is 60 miles per hour. *We have to be careful of our units here.* We don't want to mix seconds and hours. To make all our units consistent, we'll convert 8 seconds to 8/3600 of an hour. Then we can claim that $60 = s'(8/3600) = C \cdot (8/3600)$, so that we have

$$C = \frac{60(3600)}{8} = 27000 \,(\text{miles/hr}^2)$$

$$s(t) = \frac{Ct^2}{2} = 13500t^2$$

and

$$s\left(\frac{8}{3600}\right) = 13500\left(\frac{8}{3600}\right)^2 = 0.0666666\ldots \text{ mile} \approx 352 \text{ feet}$$

We have shown that in going from 0 to 60, the 1998 Olds Intrigue GL will travel approximately **352** feet.   ◆

General Solutions

If *every* solution of an *n*th-order differential equation $F(x, y, y', y'', y''', \ldots, y^{(n-1)}, y^{(n)}) = 0$ on an interval (a, b) can be obtained from an *n*-parameter family by choosing appropriate values for the *n* constants, we say that the family is the **general solution** of the differential equation. In this case, we will need *n* initial conditions or *n* boundary conditions (or a combination of *n* conditions) to determine the constants.

Sometimes, however, we can't find *every* solution somewhere among the members of an *n*-parameter family. For example, you should verify that the first-order nonlinear differential equation $2xy' + y^2 = 1$ has a one-parameter family of solutions given by $y = \dfrac{Cx - 1}{Cx + 1}$. However, for all values of *x*, the constant function $y \equiv 1$ is also a solution, but it can't be obtained from the family by choosing a particular value of the parameter C. Suppose we *could* find a value of C such that $\dfrac{Cx - 1}{Cx + 1} = 1$. Cross-multiplication gives us $Cx - 1 = Cx + 1$, so that $-1 = 1$!

Also, $y(x) = kx^2$ is a solution of $x^2 y'' - 3xy' + 4y = 0$ for any constant k and for all values of *x*, but $y(x) = x^2 \ln |x|$ is also a solution for all *x*. (*Check these claims.*) Of course, because the equation is second-order, we should realize that a one-parameter family can't be the general solution.

A solution of an *n*th-order differential equation that can't be obtained by picking particular values of the parameters in an *n*-parameter family of solutions is called a **singular solution.** We'll see in Chapter 2 that some of these singular solutions are created when we perform certain algebraic manipulations on differential equations.

SOLUTIONS OF SYSTEMS OF ODES

For a *system* of two equations with unknown functions $x(t)$ and $y(t)$, a solution on an interval (a, b) consists of a *pair* of differentiable functions $x(t)$, $y(t)$ satisfying both equations that make up the system at all points of the interval. Initial conditions are given as $x(t_0) = x_0$ and $y(t_0) = y_0$.

EXAMPLE 1.2.9   A System IVP

In Chapter 4, we shall see why the only solution of the system

$$\frac{dx}{dt} = -3x + y$$

$$\frac{dy}{dt} = x - 3y$$

satisfying the conditions $x(0) = 0$ and $y(0) = 7$ is $\left\{ x(t) = \dfrac{7}{2}e^{-2t} - \dfrac{7}{2}e^{-4t}, \; y(t) = \dfrac{7}{2}e^{-2t} + \dfrac{7}{2}e^{-4t} \right\}$. (*Verify that these are solutions.*) For now, let us accept this last statement as fact.

You can think of the solution pair as coordinates of a point $(x(t), y(t))$ in two-dimensional space, R^2. As the independent variable t changes, the points trace out a curve in the x-y plane called a **trajectory.** The *positive* direction of the curve is the direction it takes as t *increases.*

You may have seen the *parametric representation* of a curve in calculus. For example, if you let the variable t go continuously from 0 to 2π, the points $(x(t), y(t)) = (\cos t, \sin t)$ trace the *unit circle* (center $= (0, 0)$, radius $= 1$) in the plane—counterclockwise as t increases. You should know how to use your graphing calculator or computer algebra system (CAS) to graph parametrically.

The curve in the x-y plane corresponding to the system solution given above is shown in Figure 1.3a, together with arrows indicating its direction. The initial point $(x(0), y(0)) = (0, 7)$ is indicated. Looking at the solution formulas for $x(t)$ and $y(t)$, you can see that $\lim\limits_{t\to\infty} x(t) = 0 = \lim\limits_{t\to\infty} y(t)$, so that the curve tends toward the origin as t increases.

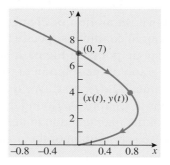

Figure 1.3a

Plot of $(x(t), y(t)) = (\frac{7}{2}e^{-2t} - \frac{7}{2}e^{-4t}, \frac{7}{2}e^{-2t} + \frac{7}{2}e^{-4t})$
in the x-y plane, $-0.1 \le t \le 4$

Figure 1.3b shows x plotted against t, and Figure 1.3c shows y plotted against t.

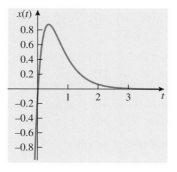

Figure 1.3b

Plot of $x(t) = \frac{7}{2}e^{-2t} - \frac{7}{2}e^{-4t}$, $-0.1 \le t \le 4$

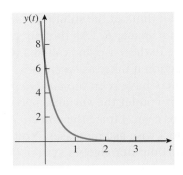

Figure 1.3c

Plot of $y(t) = \frac{7}{2}e^{-2t} + \frac{7}{2}e^{-4t}, -0.1 \le t \le 4$ ◆

A very important, *dynamical* way of looking at the situation in the last example is to think of the curve in Figure 1.3a as the path (or trajectory) of an object or quantity whose motion or change is governed by the system of differential equations. Initial conditions specify the behavior (the value, rate of change, and so on) at a single point on the path of the moving object or changing quantity. The proper graph of the solution of the system in Example 1.2.9 is a **space curve,** the set of points $(t, x(t), y(t))$. We'll see further graphical interpretations of system solutions in Chapter 4. Boundary conditions also determine certain aspects of the path of the phenomenon under study.

Similarly, each solution of the nonlinear system

$$\dot{x} = -sx + sy$$
$$\dot{y} = -xz + rx - y$$
$$\dot{z} = xy - bz$$

where b, r, and s are constants, is an ordered triple $(x(t), y(t), z(t))$, and initial conditions have the form $x(t_0) = x_0$, $y(t_0) = y_0$, and $z(t_0) = z_0$. Boundary conditions in this situation can take various forms. The trajectory in this case is a *space curve,* a path in three-dimensional space. The true graph of the solution is the set of points $(t, x(t), y(t), z(t))$ in *four-dimensional space*. These points of view, especially the idea of a trajectory, are very useful, and we'll follow up on these concepts in Chapters 4, 5, and 7.

EXERCISES 1.2

In Exercises 1–11, verify that the indicated function is a solution of the given differential equation. The letters a, b, c, and d denote constants.

1. $y'' + y = 0; \quad y = \sin x$

2. $x'' - 5x' + 6x = 0; \quad x = -\pi e^{3t} + \frac{2}{3}e^{2t}$

3. $\frac{1}{4}\left(\frac{dy}{dx}\right)^2 - x\frac{dy}{dx} + y = 0; \quad y = x^2$

4. $t\dfrac{dR}{dt} - R = t^2 \sin t; \quad R = t(c - \cos t)$

5. $\dfrac{d^4y}{dt^4} = 0; \quad y = at^3 + bt^2 + ct + d$

6. $\dfrac{dr}{dt} = at + br; \quad r = ce^{bt} - \dfrac{a}{b}t - \dfrac{a}{b^2}$

7. $xy' - 2 = 0; \quad y = \ln(x^2)$

8. $y'' = a\sqrt{1 + (y')^2}; \quad y = \dfrac{e^{ax} + e^{-ax}}{2a}$

9. $2y = xy' + \ln(y'); \quad y = \dfrac{x^2}{2} + \dfrac{x}{2}\sqrt{x^2 + 1} + \ln\sqrt{x + \sqrt{x^2 + 1}}$

10. $xy' - \sin x = 0; \quad y = \displaystyle\int_1^x \dfrac{\sin t}{t}dt$ [*Think of the Fundamental Theorem of Calculus.*]

11. $y'' + 2xy' = 0; \quad y = \displaystyle\int_3^x e^{-t^2}dt$ [*Think of the Fundamental Theorem of Calculus.*]

12. Write a paragraph explaining why $B(t)$ in Example 1.2.1—a solution of the differential equation $\dfrac{dB}{dt} = kB$—can't be a polynomial, a trigonometric, or a logarithmic function.

13. a. Why does the equation $(y')^2 + 1 = 0$ have no real-valued solution?
   b. Why does the equation $|y'| + |y| = 0$ have only one solution? What is the solution?

14. Explain why the equation $\dfrac{dx}{dt} = \sqrt{-|x - t|}$ has *no* real-valued solution.

15. If c is a positive constant, show that the two functions $y = \sqrt{c^2 - x^2}$ and $y = -\sqrt{c^2 - x^2}$ are both solutions of the nonlinear equation $y\dfrac{dy}{dx} + x = 0$ on the interval $-c < x < c$. Explain why the solutions are not valid outside the open interval $(-c, c)$.

16. Consider the equation and solution in Exercise 4. Find the particular solution that satisfies the initial condition $R(\pi) = 0$.

17. Consider the equation and solution in Exercise 5. Find the particular solution that satisfies the initial conditions $y(0) = 1$, $y'(0) = 0$, $y''(0) = 1$, and $y'''(0) = 6$. [*Hint:* Use the initial conditions one at a time, beginning at the left.]

18. Consider the equation and solution in Exercise 6. Find the particular solution that satisfies the initial condition $r(0) = 0$. (Your answer should involve only the constants a and b.)

19. Consider the equation and solution in Exercise 8. Find the particular solution that satisfies the initial conditions $y(0) = 2$, $y'(0) = 0$.

20. Let $W = W(t)$ denote your weight on day t of a diet. If you eat C calories per day and your body burns EW calories per day, where E represents calories per pound, then the equation $\dfrac{dW}{dt} = k(C - EW)$ models your change of weight.[3] (This equation says that your change of weight is proportional to the difference between calories eaten and calories burned off, with constant of proportionality k.)

   a. Show that $W = \dfrac{C}{E} + \left(W_0 - \dfrac{C}{E}\right)e^{-kEt}$ is a solution of the equation, where $W_0 = W(0)$, your weight at the beginning of the diet.

   b. Given the solution in part (a), what happens to $W(t)$ as $t \to \infty$?

   c. If $W_0 = 180$ lb, $E = 20$ cal/lb, $k = 1/3500$ lb/cal, and $C = 2500$ cal/day, then how long will it take to lose 20 lb? How long for 30 lb? 35 lb? What do your answers seem to say about the process of weight loss?

21. A particle moves along the x-axis in such a way that its velocity at any time $t \geq 0$ is given by $v(t) = 1/(t^2 + 1)$. Assuming that the particle is at the origin initially, show that it will never get past $x = \pi/2$.

22. Ayanna leaves her home at noon and drives to her aunt's house, arriving at 3:20 P.M. She started from a parked position and steadily increased her speed in such a way that when she reached her aunt's house she was driving at 60 miles per hour. (The house had been repainted recently and Ayanna didn't recognize it.) How far is it from Ayanna's home to her aunt's house?

23. A 727 jet needs to be flying 200 mph to take off. If the plane can accelerate from 0 to 200 mph in 30 seconds, how long must the runway be, assuming constant acceleration?

24. In the "Automobiles" section of the Sunday *New York Times* (November 14, 1999), we are told that a 2000 model Chevrolet Corvette convertible will go from 0 to 60 mph in 5.2 seconds.

   a. Assuming constant acceleration, how far will the car travel before it reaches 60 mph?

   b. The car's brakes are applied when the car is going 60 mph. Assuming constant deceleration, how long will it take the car to stop if it stops in 123 feet?

25. Solve the equation $EI\dfrac{d^4y}{dx^4} = -\dfrac{W}{L}$, with the boundary conditions $y(0) = 0$, $y'(0) = 0$; $y(L) = 0$, $y'(L) = 0$. (This problem arises in the analysis of the

---

3. A. C. Segal, "A Linear Diet Model," *College Mathematics Journal* 18 (1987): 44–45.

stresses on a uniform beam of length L and weight W, both ends of which are fixed in concrete. The solution y describes the shape of the beam when a certain type of load is placed on it. Here E and I are constants, and the product EI is a constant called the *flexural rigidity* of the beam.) [*Hint:* Integrate successively, introducing a constant of integration at each stage. Then use the boundary conditions to evaluate these constants of integration.]

26. Show that the first-order nonlinear equation $(xy' - y)^2 - (y')^2 - 1 = 0$ has a general solution given by $y = Cx \pm \sqrt{C^2 + 1}$ but that any function y defined implicitly by the relation $x^2 + y^2 = 1$ is also a solution—one that does not correspond to a particular value of C in the general solution formula.

27. a. Verify that the function $y = \ln(|C_1 x|) + C_2$ is a solution of the differential equation $y' = \dfrac{1}{x}$ for each value of the parameters $C_1 \ (\neq 0)$ and C_2.

   b. Show that there is only one genuine parameter needed for y. In other words, write $y = \ln(|C_1 x|) + C_2$ using only one parameter C.

28. For each function, find a differential equation satisfied by that function.

   a. $y = c + \dfrac{x}{c}$, where c is a constant

   b. $y = e^{ax} \sin bx$, where a and b are constants

   c. $y = (A + Bt)e^t$, where A and B are constants

In Exercises 29–30, the function y is defined implicitly as a function of x by the given equation, where C is a constant. In each case, use the technique of implicit differentiation to find a differential equation for which y is a solution.

29. $xy - \ln y = C$

30. $y + \arctan y = x + \arctan x + C$

31. Find a solution of $\dfrac{dy}{dx} + y = \sin x$ of the form $y(x) = c_1 \sin x + c_2 \cos x$, where c_1 and c_2 are constants.

32. Find a second-degree polynomial $y(x)$ that is a particular solution of the linear differential equation $2y' - y = 3x^2 - 13x + 7$.

33. Consider the equation $xy'' - (x + n)y' + ny = 0$, where n is a nonnegative integer.

   a. Show that $y = e^x$ is a solution.

   b. Show that $y = 1 + x + \dfrac{x^2}{2!} + \dfrac{x^3}{3!} + \cdots + \dfrac{x^n}{n!}$ is a solution.

34. The *logistic equation* $\dfrac{dy}{dt} = k\,y(t)\left(1 - \dfrac{y(t)}{M}\right)$ is used to describe the growth of certain kinds of human and animal populations. Here k and M denote constants that describe characteristics of the population being modeled.

a. Show that the function $y(t) = \dfrac{M}{1 + Ae^{-kt}}$ satisfies the logistic equation with $y(0) = \dfrac{M}{1 + A}$.

b. A study of U.S. population data[4] indicates that the solution given in part (a) provides a good fit if $M = 387.9802$, $A = 54.0812$, and $k = 0.02270347$. Using technology, plot the graph of $y(t)$ using these values of M, A, and k. (Here t denotes the time in years since 1790, the year of the first U.S. census.)

c. In 1790, the U.S. population was 3,929,214. In 1980, the figure was 226,545,805; in 1990, the population was 248,709,873. By evaluating the function plotted in part (b) at $t = 0$, 190, and 200, compare the values (in millions) given by $y(t)$ to the actual populations.

d. According to the model with parameters as given in part (b), what happens to the population of the United States as $t \to \infty$?

35. a. Show that the functions $x(t) = (A + Bt)e^{3t}$ and $y(t) = (3A + B + 3Bt)e^{3t}$ are solutions of the system

$$\begin{aligned} x' &= \quad\quad y \\ y' &= -9x + 6y \end{aligned}$$

for all values of the parameters A and B.

b. Find the solution to the system in part (a) with $x(0) = 1$ and $y(0) = 0$.

36. Show that functions $x(t) = e^{-t/10}\sin t$ and $y(t) = \dfrac{1}{10}e^{-t/10}(-10\cos t + \sin t)$ are solutions of the initial-value problem

$$\frac{dx}{dt} = -y$$

$$\frac{dy}{dt} = (1.01)x - (0.2)y; \quad x(0) = 0,\ y(0) = -1$$

37. The equations

$$\frac{dT^*}{dt} = kV_1T_0 - \delta T^*$$

$$\frac{dV_1}{dt} = -cV_1$$

are used in modeling HIV-1 infections.[5] Here $T^* = T^*(t)$ denotes the number of infected cells, $T_0 = T(0)$ is the number of potentially infected cells at

---

4. E. K. Yeargers, R. W. Shonkwiler, and J. V. Herod, *An Introduction to the Mathematics of Biology: With Computer Algebra Models* (Boston: Birkhäuser, 1996): 117.

5. A. S. Perelson, A. U. Neumann, M. Markowitz, J. M Leonard, and D. D. Ho, "HIV-1 Dynamics in Vivo: Virion Clearance Rate, Infected Cell Life-Span, and Viral Generation Time," *Science* 271 (1996): 1582–1586.

the time therapy is begun, $V_1 = V_1(t)$ is the concentration of viral particles in plasma, k is the rate of infection, c is the rate constant for viral particle clearance, and δ is the rate of loss of virus-producing cells.

a. Imitate the analysis shown in Example 1.2.1 and solve the second equation for $V_1(t)$, expressing your solution in terms of $V_0 = V_1(0)$.

b. Using the solution found in part (a), show that the solution of the differential equation for T^* can be written as

$$T^*(t) = T^*(0)e^{-\delta t} + \frac{kT_0V_0}{\delta - c}(e^{-ct} - e^{-\delta t})$$

c. What does the solution in part (a) say about the number of infected cells as $t \to \infty$?

38. A mathematical model of an idealized company consists of the equations

$$\frac{du}{dt} = kau, u(0) = A$$

$$\frac{dw}{dt} = a(1 - k)u, w(0) = 0$$

Here $u(t)$ represents the capital invested in the company at time t, $w(t)$ denotes the total dividend paid to shareholders in the period $[0, t]$, and a and k are constants with $a > 0$ and $0 \le k \le 1$.

a. Solve the first equation for $u(t)$. (See Example 1.2.1.)

b. Substitute your answer in part (a) in the differential equation for w and integrate to find $w(t)$. (Distinguish between $w(t)$ for $0 < k \le 1$ and for $k = 0$.)

39. Consider the linear equation $x^2y'' + xy' - 4y = x^3$ (*) . Let y_{GR} be the *general* solution of the "reduced" (or "complementary") equation $x^2y'' + xy' - 4y = 0$, and let y_P be a *particular* solution of (*). Show that $y_{GR} + y_P$ is the *general* solution of (*). [For this problem, define the general solution of a second-order ODE as a solution that has two arbitrary constants. A particular solution, of course, has *no* arbitrary constants.]

40. The French mathematician Pierre Simon Laplace (1749–1827), who devoted much of his time to applying Newton's laws of motion to the movement of the planets, held a rather *deterministic* view of the universe. Impressed by the power of mathematics to describe nature, in essence he believed that solving an initial-value problem always allowed one to understand the past and to predict all future states of a system. Modern physics, however, has shown that many laws of physics are in fact *stochastic* (random, depending on chance) rather than deterministic. They deal with *probabilities* rather than with certainties.

Read about the life and work of Laplace and report on his views of mathematics and the universe. Some references: *Men of Mathematics* by E. T. Bell (New York: Simon & Schuster, 1986); *The History of Mathematics* by

D. M. Burton (Boston, Mass: McGraw-Hill, 1999); *A History of Mathematics* by V. J. Katz (Reading, Mass: Addison Wesley Longman, 1998); *Calculus Gems* by G. F. Simmons (New York: McGraw-Hill, 1992).

1.3   TECHNOLOGY AND DIFFERENTIAL EQUATIONS

The users of this book are living in a marvelous age for teaching and learning. The availability and relatively low cost of calculators, computers, and computer software are making it easier than ever before to bring technology into the classroom and into the backpacks and homes of students everywhere.

In this course, you are assumed to have access to powerful graphing calculators and/or computer algebra systems (CAS). Even spreadsheet programs can perform certain calculations that would be tedious if done by hand. (See Sections 3.1, 3.3, and 3.4.) You should try to duplicate figures and tables in the text using your own technology. General-purpose mathematical software such as *Derive*®, *Macsyma*®, *Maple*®, *Mathematica*®, and *MATLAB*® is available, as well as specialized differential equation programs such as *Differential Systems, ODE Solver, Phaser, ODE Toolkit,* and *MDEP.* Your instructor may even have his or her own programs to use in class. Even without a computer, you can deal with some differential equations and systems of ODEs by using such algebraically powerful (and programmable) graphing calculators as the HP-48G/X and the TI-92.

Early in this course, you should learn how to use technology to do basic calculations and graph various kinds of equations. As you advance through the material, you will have to learn commands and procedures that pertain specifically to differential equations. Using technology will free you from the burden of tedious calculation and enable you to focus on the appropriateness of inputs and the reasonableness of outputs. A graphing calculator or CAS gives you the power to think about problems in new ways because of its numerical, graphical, and analytical capabilities. With the aid of technology, you can analyze problems of greater complexity than was possible just a college generation or two ago. However, it is important for users of graphing calculators and computers to realize that these powerful technological tools can lead them astray.

Sophisticated technological devices may *fail* to give an answer to a problem. On the other hand, calculators and computers can give *incorrect, incomplete, or misleading results,* even when all the preliminary information about a problem is entered correctly and the right keys are pressed in the proper order. For example, a popular CAS returns nothing at all when you ask it to solve the first-order equation $y' = \ln(x^2 + y^2)$. Yet the same system can produce accurate *numerical* solutions (see Chapter 3). The same program's ODE-solving procedure applied to the first-order nonlinear equation $2xy' + y^2 = 1$ returns the one-parameter family $y = \dfrac{e^{-C}x + 1}{-1 + e^{-C}x}$, which doesn't look quite the same as the solution given in Section 1.2, but the *graphs* of these families will be identical. Furthermore, the CAS misses the singular solution $y \equiv 1$. More fundamentally, because indefinite integration is important for solving many differential equations, it might be disturbing to find that computer software may give the value of $\int 1/x\, dx$ as ln x, rather than the answer we might expect, $\ln|x| + C$. In this case, the computer is correct because it is interpreting the logarithm in terms of a *complex number x*. But because we are interested in real-valued solutions in this course, we must remember to integrate $1/x$ in the usual calculus way. (*Check your own CAS on the examples in this paragraph.*)

On a more global level, we can all use the resources of the *Internet* (the *World Wide Web, WWW*) to find information. For differential equations, this exploration can take the form of finding online tools (such as *Java applets*) to graph or do numerical calculations, gathering real-life data for a modeling project, or just using tutorials or CAS worksheets provided by professors far from your own institution. In this activity, as in using a calculator or a CAS, you must be alert. The Web is notorious for providing information that may be inaccurate. Develop skill in browsing (or "surfing") the Net. Strengthen your intuition about the reliability of the data you see. Finally, neither teacher nor student should get so hung up on the technology that there is no human interaction going on. Teachers and students should speak to each other and listen carefully to each other.

The moral of this brief section is that graphing calculators and computers are wonderful, but knowledge of mathematical theory and analysis techniques is necessary. Always try to focus on the science and mathematics behind the numbers and graphs. The Internet can help us learn a great deal without leaving our classroom, library, or home, but we have to be cautious and not believe everything we see. We must use technology wisely, remembering that only human beings can think and make judgments—so far.

EXERCISES 1.3

1. If your school library (or a generous professor) has back issues of the *College Mathematics Journal,* look for reviews of some computer software (general math/scientific or specifically for ODEs), especially any that you have access to at home or at school. For example, you could find the following articles by E. Teles, H. L. Penn, and J. Wilkin in the indicated issues: "ODE Software for the IBM PC" 21 (1990), pp. 242–245, and "ODE Software for the Macintosh" 21 (1990), pp. 330–332. There's a review of a product called Differential Systems in the September 1994 issue. The May/June 1989 and April 1990 issues of the *Notices* of the American Mathematical Society (AMS) also contain some reviews of ODE software. Even if you don't understand all the math in the reviews, you can get a general idea of the strengths and weaknesses of these programs.

2. Read and report on the article "Crimes and Misdemeanors in the Computer Algebra Trade" by David R. Stoutemyer in the *Notices* of the American Mathematical Society (AMS), 38/7 (September 1991), pp. 778–785.

3. Learn the commands necessary to *solve* simple ODEs with the software you have available. Try this knowledge on the first-order nonlinear ODE $y' + y = y^3 \sin x$, which has the family of solutions $y(x) = \left[Ce^{2x} + \dfrac{2}{5}(\cos x + 2 \sin x) \right]^{-1/2}$, where C is a constant, and has $y \equiv 0$ as a singular solution. (*Verify this by hand.*) Does your computer output look like this? If not, use some algebra to get it to. Do you get the singular solution from your computer? Also, see what your computer makes of the equation $|y'| + |y| = 0$.

1.4   SUMMARY

The study of differential equations is as old as the development of calculus by Newton and Leibniz in the late seventeenth century. Motivation was provided by important questions about change and motion on earth and in the heavens.

An **ordinary differential equation** (**ODE**) is an equation that involves an unknown function, its independent variable, and one or more of its derivatives:

$$F(x, y, y', y'', y''', \dots, y^{(n-1)}, y^{(n)}) = 0$$

Such an equation can be described in terms of its **order,** the order of the highest derivative of the unknown function in the equation.

Differential equations can also be classified as either **linear** or **nonlinear. Linear equations** can be written in the form

$$a_n(x)y^{(n)} + a_{n-1}(x)y^{(n-1)} + \cdots + a_2(x)y'' + a_1(x)y' + a_0(x)y = f(x)$$

where each coefficient function $a_i(x)$ depends on x alone and doesn't involve y or any of its derivatives. **Nonlinear equations** usually contain products, quotients, or more elaborate combinations of the unknown function and its derivatives.

A **solution** of an ODE is a real-valued function that, when substituted in the equation, makes the equation valid on some interval. A given nth-order ODE may have *no* solutions, *only one* solution, or *infinitely many solutions. An infinite family of solutions* may be characterized by n constants (parameters). These arbitrary constants, if present, may be evaluated by imposing appropriate **initial conditions** (usually n of them, involving behavior of the solution at a single point of its domain) or **boundary conditions** (at two or more points). Solving a differential equation with initial conditions is referred to as solving an **initial-value problem** (**IVP**). Solving a differential equation with boundary conditions is referred to as solving a **boundary-value problem** (**BVP**). In general, BVPs are harder to solve than IVPs. The result of solving either an IVP or a BVP is called a **particular solution** of the equation or an **integral** of the equation. The graph of a particular solution is called an **integral curve** or **solution curve**. In Chapters 2 and 3, we'll discuss the question of *existence and uniqueness* for IVPs: Does the equation or system have a solution that satisfies the initial conditions? If so, is there only one solution?

If *every* solution of an *n*th-order ODE on an interval can be obtained from an *n*-parameter family by choosing appropriate values for the *n* constants, then we say that the family is the **general solution** of the differential equation. In this case we need *n* initial conditions or *n* boundary conditions to determine the constants. However, sometimes there are **singular solutions** that can't be found just by choosing particular values of the constants.

Just as high school or college algebra introduces systems of algebraic equations, the study of certain problems often leads to **systems of differential equations.** These, in turn, can be classified as either **linear systems** or **nonlinear systems.** We can specify initial or boundary conditions for systems. Whether we're considering single equations or systems of equations, we are dealing with *dynamical* situations—situations in which objects and quantities are moving and changing. In such a dynamical situation, it is often useful to focus on a **trajectory**—for a single equation, the curve made up of points $(x(t), x'(t))$, where *x* is a solution; for a system of two equations, the set of points $(x(t), y(t))$, where *x* and *y* are solutions of the system.

Finally, in studying differential equations using technology, we must always be aware that calculators and computers are not infallible. In addition to human data entry errors and mistakes in entering commands, there is the fact that a technological device may yield *no answer* or an *incorrect answer* to a given problem. The Internet can help us to understand differential equations and their applications, but this tool must be used carefully. The really powerful combination is technology together with human intelligence and mathematical understanding.

PROJECT 1-1

Draw Your Own Conclusions

Even before you learn techniques for solving differential equations, you may be able to analyze equations *qualitatively*. As an example, look at the nonlinear equation $\dfrac{dy}{dt} = y(1 - y)$. You are going to analyze the solutions, *y*, of this equation without actually finding them.

In what follows, picture the *t*-axis running horizontally and the *y*-axis running vertically.

a. For what values of *y* is the graph of *y* as a function of *t* increasing? For what values of *y* is it decreasing?

b. For what values of *y* is the graph of *y* concave up? For what values of *y* is it concave down? (What information do you need to answer a question about concavity? Remember that *y* is an implicit function of *t*.)

c. Say you are given the initial condition $y(0) = 0.5$. Use the information found in parts (a) and (b) to sketch the graph of y. What is the *long-term* behavior of $y(t)$? That is, what is $\lim_{t \to \infty} y(t)$?

d. Say you are given the initial condition $y(0) = 1.5$. Use the information found in parts (a) and (b) to sketch the graph of y. What is the *long-term* behavior of $y(t)$? That is, what is $\lim_{t \to \infty} y(t)$?

e. Sketch the graph of y if $y(0) = 1$. (Look at the original equation.)

f. If $y(t)$ represents the population of some animal species, and if units on the y-axis are in thousands, interpret the results of parts (c), (d), and (e).

2 | First-Order Differential Equations

2.0 INTRODUCTION

The various examples in the last chapter should have convinced you that there are different possible answers to the question of what the solution or solutions to a differential equation look like. In this chapter, we'll examine first-order differential equations from both the analytic and the qualitative point of view. Chapter 3 will focus on some numerical solution methods concerned with calculating *approximate* values of solutions.

First we'll learn *analytic* solution techniques for two important types of first-order equations. For these kinds of equations, we'll come up with explicit or implicit formulas for their solution curves. This technique is often referred to as *integrating* a differential equation.

Next there is a *qualitative* way of viewing differential equations. This is a neat geometrical way of studying the behavior of solutions without actually solving the differential equation. The idea is to examine certain pictures or graphs derived from differential equations. Although we can do some of this work by hand, most computer algebra systems and many graphing calculators can produce these graphs, and you'll be expected to use technology when appropriate. There are also specialized programs just for doing this sort of thing. (Follow your instructor's guidance in using technology.)

Both qualitative and numerical methods are necessary, because it is often impossible to represent the solutions of differential equations—even first-order equations—by formulas involving elementary functions.[1]

---

1. In general, "elementary functions" are finite combinations of integer powers of the independent variable, roots, exponential functions, logarithmic functions, trigonometric functions, and inverse trigonometric functions.

2.1   SEPARABLE EQUATIONS

The simplest type of differential equation to solve is one in which the variables are *separable*. For example, look at the form of the equation $\dfrac{dy}{dt} = -2y$, which could be expressing the fact that a certain radioactive substance is decaying at a rate that is proportional to the quantity y present at any time t. (Note the negative growth rate if $y > 0$.)

If we consider the left-hand side of the equation as a *fraction* rather than as a single symbol, we can separate the "numerator" and "denominator" and multiply both sides of the equation by the "denominator" dt: $dy = -2y\,dt$. Now we can divide both sides by $-2y$ (assuming that $y \neq 0$) to get $\dfrac{dy}{-2y} = dt$. Integrating both sides, we have $\displaystyle\int \dfrac{dy}{-2y} = \int dt$, or $-\tfrac{1}{2}\ln|-2y| = t + C_1$. (There is a constant of integration on *both* sides, but it is customary to combine constants on the right-hand side of the solution equation.) This simplifies to $\ln|y| = -2t + C_2$, so that (exponentiating) we find that $|y| = e^{-2t + C_2} = C_3 e^{-2t}$, or $y = Ce^{-2t}$, where C is any nonzero constant. Note that C_3 denotes e^{C_2} (and is therefore a *positive* constant) and that the final constant C could be positive or negative once we removed the absolute value symbol around y.

Because we divided by y, we must examine the case $y \equiv 0$ separately. We see that $y \equiv 0$ is a solution. Note that this is not a singular solution because we can get this solution by letting $C = 0$ in our one-parameter family of solutions. C was described as a nonzero constant in the previous paragraph, so allowing C to be zero amounts to extending the set of admissible values of C in order to include all possible solutions under the umbrella of a one-parameter family.

In using Leibniz's $\dfrac{dy}{dx}$ notation in calculus, we usually think of this as $\dfrac{d}{dx}(y)$, the operation or process of taking the derivative of y with respect to x, and we don't give meaning to the symbols dy and dx separately. However, you may have seen the form of the Chain Rule that looks like this: $\dfrac{dy}{dt} = \dfrac{dy}{dx} \cdot \dfrac{dx}{dt}$, where y is a function of x, and x in turn depends on t. (See Appendix A.2 if necessary.) Leibniz's notation suggests that some "cancellation" of fractions is occurring and helps us to remember the form of the Chain Rule for differentiating composite functions. Although you may have seen the mathematically precise way to deal with this under the name *differentials*,[2] we will be content to manipulate these symbols in an obvious (intuitive) way.

---

2. If $y = f(x)$, where f is a differentiable function and dx is an arbitrary number, then the *differential* dy is defined in terms of dx by the equation $dy = f'(x)dx$. Note that dy is a dependent variable, a function of the two variables x and dx. (In this context, the number dx is also called a differential.)

Formally, a first-order differential equation $\dfrac{dy}{dx} = F(x, y)$ is called **separable** if it can be written in the form $\dfrac{dy}{dx} = f(x)g(y)$ where f denotes a function of x alone and g denotes a function of y alone. In this situation, we have separated the independent and dependent variables. A separable equation can then be rewritten as $\dfrac{dy}{g(y)} = f(x)\,dx$ (a process called **separating variables**) and can be solved by integrating both sides with respect to x:

$$\int \frac{1}{g(y)} \frac{dy}{dx} dx = \int \frac{dy}{g(y)} = \int f(x)\,dx.$$

In the example $\dfrac{dy}{dt} = -2y$ that we've already seen, $f(x) \equiv 1$ and $g(y) = -2y$.

There are three things to be careful about: (1) Not every first-order differential equation is separable. (2) Even after you have separated the variables and integrated, it may not be possible to solve for one variable (say y) in terms of the other (say x); you may have to express your answer *implicitly*. (3) You may not be able to carry out the integration(s) in terms of elementary functions. We'll see examples of these situations.

Also, note that in a separable equation $y' = f(x)g(y)$, a solution of $g(y) \equiv 0$ (a constant solution) is also a solution of the differential equation—possibly a *singular* solution (see Section 1.2). If $g(y) = 0$, then $y' = 0$, implying that y is a constant. Conversely, if $y(x) = c$ is a constant solution, then $y' = 0$, which implies that $g(y) = 0$ because $f(x) = 0$ is unlikely in a physical problem. This says that the zeros of g are constant solutions; and, in general, they are the only constant solutions.

The next example—so basic, yet so important in many applications—is one we've seen before. Back in Chapter 1, we *guessed* at the solution and then verified that our guess was correct.

EXAMPLE 2.1.1 Solving a Separable Equation—Example 1.2.1 Revisited

The way the balance $B(t)$ of a bank account grows under continuous compounding demonstrates the "snowball effect": The larger the balance at a given time, the more rapid the growth—that is, the greater rate of growth. (You may have heard the expression "them that has, gets.") In the language of differential equations, this becomes $\dfrac{dB}{dt} = rB$, where r, the constant of proportionality, is the annual interest rate (expressed as a decimal). If the initial balance (the principal at $t = 0$) is positive, we want to find the balance at time t.

Separating the variables, we can write $\dfrac{dB}{B} = r\,dt$, so that $\int \dfrac{dB}{B} = \int r\,dt$ and $\ln|B| = rt + C$. Then we exponentiate: $e^{\ln|B|} = e^{rt+C} = e^{rt}e^C$, or $|B| = Ke^{rt}$, where $K = e^C$, a positive constant. Given that the initial balance was positive, we realize that B must be positive, so that we can just write $B(t) = Ke^{rt}$.

Finally, we can bring in the positive initial balance to write $B(0) = Ke^0 = K$, so that our final formula is $B = B(t) = B(0)e^{rt}$. For example, if we invest $1000 at 4% interest compounded continuously for 6 years, we will have $(1000)e^{(0.04)6} \approx \$1271.25$ in our account. (You should use your calculator or CAS to verify this.) ◆

Of course, there is another way to do the previous problem, but you have to know the formula for the balance $B_n(t)$ if you invest P dollars at interest rate r compounded n times a year for t years. This formula is

$$B_n(t) = P\left(1 + \frac{r}{n}\right)^{nt}.$$

Continuous compounding involves compounding infinitely often, compounding at *every instant* of the year. Mathematically, we want

$$\lim_{n\to\infty} B_n(t) = \lim_{n\to\infty} P\left(1 + \frac{r}{n}\right)^{nt} = P \cdot \lim_{n\to\infty}\left(1 + \frac{r}{n}\right)^{nt} = P \cdot \left\{\lim_{n\to\infty}\left(1 + \frac{r}{n}\right)^{n}\right\}^{t}$$
$$= P \cdot \{e^r\}^t = Pe^{rt} = B(0)e^{rt}.$$

You may have seen this derivation in calculus class, as well as the fact that if you invest $1 for 1 year at 100% interest compounded continuously, you will have e ($\approx \$2.72$) at the end of the year.

The next problem continues the analysis of separable equations and adds the use of technology.

EXAMPLE 2.1.2   A Separable Equation and the Graph of a Solution
Suppose that an insect population P shows seasonal growth modeled by the differential equation $\dfrac{dP}{dt} = kP\cos(\omega t)$, where k and ω are positive constants. (The cosine factor suggests periodic fluctuation.)

You should be able to see that the equation is separable: $\dfrac{dP}{dt} = f(P)g(t)$, where $f(P) = P$ and $g(t) = k\cos(\omega t)$. (We could have stuck the constant k with the factor P, but if you think ahead, you'll realize that there's one less algebraic step if we keep the constant with the cosine term.) Separating the variables, we get $\dfrac{dP}{P} = k\cos(\omega t)dt$, so that $\displaystyle\int \dfrac{dP}{P} = k\int \cos(\omega t)dt$, or $\ln|P| = \dfrac{k}{\omega}\sin(\omega t) + C$.

Exponentiating, we see that $P(t) = Re^{\frac{k}{\omega}\sin(\omega t)}$, where $R > 0$. (This is a population problem, so $R > 0$ is a realistic assumption.) Letting $P_0 = P(0)$ denote the initial insect population, we have $P(t) = P_0 e^{\frac{k}{\omega}\sin(\omega t)}$ as the solution.

Graphing the solution curve for $P_0 = 100$, $k = 2$, and $\omega = \pi$ (Figure 2.1), we can see that the population varies periodically, fluctuating from a minimum value of $100e^{-\frac{2}{\pi}}$ (approximately 53) to a maximum value of $100e^{\frac{2}{\pi}}$ (approximately 189).

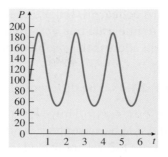

Figure 2.1

The solution of the IVP $\dfrac{dP}{dt} = 2P\cos(\pi t)$; $P(0) = 100$ ◆

Sometimes, as the next example shows, it may not be easy to find an explicit solution for a separable differential equation. (See the second concern we mentioned just before Example 2.1.1.)

EXAMPLE 2.1.3 A Separable Equation with Implicit Solutions

The equation $\dfrac{dy}{dx} = \dfrac{x^2}{1 + y^2}$ can be written as $\dfrac{dy}{dx} = f(x)g(y)$, where $f(x) = x^2$ and

and $g(y) = \dfrac{1}{1 + y^2}$. Separating the variables, we get $(1 + y^2)dy = x^2 dx$. Integrating both sides, we find that $y + \dfrac{y^3}{3} = \dfrac{x^3}{3} + C$, or $\dfrac{x^3}{3} - \left(y + \dfrac{y^3}{3}\right) = C$. This

gives the solution *implicitly*. (See Chapter 1, right after Example 1.2.2.) To get an *explicit* solution, we must solve this last equation for y in terms of x or for x in terms of y. Either way is acceptable, although solving for x as a function of y is easier algebraically. But even if we don't find an explicit solution, we can plot solution curves for different values of the constant C. (This may be a good time to find out how to graph implicit functions using your available technology.) In Figure 2.2 we use (from top to bottom) $C = -7, -5, -3, 0, 3, 5$, and 7.

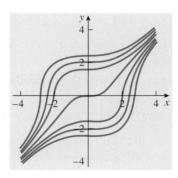

Figure 2.2

Implicit solutions of $\dfrac{dy}{dx} = \dfrac{x^2}{1 + y^2}$: the curves $\dfrac{x^3}{3} - \left(y + \dfrac{y^3}{3}\right) = C$

$C = -7, -5, -3, 0, 3, 5$, and 7; $-4 \le x \le 4$, $-4 \le y \le 4$ ◆

The third concern we mentioned just before Example 2.1.1 is that you may not be able to integrate one or both of the sides after you have separated the variables. We will address this problem next.

EXAMPLE 2.1.4 The differential equation $\dfrac{dy}{dt} = e^{y^2}t$ is clearly separable—we can write $e^{-y^2}dy = t\,dt$. However, we can't carry out the integration $\int e^{-y^2}dy$ on the left-hand side because there is no combination of elementary functions whose derivative is e^{-y^2}, so we are forced to write the family of solutions as

$$\int e^{-y^2}dy = \frac{t^2}{2} + C, \quad \text{or} \quad 2\int e^{-y^2}dy = t^2 + K,$$

where $K = 2C$.

Integrals of the form $\displaystyle\int_a^b e^{-y^2}dy$ have many applications in mathematics and science, especially in problems dealing with probability and statistics. For instance, the ***error function*** $erf(x) = \dfrac{2}{\sqrt{\pi}}\displaystyle\int_0^x e^{-y^2}dy$ appears in many applied problems and can be evaluated easily by any CAS. ◆

Dealing with separable equations often requires some algebraic skills and some integration smarts, although technology can help in tough situations. The next example introduces a common algebraic problem.

EXAMPLE 2.1.5 **Using Partial Fractions**

The equation $\dfrac{dz}{dt} + 1 = z^2$ looks simple enough but requires some algebraic manipulation to get a neat solution. Separating variables, we get $\dfrac{dz}{z^2 - 1} = dt$.

Using the method of *partial fractions* (see Appendix A.5), we can write $\dfrac{1}{z^2 - 1}$ as $\dfrac{1}{2}\left(\dfrac{1}{z - 1} - \dfrac{1}{z + 1}\right)$, so integration gives us $\displaystyle\int \dfrac{1}{2}\left(\dfrac{1}{z - 1} - \dfrac{1}{z + 1}\right)dz = \int 1\,dt$, or $\dfrac{1}{2}(\ln|z - 1| - \ln|z + 1|) = t + C_1$. Multiplying both sides of this last equation by 2 and then simplifying the logarithmic expression, we get $\ln\left|\dfrac{z - 1}{z + 1}\right| = 2t + C_2$.

Exponentiating, we find that $\dfrac{z - 1}{z + 1} = Ke^{2t}$. Finally, solving this last equation for z

(*a bit tricky—so do it*), we conclude that $z = \dfrac{1 + Ke^{2t}}{1 - Ke^{2t}}$, a one-parameter family of solutions.

Note that in going through the process of separating variables, we divided by $z^2 - 1$, implicitly assuming that this expression was not zero. Going back to this, we see that the constant function $z \equiv 1$ corresponds to $K = 0$ in our general solution, whereas $z \equiv -1$ is a *singular* solution (*Why?*).   ◆

As simple as separable equations may seem, they have some very important applications. The calculations and manipulations involved in the next example may seem tedious, but they should remind you of things you have seen in previous classes. The analysis at the end of the example should convince you of the usefulness of a graphical approach.

EXAMPLE 2.1.6   A Model of a Bimolecular Chemical Reaction
Most chemical reactions can be viewed as interactions between two molecules that undergo a change and result in a new product. The rate of a reaction, therefore, depends on the number of interactions or collisions, which in turn depends on the concentrations (in moles per liter) of both types of molecules. Consider the simple (*bimolecular*) reaction $A + B \rightarrow X$, in which molecules of substance A collide with molecules of substance B to create substance X.

Let's designate the concentrations at time 0 of A and B by α and β, respectively. We'll assume that the concentration of X at the beginning is 0 and that at time t it is $x = x(t)$. The concentrations of A and B at time t are, correspondingly, $\alpha - x$ and $\beta - x$. Note that $\alpha - x > 0$ and $\beta - x > 0$ (*Why?*). The rate of formation (the *velocity of reaction* or *reaction rate*) is given by the differential equation $\dfrac{dx}{dt} = k(\alpha - x)(\beta - x)$, where k is a positive constant (called the *velocity constant*). The product on the right-hand side of the equation reflects the interactions or collisions between the two kinds of molecules. We want to determine $x(t)$.

Separating variables and integrating, we get

$$\int \frac{dx}{(\alpha - x)(\beta - x)} = \int k \, dt.$$

To simplify the integrand $\dfrac{1}{(\alpha - x)(\beta - x)}$, we use the technique of partial fractions so that we can write

$$\int \frac{dx}{(\alpha - x)(\beta - x)} = \frac{1}{\beta - \alpha} \int \frac{dx}{\alpha - x} + \frac{1}{\alpha - \beta} \int \frac{dx}{\beta - x} = \int k \, dt$$

or

$$-\frac{1}{\beta - \alpha}\ln(\alpha - x) - \frac{1}{\alpha - \beta}\ln(\beta - x) = kt + C$$

which simplifies to

$$\frac{1}{\alpha - \beta}\ln\left(\frac{\alpha - x}{\beta - x}\right) = kt + C.$$

The initial condition $x(0) = 0$ leads us to conclude that

$$C = \frac{1}{\alpha - \beta}\ln\left(\frac{\alpha}{\beta}\right).$$

Then

$$\frac{1}{\alpha - \beta}\ln\left(\frac{\alpha - x}{\beta - x}\right) = kt + \frac{1}{\alpha - \beta}\ln\left(\frac{\alpha}{\beta}\right),$$

so

$$\ln\left(\frac{\alpha - x}{\beta - x}\right) = (\alpha - \beta)kt + \ln\left(\frac{\alpha}{\beta}\right)$$

or

$$\frac{\alpha - x}{\beta - x} = \frac{\alpha}{\beta}e^{(\alpha - \beta)kt}.$$

A few more algebraic manipulations lead to the solution

$$x = x(t) = \frac{\alpha\beta(1 - e^{(\alpha - \beta)kt})}{\beta - \alpha e^{(\alpha - \beta)kt}}. \tag{2.1.1}$$

Formula (2.1.1) does not seem very informative as far as understanding the nature of the chemical reaction goes, but Exercise 33 suggests some useful ways of analyzing the formula. A CAS-generated graph of a solution (Figure 2.3) of the equation with $\alpha = 250$, $\beta = 40$, and $k = 0.0006$ is more informative, showing the steady rise in the concentration of molecule X to what is called an *equilibrium value* of 40. (We'll explore the idea of an equilibrium value in Section 2.5.) The particular solution shown corresponds to $x(0) = 0$.

Of course, as we've noted above, the right-hand side of the original differential equation is positive, so we know ahead of time that the concentration function is increasing. Also, you can calculate $\frac{d^2x}{dt^2}$ from the original differential equation to see why the graph of x is concave down. Remember that $k > 0$ and $0 \le x < \alpha$, $0 \le x < \beta$.   ◆

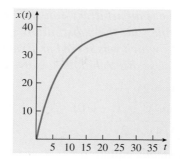

Figure 2.3
Solution of the

$$\text{IVP } \frac{dx}{dt} = 0.0006(250 - x)(40 - x); x(0) = 0$$

$$0 \le t \le 35, 0 \le x \le 40$$

EXERCISES 2.1

Solve the equations or IVPs in Exercises 1–9 by separating variables. Be sure to describe any singular solutions where appropriate.

1. $\dfrac{dy}{dx} = \dfrac{A - 2y}{x}$, where A is a constant

2. $\dfrac{dy}{dx} = \dfrac{-xy}{x + 1}$

3. $y' = 3\sqrt[3]{y^2};\quad y(2) = 0$

4. $\dfrac{dy}{dx} = \dfrac{(y - 1)(y - 2)}{x}$

5. $(\cot x)y' + y = 2;\quad y(0) = -1$

6. $x' = -\dfrac{\sin t \cos^2 x}{\cos^2 t};\quad x(0) = 0.$

7. $x^2 y^2 y' + 1 = y$

8. $xy' + y = y^2;\quad y(1) = 0.5$

9. $z' = 10^{x+z}$

10. Solve the equation $y' = 1 + x + y^2 + xy^2$. [*Hint:* Factor cleverly.]

11. Solve the equation $(y')^2 + (x + y)y' + xy = 0$. [*Hint:* Solve this quadratic equation for y' by factoring or by using the quadratic formula. Then solve the two resulting differential equations separately.]

An equation of the form dy/dx = f(ax + by) can be transformed into an equation with separable variables by making the substitution z = ax + by or z = ax + by

$+ c$, where c is an arbitrary constant. For example, the equation $y' = (y - x)^2$ is not separable, but the substitution $z = y - x$ leads to the separable equation $z' + 1 = z^2$, which was solved as Example 2.1.5. Then substitute the original variables for z. Use this technique to solve the equations in Exercises 12–14.

12. $y' - y = 2x - 3$

13. $(x + 2y)y' = 1$; $y(0) = -1$

14. $y' = \sqrt{4x + 2y - 1}$

A **homogeneous** equation has the form $dy/dx = f(x, y)$, where $f(x, y)$ can be expressed in the form $g(y/x)$ or $g(x/y)$—that is, as a function of the quotient y/x or the quotient x/y alone. For example, by dividing numerator and denominator by x^2, the equation $\dfrac{dy}{dx} = \dfrac{2x^2 - y^2}{3xy}$ can be written in the form $\dfrac{dy}{dx} = \dfrac{2 - (y/x)^2}{3(y/x)} = g\left(\dfrac{y}{x}\right)$.

Any such equation can be changed into a separable equation by making the substitution $z = y/x$ (or $z = x/y$). Making the substitution $z = y/x$ in our example, we have

$$\frac{dy}{dx} = \frac{d}{dx}(xz) \overset{\text{(Product Rule)}}{=} 1 \cdot z + x\left(\frac{dz}{dx}\right)$$

so our equation becomes $x\dfrac{dz}{dx} + z = \dfrac{2 - z^2}{3z}$,   or, separating variables,

$\left(\dfrac{3z}{2 - 4z^2}\right)dz = \dfrac{1}{x}dx$. After integrating, remember to replace z by y/x (or x/y). Use this technique to solve the equations in Exercises 15–18.

15. $y' = \dfrac{x + y}{x - y}$

16. $\dot{x} = \dfrac{t - 3x}{3t + x}$

17. $y' = \dfrac{x}{y} + \dfrac{y}{x}$

18. $\dfrac{dy}{dx} = \dfrac{y^2 + 2xy - x^2}{x^2 + 2xy - y^2}$

19. Suppose that f is a function such that $f(x) = \displaystyle\int_0^x f(t)\,dt$ for all real numbers x. Show that $f(x) \equiv 0$. [*Hint:* Use the Fundamental Theorem of Calculus to get a differential equation. Then think of an appropriate initial condition.]

20. Consider the equation $\dot{x} = \dfrac{x^2 + x}{t}$.

    a. Find a one-parameter family of solutions.

    b. Can you find a solution that satisfies the initial condition $x(0) = -1$? If so, give it. If not, give a reason.

    c. Find a singular solution.

21. a. Solve the initial-value problem $\dot{x} = x^2$, $x(1) = 1$.

    b. If the solution in part (a) is valid over an interval I, how large can I be?

    c. Use technology to draw the graph of the solution $x(t)$ found in part (a).

    d. Solve the initial-value problem $\dot{x} = x^2$, $x(0) = 0$.

22. Solve the initial-value problem $\dfrac{dQ}{dt} = \dfrac{Q^3 + 2Q}{t^2 + 3t}$, $Q(1) = 1$ explicitly for $Q(t)$.

23. A quantity y varies in such a way that $\dfrac{dy}{dt} = -\dfrac{\ln 2}{30}(y - 20)$. If $y = 60$ when $t = 30$, find the value of t for which $y = 40$.

24. The volume V of water in a particular container is related to the depth h of the water by the equation $\dfrac{dV}{dh} = 16\sqrt{4 - (h - 2)^2}$. If $V = 0$ when $h = 0$, find V when $h = 4$.

25. The slope m of a curve is 0 where the curve crosses the y-axis, and $\dfrac{dm}{dx} = \sqrt{1 + m^2}$. Find m as a function of x.

26. A Police Department forensics expert checks a gun by firing a bullet into a bale of cotton. The friction force resulting from the passage of the bullet through the cotton causes the bullet to slow down at a rate proportional to the square root of its velocity. It stopped in 0.1 second and penetrated 10 feet into the bale of cotton. How fast was the bullet going when it hit?

27. The relationship between the velocity v of a rifle bullet and the distance L that it traveled in the barrel of the gun is established in ballistics by the equation

$$v = \frac{aL^n}{b + L^n},$$

where $v = \dfrac{dL}{dt}$ and $n < 1$. Find the relationship between the time t during which the bullet moves in the barrel and the distance L that it covers in the barrel.

28. In trying to determine the shape of a flexible, nonstretching cable suspended between two points A and B of equal height, we can analyze the forces acting on the cable and get the differential equation

$$\frac{d^2y}{dx^2} = k\left[1 + \left(\frac{dy}{dx}\right)^2\right]^{1/2},$$

where $k > 0$ is a constant.

    a. Use the substitution $p(x) = dy/dx$ to reduce the second-order equation to a separable first-order equation.

    b. Express the general solution of the equation in terms of exponential functions. (You may need a table of integrals here. Your CAS may evaluate the more difficult integral in an awkward way.)

29. When the drug Theophylline is administered for asthma, a concentration in the blood below 5 mg/liter of blood has little effect, but undesirable side effects appear when the concentration exceeds 20 mg/liter. Suppose a dose corresponding to 14 mg/liter of blood is administered initially. The concentration satisfies the differential equation $\frac{dC}{dt} = -\frac{C}{6}$, where the time t is measured in hours.

    a. Find the concentration at time t.

    b. Show that a second injection will need to be given after about 6 hours to prevent the drug from becoming ineffective.

    c. Given that the second injection also increases the concentration by 14 mg/liter, how long will it be before another injection is necessary?

    d. What is the shortest safe time after which a second injection may be given so that side effects do not occur?

    e. Sketch graphs of the situations in parts (b), (c), and (d).

30. One method of administering a drug is to feed it continuously into the bloodstream by a process called *intravenous infusion*. This process may be modeled by the separable (and linear) differential equation $\frac{dC}{dt} = -\mu C + D$, where C is the concentration in the blood at time t, μ is a positive constant, and D is also a positive constant, the rate at which the drug is administered.

    a. Find the *equilibrium solution* of the differential equation, the solution such that $\frac{dC}{dt} = 0$.

    b. Given that $C = C_0$ when $t = 0$, find the concentration at time t. What limit does the concentration approach as $t \to \infty$? Compare with your answer to part (a).

    c. Sketch the graph of a typical solution.

31. Let $\frac{dP}{dt} = P(1 - P)$.

    a. Find all solutions by separating variables. (You will have to integrate by using partial fractions.)

    b. Let $P(0) = P_0$. Suppose that $0 < P_0 < 1$. What happens to $P(t)$ as $t \to \infty$?

    c. Let $P(0) = P_0$. Suppose that $P_0 > 1$. What happens to $P(t)$ as $t \to \infty$?

32. Consider each of the problems in parts (a)–(e). If a problem makes sense, solve it. If a problem does not seem to make sense, explain why.

    a. Find the general solution of $y' = y$, $y(0) = 1$.

    b. Find the unique solution of the initial-value problem $y' = y$, $y(0) = 1$.

    c. Find a nontrivial solution (that is, *not* $y \equiv 0$) of $y' = y$, $y(0) = 0$.

    d. Find the unique solution of $y' = y$.

    e. Find the unique solution of $y' = y$, $y(0) = 1$, $y(1) = 0$.

33. Consider formula (2.1.1), the solution to Example 2.1.6.

    a. If $\alpha > \beta$, factor $e^{(\alpha - \beta)kt}$ from the numerator and denominator and show that $x(t) \to \beta$ as $t \to \infty$.

    b. If $\alpha < \beta$, explain what happens to $e^{(\alpha - \beta)kt}$ as $t \to \infty$ and show that $x(t) \to \alpha$ as $t \to \infty$.

2.2   LINEAR EQUATIONS

We introduced the idea of a linear differential equation in Section 1.1. Now let's see what we can do when the order of the differential equation is 1. A **linear first-order differential equation** has the form

$$a_1(x)\frac{dy}{dx} + a_0(x)y = f(x),$$

where $a_1(x)$, $a_0(x)$, and $f(x)$ are functions of the independent variable x alone. After dividing through by $a_1(x)$—being careful to note where this function is zero—we can write the equation in the *standard form*

$$\frac{dy}{dx} + P(x)y = Q(x), \tag{2.2.1}$$

where P and Q are functions of x alone. In this standard form, if the function $Q(x)$ is the zero function, we call equation (2.2.1) **homogeneous.** Otherwise, we say that the equation is **nonhomogeneous.** (Don't confuse this terminology with the use of the term *homogeneous* as explained before problems 15–18 in Exercises 2.1.) In certain applied problems, $Q(x) \neq 0$ may be referred to as the **forcing term,** the **driving term,** or the **input,** as we'll see in Example 2.2.5, for instance. The solution y can be called the **output.**

For example, $\dfrac{dy}{dx} + \sin(x)y = e^{-x}$ is linear with $P(x) = \sin x$ and $Q(x) = e^{-x}$. The equation $x\dfrac{dy}{dx} + y^2 = 0$ is *not* linear, because even when we divide by x (assuming that x is nonzero), we get $\dfrac{dy}{dx} + \left(\dfrac{y}{x}\right)y = 0$. The function $Q(x)$ can be taken as $Q(x) \equiv 0$, but the coefficient of y, $\dfrac{y}{x}$, is not a function of x alone.

However, even the complicated-looking equation $2tz^3 + 3t^2z^2\dfrac{dz}{dt} = t^5z^2$ can be made linear. Just divide by $3t^2z^2$ to get $\dfrac{dz}{dt} + \left(\dfrac{2}{3t}\right)z = \dfrac{1}{3}t^3$, so that $P(t) = \dfrac{2}{3t}$ and $Q(t) = \dfrac{1}{3}t^3$. Of course, we must consider the cases $t = 0$ and $z \equiv 0$ separately.

THE SUPERPOSITION PRINCIPLE

In some applications, it is useful to think of a linear first-order equation in terms of an **operator,** or **transformation,** L, that changes a differentiable function y into the left-hand side of equation (2.2.1): $L(y) = \dfrac{dy}{dx} + P(x)y$. Then the equation (2.2.1) can be expressed simply as $L(y) = Q(x)$. For example, if the nonhomogeneous linear equation in standard form is $\dfrac{dy}{dx} - y = x$, then we have the operator L defined as $L(y) = \dfrac{dy}{dx} - y$. If $y(x) = x^2$, for instance, then $L(y) = 2x - x^2$. A solution y of the differential equation $\dfrac{dy}{dx} - y = x$ would have to satisfy $L(y) = x$. (We can see that $y = x^2$ is not a solution.)

In this general context, if y_1 is a solution of $L(y) = Q_1(x)$ and y_2 is a solution of $L(y) = Q_2(x)$, then

$$L(y_1 + y_2) = \frac{d}{dx}(y_1 + y_2) + P(x)(y_1 + y_2)$$

$$= \frac{d}{dx}y_1 + \frac{d}{dx}y_2 + P(x)y_1 + P(x)y_2$$

$$= \left(\frac{d}{dx}y_1 + P(x)y_1\right) + \left(\frac{d}{dx}y_2 + P(x)y_2\right)$$

$$= Q_1(x) + Q_2(x) = L(y_1) + L(y_2).$$

The result $L(y_1 + y_2) = Q_1(x) + Q_2(x) = L(y_1) + L(y_2)$ is usually called the **Superposition Principle.** We can describe this by saying that adding two inputs (Q_1 and Q_2) of a linear equation gives us an output that is the sum ($y_1 + y_2$) of the individual outputs. In particular, if both $Q_1(x)$ and $Q_2(x)$ are zero, we can conclude that **the sum of two solutions of a homogeneous linear equation is again a solution.** (Be sure that you understand the statements in the last two sentences.)

Now, as in the previous paragraph, assume that y_1 is a solution of $L(y) = Q_1(x)$ and y_2 is a solution of $L(y) = Q_2(x)$, where L is an operator defined by a linear first-order differential equation $\dfrac{dy}{dx} + P(x)y = Q(x)$. We have the following more general result: If c_1 and c_2 are constants, then $L(c_1 y_1 + c_2 y_2) = c_1Q_1(x) + c_2Q_2(x) = c_1L(y_1) + c_2L(y_2)$. Any operator that satisfies this last condition is called a **linear operator.** Otherwise, it is called **nonlinear.**

The next two examples should clarify the difference between linear and nonlinear operators.

EXAMPLE 2.2.1   A Linear Operator
We can check that $y_1 = e^{-x}$ is a solution of the homogeneous linear equation $L(y) = y' + y = 0 = Q_1$ and that $y_2 = \sin x$ is a solution of $L(y) = y' + y = \cos x + \sin x = Q_2$. (*Note*: the same left-hand side, but different right-hand

sides.) You should see that $y_1 + y_2 = e^{-x} + \sin x$ is a solution of the equation $y' + y = Q_1 + Q_2 = 0 + \cos x + \sin x = \cos x + \sin x$—that is, that $L(y_1 + y_2) = Q_1 + Q_2 = L(y_1) + L(y_2)$. ◆

However, not every operator defined by a first-order equation is linear.

EXAMPLE 2.2.2   A Nonlinear Operator

Now consider the operator defined as $T(y) = xy' + y^2$ and suppose that $T(y_1) = 0$ and $T(y_2) = 0$. Then

$$T(y_1 + y_2) = x(y_1 + y_2)' + (y_1 + y_2)^2$$
$$= xy_1' + xy_2' + y_1^2 + y_2^2 + 2y_1y_2$$
$$= \overbrace{(xy_1' + y_1^2)}^{T(y_1)} + \overbrace{(xy_2' + y_2^2)}^{T(y_2)} + 2y_1y_2$$
$$= 0 + 0 + 2y_1y_2 = 2y_1y_2 \neq T(y_1) + T(y_2).$$

The equation $T(y) = xy' + y^2 = 0$ is nonlinear, and the operator T is not a linear operator. ◆

THE INTEGRATING FACTOR

Note that if the equation $\dfrac{dy}{dx} + P(x)y = Q(x)$ is homogeneous, then the equation is separable. (*Do you see why?*) Clearly, the more interesting problems are those for which $Q(x)$ is not the zero function. In addition to their applicability to significant problems, linear first-order equations are nice because you can always solve them explicitly and find the general solution. This is done by a clever technique, the use of something called an **integrating factor**—a special multiplier function that has been used to solve first-order linear equations since the late 1600s.

We will demonstrate the method in the next example, at the same time explaining its effectiveness.

EXAMPLE 2.2.3   Using an Integrating Factor

Suppose we want to solve the linear nonhomogeneous equation $y' + xy = 2x$. One way would be to reach into our sleeve and pluck out the magic function $\mu(x) = e^{x^2/2}$, an integrating factor for this equation. Now we get an equivalent differential equation by multiplying each side of the original equation by $\mu(x)$:

$$e^{x^2/2}y' + xe^{x^2/2}y = 2xe^{x^2/2}.$$

"Why would we do such a crazy thing?" you're probably asking yourself. Well, just notice that if we assume that $y = y(x)$, an implicit function of x, the Product Rule gives us $(e^{x^2/2}y)' = xe^{x^2/2}y + e^{x^2/2}y'$, the left-hand side of our new differential equation. This observation tells us that the left side is an *exact derivative* and enables us to write the differential equation in a more compact form: $(e^{x^2/2}y)' = 2xe^{x^2/2}$. (*Be sure that you see this.*) Now we can integrate each side with respect to x to get

$e^{x^2/2}y = \int 2xe^{x^2/2}dx = 2e^{x^2/2} + C$. Solving for y by multiplying each side of this last equation by $e^{-x^2/2}$, we get $y(x) = 2 + Ce^{-x^2/2}$, valid for $-\infty < x < \infty$.

We see from the closed-form solution that all solutions approach 2 as $x \to \pm\infty$: If $y(x)$ is any solution of the differential equation $y' + xy = 2x$, then

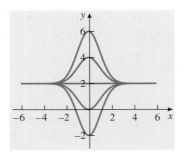

Figure 2.4
Solutions of the IVP $y' + xy = 2x$;
$y(0) = -2, 0, 2, 4,$ and 6
$-6 \le x \le 6, -2 \le y \le 6$

$\lim\limits_{x\to\infty} y(x) = 2 = \lim\limits_{x\to-\infty} y(x)$. Figure 2.4 shows five solutions of this linear equation. From top to bottom, the five particular solutions plotted correspond to $C = 4, 2, 0, -2,$ and -4, respectively. The choice $C = 0$ gives us the asymptotic solution $y \equiv 2$.

Finally, you may have recognized that our original equation is actually a *separable* equation. You should solve by separating variables and then compare your solution to the one we have given.   ◆

Rationale

Now let's step back and look at this integrating-factor technique in more generality. Suppose we have written a linear first-order differential equation in the standard form $\dfrac{dy}{dx} + P(x)y = Q(x)$. Let's take $P(x)$, the coefficient of y in the equation, and form the new function $\mu(x) = e^{\int P(x)dx}$. (Note that in Example 2.2.3, $P(x) = x$ and $e^{\int P(x)dx} = e^{\int xdx} = e^{x^2/2+K}$, where we chose $K = 0$ for convenience.) The Chain Rule and the Fundamental Theorem of Calculus tell us that

$$\frac{d}{dx}(e^{\int P(x)dx}) = e^{\int P(x)dx} \cdot \frac{d}{dx}(\int P(x)\,dx) = e^{\int P(x)dx}P(x).$$

Then, if we multiply each side of the standard form equation by $\mu(x) = e^{\int P(x)dx}$, we have

$$e^{\int P(x)dx}\frac{dy}{dx} + e^{\int P(x)dx}P(x)y = e^{\int P(x)dx}Q(x)$$

and we can rewrite the last line as

$$\frac{d}{dx}(e^{\int P(x)dx}y) = e^{\int P(x)dx}Q(x).$$

If we integrate both sides of this equation, we get

$$e^{\int P(x)dx}y = \int e^{\int P(x)dx}Q(x)dx + C$$

so we can multiply each side by $e^{-\int P(x)dx}$ to find that

$$y = e^{-\int P(x)dx} \cdot \int e^{\int P(x)dx}Q(x)dx + Ce^{-\int P(x)dx}. \qquad (2.2.2)$$

This is an explicit formula for the general solution of any first-order linear differential equation in standard form. Even if the integrals involved can't be evaluated in closed form, they can still be approximated by numerical methods usually learned in a calculus course. (Try the formula out on Example 2.2.3.) Do *not* bother memorizing this formula. Just remember that *any linear first-order equation has an explicit general solution* and understand how to find the appropriate integrating factor.

Technically, there is a whole family of integrating factors for a given linear equation (so we should say *an* integrating factor, rather than *the* integrating factor), but we can always take the family member with $K = 0$: If $R(x)$ is an antiderivative of $P(x)$ (so that $R'(x) = P(x)$), then $\mu(x) = e^{\int P(x)dx} = e^{R(x)+K} = e^{R(x)}e^{K}$, and $\mu(x)\dfrac{dy}{dx} + \mu(x)P(x)y = \mu(x)Q(x)$ has the form $e^{K}\left\{e^{R(x)}\dfrac{dy}{dx} + e^{R(x)}P(x)y\right\} = e^{K}\{e^{R(x)}Q(x)\}$. Because e^{K} is always positive, we can cancel it out in the last equation to get $\dfrac{d}{dx}(e^{R(x)}y) = e^{R(x)}Q(x)$, and we can continue to the solution as before. The constant K disappears and plays no part in the final stage of the solution, so we could take its value to be 0 from the very beginning.

Now that we know how to choose an integrating factor and find an explicit solution, let's practice.

EXAMPLE 2.2.4 **Using an Integrating Factor**

Let's try the integrating-factor technique on the linear equation $x\dfrac{dy}{dx} - 2y = x^3e^{-2x}$. The standard form of the equation is $\dfrac{dy}{dx} - \left(\dfrac{2}{x}\right)y = x^2e^{-2x}$. Our integrating factor is $\mu(x) = e^{\int -\frac{2}{x}dx} = e^{-2\ln|x|} = x^{-2}$.

Multiplying both sides of the equation in standard form by this factor, we get $x^{-2}\dfrac{dy}{dx} - 2x^{-3}y = e^{-2x}$. Recognizing the left side as the derivative of the product

$\mu(x)y = x^{-2}y$, we can write the differential equation as $\dfrac{d}{dx}(x^{-2}y) = e^{-2x}$.
Integrating both sides, we find that $x^{-2}y = \int e^{-2x}dx = -\frac{1}{2}e^{-2x} + C$. Now we can solve for y and see that $y = -\frac{1}{2}x^2 e^{-2x} + Cx^2$.   ◆

The next example, an important application of linear differential equations to electrical network theory, shows that the details of using an integrating factor may get messy.

EXAMPLE 2.2.5   A Circuit Problem

As a consequence of one of *Kirchhoff's laws* in physics, suppose we know that the current I flowing in a particular electrical circuit satisfies the first-order linear differential equation $L\dfrac{dI}{dt} + RI = v_0 \sin(\omega t)$, where L, R, v_0, and ω are positive constants that give information about the circuit. (See Exercises 25–27 for related problems.) Let's try to find the current $I(t)$ at time t, for $t > 0$, given that $I(0) = 0$. This initial condition says that at the beginning ($t = 0$) of our analysis, there is no current flowing in the circuit.

First we divide both sides of the differential equation by L to get our equation in standard form: $\dfrac{dI}{dt} + \left(\dfrac{R}{L}\right)I = \left(\dfrac{v_0}{L}\right)\sin(\omega t)$. Now, in terms of the standard form [equation (2.2.1)], we make the identifications $P(t) \equiv R/L$, a constant function, and $Q(t) = (v_0/L)\sin(\omega t)$. In this problem, the forcing term $Q(t)$ represents an (alternating) electromotive force supplied by a generator. Next we compute the integrating factor

$$\mu(t) = e^{\int P(t)dt} = e^{\int \frac{R}{L}dt} = e^{\frac{R}{L}t}.$$

Multiplying each side of the equation in standard form by $\mu(t)$, we get

$$e^{\frac{R}{L}t}\dfrac{dI}{dt} + e^{\frac{R}{L}t}\left(\dfrac{R}{L}\right)I = \left(\dfrac{v_0}{L}\right)e^{\frac{R}{L}t}\sin(\omega t), \quad \text{or} \quad \dfrac{d}{dt}\left(e^{\frac{R}{L}t}I\right) = \left(\dfrac{v_0}{L}\right)e^{\frac{R}{L}t}\sin(\omega t).$$

Integrating each side yields $e^{\frac{R}{L}t}I = \left(\dfrac{v_0}{L}\right)\displaystyle\int e^{\frac{R}{L}t}\sin(\omega t)dt$.

To evaluate this last integral, we have three choices: (1) Integrate by parts twice, (2) use a table of integrals, or (3) feed the integral to a computer algebra system capable of integration. In any case, we get

$$e^{\frac{R}{L}t}I = \left(\dfrac{v_0}{L}\right)\left[\dfrac{Re^{\frac{R}{L}t}\sin(\omega t)}{L\left(\dfrac{R^2}{L^2} + \omega^2\right)} - \dfrac{\omega e^{\frac{R}{L}t}\cos(\omega t)}{\left(\dfrac{R^2}{L^2} + \omega^2\right)}\right] + C$$

$$= \left(\dfrac{v_0}{L}\right)\dfrac{e^{\frac{R}{L}t}}{\left(\dfrac{R^2}{L^2} + \omega^2\right)} \cdot \left[\dfrac{R}{L}\sin(\omega t) - \omega\cos(\omega t)\right] + C.$$

To find the general solution, we multiply each side of this last equation by $e^{-\frac{R}{L}t}$ to get

$$I(t) = \frac{\left(\dfrac{v_0}{L}\right)}{\left(\dfrac{R^2}{L^2} + \omega^2\right)} \cdot \left[\frac{R}{L}\sin(\omega t) - \omega\cos(\omega t)\right] + Ce^{-\frac{R}{L}t}.$$

Now we use the initial condition $I(0) = 0$:

$$0 = I(0) = \frac{\left(\dfrac{v_0}{L}\right)}{\left(\dfrac{R^2}{L^2} + \omega^2\right)} \cdot \left[\frac{R}{L}\sin(\omega \cdot 0) - \omega\cos(\omega \cdot 0)\right] + Ce0$$

$$= \frac{-\omega\left(\dfrac{v_0}{L}\right)}{\left(\dfrac{R^2}{L^2} + \omega^2\right)} + C$$

so that $C = \dfrac{\omega\left(\dfrac{v_0}{L}\right)}{\left(\dfrac{R^2}{L^2} + \omega^2\right)}$, and we have (*finally!*)

$$I(t) = \frac{\left(\dfrac{v_0}{L}\right)}{\left(\dfrac{R^2}{L^2} + \omega^2\right)} \cdot \left[\frac{R}{L}\sin(\omega t) - \omega\cos(\omega t)\right] + \frac{\omega\left(\dfrac{v_0}{L}\right)}{\left(\dfrac{R^2}{L^2} + \omega^2\right)}e^{-\frac{R}{L}t}$$

$$= \frac{\left(\dfrac{v_0}{L}\right)}{\left(\dfrac{R^2}{L^2} + \omega^2\right)} \cdot \left[\frac{R}{L}\sin(\omega t) - \omega\cos(\omega t) + \omega e^{-\frac{R}{L}t}\right].$$

In this kind of problem, we call the term $\omega e^{-\frac{R}{L}t}$ (or its constant multiple) a **transient** term because it eventually goes to 0. "Eventually" means as $t \to \infty$. The trigonometric terms make up the **steady-state** part of the solution and have the same period as the original forcing term. (*Can you see that this last claim is true?*) ◆

COMPARTMENT PROBLEMS

In analyzing certain systems in biology and chemical engineering, researchers encounter a class of problems called **mixing problems** or **compartment problems.**

Suppose we have a single container, or compartment, containing some substance. Now think of some other substance entering the compartment at a certain rate, and imagine that a mixture of the two substances leaves the compartment at another rate. For example, we could be talking about a tank of water into which some chemical is introduced via a pipe. What emerges from the tank through another pipe will be some mixture of the water and the chemical. In biology and physiology, the compartment may be the bloodstream or a particular organ, such as the kidneys. In fact, mathematical analyses in these research fields often regard the organism under study as a whole collection of individual components (compartments). In the human body, these could be different organs or groups of cells, for example.

To get our bearings, we can start with a simple one-compartment model (Figure 2.5).

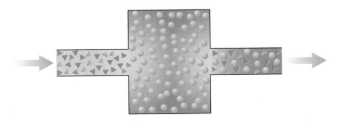

Figure 2.5
A one-compartment model

We have a single tank with a certain amount of material in it. The amount of substance (or the concentration of the substance) that is added to the tank is called the *inflow*, and the amount (or concentration) of substance leaving the tank is called the *outflow*. We assume that there is a thorough mixing process taking place in the tank—an almost instantaneous uniform blending of the two substances. To model this process using a differential equation, it is important to focus on three different *rates* associated with this situation: (1) the rate of inflow, (2) the rate at which some aspect of the mixture in the tank is changing, and (3) the rate of outflow of the mixture.

As our first example, let's look at a simple model of medicine in the bloodstream.

EXAMPLE 2.2.6   Medicine in the Bloodstream

Intravenous infusion is the process of administering a substance into the veins at a steady rate. (See Exercise 30 of Exercises 2.1.) Suppose a patient in a hospital is receiving medication through an intravenous tube that drips the substance into the bloodstream at a constant rate of I milligrams (mg) per minute. Also suppose that the medication is dispersed through the body and eliminated at a rate proportional to the concentration of the medication at the time. In this problem, *concentration* is defined to be

$$\frac{\text{Quantity of medication}}{\text{Volume of blood plus medication}}$$

where we assume that the volume V of blood plus medication remains constant. The problem is to find the concentration of the medication in the body at any time t. To do this, we can consider the bloodstream as a single compartment and examine a differential equation that models the process.

If we let $C = C(t)$ denote the concentration of the medication at time t (in mg/cm³), then the conditions of this type of problem lead us to the relation

Net rate of change = rate of inflow − rate of outflow

or

$$V\frac{dC}{dt} = I - kC,$$

where k is a positive constant of proportionality that depends on the specific medication and the physiological characteristics of the patient.

Note that the left-hand side of the differential equation is in units of

$$cm^3 \times \frac{\frac{mg}{cm^3}}{min} = \frac{mg}{min}$$ and that the right-hand term I is also in mg/min. Because C is

expressed in units of $\frac{mg}{cm^3}$, we see that the units for k, representing a removal rate, must be cm³/min. This seems appropriate. (Be sure that you understand this "dimensional analysis.")

This is a linear equation that we can write in the standard form

$$\frac{dC}{dt} + \left(\frac{k}{V}\right)C = \frac{I}{V}.$$

An integrating factor for this equation is $\mu = e^{\int \frac{k}{V}dt} = e^{\frac{kt}{V}}$. Multiplying each side of this last differential equation by μ gives us

$$e^{\frac{kt}{V}}\frac{dC}{dt} + \frac{k}{V}e^{\frac{kt}{V}}C = \left(\frac{I}{V}\right)e^{\frac{kt}{V}}$$

or

$$\frac{d}{dt}\left(e^{\frac{kt}{V}}C\right) = \left(\frac{I}{V}\right)e^{\frac{kt}{V}}$$

so that integrating each side gives us

$$e^{\frac{kt}{V}}C = \int\left(\frac{I}{V}\right)e^{\frac{kt}{V}}dt$$

and

$$C(t) = e^{-\frac{kt}{V}}\int\left(\frac{I}{V}\right)e^{\frac{kt}{V}}dt = e^{-\frac{kt}{V}}\left[\frac{V}{k}\left(\frac{I}{V}\right)e^{\frac{kt}{V}} + \alpha\right] = \frac{I}{k} + \alpha e^{-\frac{kt}{V}}.$$

Using the implied initial condition $C(0) = 0$, we find that $\alpha = -\frac{I}{k}$, so that we can write our solution as

$$C(t) = \frac{I}{k} - \frac{I}{k}e^{-\frac{kt}{V}} = \frac{I}{k}\left(1 - e^{-\frac{kt}{V}}\right).$$

Note what happens as time goes by. Analytically, $\lim\limits_{t\to\infty} C(t) = \dfrac{I}{k}$. This says that the concentration of medication in the patient's body reaches a *threshold*, or *saturation level*, of $\dfrac{I}{k}$. Figure 2.6 is a graph of the concentration when $I = 4$, $V = 1$, and $k = 0.2$, showing a saturation level of 20 mg/cm^3.

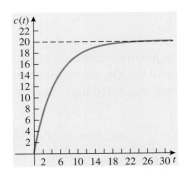

Figure 2.6

$C(t) = 20(1 - e^{-0.2t}), 0 \le t \le 30, 0 \le C \le 20$ ◆

In compartment model problems, it is often important to determine *how long* it may take for a certain result to occur.

EXAMPLE 2.2.7   Air Pollution

By 10:00 P.M. on a lively Friday night, a club of dimensions 30 feet by 50 feet by 10 feet is full of customers. Sadly, most of these customers are smokers, so cigarette smoke containing 4% carbon monoxide is introduced into the room at a rate of 0.15 cubic foot per minute. Suppose that this rate does not vary significantly during the evening. Before 10:00 there is no trace of carbon monoxide in the club, and fortunately, this club is equipped with good ventilators. These ventilators allow the formation of a uniform smoke-air mixture in the room, and they provide for the ejection of this mixture to the outside at the rate of 1.5 cubic feet per minute—that is, at a rate 10 times greater than that of the arrival of pollutants.

You want to dance and socialize, but you also want to preserve your health. A prolonged exposure to a concentration of carbon monoxide greater than or equal to 0.012% is considered dangerous by the health department. Knowing that the club closes its doors at 3 A.M., will you allow yourself to stay until the end? To be more precise, you want to find the time when the concentration of carbon monoxide reaches the critical concentration of 0.012%.

The key to this type of single-compartment problem is the fundamental relation we saw in the last example:

Net rate of change = rate of inflow − rate of outflow   (2.2.3)

Let $C(t)$ be the concentration of carbon monoxide in the club (the grams of carbon monoxide per cubic foot of air, abbreviated g/ft³) at any time t, where $t = 0$ represents 10 P.M. Then $Q(t)$, the *amount* of pollutant in the room at time t, is described by the equation $Q(t)$ = (volume of room) $\times C(t)$. Because the room is $30 \times 50 \times 10 = 15,000$ cubic feet, this expression for the amount of monoxide in the room at time t becomes $Q(t) = 15,000\, C(t)$.

Now the *rate* at which carbon monoxide is *entering* the room is given by

$$\left(0.15\,\frac{\text{ft}^3}{\text{min}}\right)\left(0.04\,\frac{\text{g}}{\text{ft}^3}\right) = 0.006\,\frac{\text{g}}{\text{min}}.$$

Similarly, the *rate* at which carbon monoxide is *leaving* the room (via the ventilators) is $\left(1.5\,\dfrac{\text{ft}^3}{\text{min}}\right) \cdot C(t)$

The relationship (2.2.3) tells us that the rate of change of the amount of carbon monoxide in the room is equal to the rate at which the pollutant is introduced minus the rate at which it leaves:

$$\frac{dQ(t)}{dt} = \frac{d}{dt}\{15,000\,C(t)\} = \text{rate of inflow} - \text{rate of outflow}$$

$$= \left(0.15\,\frac{\text{ft}^3}{\text{min}}\right)\left(0.04\,\frac{\text{g}}{\text{ft}^3}\right) - \left(1.5\,\frac{\text{ft}^3}{\text{min}}\right)C(t)$$

$$= 0.006 - 1.5\,C(t)\ \text{g/min}$$

so we have the differential equation

$$15,000\,\frac{d}{dt}C(t) = 0.006 - 1.5\,C(t).$$

This is a linear equation, and we can write it in the form

$$\frac{d}{dt}C(t) + (0.0001)C(t) = (4 \times 10^{-7}).$$

An integrating factor is $\mu(t) = e^{\int (0.0001)dt} = e^{0.0001t}$, so the last equation has the form

$$\frac{d}{dt}\{e^{0.0001t}C(t)\} = (4 \times 10^{-7})e^{0.0001t}.$$

Integrating, we find that

$$C(t) = (4 \times 10^{-7})e^{-0.0001t}\int e^{0.0001t}dt$$

$$= (4 \times 10^{-7})e^{-0.0001t}\left(\frac{e^{0.0001t}}{0.0001} + k\right)$$

$$= 0.004 + \alpha \cdot e^{-0.0001t}$$

where $\alpha = (4 \times 10^{-7}) \cdot k$.

Because we are told that $C(0) = 0$, we have $0 = C(0) = 0.004 + \alpha$, which gives us the information that $\alpha = -0.004$. Therefore, we can write the solution of our differential equation as

$$C(t) = 0.004(1 - e^{-0.0001t}).$$

Because we want to know the time t at which the concentration equals 0.012%, we must solve the equation $C(t) = 0.00012$ for t. Hence we must have

$$0.00012 = 0.004(1 - e^{-0.0001t})$$
$$0.03 = 1 - e^{-0.0001t}$$
$$e^{-0.0001t} = 1 - 0.03 = 0.97$$
$$-0.0001t = \ln(0.97)$$
$$t = \frac{\ln(0.97)}{(-0.0001)}$$

so $t = 304.59$ minutes ≈ 5.08 hours ≈ 5 hours, 5 minutes. Therefore, the critical concentration of carbon monoxide is reached at 3:05 A.M. *Too close!*   ◆

EXERCISES 2.2

For Exercises 1–18, solve each equation or initial-value problem.

1. $y' + 2y = 4x$

2. $y' + 2xy = xe^{-x^2}$

3. $\dot{x} + 2tx = t^3$

4. $y' + y = \cos x$

5. $ty' = -3y + t^3 - t^2$

6. $\dfrac{dx}{ds} = \dfrac{x}{s} - s^2$

7. $y = x(y' - x\cos x)$

8. $(1 + x^2)y' - 2xy = (1 + x^2)^2$

9. $t(x' - x) = (1 + t^2)e^t$

10. $Q' - (\tan t)Q = \sec t; \quad Q(0) = 0$

11. $xy' + y - e^x = 0; \quad y(a) = b$ [a and b are constants.]

12. $(xy' - 1)\ln x = 2y$

13. $y' + ay = e^{mx}$ [Consider two cases: $m \neq -a$ and $m = -a$.]

14. $y' + \left(\dfrac{1 - 2x}{x^2}\right)y = 1$

15. $tx' - \left(\dfrac{x}{t + 1}\right) = t; \quad x(1) = 0$

16. $y = (2x + y^3)y'$ [*Hint:* Think of y as the independent variable and of x as the dependent variable, and rewrite the equation in terms of dx / dy.]

17. $x(e^y - y') = 2$ [Use the hint from Exercise 16.]

18. $y(x) = \int_0^x y(t)\,dt + x + 1$ [Use the Fundamental Theorem of Calculus to get an ODE.]

An equation of the form $y' + a(x)y = b(x)y^n$ is called a Bernoulli equation (named for Jakob Bernoulli, 1654–1705, one of a family of noted Swiss scientists/ mathematicians). Note that if $n = 0$ or $n = 1$, we just have a linear equation. Now if n is not equal to 0 or 1, and if we divide both sides of the equation by y^n, we can let $z = y^{1-n}$ and get a linear equation in the variable z. We solve the linear equation for z in terms of x and then return to the original variables x and y. This substitution method was devised by Leibniz in 1696. For example, $y' - y = xy^2$ is a Bernoulli equation with $a(x) \equiv -1$, $b(x) = x$, and $n = 2$. Divide by y^2 and the equation becomes $y^{-2}y' - y^{-1} = x$. Letting $z = y^{-1}$, we get the linear equation $-z' - z = x$, or $z' + z = -x$. Solving for z, we find that $z = 1 - x + ce^{-x}$. Because $z = y^{-1}$, we conclude that $y = (1 - x + ce^{-x})^{-1}$. Note that we divided by y^2 and that $y \equiv 0$ is a singular solution. Find all solutions of each Bernoulli equation in Exercises 19–22.

19. $y' = \dfrac{4}{t}y - 6ty^2$

20. $\dot{x} = \dfrac{1}{t}x + \sqrt{x}$

21. $\dfrac{dy}{dx} + y = xy^3$

22. $y' + xy = \sqrt{y}$

23. In trying to regulate fishing in the oceans, international commissions have been set up to implement controls. To understand the effect of such controls, mathematical models of fish populations have been constructed. One stage in this modeling effort involves predicting the growth of an individual fish. The *von Bertalanffy growth model* is reflected in the Bernoulli equation (see above):

$$\frac{dW}{dt} = \alpha W^{2/3} - \beta W$$

where $W = W(t)$ denotes the weight of a fish and α and β are positive constants.

   a. Find the general solution of the equation.

   b. Calculate $W_\infty = \lim_{t \to \infty} W(t)$, the limiting weight of the fish.

   c. Using the answer to part (b) and the initial condition $W(0) = 0$, write the formula for $W(t)$ free of any arbitrary constants.

   d. Sketch a graph of W against t.

24. Show that if a linear first-order differential equation is homogeneous, then the equation is separable.

25. When a switch is closed in a circuit containing a *resistance R*, an *inductance L*, and a battery that supplies a constant *voltage E*, the *current I* builds up at a rate described by the equation $L\dfrac{dI}{dt} + RI = E$. [In Example 2.2.5, the **electromotive force** on the right-hand side of the equation is not constant. Instead of a battery, there is a generator supplying an **alternating voltage** equal to $(v_0/L)\sin(\omega t)$.]

   a. Find the current I as a function of time.

   b. Evaluate $\lim_{t \to \infty} I(t)$.

   c. How long will I take to reach one-half its "final" value?

   d. Find I if $I_0 = I(0) = E/R$.

26. In an electrical circuit, when a capacitor of capacitance C is being charged through a resistance R by a battery that supplies a constant voltage E, the instantaneous charge Q on the capacitor satisfies the differential equation

$$R\frac{dQ}{dt} + \frac{Q}{C} = E$$

   a. Find Q as a function of time if the capacitor is initially uncharged—that is, if $Q_0 = Q(0) = 0$.

   b. How long will it be before the charge on the capacitor is one-half its "final" value?

27. In Exercise 26, determine Q if $Q_0 = 0$ and if the battery is replaced by a generator that supplies an alternating voltage equal to $E_0 \sin(\omega t)$.

28. A study of the population of Botswana from 1975 to 1990 leads to the following model for the country's growth rate: $\dfrac{dP}{dt} = kP - \alpha t$, where t denotes time in years with 1990 corresponding to $t = 0$, $P(0) = 1.285$ (million), $k = 0.0355$, and $\alpha = 1.60625 \times 10^{-3}$. (The term kP reflects births and immigration, and the term αt expresses deaths and emigration.)

   a. Find a formula for $P(t)$.

   b. Estimate Botswana's population in the year 2010.

29. In analyzing the effect of advertising on the sales of a product, the following model can be extracted from work done by the economists Vidale and Wolf:[3]

$$\frac{dS}{dt} + \left(\frac{rA}{M} + \lambda\right)S = rA$$

Here $S = S(t)$ denotes sales, $A = A(t)$ indicates the amount of advertising, M is the saturation level of the product (the practical limit of sales that can be generated), and r and λ are positive constants. Clearly, the solution of this linear equation depends on the form of the advertising function A.

   a. Solve the equation if A is constant over a particular time interval and is zero after this:

$$A(t) = \begin{cases} \overline{A} & \text{for } 0 < t < T \\ 0 & \text{for } t > T \end{cases}$$

(You really have to solve *two* equations and then combine the solutions appropriately.)

─────────

3. M. L. Vidale and H. B. Wolfe, "Response of Sales to Advertising, " in *Mathematical Models in Marketing,* ed. Robert G. Murdick (Scranton, Pa.: Intext Educational Publishers, 1971): 249–256.

b. Sketch a typical graph of S against t. (Choose reasonable values for any constants in your solution.)

30. In the study of population genetics, biological units called *genes* determine what characteristics living things inherit from their parents. Suppose we look at a gene with two "flavors," A and a, that occur in the proportions $p(t)$ and $q(t) = 1 - p(t)$, respectively, at time t in a particular population. Suppose that we have the relation

$$\frac{dp}{dt} = \nu - (\mu + \nu)p$$

where μ is a constant describing a "forward mutation rate" and ν is another constant representing the "backward mutation rate."

a. Determine $p(t)$ and $q(t)$ in terms of $p(0)$, $q(0)$, μ, and ν.

b. Show that $\lim_{t \to \infty} p(t) = \nu/(\mu + \nu)$ and $\lim_{t \to \infty} q(t) = \mu/(\mu + \nu)$. These are called the *equilibrium* gene frequencies.

31. A tank with a capacity of 100 gallons is half full of fresh water. A pipe is opened that lets treated sewage enter the tank at the rate of 4 gallons per minute. At the same time, a drain is opened to allow the mixture to leave the tank at the rate of 2 gallons per minute. If the treated sewage contains 10 grams of usable potassium per gallon, what is the concentration of potassium in the tank when it is full? [*Be careful of your units!*]

32. A tank having a capacity of 100 gallons is initially full of water. Pure water is allowed to run into the tank at the rate of 1 gallon per minute. At the same time, brine (a mixture of salt and water) containing $\frac{1}{4}$ pound of salt per gallon flows into the tank at the rate of 1 gallon per minute. (Assume that there is perfect mixing.) The mixture flows out at the rate of 2 gallons per minute. Find the amount of salt in the tank after t minutes.

33. Suppose you have a 200-gallon tank full of fresh water. A drain is opened that removes 3 gallons per second from the tank, and at the same moment, a valve is opened that lets in a 1% solution (a 1% *concentration*) of chlorine at the rate of 2 gallons per second.

a. When is the tank half full and what is the concentration of chlorine then?

b. If the drain is closed when the tank is half full and the tank is allowed to fill, what will be the final concentration of chlorine in the tank?

34. Suppose that the maximum concentration of a drug present in a given organ of constant volume V must be c_{max}. Assuming that the organ does not contain the drug initially, that the liquid carrying the drug into the organ has constant concentration $c > c_{max}$, and that the inflow and outflow rates are both equal to r, show that the liquid must not be allowed to enter for a time longer than

$$\frac{V}{r} \ln\left(\frac{c}{c - c_{max}}\right).$$

35. A government agency has a current staff of 6000, of whom 25% are women. Employees are quitting randomly at the rate of 100 per week. Replacements are being hired at the rate of 50 per week, with the requirement that half be women.

    a. What is the size of the agency staff in 40 weeks, and what percentage is then female?

    b. What would have been the percentage if it had been required that *all* new employees be women?

36. If $V = V(t)$ represents the value of a bond at time t, $r(t)$ is the interest rate, and $K(t)$ is the coupon payment, then $\dfrac{dV}{dt} + K(t) = r(t)V$ describes the value of the bond at a time before maturity.

    a. If T is the time of the bond's maturity and $V(T) = Z$, show that

$$V(t) = e^{-\int_t^T r(x)dx}\left(Z + \int_t^T K(u)e^{\int_t^T r(x)dx}du\right)$$

    b. What does $V(t)$ look like if you have a zero-coupon bond—that is, if $K(t) \equiv 0$?

37. Suppose you have a linear first-order differential equation in the standard form

$$\frac{dy}{dx} + P(x)y = Q(x)$$

where $Q(x)$ is not the zero function.

    a. Looking at the general solution given by equation (2.2.2), show that the term $Ce^{-\int P(x)dx}$ is the general solution, y_{GH}, of the homogeneous equation you get by setting $Q(x) \equiv 0$.

    b. Show that the term $e^{-\int P(x)dx} \cdot \int e^{\int P(x)dx}Q(x)dx$ is a particular solution, y_{PNH}, of the original nonhomogeneous equation. (Thus we can express the general solution, y_{GNH}, of the nonhomogeneous equation as follows: $y_{GNH} = y_{GH} + y_{PNH}$. See Problem 39 of Exercises 1.2.)

    c. Examine the result of part (b) in light of the *Superposition Principle*.

2.3 SLOPE FIELDS

Now that we have become familiar with the basic concepts of ordinary differential equations (ODEs) and have learned how to solve separable and linear equations, we can consider a *qualitative* approach to understanding solutions of first-order equations. This is a graphical approach to an equation that provides insights into the behavior of solutions, even when we don't know the techniques for solving the equation.

Let's look at first-order equations of the general form

$$\frac{dy}{dx} = y' = f(x, y).$$

Many equations can be written in this way, with the derivative isolated. For example, we could have $\dfrac{dy}{dx} = f(x, y) = 3y - 4x$, $y' = g(x, y) = \sqrt{xy}$, $y' = F(x) = 2x^3 - 1$, or $y' = G(y) = 2 - y^2$. Now remember what a first derivative tells us. One interpretation of a derivative is as the slope of the tangent line drawn to a curve at a particular point. The equation $y' = f(x, y)$ means that at the point (x, y) of any solution curve of the differential equation, the slope of the tangent line is given by the value of the function f at that point—that is, the slope is given by $f(x, y)$. Remember that there may be a whole family of solution curves as well as singular solutions.

For a first-order differential equation, a set of possible tangent line segments (sometimes called **lineal elements**), whose slopes at (x, y) are given by $f(x, y)$, is called a **slope field** (or **direction field**) of the equation. Visually, this establishes a flow pattern for solutions of the equation. A slope field includes tangent line segments for many solutions of the equation, but the general shapes of the integral curves should be clear. You can think of these outlines as the "ghosts" of solution curves, and they may reveal certain *qualitative* aspects of the solutions, even if a closed form solution is difficult or impossible to find.

Our first example will indicate how to generate a slope field.

EXAMPLE 2.3.1   A Slope Field

To get a feeling for these ideas, let's get a piece of graph paper and plot some tangent line segments for the first-order linear equation $y' - y = x$, which we can write as $y' = f(x, y) = x + y$. To make things a bit easier, we can construct a table (Table 2.1).

TABLE 2.1   Slopes at Points (x, y) for $y' = x + y$

x	y	Point (x, y)	$y' = x + y$ Slope at (x, y)
-3	3	$(-3, 3)$	0
1	-1	$(1, -1)$	0
0	0	$(0, 0)$	0
0	1	$(0, 1)$	1
1	0	$(1, 0)$	1
2	-1	$(2, -1)$	1
-1	2	$(-1, 2)$	1
0	-1	$(0, -1)$	-1
-1	0	$(-1, 0)$	-1
2	-3	$(2, -3)$	-1
$\vdots$	$\vdots$	$\vdots$	$\vdots$

We've made things even easier for ourselves by choosing points at which the slopes are 0, 1, and -1. Now we can draw some tangent line segments corresponding to these slopes (Figure 2.7a).

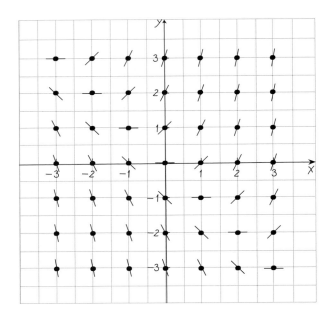

Figure 2.7a
Some lineal elements for $y' - x + y$

Note that we have drawn the little tangent line segments so that the midpoint of each segment is the point (x, y). We have used portions of the slope field given by $f(x, y) = 0$ and $f(x, y) = \pm 1$. Figure 2.7b is a computer-drawn direction field for the same ODE, with some solution curves superimposed on the slope field.

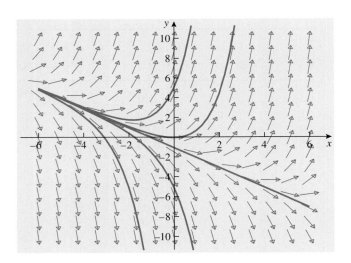

Figure 2.7b
Slope field for $y' = x + y$, $-6 \leq x \leq 6$, $-10 \leq y \leq 10$
and five computer-generated solution curves

Note that as $x \to -\infty$, the solution curves seem to be approaching a straight line as an asymptote. The solution curves seem to be veering *away* from this line as $x \to +\infty$. If you look very closely, you may be able to guess that the straight line is $x + y = -1$, or $y = -1 - x$. In Section 2.2 we learned how to find the general solution, $y = -x - 1 + Ce^x$, for this linear equation. The straight line $y = -1 - x$ is the particular solution of the ODE corresponding to $C = 0$, a solution of the initial-value problem (IVP) $y' - y = x$, $y(0) = -1$. Also note that if y is the general solution and $C \neq 0$, then $\lim_{x \to +\infty} y(x) = \infty$ if $C > 0$ and $\lim_{x \to +\infty} y(x) = -\infty$ if $C < 0$. ◆

Although the slope field suggests some features of the solution curves, we have to be careful not to read too much into it. In the last example, without the analytic form of the general solution or some sound numerical evidence, we can't be sure that y doesn't have vertical asymptotes, so that $y \to \pm \infty$ as x approaches some *finite* value x_0.

Note that in Example 2.3.1 we used portions of the slope field given by $f(x, y) = 0$ and $f(x, y) = \pm 1$. For any first-order equation $y' = f(x, y)$, if we look at the set of points (x, y) such that $f(x, y) = C$, a constant, we get an **isocline**—a curve or curves along which the slopes of the tangent lines are all the same. (The word *isocline* is made up of parts that mean "equal inclination" or "equal slope.") Isoclines are used to simplify the construction of a slope field because once you draw the isoclines, you can quickly and easily draw, for each C, a series of parallel line segments of slope C, all having their midpoints on the curve $f(x, y) = C$. In Example 2.3.1 the isoclines are the curves $x + y = C$, which are straight lines through $(0, C)$ and $(C, 0)$ with slope -1. (Be sure that you understand the previous sentence.)

It is important to realize that *an isocline is usually not a solution curve* but that through any point on an isocline, a solution to the differential equation passes with slope C. However, as we'll see in Section 2.5, isoclines corresponding to $C = 0$—called **nullclines**—turn out to be important solutions (*equilibrium solutions*) of equations in which the independent variable does not appear explicitly—that is, equations of the form $y' = f(y)$.

The next example has something important to say about the difference between equations of the general form $y' = f(x, y)$ and equations of the special form $y' = f(y)$.

EXAMPLE 2.3.2   The slope field (Figure 2.8) corresponding to the equation $x' = f(x) = -2x$ reveals something interesting about certain kinds of equations and their corresponding slope fields. (*Don't be confused by the labeling of the axes.* Here we are assuming that t is the independent variable and x is the dependent variable: $x = x(t)$.) First of all, note that algebraically we can write the equation in the form $F(x, x') = x' + 2x = 0$, or $x' = f(x) = -2x$. In other words, we have a first-order equation in which the independent variable t does not appear explicitly. This says that the slopes of the tangent line segments making up the slope field of this equation depend only on the values of x. In the slope field plot given in Figure 2.8, if you fix the value of x by drawing a horizontal line $x = C$ for

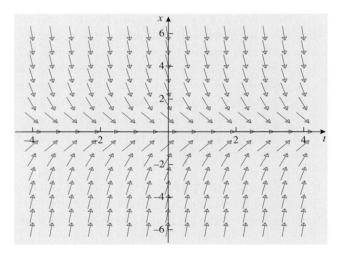

Figure 2.8
Slope field for $x' = -2x$, $-4 \leq t \leq 4$, $-6 \leq x \leq 6$

any constant C, you will see that all the tangent line segments along this line have the same slope, no matter what the value of t. Another way to look at this is to realize that you can generate infinitely many solutions by taking any one solution and translating (shifting) its graph left or right. (See Exercise 21.)   ◆

AUTONOMOUS AND NONAUTONOMOUS EQUATIONS

A differential equation, such as the one in the last example, in which the independent variable does not appear explicitly is called an **autonomous** equation. If the independent variable *does* appear, the equation is called **nonautonomous.** This definition is valid for an equation of any order.

For example, $y' = y^2 - t^2$ is nonautonomous because the independent variable t appears explicitly, whereas $y' = 3y^4 + 2\sin(y)$ is autonomous because the independent variable (t, x, or whatever) is missing. Note that the independent variable is always present *implicitly* (in the background), but if you don't see it "up front," the equation is autonomous. Example 2.3.1 discusses a nonautonomous equation. If we look carefully at its slope field (Figure 2.7b), we see that the slopes change as we move along any horizontal line.

Autonomous equations arise frequently in physical problems because the physical components generally depend on the *state* of the system, but not on the actual time. For example, according to *Newton's Second Law of Motion* (to be discussed and applied to spring-mass problems in Chapter 4), an object of mass m falling under the influence of gravity satisfies the autonomous equation $\ddot{x} = -g$, where $x(t)$ is the position of the mass measured from the earth's surface and g is the acceleration due to gravity. Gravity is considered time-independent because the mass follows the same path no matter when the mass is dropped.

Now let's see how we can recognize the correspondence between first-order differential equations and their slope fields.

EXAMPLE 2.3.3   Looking at the two differential equations

$$\text{(A) } \frac{dx}{dt} = x^2 - t^2 \quad \text{and} \quad \text{(B) } \frac{dx}{dt} = x^2 - 1$$

and the accompanying slope fields 1–4, let's try to match each equation with exactly one of the slope fields.

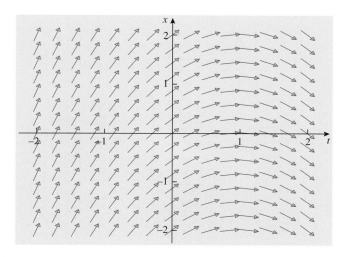

Figure 2.8a
Slope field 1

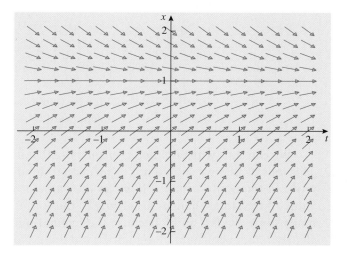

Figure 2.8b
Slope field 2

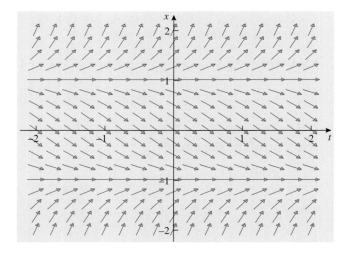

Figure 2.8c
Slope field 3

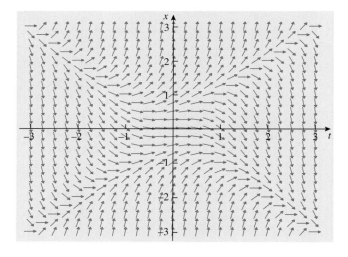

Figure 2.8d
Slope field 4

We can start with equation (A) and note that it is a *nonautonomous* equation. This tells us that we should not expect equal slopes along horizontal lines. As we move horizontally—that is, if we fix the value of x and vary the value of t—the value of the slope changes according to the formula $x^2 - t^2$. This analysis eliminates slope fields 2 and 3 because the inclinations of the tangent line segments clearly remain constant along horizontal lines. Now if we write equation (A) in factored form, $\dfrac{dx}{dt} = (x + t)(x - t)$, we can see that the tangent line segments

must be horizontal where $x = t$ or $x = -t$ because that's where the slope $\dfrac{dx}{dt}$

equals 0. (These are the *nullclines*—isoclines corresponding to $C = 0$.) Looking carefully at slope fields 1 and 4, we see that field 4 exhibits a series of horizontal "steps" forming an X through the origin. If you look closely, it seems that these horizontal line segments lie on the lines $x = t$ and $x = -t$, so we conclude that equation A corresponds to slope field 4.

Equation B is *autonomous* because the independent variable t does not appear explicitly. The corresponding slope field must show equal slopes along any horizontal line. Only fields 2 and 3 exhibit this behavior. What else can we look for?

Well, if we factor equation (B) to get $\dfrac{dx}{dt} = (x + 1)(x - 1)$, we realize that the

slope field must show horizontal line segments when $\dfrac{dx}{dt}$ equals 0—that is, where

$x = 1$ or $x = -1$. Slope field 2 has horizontal tangents at $x = 1$ but doesn't have them at $x = -1$. Only slope field 3 shows zero slopes along both horizontal lines $x = 1$ and $x = -1$, so we conclude that equation (B) must match up with slope field 3.   ◆

The next example shows an advantage—and a possible drawback—of using slope fields.

EXAMPLE 2.3.4   A Slope Field for an Autonomous Equation
The first-order nonlinear autonomous equation $y' = y^4 + 1$ looks innocent, but (*surprise!*) it has the one-parameter family of solutions

$$\frac{\sqrt{2}}{8}\ln\left(\frac{y^2 + \sqrt{2}y + 1}{y^2 - \sqrt{2}y + 1}\right) +$$

$$\frac{\sqrt{2}}{4}\{\arctan(y\sqrt{2} + 1) + \arctan(y\sqrt{2} - 1)\} = t + C$$

(The equation is separable, but the integration required to solve it is tricky. Use your CAS to evaluate the integral, but don't be surprised if your answer doesn't look exactly like the one given here.) Without looking at the solution formula, you can see immediately that the differential equation has no constant function as a solution: If y is constant, then $y' = 0$; but the right-hand side of the differential equation, $y^4 + 1$, can never be zero. In fact, this simple analysis shows that any solution y must be an *increasing* function. (*Why?*)

The fearsome formula describing a family of implicit solutions gives little useful information. However, let's take a look at the equation's slope field (Figure 2.9). First of all, the autonomous nature of the equation is clear from the fact that

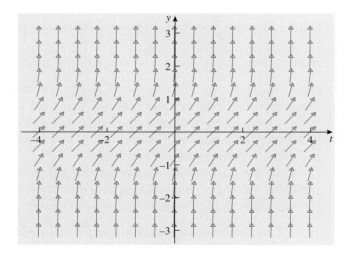

Figure 2.9
Slope field for $y' = y^4 + 1$, $-4 \le t \le 4$, $-3 \le y \le 3$

along any horizontal line, the inclinations of the tangent line segments are equal. Furthermore, it should be evident that any solution curve is increasing. In fact, any solution curve has vertical asymptotes holding it in on the left and on the right. Of course, we can't tell whether this last statement is true by merely looking at the slope field. A purely graphical analysis can't reveal this. But the slope field does give us an idea of what to expect when we try to solve the equation analytically or to approximate a solution numerically.   ◆

As we shall see in later chapters, a type of slope field can help us analyze certain *systems* of differential equations as well.

EXERCISES 2.3

In Exercises 1–15, first sketch the slope field for the given equation by hand, and then try using a computer or graphing calculator to generate the slope field. Sketch several possible solution curves for each equation. (Your graphing calculator or CAS may have trouble with Exercises 5 and 10.)

1. $y' = x$

2. $\dfrac{dx}{dt} = t$

3. $\dfrac{dr}{dt} = t - 2r$

4. $\dfrac{dx}{dt} = 1 - 0.01x$

5. $Q' = |Q|$

6. $y' = y - x$

7. $r\dfrac{dr}{dt} = -t$

8. $\dfrac{dy}{dx} = \dfrac{1}{y}$

9. $\dfrac{dy}{dx} = \dfrac{2y}{x}$

10. $y' = \max(x, y)$, the larger of the two values x and y

11. $y' = x^2 + y^2$ **12.** $x' = 1 - tx$ **13.** $\dfrac{dy}{dt} = \dfrac{ty}{t^2 - 1}$

14. $\dfrac{dP}{dt} = 2P(1 - P)$ **15.** $y' = \dfrac{\cos x}{\cos y}$

16. The German physiologist Gustav Fechner (1801–1887) devised the model expressed as $\dfrac{dR}{dS} = \dfrac{k}{S}$, where k is a constant, to describe the response, R, to a stimulus, S. Use technology to sketch the slope field for $k = 0.1$.

17. A *bimolecular chemical reaction* is one in which molecules of substance A collide with molecules of substance B to create substance X. The rate of formation (the *velocity of reaction*) is given by a differential equation of the form $\dfrac{dx}{dt} = k(\alpha - x)(\beta - x)$, where α and β represent the initial amounts of substances A and B, respectively, and $x(t)$ denotes the amount of substance X present at time t. (See Example 2.1.6.)

   a. Use technology to plot the slope field when $\alpha = 250$, $\beta = 40$, and $k = 0.0006$.

   b. If $x(0) = 0$, what seems to be the behavior of x as $t \to \infty$?

18. The one-parameter family $y = \dfrac{c}{t}$ represents a solution of $\dfrac{dy}{dt} = f(t, y)$. Sketch (by hand) the slope field of the differential equation.

19. Describe the isoclines of the equation $\dfrac{dy}{dt} = \dfrac{y + t}{y - t}$.

20. Which of the equations in Exercises 1–15 are *autonomous*? If you have done some of these problems, look at their slope fields to confirm your answers.

21. If $\varphi(t)$ is a solution of an autonomous differential equation $x' = f(x)$ and k is any real number, show that $\varphi(t + k)$ is also a solution. [*Hint:* Use the Chain Rule.]

22. Use the result of Exercise 21 to show that if $\sin t$ is a solution of an autonomous differential equation $x' = f(x)$, then $\cos t$ is also a solution. [*Hint:* How are the graphs of sine and cosine related?]

23. In your own words, explain the important differences in the slope fields for the following forms of differential equations:

   a. $y' = f(t, y)$ **b.** $y' = f(t)$ **c.** $y' = f(y)$

24. In any way your instructor tells you, manually or using technology, sketch the slope field for each of the following equations, and then sketch the solution curve that passes through the given point (x_0, y_0).

   a. $\dfrac{dy}{dx} = x^2$; $(x_0, y_0) = (0, -2)$ **b.** $\dfrac{dy}{dx} = -xy$; $(x_0, y_0) = (0, 3)$

25. Match each of these equations with one of the accompanying slope fields.

  a. $\dfrac{dy}{dt} = y + 1$ **b.** $\dfrac{dy}{dt} = y - t$ **c.** $\dfrac{dy}{dt} = t + 1$

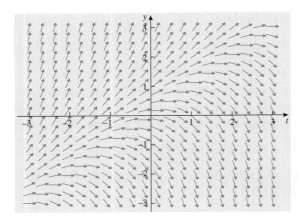

Figure 2.9a

Slope field 1

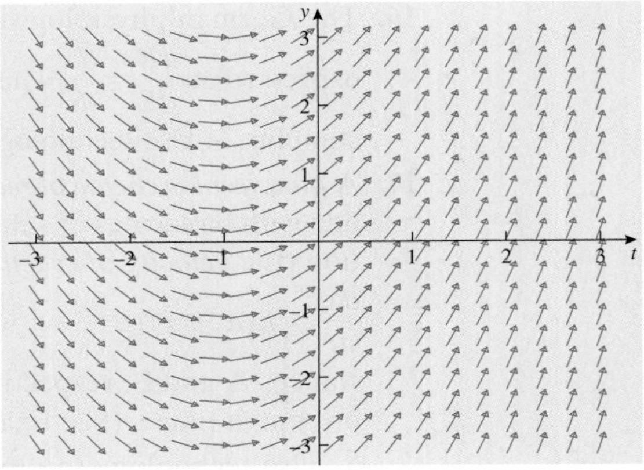

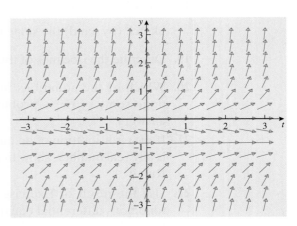

Figure 2.9b

Slope field 2

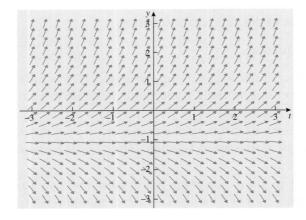

Figure 2.9c

Slope field 3

Figure 2.9d

Slope field 4

26. By looking at the slope field for each of the following equations, describe the behavior of solutions of each equation as $t \to \infty$. How do your answers seem to depend on initial conditions in each case?

a. $y' = 3y$

b. $\dfrac{dP}{dt} = P(1 - P)$

c. $y' = e^{-t} + y$

d. $y' = 3 \sin t + 1 + y$

27. Examine the slope field for the first-order nonlinear equation $\dfrac{dy}{dx} = e^{-2xy}$. On the basis of your examination, what can you say about the solutions to this equation? (You may want to look at parts of various quadrants more closely.) In particular, what can you say about the behavior of the solutions as $x \to \pm\infty$? (*Be careful: Some solutions become infinite as x approaches finite values.*)

2.4   PHASE LINES AND PHASE PORTRAITS

THE LOGISTIC EQUATION

When we are dealing with an *autonomous* first-order equation, the qualitative analysis can be strengthened quite a bit to provide useful information about solution curves.

We'll begin to examine this new analysis technique by using an important population growth model, first studied by the Belgian mathematician Pierre Verhulst in 1838 and later rediscovered independently by the American scientists Raymond Pearl and Lowell Reed in the 1920s.

EXAMPLE 2.4.1   The Qualitative Analysis of the Logistic Equation

The autonomous differential equation $\dfrac{dP}{dt} = P(1 - P)$, a particular example of something called a **logistic equation,** is useful, for instance, in analyzing such phenomena as epidemics. (We dealt with this equation in another way in part (b) of Problem 26 in Exercises 2.3.) In an epidemiological situation, P could represent the infected population as a function of time. We'll work more with this kind of model later, but for now let's ignore the fact that this is a separable equation that we can solve explicitly and see what basic calculus can tell us.

First of all, the right-hand side represents a derivative, the instantaneous rate at which P is changing with respect to time. From calculus, we know that if the derivative is positive, then P is increasing, and if the derivative is negative, then P is decreasing. Now when is dP/dt positive? The answer is when $P(1 - P)$ is greater than zero. Similarly, dP/dt is negative when $P(1 - P)$ is less than zero. Finally, we see that $dP/dt = 0$ when $P(1 - P) = 0$—that is, when $P = 0$ or $P = 1$.

These two critical points split a *P*-axis into three pieces (Figure 2.10): $-\infty < P < 0$, $0 < P < 1$, and $1 < P < \infty$.

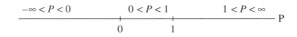

Figure 2.10
P-axis divided by critical points

What is the sign of *dP/dt* when *P* satisfies $-\infty < P < 0$? Well, for these values of *P*, *P* is negative and $1 - P$ is positive, making the product $dP/dt = P(1 - P)$ *negative*. This means that *P* is *decreasing*. When *P* is between 0 and 1, we see that *P* is positive and $1 - P$ is positive, so *dP/dt* is *positive* and *P* is *increasing*. Finally, when *P* is greater than 1, we see that *P* is positive and $1 - P$ is negative, so *dP/dt* is *negative* and *P* is *decreasing*.

We can redraw Figure 2.10 with arrows indicating whether *P* is increasing or decreasing on a particular interval for *P*. The direction of any arrow shows the algebraic sign of *dP/dt* in a subinterval and so indicates whether *P* is increasing or decreasing: → means "positive derivative/increasing *P*" and ← means "negative derivative/decreasing *P*." Figure 2.11 is called the **(one-dimensional) phase portrait** of the differential equation $\dfrac{dP}{dt} = P(1 - P)$. The horizontal line itself is called the **phase line**.

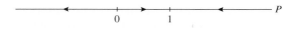

Figure 2.11
Phase portrait of $\dfrac{dP}{dt} = P(1 - P)$

We can actually do a little more in this situation. If we differentiate each side of our original differential equation with respect to *t*, we get

$$\frac{d^2P}{dt^2} = \frac{dP}{dt} \cdot (1 - 2P) = P(1 - P)(1 - 2P)$$

where we have replaced *dP/dt* by the right-hand side of the original differential equation. (*Check all this!*) Remember that the second derivative of a function tells us about the *concavity* of the function: *P* is concave up when $d^2P/dt^2 > 0$ and *P* is concave down when $d^2P/dt^2 < 0$. Using the critical points $0, \frac{1}{2}$, and 1 of $\dfrac{d^2P}{dt^2}$ as a guide, we can construct the following table of signs (Table 2.2).

TABLE 2.2   Table of Signs

P Interval	P	$1 - P$	$1 - 2P$	$P'' = P(1 - P)(1 - 2P)$	Concavity
$(-\infty, 0)$	$-$	$+$	$+$	$-$	Down
$(0, \frac{1}{2})$	$+$	$+$	$+$	$+$	Up
$(\frac{1}{2}, 1)$	$+$	$+$	$-$	$-$	Down
$(1, \infty)$	$+$	$-$	$-$	$+$	Up

We have to remember that t is the independent variable in this problem and P is the dependent variable. It's easy to lose sight of this because the (autonomous) form of this differential equation makes us focus on P alone.

On the basis of our analysis of $\dfrac{dP}{dt}$ and $\dfrac{d^2P}{dt^2}$, let's take a look at what the graph of P could look like in the t-P coordinate plane (Figure 2.12). We'll focus on the first quadrant because $t \geq 0$ and $P \geq 0$ are realistic assumptions when one is dealing with a population growth model. Note that the phase line (representing the P-axis) is now drawn vertically and placed next to the graph and that we've marked the important values from our previous investigation of $\dfrac{dP}{dt}$ and $\dfrac{d^2P}{dt^2}$.

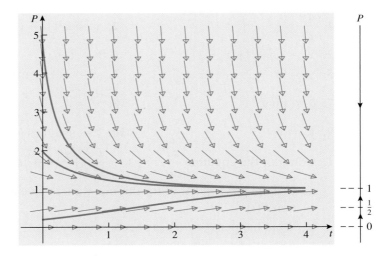

Figure 2.12

Sketch of three solutions of $\dfrac{dP}{dt} = P(1 - P)$, based on the phase portrait and concavity.

Initial conditions are $P(0) = 0.2, 2,$ and $5; 0 \leq t \leq 4, 0 \leq P \leq 5$.

The graph indicates the change of concavity at $P = \frac{1}{2}$. Note how the three solutions we have sketched seem to approach $P = 1$ as an asymptote as t increases. In terms of a realistic scenario, this says that if the initial population is below 1 (the unit could be thousands or millions), the population will increase to 1 asymptotically. On the other hand, any population starting above 1 will eventually decrease toward 1. If we had drawn the rest of the phase line (for $P < 0$) and solutions in the fourth quadrant ($t \geq 0$, $P < 0$), we would have seen these solutions moving away from the t-axis. We'll say more about this phenomenon in the next section.   ◆

The *logistic equation,* which is commonly used to model population growth when resources (such as food) are limited, is usually written as $\dfrac{dP}{dt} = rP\left(1 - \dfrac{P}{k}\right)$, where r is a per capita growth rate balancing births and deaths and k represents the theoretical maximum population that a given environment (forest, petri dish, etc.) can sustain. The value k is called the *carrying capacity*. This model will reappear from time to time in this text.

EXERCISES 2.4

Draw phase portraits for each of the equations in Exercises 1–10.

1. $\dfrac{dy}{dt} = y^2 - 1$

2. $y' = y^2(1 - y)^2$

3. $x' = (x + 1)(x - 3)$

4. $\dot{x} = \cos x$

5. $y' = e^y - 1$

6. $y' = y(1 - y)(2 - y)$

7. $\dot{y} = \sin y$

8. $x' = 1 - \dfrac{x}{1 + x}$

9. $y' = ye^{y-1}$

10. $\dot{y} = \sin y \cos y$

11. Consider the equation $\dfrac{dP}{dt} = \left(1 - \dfrac{P}{15}\right)^3\left(\dfrac{P}{7} - 1\right)P^5$, with $P(0) = 3$.

    a. Use the phase portrait for this equation to give a rough sketch of the solution $P(t)$.

    b. What happens to $P(t)$ as t becomes very large?

12. Using one of *Kirchhoff's laws* in physics, it is found that the current, I, flowing in a particular electrical circuit satisfies the equation $0.5\dfrac{dI}{dt} + 10I = 12$.

    (The resistance is 10 ohms, the inductance is 0.5 henry, and there is a 12-volt battery.)

    a. Sketch the phase portrait of the equation.

    b. If the initial current, $I(0)$, is 3 amps, use the sketch you made in part (a) to describe the behavior of I for large values of t.

13. Example 2.1.6 and Problem 17 of Exercises 2.3 indicated that a type of chemical reaction can be modeled by the equation $\dfrac{dx}{dt} = k(\alpha - x)(\beta - x)$.

    a. If $\alpha = 250$, $\beta = 40$, and k is a positive constant, produce the phase portrait of the equation.

    b. If $x(0) = 0$, how does x behave as $t \to \infty$?

14. Look at the equation $\dfrac{dy}{dt} = (1 + y)^2$.

    a. What happens to solutions with initial conditions $y(0) > -1$ as t increases?

    b. Describe the behavior of solutions with initial conditions $y(0) < -1$ as t increases.

15. Given the following phase portrait for $\dfrac{dy}{dt} = f(y)$, make a rough sketch of the graph of $f(y)$, assuming that $y = 0$ is in the center of the phase line.

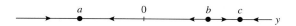

16. Given the following phase portrait, find a first-order ODE that is consistent with this phase portrait.

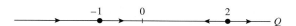

17. The *Landau equation* arises in the analysis of the dynamics of fluid flow. It is $\dfrac{dx}{dt} = ax - bx^3$, where a and b are positive real constants.

    a. Draw the phase portrait of the Landau equation.

    b. What happens to x as t increases if $x(0) = \sqrt{\dfrac{a}{b}} + \varepsilon$, where ε is a small positive quantity?

    c. What happens to x as t increases if $x(0) = 0$?

    d. How does x behave as t increases if $x(0) = \sqrt{\dfrac{a}{b}} - \varepsilon$?

2.5   EQUILIBRIUM POINTS: SINKS, SOURCES, AND NODES

Let's take another look at Figure 2.12 and focus on the **critical points**—the places where $\dfrac{dP}{dt} = 0$. Geometrically, these are the horizontal lines $P = 0$ and $P = 1$, which represent the functions $P(t) \equiv 0$ and $P(t) \equiv 1$, constant solutions of the differential equation $\dfrac{dP}{dt} = P(1 - P)$. These values of P are called **equilibrium points** or **stationary points** of the autonomous differential equation. We also say that $P(t) \equiv 0$ and $P(t) \equiv 1$ are **equilibrium solutions** or **stationary solutions** of the equation. Assuming that the solutions of an autonomous differential equation describe some physical, economic, or biological system, we can conclude that if the system actually reaches an equilibrium point P, it must always have been at P—and will always remain at P. (*Think about this. You have* $\dfrac{dP}{dt} = 0$ *at an equilibrium point.*)

We can go further in this analysis and classify equilibrium points for autonomous first-order differential equations. It turns out that there are only three basic kinds of equilibrium points: **sinks, sources,** and **nodes.**

If we look at the equilibrium point $P = 1$ in Example 2.4.1, we see from Figure 2.12 that the solution curves near the line $P = 1$ seem to swarm into (or converge to) the horizontal line. We call $P = 1$ a **sink.** A little more accurately, an equilibrium solution $P \equiv k$ is a **sink** if solutions with initial conditions sufficiently close to $P \equiv k$ are asymptotic to $P \equiv k$ as $t \to \infty$. This idea of being "sufficiently close" can be made mathematically precise, but we shall just consider the situation intuitively. Sinks are also called **attractors** or **asymptotically stable solutions.** The term *sink* is meant to suggest the drain of a bathroom or kitchen sink: Along the sides, water that is close enough will flow into the drain.

On the other hand, we see different behavior near $P = 0$. As we look along solution curves from left to right, they seem to be moving *away* from the line $P = 0$. The equilibrium point $P = 0$ is called a **source.** In other words, an equilibrium solution $P \equiv k$ is a **source** if solutions with initial conditions sufficiently close to $P \equiv k$ are asymptotic to $P \equiv k$ as $t \to -\infty$—that is, as we go backward in time. A source is also called a **repeller** or an **unstable equilibrium solution.** Here we can think of a faucet pouring forth water or a hand-held hair dryer putting out streams of hot air.

If we had another equilibrium point such that nearby solutions showed any other kind of behavior—perhaps somewhat like a sink and somewhat like a source at the same time—we would call that equilibrium point a **node.** (See Example 2.5.2 and Figure 2.15.) More technically, we can refer to a node as a **semistable equilibrium solution.**

A Test for Equilibrium Points

We can test equilibrium points/solutions for autonomous first-order equations using a criterion that should remind you of the *Second Derivative Test* from calculus:

Derivative Test

> If x^* is an equilibrium point for the equation $\dfrac{dx}{dt} = f(x)$, then it is true that (1) if $f'(x^*) > 0$, then x^* is a source; (2) if $f'(x^*) < 0$, then x^* is a sink; and (3) if $f'(x^*) = 0$, then we can't tell, without further investigation, what sort of equilibrium point x^* may be.

In Example 2.4.1, we saw that $P = 0$ and $P = 1$ were equilibrium points. Because $f(P) = P(1 - P)$, we have $f'(P) = 1 - 2P$, so $f'(0) = 1 > 0$ indicates that $P = 0$ is a *source* and $f'(1) = -1 < 0$ shows that $P = 1$ is a *sink*.

We can understand why the Derivative Test works by using the concept of *local linearity*: Near an equilibrium solution x^*, we can approximate $f(x)$ by the equation of its *tangent line* at x^*. (See Appendix A.1 if necessary.) Therefore, if x is close enough to x^*, we can write

$$\frac{dx}{dt} = f(x) \approx f(x^*) + f'(x^*)(x - x^*) = f'(x^*)(x - x^*)$$

because $f(x^*) = 0$ when x^* is an equilibrium solution. Now use Table 2.3 to compare the signs of $f'(x^*)$ and $(x - x^*)$:

TABLE 2.3 Signs of $\dfrac{dx}{dt}$

$(x - x^*)$	$f'(x^*)$	$\dfrac{dx}{dt}$
+	+	+
+	−	−
−	+	−
−	−	+

The first row of signs in Table 2.3, for example, tells us that if $(x - x^*)$ is positive—so that a solution x is slightly *above* the equilibrium solution—and $f'(x^*) > 0$, then $\dfrac{dx}{dt} > 0$, which means that the solution x is moving *away* from

x^*. This last statement says that x^* must be a *source*. Similarly, the third row of signs indicates that if x starts out *below* x^* and $f'(x^*) > 0$, then x falls away from x^* as t increases, so x^* is a source. The remaining two rows describe a sink.

The next two examples show us the power and the limitations of the Derivative Test.

EXAMPLE 2.5.1   Using the Derivative Test

If we examine the autonomous equation $\dfrac{dx}{dt} = x - x^3 = x(1 - x^2)$, we see that the equilibrium points are $x = 0$, $x = -1$, and $x = 1$. Can we determine what kinds of equilibrium points these are without any kind of graph?

Yes, we just apply the Derivative Test given above. First of all, we have $\dfrac{dx}{dt} = f(x)$, where $f(x) = x - x^3$, so $f'(x) = 1 - 3x^2$. Because $f'(0) = 1 > 0$, we know that $x = 0$ is a *source*. The fact that $f'(-1) = -2 < 0$ tells us that $x = -1$ is a *sink*. Finally, because $f'(1) = -2 < 0$, we see that $x = 1$ is another *sink*.

The phase portrait shown in Figure 2.13 reflects this information. Finally, the slope field (Figure 2.14) confirms our analysis.

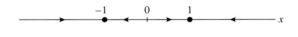

Figure 2.13

Phase portrait of $\dfrac{dx}{dt} = x - x^3$

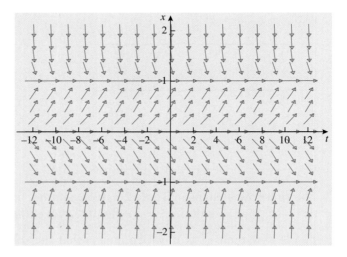

Figure 2.14

Slope field for $\dfrac{dx}{dt} = x - x^3$

EXAMPLE 2.5.2   **Failure of the Derivative Test**

Let's look at the first-order nonlinear equation $\dfrac{dx}{dt} = f(x) = (1 - x)^2$. The only equilibrium solution is $x \equiv 1$, and we have $f'(x) = 2(1 - x)(-1) = 2(x - 1)$. Because $f'(1) = 0$, our test doesn't allow us to draw any conclusion. However, we can examine the behavior of $f'(x)$ near $x = 1$ to get an idea of what's going on.

We can see that $f'(x)$ is greater than zero for values of x greater than 1, so $x \equiv 1$ looks like a source; but values of x just below 1 give us *negative* values of the derivative, so $x \equiv 1$ looks like a sink. This ambivalent behavior enables us to conclude that $x \equiv 1$ is a *node*. Figure 2.15 shows the phase portrait of this equation. Figure 2.16 shows the slope field with some particular solutions superimposed.

Figure 2.15

Phase portrait of $\dfrac{dx}{dt} = (1 - x)^2$

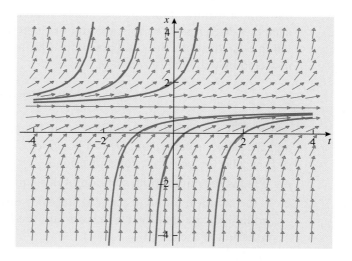

Figure 2.16

Solutions of $\dfrac{dx}{dt} = (1 - x)^2 : x(0) = -\frac{1}{2}, \frac{1}{2},$ and 2

Note that solution curves starting *above* the line $x = 1$ seem to flow *away* from the line $x = 1$, whereas those starting *below* the equilibrium solution flow *toward* the line $x = 1$. In other words, the point $x = 1$ is neither a sink nor a source. It is a *node*.   ◆

Equilibrium solutions and their nature will be particularly useful when we discuss qualitative aspects of *systems* of linear and nonlinear equations in Chapters 4, 5, and 7.

EXERCISES 2.5

For Exercises 1–12, find the equilibrium point(s) of each equation and classify them as sinks, sources, *or* nodes.

1. $y' = y^2(1 - y)^2$

2. $\dot{x} = \cos x$

3. $y' = e^y - 1$

4. $y' = y^2(y^2 - 1)$

5. $\dot{x} = ax + bx^2, \quad a > 0, b > 0$

6. $\dot{x} = x^2 - x^3$

7. $\dot{y} = 10 + 3y - y^2$

8. $\dot{x} = x(2 - x)(4 - x)$

9. $\dot{x} = -x^3$

10. $\dot{x} = x^3$

11. $\dot{y} = y \ln(y + 2)$

12. $\dot{x} = x - \cos x$ (*Hint:* Either use technology to find the equilibrium point explicitly (via a **solve** command) or graph $y = x$ and $y = \cos x$ separately to estimate the graphs' point of intersection.)

13. A lake has two rivers flowing into it, one that each day discharges a certain amount of water containing a concentration of pollutant and the other that each day discharges a certain amount of clean water. Assuming that the lake volume is constant, the total amount of pollution in the lake, $Q(t)$, can be modeled by the *balance equation*

$$\frac{dQ}{dt} = D(Q^* - Q)$$

where D is a positive constant involving the two rates of flow into the lake and the lake's volume, and Q^* is a positive constant involving volume, rates of flow, and the pollutant concentration.

a. What is the equilibrium solution of this equation?

b. Is the solution found in part (a) stable or unstable? (For example, a *sink* would indicate that the clean river input *reduces* the long-term amount of pollution in the lake.)

14. The following equation has been proposed for determining the speed of a rowing boat:[4]

$$M\frac{du}{dt} = \frac{8P}{u} - bSu^2$$

where $u(t)$ denotes the speed of the boat at time t, M is its mass, and P, S, and b are positive constants describing various other aspects of the boat and the person rowing it.

---

4. Michael Mesterton-Gibbons, *A Concrete Approach to Mathematical Modelling* (New York: John Wiley & Sons, 1995): 32 ff.

    a. Determine the equilibrium speed of the boat.

    b. Determine whether the speed found in part (a) is a sink or a source.

    c. Interpret the result of part (b) physically.

15. Given $\dfrac{dx}{dt} = f(x)$ and the graph of $f(x)$ that follows,

    a. Sketch the phase portrait of the equation.

    b. Identify all equilibrium points and classify each as a *sink*, a *source*, or a *node*.

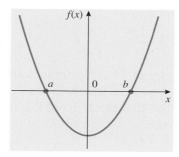

16. A population growth model that is fairly simple yet amazingly accurate in predicting tumor growth is described by the *Gompertz equation*,

$$\frac{dN}{dt} = -aN\ln(bN)$$

where $N(t)$ is proportional to the number of cells in the tumor and $a, b > 0$ are parameters that are determined experimentally. [Benjamin Gompertz (1779–1865) was an English mathematician/actuary.]

    a. Sketch the phase portrait for this equation.

    b. Sketch the graph of $f(N)$ against N.

    c. Find and classify all equilibrium points for this equation.

    d. For $0 < N \le 1$, determine where the graph of $N(t)$ against t is concave up and where it is concave down. (You may want to review Example 2.4.1.)

    e. Sketch $N(t)$.

17. Find an equation $\dot{x} = f(x)$ with the property that there are exactly three equilibrium points and all of them are *sinks*.

18. A population of animals following the *logistic* growth pattern (see Section 2.4) is harvested at a constant rate; that is, as long as the population size, P, is positive, a fixed number, h, of animals is removed per unit of time. The equation modeling the dynamics of this situation is $\dfrac{dP}{dt} = rP\left(1 - \dfrac{P}{k}\right) - h$

for $P > 0$.

a. Show that if $h < \dfrac{rk}{4}$, there are two nonzero equilibrium solutions.

b. Show that the *smaller* of the equilibrium solutions in part (a) is a *source*, whereas the *larger* of the two is a *sink*.

*2.6 BIFURCATIONS

BASIC CONCEPTS

To get an idea of what this topic is all about, let's go back to elementary algebra and look at the quadratic function $f(x) = x^2 + x + c$, where c is a constant. We should realize that the zeros of this function depend on the parameter c. To see this, let's write

$$x^2 + x + c = \left(x + \frac{1}{2}\right)^2 + \left(c - \frac{1}{4}\right) \tag{2.6.1}$$

Clearly the term $(x + \frac{1}{2})^2$ is always nonnegative, so that if $c > \frac{1}{4}$, the expression (2.6.1) is always *strictly greater than zero,* and the quadratic equation $x^2 + x + c = 0$ has *no* real solutions. If $c = \frac{1}{4}$, then the equation has $x = -\frac{1}{2}$ as its only root, a repeated root. Finally, if $c < \frac{1}{4}$, we have *two* solutions, $x = -\frac{1}{2} \pm \sqrt{\frac{1}{4} - c}$. *Verify all the assertions in this paragraph.* Figure 2.17 shows the graph of $y = x^2 + x + c$ for three values of c.

The important thing in this example is that $\frac{1}{4}$ is the value of the parameter c at which the nature of the solutions of the quadratic equation changes. We say that

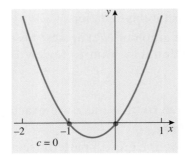
$c = 0$

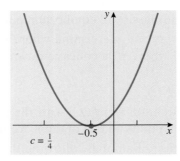
$c = \frac{1}{4}$

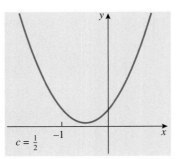
$c = \frac{1}{2}$

Figure 2.17
Graphs of $y = x^2 + x + c$

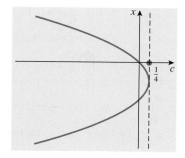

Figure 2.18
Bifurcation diagram for $x^2 + x + c = 0$

$c = \frac{1}{4}$ is a **bifurcation point** because as c decreases through $\frac{1}{4}$, the solution $x = 0$ bifurcates into two solutions. (The word *bifurcation* refers to a splitting or branching.)

We can see the effect of the bifurcation most clearly by plotting the solution x against the parameter c—in our example, showing the graph of the relationship $x = -\frac{1}{2} \pm \sqrt{\frac{1}{4} - c}$ (Figure 2.18). This graph showing the dependence of a variable on a parameter is called the **bifurcation diagram** for the equation $x^2 + x + c = 0$. Be sure you understand what this diagram tells you. Note, in particular, what happens as c passes through the value $\frac{1}{4}$.

APPLICATION TO DIFFERENTIAL EQUATIONS

This sort of qualitative change caused by a change of parameter value is particularly interesting when we observe it in an autonomous differential equation. What changes for such an ODE at a bifurcation point is the number and/or nature of the equilibrium solutions.

The real significance of bifurcations was first revealed in Euler's 1744 work on the buckling of an elastic straight beam or column under a compressive force. [The great Swiss mathematician Leonhard Euler (1707–1783) has been called "the Shakespeare of mathematics."] The normal upright position represents an equilibrium position. The parameter here is the force F exerted on the top of the column. For certain values of F, say $F < F^*$, the column maintains its vertical position; but if the force is too great, say $F > F^*$, the vertical equilibrium position becomes unstable, and the column may buckle. The critical force F^* is the bifurcation point. The equilibrium situation changes as the size of the force passes through F^*.

The next example reveals the bifurcation point for a simple first-order equation of a type we've discussed before (Example 1.2.1).

EXAMPLE 2.6.1   A Bifurcation Point for a Linear Equation

The equation $\dfrac{dy}{dt} = ay$ expresses the fact that at any time t, some quantity y grows at a rate proportional to its size at time t. The parameter a is the constant of proportionality that captures some growth characteristic of the quantity.

Setting $\dfrac{dy}{dt} = 0$, we find that the equilibrium solutions are described by $ay = 0$. If $a = 0$, then *every* value of y is an equilibrium point. If $a \neq 0$, then $y \equiv 0$ is the only equilibrium point. For the equation $\dfrac{dy}{dt} = ay = f(y)$, we have $f'(y) \equiv a$; and we use the Derivative Test of Section 2.5 to conclude that if $a > 0$, then $y \equiv 0$ is a *source*, and if $a < 0$, then $y \equiv 0$ is a *sink*. Clearly $a = 0$ is a bifurcation point, because the number and nature of the equilibrium solutions change as a passes through 0. Figure 2.19 shows graphs of $f(y)$ against y for the three possibilities for a and corresponding phase portraits.

We can show the dependence of the equilibrium points on a by drawing a bifurcation diagram, plotting $y(t)$ against a (Figure 2.20). The y-axis itself repre-

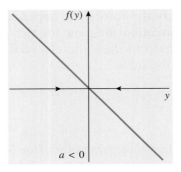
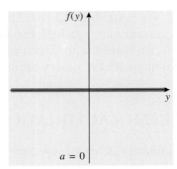
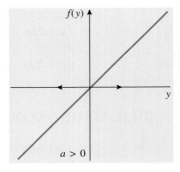

Figure 2.19

$$\frac{dy}{dx} = f(y) = ay \text{ vs. } y$$

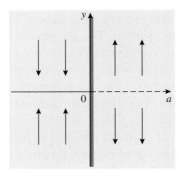

Figure 2.20

Bifurcation diagram for $\dfrac{dy}{dt} = ay$

sents all the solutions $y = C$, where C is any constant, for $a = 0$. It is usual in bifurcation diagrams to use solid curves to indicate stable equilibrium solutions (sinks) and dashed lines to denote unstable solutions (sources). Arrows indicate the directions of change of some solutions with time. ◆

EXAMPLE 2.6.2   A Bifurcation Point for a Nonlinear Equation

Now let's look at the first-order nonlinear equation $\dfrac{dy}{dx} = \alpha y - y^3 = f(y)$. This is the Landau equation, which we encountered in Problem 17 of Exercises 2.4; it arises in the study of one-dimensional patterns in fluid systems. Here $y = y(t)$ gives the amplitude of the patterns, and α is a small, dimensionless parameter that measures the distance from the bifurcation. [L. D. Landau (1908–1968) was a Russian physicist who won the Nobel Prize in 1962.]

We see that $\dfrac{dy}{dx} = 0$ implies that $\alpha y - y^3 = y(\alpha - y^2) = 0$, so $y = 0$, $y = \sqrt{\alpha}$, and $y = -\sqrt{\alpha}$ are the only equilibrium points. Looking at these points, we can see (because of the radical sign) that we have three cases to consider: (1) $\alpha = 0$, (2) $\alpha > 0$, and (3) $\alpha < 0$.

When $\alpha = 0$, there is the single equilibrium point $y = 0$. Then we have $f'(y) = \alpha - 3y^2 = -3y^2$, so $f'(0) = 0$ and we can't determine the nature of the equilibrium solution $y = 0$ from the Derivative Test. However, we can see that when $\alpha = 0$, the differential equation is $\dfrac{dy}{dx} = -y^3$, whose solution tends to zero as x becomes infinite in the positive direction. (*Look at the slope field or a phase portrait.*) Thus $y = 0$ is a *sink*.

If α is less than zero, then $y = 0$ is the only equilibrium point because $\sqrt{\alpha}$ and $-\sqrt{\alpha}$ are imaginary numbers. For this case, we see that $f'(0) = \alpha < 0$, so $y = 0$ is a *sink*.

However, if α is greater than zero, then the equation has three distinct equilibrium points: $y = 0$, $y = \sqrt{\alpha}$, and $y = -\sqrt{\alpha}$. We see that $f'(0) = \alpha > 0$, so $y = 0$ is a *source*; $f'(\sqrt{\alpha}) = -2\alpha < 0$, so $y = \sqrt{\alpha}$ is a *sink*; and $f'(-\sqrt{\alpha}) = -2\alpha < 0$, so $y = -\sqrt{\alpha}$ is also a *sink*.

Note how the value of α determines the number and the nature of the equilibrium solutions of our equation. Clearly $\alpha = 0$ is the only bifurcation point for our original equation. Figure 2.21 shows two representations of our situation: graphs of $f(y)$ against y for the three descriptions of α considered above and corresponding phase portraits.

We can also show this dependence of the equilibrium points on α by means of a bifurcation diagram, in which we plot y against α (Figure 2.22).

There are different kinds of bifurcations. Figure 2.22 shows a **pitchfork bifurcation,** named for obvious reasons. (See Exercise 7 for a generalization of this example.)

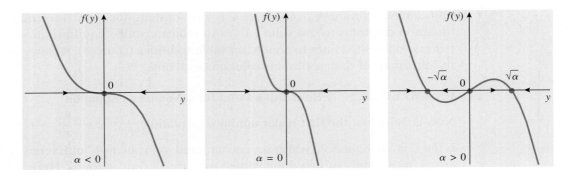

Figure 2.21

$f(y) = \alpha y - y^3$ vs. y

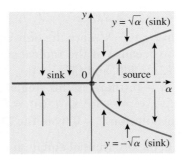

Figure 2.22

Bifurcation diagram for $\dfrac{dy}{dx} = \alpha y - y^3$ ◆

A laser—the word stands for **l**ight **a**mplification by **s**timulated **e**mission of **ra**diation—is a marvelous device that produces a beam of intense, concentrated pure light that can be used to cut diamonds, destroy cancerous cells, perform eye surgery, and enhance telecommunications when used in fiber optics. Basically, an external energy source is used to excite atoms and produce photons (light particles) that have the same frequency and phase. A. Schawlow and C. Townes received a patent for the invention of the laser in 1960, and the first laser was built by the American physicist T. H. Maiman in the same year. The mathematical model of a laser that follows is an important scientific example that illustrates another type of bifurcation. It is more complicated than the previous two examples because the bifurcation behavior depends on the values of *two* parameters.

EXAMPLE 2.6.3   A Laser Model That Has a Transcritical Bifurcation

A simplified model of the basic physics behind a laser is given by the equation

$$\dot{n} = f(n) = Gn(N_0 - n) - kn = (GN_0 - k)n - Gn^2.$$

In this equation, $n = n(t)$ represents the number of photons at time t, N_0 is the (constant) number of "excited" atoms (in the absence of laser action), and G and

k are positive parameters related to the gain and loss, respectively, of photons that have the same frequency and phase. We emphasize that we have *two* parameters in our equation, and we shall see that our bifurcation analysis depends on the value of N_0 in relationship to them.

We can write the equation as $\dot{n} = n(GN_0 - k - Gn)$, so setting $\dot{n}$ equal to zero gives us $n \equiv 0$ or $GN_0 - k - Gn \equiv 0$. This tells us that the equilibrium solutions are $n \equiv 0$ and $n = (GN_0 - k)/G = N_0 - (k/G)$, where $N_0 \neq k/G$.

Looking at the first equilibrium solution, $n \equiv 0$, we will see that this equilibrium solution is a sink when $N_0 < k/G$—that is, when $GN_0 - k < 0$. From the original equation, we have $f(n) = (GN_0 - k)n - Gn^2$, so $f'(n) = (GN_0 - k) - 2Gn$. Then $f'(0) = GN_0 - k < 0$, so $n \equiv 0$ is indeed a sink by the Derivative Test. Physically, this means that there is no stimulated emission and no photons are produced that have the same frequency and phase. The laser device functions like a light bulb. Similarly, we can determine that $n \equiv 0$ is a source when $N_0 > k/G$.

Focusing on the second equilibrium point, $n = N_0 - (k/G)$, where $N_0 \neq k/G$, we see that

$$f'\left(N_0 - \frac{k}{G}\right) = (GN_0 - k) - 2G\left(N_0 - \frac{k}{G}\right) = -GN_0 + k.$$

If $N_0 < k/G$, then $-GN_0 + k > 0$ and therefore $n = N_0 - (k/G)$ is a source. If $N_0 > k/G$, then $-GN_0 + k < 0$ and therefore $n = N_0 - (k/G)$ is a sink. The physical interpretation of this last fact is that the external energy source has excited the atoms enough so that some atoms produce photons that have the same frequency and phase. The device is now producing coherent light.

Finally, if $N_0 = k/G$, then our original equation reduces to $\dot{n} = -Gn^2$, so we get only one equilibrium solution, $n \equiv 0$, which is a sink if we consider only positive values of n. Because of this change in the nature and number of equilibrium solutions, we can interpret $N_0 = k/G$ as our bifurcation point (called the *laser threshold*). The bifurcation diagram (Figure 2.23) summarizes this model.

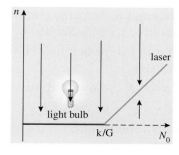

Figure 2.23

Bifurcation diagram for the laser model[5]

5. Adapted from Figure 3.3.3 in Steven H. Strogatz, *Nonlinear Dynamics and Chaos: With Applications to Physics, Biology, Chemistry, and Engineering* (Reading, Mass.: Addison-Wesley, 1994): 55.

The physical interpretation is that when the amount of energy supplied to the laser exceeds a certain threshold—that is, when $N_0 > k/G$—the "light bulb" has turned into a laser. Notice that at the bifurcation value $N_0 = k/G$, the two equilibrium solutions merge, and when they split apart, they have interchanged stability. Such a bifurcation is called a **transcritical bifurcation**. For $N_0 > k/G$, the equilibrium solution $n \equiv 0$ becomes unstable by transferring its stability to another equilibrium solution, $n = N_0 - (k/G)$, the straight line with slope 1 in Figure 2.23.   ◆

EXERCISES 2.6

For each of the equations in Exercises 1–4, (a) sketch all the qualitatively different graphs of $f(x)$ against x as the parameter c is varied; (b) determine the bifurcation point(s); and (c) sketch the bifurcation diagram of equilibrium solutions against c.

1. $\dfrac{dx}{dt} = 1 + cx + x^2$ **2.** $\dfrac{dx}{dt} = x - cx(1 - x)$

3. $\dfrac{dx}{dt} = x^2 - 2x + c$ **4.** $\dfrac{dx}{dt} = x + rx^3$

5. a. Determine the bifurcation value for the equation $\dot{x} = x(\alpha - x^2)$, where α is a real number, and construct a bifurcation diagram.

   b. Use curves to connect the appropriate critical points and give their equations.

6. Consider the *logistic equation* (see Example 2.4.1) with a constant *harvesting* (hunting, fishing, reaping, etc.) rate h:

$$\frac{dP}{dt} = P(5 - P) - h.$$

Does there exist a maximum harvest rate h^* beyond which the population will become extinct for *every* initial population $P_0 = P(0)$?

7. The *Landau equation*, $\dot{x} = (R - R_c)x - kx^3$, where k and R_c are positive constants and R is a parameter that may take on various values, is important in the field of fluid mechanics.

   a. If $R < R_c$, show that there is only the equilibrium solution $x = 0$ and that it is a sink.

   b. If $R > R_c$, show that there are *three* equilibrium solutions, $x = 0$, $x = \sqrt{(R - R_c)/k}$, and $x = -\sqrt{(R - R_c)/k}$, and that the first solution is a source whereas the other two are sinks.

   c. Sketch a graph in the R-x plane showing all equilibrium solutions, and label each one as a sink or a source. How would you describe the bifurcation point $R = R_c$?

2.7 EXISTENCE AND UNIQUENESS OF SOLUTIONS

This is the time to acknowledge that we have been avoiding a very important question: When we're trying to solve a differential equation, how do we know whether there *is* a solution? We could be looking for something that doesn't exist—a waste of time, effort, and (these days) computer resources.

We've already noted in Section 1.2 that the equation $(y')^2 + 1 = 0$ has *no* real-valued solution. You can easily check that the IVP $y' = \frac{3}{2}y^{1/3}$, $y(0) = 0$ has *three* distinct solutions: $y \equiv 0$, $y = -x^{3/2}$, and $y = x^{3/2}$.

Back in Section 1.3, you were warned that calculators and computers can mislead. They may present you with a solution where there is none. If there are several possible solutions, your user-friendly device may make its own selection, whether or not it is the one that you want for your problem. A skeptical attitude and a knowledge of mathematical theory will protect you against inappropriate answers.

First let's look at what can happen when we try to solve first-order initial-value problems. Then we'll discuss an important result guaranteeing when such IVPs have one and only one solution.

EXAMPLE 2.7.1   An IVP with a Unique Solution on a Restricted Domain
We'll see that for each value of x_0, the initial-value problem $x' = 1 + x^2$, $x(0) = x_0$ has a unique solution but that this solution does not exist for *all* values of the independent variable t. The slope field for this equation (Figure 2.24) gives us some clues.

To see things clearly, we can focus on the initial condition $x(0) = 0$. There seems to be only one solution satisfying this condition, but the direction field

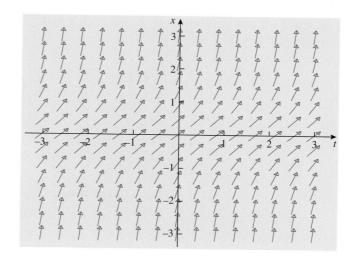

Figure 2.24
Slope field for $x' = 1 + x^2$; $-3 \le t \le 3$, $-3 \le x \le 3$

suggests that the solution curve may have vertical asymptotes. Separating variables, we see that

$$\int \frac{dx}{1 + x^2} = \int dt$$

which gives us arctan $x = t + C$, or $x(t) = \tan(t + C)$. The initial condition $x(0) = 0$ implies that $C = 0$, so that the solution of the IVP is $x(t) = \tan t$. But this solution's domain is the open interval $(-\pi/2, \pi/2)$. Recall that the function approaches $\pm \infty$ as $t \to \pm \pi/2$. (We can say the function "blows up in finite time.") Therefore, the unique solution of our IVP doesn't exist outside the (time) interval $(-\pi/2, \pi/2)$.   ◆

Now even if we have determined that a given equation *has* a solution, a second important concern is whether there is *only one* solution. This question is usually asked about solutions to initial-value problems.

EXAMPLE 2.7.2   An IVP with Infinitely Many Solutions

The nonlinear separable differential equation $x' = x^{2/3}$ has *infinitely many* solutions satisfying $x(0) = 0$ on every interval $[0, \beta]$. To show the truth of this claim, we actually construct the family of solutions of the IVP.

For each number c such that $0 < c < \beta$, we can define the function

$$x_c(t) = \begin{cases} 0 & \text{for } 0 \leq t \leq c \\ \dfrac{1}{27}(t - c)^3 & c \leq t \leq \beta \end{cases}$$

You should verify that each such function satisfies the differential equation with $x(0) = 0$. (You should even be able to show that such a function is differentiable at the break point c.) Because there are infinitely many values of the parameter c, our IVP has infinitely many solutions. Figure 2.25 shows a few of these solutions with $\beta = 7$.

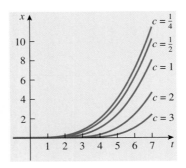

Figure 2.25

Solutions of the IVP $x' = x^{2/3}, x(0) = 0$
The functions $x_c(t)$ for $c = \frac{1}{2}, \frac{1}{4}, 1, 2,$ and 3
$\beta = 7, -1 \leq t \leq 7$ ◆

AN EXISTENCE AND UNIQUENESS THEOREM

For first-order differential equations, the answers to the existence and uniqueness questions we have just posed are fairly easy. We have an **Existence and Uniqueness Theorem**—simple conditions that guarantee one and only one solution of an initial-value problem.

Existence and Uniqueness Theorem

Let R be a rectangular region in the x-y plane described by the two inequalities $a \le x \le b$ and $c \le y \le d$. Suppose that the point (x_0, y_0) is inside R. Then if $f(x, y)$ and the partial derivative $\dfrac{\partial f}{\partial y}(x, y)$ are continuous functions on R, there is an interval I centered at $x = x_0$ and a unique function $y(x)$ defined on I such that y is a solution of the initial-value problem $y' = f(x, y), y(x_0) = y_0$.

This last statement may look a bit abstract, but it is the easiest and probably the most widely used result that guarantees the existence and uniqueness of a solution of a first-order initial-value problem. Using this theorem is simple. Take your IVP, write it in the form $y' = f(x, y), y(x_0) = y_0$, and then examine the functions $f(x, y)$ and $\dfrac{\partial f}{\partial y}$, the *partial derivative* of f with respect to the dependent variable y. (If you don't know about partial derivatives, see Appendix A.7 for a quick introduction.)

Figure 2.26 gives an idea of what the region R and the interval I in the Existence and Uniqueness Theorem may look like.

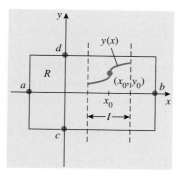

Figure 2.26

Region of existence and uniqueness:
$R = \{(x, y) \mid a \le x \le b, c \le y \le d\}$

It's important to make the following comments about this fundamental theorem:

1. If the conditions of our result are satisfied, then solution curves for the IVP can never intersect. (*Do you see why?*)
2. If $f(x, y)$ and $\partial f / \partial y$ happen to be continuous for *all* values of x and y, our result does *not* say that the unique solution must be valid for *all* values of x and y.
3. The continuity of $f(x, y)$ and $\partial f / \partial y$ are *sufficient* for the existence of solutions, but they may not be *necessary* to guarantee existence. This means that you may have solutions even if the continuity condition is not satisfied.
4. Note that this is an *existence theorem,* which means that if the right conditions are satisfied, you can find a solution, *but you are not told how to find it.* In particular, you may not be able to describe the interval I without actually solving the differential equation.

The significance of these remarks will be explored in some of the following examples and in some of the problems in Exercises 2.7. First let's apply the Existence and Uniqueness Theorem to IVPs involving first-order linear ODEs.

EXAMPLE 2.7.3   Any "Nice" Linear IVP Has a Unique Solution

Because linear equations model many important physical situations, it's important to know when such equations have unique solutions. We show that if $P(x)$ and $Q(x)$ are continuous ("nice") on an interval (a, b) containing x_0, then any IVP of the form $\dfrac{dy}{dx} + P(x)y = Q(x)$, $y(x_0) = y_0$, has one and only one solution on (a, b).

In terms of the Existence and Uniqueness Theorem, we have

$$f(x, y) = -P(x)y + Q(x) \qquad \text{and} \qquad \frac{\partial f}{\partial y}(x, y) = -P(x)$$

But both $P(x)$ and $Q(x)$ are assumed continuous on the rectangle $R = \{(x, y) \,|\, a \le x \le b, c \le y \le d\}$ for any values of c and d, and $f(x, y)$ is a combination of continuous functions. (There are no values of x and y that give us division by zero or an even root of a negative number, for example.) The conditions of the theorem are satisfied, and so any IVP of the form described above has a unique solution.

In Section 2.2 we showed how to find a solution of a linear differential equation explicitly. Now we see that, given an appropriate initial condition, we have learned how to find the *unique* solution.   ◆

Now let's go back to re-examine examples we discussed earlier.

EXAMPLE 2.7.4   Example 2.7.1 Revisited

Assume that x is a function of the independent variable t. If we look at the IVP $x' = 1 + x^2$, $x(0) = x_0$, in light of the theorem, we see that $f(t, x) = 1 + x^2$, a function of x alone that is clearly continuous at all points (t, x), and $\dfrac{\partial f}{\partial x} = 2x$, also continuous for all (t, x).

The conditions of the theorem are satisfied, and so the IVP has a unique solution. But even though both $f(t, x)$ and $\dfrac{\partial f}{\partial x}$ are continuous for *all* values of t and x, we know that any unique solution is limited to an interval

$$\left(\frac{(2n-1)\pi}{2}, \frac{(2n+1)\pi}{2} \right), \quad n = 0, \pm 1, \pm 2, \pm 3, \ldots,$$

separating consecutive vertical asymptotes of the tangent function. (Go back to look at the one-parameter family of solutions for the equation, and see comment 2 that follows the statement of the Existence and Uniqueness Theorem.)   ◆

Next we scrutinize Example 2.7.2 in light of the Existence and Uniqueness Theorem.

EXAMPLE 2.7.5   Example 2.7.2 Revisited

Here we have the form $x' = x^{2/3} = f(x)$, with $x(0) = 0$, so we must look at $f(x)$ and $\dfrac{\partial f}{\partial x}$. But $\dfrac{\partial f}{\partial x} = f'(x) = \dfrac{2}{3}x^{-1/3} = \dfrac{2}{3\sqrt[3]{x}}$, which is not continuous in any rectangle in the t-x plane that includes $x = 0$ (that is, any part of the t-axis). Therefore, we shouldn't expect to have both existence and uniqueness on an interval of the form $[0, \beta]$—and in fact we don't have uniqueness, as we saw.

However, if we avoid the t-axis—that is, if we choose an initial condition $x(t_0) = x_0 \neq 0$—then the existence and uniqueness theorem guarantees that there *will* be a unique solution for the IVP. Figure 2.27a shows the slope field for the autonomous equation $x' = x^{2/3}$ in the rectangle $-1 \leq t \leq 5, 0 \leq x \leq 3$. This rectangle includes part of the t-axis, and it is easy to visualize many solutions

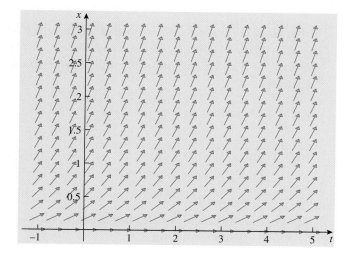

Figure 2.27a
Slope field for $x' = x^{2/3}, -1 \leq t \leq 5, 0 \leq x \leq 3$

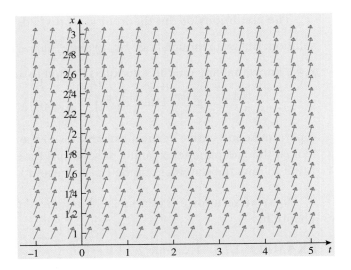

Figure 2.27b

Slope field for $x' = x^{2/3}$, $-1 \leq t \leq 5$, $1 \leq x \leq 3$

starting at the origin, gliding along the t-axis for a little while, and then taking off. Figure 2.25 shows some of these solution curves.

Figure 2.27b, on the other hand, shows what happens if we choose a rectangle that avoids the t-axis. It should be clear that if we pick any point (t_0, x_0) in this rectangle, there will be one and only one solution of the equation that passes through this point. ◆

EXERCISES 2.7

For each of the initial-value problems in Exercises 1–6, determine a rectangle R in the appropriate plane (x-y, t-x, etc.) for which the given differential equation would have a unique solution through a point in the rectangle. Do not solve the equations.

1. $\dfrac{dx}{dt} = \dfrac{1}{x}$, $x(0) = 3$ **2.** $\dfrac{dy}{dt} = \dfrac{5}{4}y^{1/5}$, $y(0) = 0$

3. $t\dfrac{dx}{dt} = x$, $x(0) = 0$ **4.** $y' = -\dfrac{t}{y}$, $y(0) = 0.2$

5. $y' = \dfrac{t}{1+t+y}$, $y(-2) = 1$ **6.** $x' = \tan x$, $x(0) = \dfrac{\pi}{2}$

7. What is the length of the largest interval I on which the IVP $y' = 1 + y^2$, $y(0) = 0$, has a solution?

8. Show that $y \equiv -1$ is the only solution of the IVP $y' = t(1 + y)$, $y(0) = -1$.

9. What is the solution to the IVP $\dfrac{dx}{dt} = x^{2/3}$, $x(0) = x_0$ if $x_0 < 0$? Compare your answer to the answer(s) in Example 2.7.2. What has changed?

10. Look at the IVP $Q' = |Q - 1|$, $Q(0) = 1$.

   a. Explain why the conditions of the Existence and Uniqueness Theorem do not hold for this equation.

   b. Guess at a solution of this initial-value problem.

   c. Can you see that the solution you found in part (b) is unique?

11. Consider the equation $\dot{y} = \sqrt{|y|} + k$, where k is a positive constant.

   a. Solve the equation. (You will get an implicit solution.)

   b. For what initial values (t_0, y_0) does the equation have a unique solution?

   c. For what values of $k \leq 0$ does the equation have unique solutions?

12. The parabola $y = x^2$ and the line $y = 2x - 1$ are both solutions of the equation $y' = 2x - 2\sqrt{x^2 - y}$, and both satisfy the initial condition $y(1) = 1$. Does this contradict the Existence and Uniqueness Theorem?

13. Consider the initial-value problem $\dfrac{dy}{dx} = P(x)y^2 + Q(x)y$, $y(2) = 5$, where $P(x)$ and $Q(x)$ are third-degree polynomials in x. Does this problem have a unique solution on some interval $|x - 2| \leq h$ around $x = 2$? Explain why or why not.

14. Consider the nonlinear equation $\dfrac{dx}{dt} = (\alpha - x)(\beta - x)$, where α and β are positive constants. (See Example 2.1.6.) Without solving the equation, show that the solution of any IVP involving this equation is unique.

15. Consider the equilibrium solution $P \equiv b$ of the logistic equation (Section 2.4) $\dfrac{dP}{dt} = kP(b - P)$, where k and b are positive constants. Is it possible for a solution near $P \equiv b$ to *reach* (that is, to equal) this solution for a finite value of x? [*Hint:* Use the uniqueness part of the Existence and Uniqueness Theorem.]

16. Why can't the following family of curves be the solution curves for the differential equation $y' = f(t, y)$, where f is a polynomial in t and y?

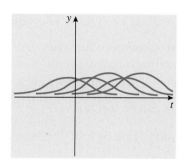

17. Consider the equation $\dfrac{dy}{dx} + x^2 y^3 = \cos x$.

   a. Does this equation have a unique solution passing through any point (x_0, y_0)?

   b. Try to solve the equation using the ODE solver in your CAS. Comment on the result.

18. Consider the nonseparable, nonlinear equation

$$\frac{dx}{dt} = \begin{cases} \dfrac{4t^3 x}{x^2 + t^4} & \text{if } xt \neq 0 \\ 0 & \text{if } xt = 0 \end{cases}$$

Learn how to enter this equation involving a piecewise-defined function into your CAS and then use your ODE solver to find the infinite number of solutions satisfying $x(0) = 0$.

19. Suppose that a differential equation is a model for a certain type of chemical reaction. Could the fact that the equation does *not* have a solution indicate that the reaction cannot take place? Would the fact that the equation *has* a solution guarantee that the reaction *does* take place?

2.8 SUMMARY

Perhaps the easiest type of first-order ODE to solve is a **separable equation,** one that can be written in the form $\dfrac{dy}{dx} = f(x)g(y)$, where f denotes a function of x alone and g denotes a function of y alone. "Separating the variables" leads to the equation $\displaystyle\int \frac{dy}{g(y)} = \int f(x)\,dx$. It is possible that you cannot carry out one of the integrations in terms of elementary functions or you may wind up with an *implicit* solution. Furthermore, the process of separating variables may introduce singular solutions.

Another important type of first-order ODE is a **linear equation,** one that can be written in the form $a_1(x)y' + a_0(x)y = f(x)$, where $a_1(x)$, $a_0(x)$, and $f(x)$ are functions of the independent variable x alone. The standard form of such an equation is $\dfrac{dy}{dx} + P(x)y = Q(x)$. The equation is called **homogeneous** if $Q(x) \equiv 0$ and **nonhomogeneous** otherwise. Any homogeneous linear equation is separable.

After writing a linear equation in standard form, we introduce an **integrating factor,** $\mu(x) = e^{\int P(x)dx}$, multiply each side of the equation by $\mu(x)$, and see that the equation can be written as $\dfrac{d}{dx}(e^{\int P(x)dx}y) = e^{\int P(x)dx}Q(x)$. Integrating each side and then multiplying by $e^{-\int P(x)dx}$, we get an explicit formula:

$$y = e^{-\int P(x)dx} \cdot \int e^{\int P(x)dx}Q(x)\,dx + Ce^{-\int P(x)dx}$$

A typical first-order differential equation can be written in the form $\frac{dy}{dx} = f(x, y)$. Graphically, this tells us that at any point (x, y) on a solution curve of the equation, the slope of the tangent line is given by the value of the function f at that point. We can outline the solution curves by using possible tangent line segments. Such a collection of tangent line segments is called a **direction field** or **slope field** of the equation. The set of points (x, y) such that $f(x, y) = C$, a constant, defines an **isocline,** a curve along which the slopes of the tangent lines are all the same (namely, C). A differential equation in which the independent variable does not appear explicitly is called an **autonomous** equation. If the independent variable *does* appear, the equation is called **nonautonomous.** For an autonomous equation, the slopes of the tangent line segments that make up the slope field depend only on the values of the dependent variable. Graphically, if we fix the value of the dependent variable, say x, by drawing a horizontal line $x = C$ for any constant C, we see that all the tangent line segments along this line have the same slope, no matter what the value of the independent variable, say t. Another way to look at this is to realize that we can generate infinitely many solutions by taking any one solution and translating (shifting) its graph left or right. Even when we can't solve an equation, an analysis of its slope field can be very instructive. However, such a graphical analysis may miss certain important features of the integral curves, such as vertical asymptotes.

An *autonomous* first-order equation can be analyzed qualitatively by using a **phase line** or **phase portrait.** For an autonomous equation, the points x such that $\frac{dy}{dx} = f(x) = 0$ are called **critical points.** We also use the terms **equilibrium points, equilibrium solutions,** and **stationary points** to describe these key values. There are three kinds of equilibrium points for an autonomous first-order equation: **sinks, sources,** and **nodes.** An equilibrium solution y is a **sink** (or **asymptotically stable solution**) if solutions with initial conditions "sufficiently close" to y are asymptotic to y as the independent variable tends to infinity. On the other hand, if solutions "sufficiently close" to an equilibrium solution y are asymptotic to y as the independent variable tends to negative infinity, then we call y a **source** (or **unstable equilibrium solution**). An equilibrium solution that shows any other kind of behavior is called a **node** (or **semistable equilibrium solution**). A simple (but not always conclusive) test is as follows:

> If x^* is an equilibrium point for the equation $\frac{dx}{dt} = f(x)$, then it is true that (1) if $f'(x^*) > 0$, then x^* is a source; (2) if $f'(x^*) < 0$, then x^* is a sink; and (3) if $f'(x^*) = 0$, then we can't tell what sort of equilibrium point x^* may be without further investigation.

Suppose that we have an autonomous differential equation with a parameter α. A **bifurcation point** α_0 is a value that causes a change in the nature of the equation's equilibrium solutions as α passes through the value α_0.

In trying to solve a differential equation, especially an initial-value problem, it is important to understand whether the problem *has* a solution and whether any solution is *unique*. There are simple conditions that guarantee that there is one and only one solution of an initial-value problem:

> Let R be a rectangular region in the x-y plane described by the two inequalities $a \le x \le b, c \le y \le d$. Suppose that the point (x_0, y_0) is inside R. Then, if $f(x, y)$ and $\dfrac{\partial f}{\partial y}(x, y)$ are continuous functions on R, there is an interval I centered at $x = x_0$ and a unique function $y(x)$ defined on I such that y is a solution of the initial-value problem $y' = f(x, y)$, $y(x_0) = y_0$.

PROJECT 2-1

The Price Is Right

Let $p(t)$, $s(t)$, and $d(t)$ denote the price, supply, and demand of a commodity at time t. **Allen's Speculative Model** in economics assumes that s and d are linear functions in $p(t)$ and $p'(t)$:

$$s(t) = a_1 p(t) + a_2 p'(t) + a_3 \qquad (*)$$
$$d(t) = b_1 p(t) + b_2 p'(t) + b_3, \qquad (**)$$

where the a_i's and b_i's are constants.

The **Economic Principle of Supply and Demand,** which guarantees a state of dynamic equilibrium, is

$$d(t) = s(t) \qquad (***)$$

a. By combining (*), (**), and (***), find a single linear differential equation involving $p(t)$.

b. Assuming that $a_1 \ne b_1$, $a_2 \ne b_2$, and $a_3 \ne b_3$, solve the equation you found in part (a) with the initial condition $p(0) = p_0$.

c. Interpret the solution in economic terms if $p_0 = \dfrac{a_3 - b_3}{b_1 - a_1}$.

d. Suppose that $\dfrac{b_1 - a_1}{b_2 - a_2} > 0$. What happens to the price as t increases without bound?

e. Suppose that $\dfrac{b_1 - a_1}{b_2 - a_2} < 0$. Now what happens to the price as t increases without bound?

f. Suppose that $s(t) = 30 + p(t) + 4p'(t)$ and $d(t) = 48 - 2p(t) + 3p'(t)$, where s and t are given in thousands of units. If $p(0) = 10$ monetary units, find the price at any later time t. What happens to the price as t increases?

PROJECT 2-2

Cultured Perils

A continuous culture device, or **chemostat,** is a well-stirred vessel that contains microorganisms and into which fresh medium (nutrient) is pumped at a constant rate F. The contents of the growth vessel are pumped out at the same rate, so the volume V remains constant. Microbiologists and ecologists use the chemostat as a laboratory simulation of an aquatic environment, and it has also been used to model the waste water treatment process. (See the accompanying schematic diagram.)

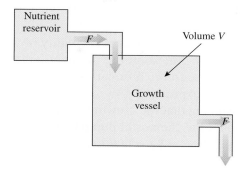

Around 1950, the biologist Jacob Monod developed a mathematical model[6] for the continuous culture of a single species of microorganism whose growth is dependent solely on a single nutrient supplied at a constant rate via the input to the growth vessel.

Chemostat experiments[7] with a certain strain of *Escherichia coli* led to the particular equation

$$\frac{dx}{dt} = \left(\frac{0.81\,(10-x)}{3 + (10-x)} - D \right)x \tag{$*$}$$

where $x(t)$ denotes the concentration of the organism at time t, and $D = \dfrac{F}{V}$, the pump rate divided by the volume, is a parameter under control of the experimenter.

We want to study the effects of varying D. Intuition suggests that if the pump is allowed to run too fast, then the *E. coli* will eventually approach extinction in the chemostat because they are being pumped out at a faster rate than they can grow and reproduce. On the other hand, if the pump is run slowly enough, the *E. coli* should be able to grow at a rate sufficient to overcome washout and should be able to thrive in the growth vessel indefinitely.

Our problem is to determine which values of D result in extinction and which result in survival. This can be done by studying equation ($*$), treating D as a bifurcation parameter (see Section 2.6).

a. Using technology, study solutions to equation ($*$) for parameter values of 0.8, 0.7, 0.4, and 0.5. For each choice of D, use several different initial conditions x_0. What are your observations?

6. J. Monod, "La technique de culture continue: Théorie et applications," *Annales de L'Institut Pasteur* 79 (1950): 390–410.
7. S. R. Hansen and S. P. Hubbell, "Single nutrient microbial competition: Agreement between experimental and theoretically forecast outcomes," *Science* 20 (1980): 1491–1493.

b. Find the equilibrium solutions $\hat{x}$ as a function of the parameter D and determine whether they are sinks, sources, or nodes. (Assume that the only meaningful equilibrium solutions are those for which $0 \le \hat{x} \le 10$.) Construct the bifurcation diagram for equation (*). (See Section 2.6.) At what value of D does a bifurcation occur? Explain the significance of this bifurcation with regard to the fate of the *E. coli*.

c. Use technology to test the validity of your bifurcation analysis in part (b) by examining the solutions of equation (*) again, choosing various values of D very close to the bifurcation value you found in part (b). Are your observations as expected?

d. Assuming that the growth vessel is kept at a volume of 20 liters, at what speed should the chemostat pump be run in order to maintain a steady-state *E. coli* population of 8 μg/liter? A population of 4.5 μg/liter?

3 | The Numerical Approximation of Solutions

3.0 INTRODUCTION

Historically, numerical methods of working with differential equations were developed when some equations could not be solved analytically—that is, with their solutions expressed in terms of elementary functions. Over the past 300 years, mathematicians and scientists have learned to solve more and more types of differential equations. However, today there are still equations that are impossible to solve in closed form (for instance, Example 2.1.4). In fact, very few differential equations that arise in applications can be solved exactly; and, perhaps more important, even solution formulas often express the solutions *implicitly* via complicated combinations of the solution and the independent variable that are difficult to work with. Take a look back at the solution in Example 2.3.4 for instance. What we will do in this chapter is describe some ways of getting an *approximate numerical solution* of a first-order IVP $y' = f(x, y)$, $y(x_0) = y_0$. This means being able to calculate approximate values of the solution function y by some process requiring a finite number of steps, so that at the end of this step-by-step process we are reasonably close to the "true" answer. Graphically, we are trying to approximate the solution curve with a simpler curve, usually a curve made up of straight line segments.

The very nature of what we will be trying to do contains the notion of *error,* the discrepancy between a true value and its approximate value. Error is what stands between reality and perfection. It is the static in our telephone line, the wobble in a kitchen chair, a slip of the tongue. Although there are various ways to measure error, we will focus on **absolute error,** which is defined by the quantity |true value − approximation|, the absolute value of the difference between the true value and the approximation. We'll have more to say about error in Section 3.2.

Let's see how all this applies to a first-order IVP $y' = f(x, y)$, $y(x_0) = y_0$.

3.1 EULER'S METHOD

One of the easiest methods of obtaining an approximation to a solution curve is attributed to the mathematician Euler. He used this approach to solve differential equations around 1768. A modern way of expressing his idea is to say that he used *local linearity*. Geometrically, this simply refers to the fact that if a function F is differentiable at $x = x_0$ and we "zoom" in on the point $(x_0, F(x_0))$ lying on the curve $y = F(x)$, then we will think we're looking at a straight line segment. Numerically, we're saying that if we have a straight line tangent to a curve $y = F(x)$ at a point $(x_0, F(x_0)) = (x_0, y_0)$, then for a value of x close to x_0, the corresponding value on the tangent line is approximately equal to the value on the curve. In other words, we can avoid the complexity of dealing with values on what may be a complicated curve by dealing with values on a straight line. (See Appendix A.1 for more on this topic.)

Using the familiar "point-slope" formula for the equation of a straight line, we can derive the equation of the line T tangent to the curve $y = F(x)$ at the point (x_0, y_0):

$$T(x) = F'(x_0)(x - x_0) + F(x_0) = y'(x_0)(x - x_0) + y_0. \qquad (3.1.1)$$

Now we can express the idea of local linearity by writing

$$\underbrace{y(x)}_{\text{value on the curve}} \approx \underbrace{y'(x_0)(x - x_0) + y_0}_{\text{value on the tangent line}}$$

where the symbol $\approx$ means "is approximately equal to." Figure 3.1 shows graphically what we are saying.

Note that because the curve we're using as an illustration is concave down near the point (x_0, y_0), the tangent line lies *above* the curve here, so the value $T(x)$ given by the tangent line is *greater than* the true value $y(x)$ for x near x_0.

Now let's look at an IVP $y' = f(x, y)$, $y(x_0) = y_0$ so that we can write $y_0' = y'(x_0) = f(x_0, y_0)$. **In what follows, we assume that there is a unique solution $y = \varphi(x)$ in some interval containing x_0.** Suppose we want to know the height of the solution curve corresponding to a value x_1 that is close to x_0, but to the

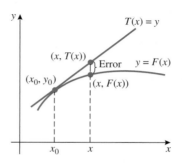

Figure 3.1
Local linearity

right of x_0. We can describe such a new value of the independent variable as $x_1 = x_0 + h$, where $h > 0$ is the size of a small "step." Now let's try to approximate $\varphi(x_1)$, a value on the actual solution curve, by some value y_1 on the tangent line to $y = \varphi(x)$ at x_0:

$$\varphi(x_1) \approx y_1 = \varphi'(x_0)(x_1 - x_0) + y_0$$
$$= f(x_0, y_0)(x_0 + h - x_0) + y_0$$
$$= f(x_0, y_0) \cdot h + y_0.$$

Therefore, we can write

$$\varphi(x_1) \approx y_1 = f(x_0, y_0) \cdot h + y_0$$

and we have a good *local linear approximation* of $\varphi(x)$ at $x = x_1$ if we choose h small enough. Figure 3.2 illustrates graphically what's going on.

We can repeat the process using (x_1, y_1) as our jumping-off point, realizing that the value y_1 is only an approximation. Using equation (3.1.1) again, with (x_0, y_0) replaced by (x_1, y_1), we see that the line through (x_1, y_1) with slope equal to $f(x_1, y_1)$ has y values given by

$$f(x_1, y_1)(x - x_1) + y_1. \tag{3.1.2}$$

We should realize that the point (x_1, y_1) is not expected to be on the actual solution curve, so in general $f(x_1, y_1) \neq f(x_1, \varphi(x_1))$, the slope of the actual solution at x_1.

For convenience, suppose that we want to approximate the solution curve's height corresponding to a value x_2 that is the same distance from x_1 as x_1 is from x_0. That is, we take a step to the right of size h: $x_2 = x_1 + h = (x_0 + h) + h = x_0 + 2h$. We can approximate $\varphi(x_2)$, the actual value of the solution function at $x = x_2$, by using equation (3.1.2):

$$\varphi(x_2) \approx y_2 = f(x_1, y_1) \cdot h + y_1.$$

Similarly, for $x_3 = x_2 + h = (x_0 + 2h) + h = x_0 + 3h$, we approximate $\varphi(x_3)$ as follows:

$$\varphi(x_3) \approx y_3 = f(x_2, y_2) \cdot h + y_2.$$

Figure 3.3 shows what we are doing.

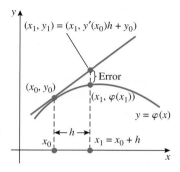

Figure 3.2
A local linear approximation of a solution

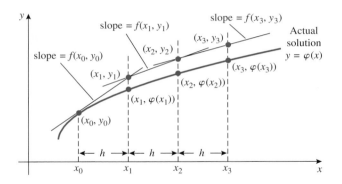

Figure 3.3
A three-step linear approximation

Continuing in this way, we generate a sequence of approximate values y_1, y_2, $y_3, \ldots, y_n$ for the solution function φ at various equally spaced points x_1, x_2, $x_3, \ldots, x_n$:

$$\underbrace{y_{k+1}}_{\text{new approx. value}} = \underbrace{y_k}_{\text{old approx. value}} + \underbrace{h \cdot f(x_k, y_k)}_{\text{step size}}, \qquad (3.1.3)$$

where $x_k = x_0 + kh$, $k = 0, 1, \ldots, n$. If you go back through the derivation, you'll realize that formula (3.1.3) is valid for $h < 0$ also. Note that if the points x_k are equally spaced with step size h and we want to get from (x_0, y_0) to (x^*, y^*) along the approximating polygonal curve, then we must have $n = \dfrac{x^* - x_0}{h}$ steps. For example, if we start at $x_0 = 2$ and want to approximate $\varphi(2.7)$ using steps of size $h = 0.1$, we can reach $x^* = 2.7$ by taking $n = \dfrac{2.7 - 2}{0.1} = 7$ steps. In practice, once we have chosen the step size h, the number of steps needed, n, will be obvious.

If we stand back from all these equations and look at Figures 3.2 and 3.3 again, we can see that what we are doing is using the slope field for our IVP as a set of stepping stones. We "walk" on a tangent line segment for a short distance, stop to look forward for the next step, jump to that step, and so on. We are approximating the flow of the solution curve by using flat rocks set into the "stream." If you play "connect the dots" with the points (x_0, y_0), (x_1, y_1), $(x_2, y_2), \ldots, (x_n, y_n)$, you see a polygonal line (called the **Euler polygon** or the **Cauchy-Euler polygon**) that approximates the actual solution curve. Figure 3.4 shows this for the initial-value problem $y' = x^2 + y$, $y(0) = 1$, where we try to approximate $y(1)$ with different step sizes $h = 1, 0.5$, and 0.25.

We can look at this approximation process, formula (3.1.3), in another geometrical way. Suppose we have the differential equation $y' = f(x, y)$. Then the Fundamental Theorem of Calculus tells us that

$$y_{k+1} - y_k \approx y(x_{k+1}) - y(x_k) = \int_{x_k}^{x_{k+1}} y'(x)\, dx = \int_{x_k}^{x_{k+1}} f(x, y)\, dx.$$

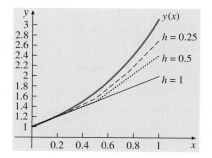

Figure 3.4
The actual solution of the IVP $y' = x^2 + y$, $y(0) = 1$, and three
Euler approximations ($h = 1, 0.5,$ and 0.25) on the interval $[0, 1]$

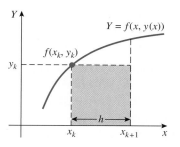

Figure 3.5
Approximation of an integral by a rectangular area

But formula (3.1.3) requires us to replace $y_{k+1} - y_k$ by $h \cdot f(x_k, y_k)$. This means
that we are approximating $\int_{x_k}^{x_{k+1}} f(x, y)\, dx$ by $h \cdot f(x_k, y_k)$. Figure 3.5 shows the
geometry of the situation in the interval $[x_k, x_{k+1}]$.

We have approximated the area under the curve $Y = f(x, y(x))$ by the area
of the shaded rectangle—a rectangle formed by using the height of the curve at
the left-hand endpoint of the interval. Thus Euler's method amounts to using a
(left-hand) *Riemann sum* approximation to the area under a curve.

Enough theory for now. Let's see how this "method of tangents" works with
a simple initial-value problem.

EXAMPLE 3.1.1   Euler's Method with Error Analysis

Suppose we're given the IVP $\dfrac{dx}{dt} = t^2 + x$, $x(1) = 3$. We want to use Euler's
method to approximate $x(1.5)$.

This is a first-order linear equation whose particular solution for the initial
condition $x(1) = 3$ is $x(t) = -t^2 - 2t - 2 + 8e^{t-1}$ (*Verify this.*) Thus the actual

value of $x(1.5)$ is $-(1.5)^2 - 2(1.5) - 2 + 8e^{(1.5)-1} = 5.939770. \ldots$ We'll use the actual value to see how good an approximation Euler's method gives us.

In our problem, $f(t,x) = t^2 + x$, so Euler's formula (3.1.3) becomes

$$x_{k+1} = x_k + h \cdot (t_k^2 + x_k), \tag{3.1.4}$$

where $t_k = t_0 + kh$, $k = 0, \ldots, n$, $t_0 = 1$, and $x_0 = 3$. (By now you should be comfortable with the switch from the traditional x-y coordinates to t-x coordinates.) Suppose we take $h = 0.1$—that is, our step size is one-tenth of a unit. Because our target $t = 1.5$ is 0.5 unit away from our initial point $t = 1$, we'll need $n = 5$ steps of size $h = 0.1$ to reach this with Euler's process (Figure 3.6).

Using formula (3.1.4), let's generate our approximate values, stepping from $t = 1$ to $t = 1.5$:

$$\begin{aligned}
x_1 &= x_0 + (0.1)(t_0^2 + x_0) = 3 + (0.1)(1^2 + 3) = 3.40 \\
x_2 &= x_1 + (0.1)(t_1^2 + x_1) = 3.40 + (0.1)(1.1^2 + 3.4) = 3.861 \\
x_3 &= x_2 + (0.1)(t_2^2 + x_2) = 3.861 + (0.1)(1.2^2 + 3.861) = 4.3911 \\
x_4 &= x_3 + (0.1)(t_3^2 + x_3) = 4.3911 + (0.1)(1.3^2 + 4.3911) = 4.99921 \\
x_5 &= x_4 + (0.1)(t_4^2 + x_4) = 4.99921 + (0.1)(1.4^2 + 4.99921) = 5.695131.
\end{aligned}$$

Thus Euler's method gives the approximation 5.695131 for the value $x(1.5)$. In this example, the *absolute error* is |true value − approximation| = |5.939770 − 5.695131| = 0.244639.

If we try again, using a step size only *half* the size of the step we used before—that is, using a step size $h = 0.05$—it will take *twice* as many steps to bridge the gap between the initial value $t = 1$ and the final value $t = 1.5$. Table 3.1 shows the result of a spreadsheet calculation of Euler's method for this new sequence of steps. If you have access to a spreadsheet program, you'll find it fairly easy to use it to set up Euler's method.

Note that the number of calculations (steps) has doubled, but the absolute error has been cut almost in half. Also note the cumulative growth of the error in the last column.

If we cut the step size in half again, working with $h = 0.025$ this time, we can see a pattern emerging (Table 3.2).

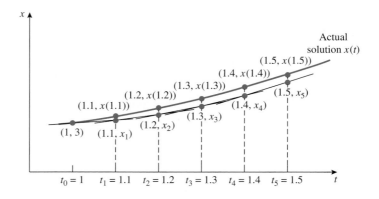

Figure 3.6
A five-step approximation

TABLE 3.1   Euler's Method for $\dfrac{dx}{dt} = t^2 + x$, $x(1) = 3$, with $h = 0.05$

k	t_k	x_k	True Value	Absolute Error
0	1	3.000000	3.000000	0.00000
1	1.05	3.200000	3.207669	0.00767
2	1.1	3.415125	3.431367	0.01624
3	1.15	3.646381	3.672174	0.02579
4	1.2	3.894825	3.931222	0.03640
5	1.25	4.161567	4.209703	0.04814
6	1.3	4.447770	4.508870	0.06110
7	1.35	4.754658	4.830040	0.07538
8	1.4	5.083516	5.174598	0.09108
9	1.45	5.435692	5.543997	0.10831
10	1.5	5.812602	5.939770	0.12717

TABLE 3.2   Euler's Method for $\dfrac{dx}{dt} = t^2 + x$, $x(1) = 3$, with $h = 0.025$

k	t_k	x_k	True Value	Absolute Error
0	1	3.000000	3.000000	0.0000
1	1.025	3.100000	3.101896	0.0019
2	1.05	3.203766	3.207669	0.0039
3	1.075	3.311422	3.317448	0.0060
4	1.1	3.423098	3.431367	0.0083
5	1.125	3.538926	3.549563	0.0106
6	1.15	3.65904	3.672174	0.0131
7	1.175	3.783578	3.799345	0.0158
8	1.2	3.912683	3.931222	0.0185
9	1.225	4.046500	4.067957	0.0215
10	1.25	4.185178	4.209703	0.0245
11	1.275	4.32887	4.356620	0.0277
12	1.3	4.477733	4.508870	0.0311
13	1.325	4.631926	4.666620	0.0347
14	1.35	4.791615	4.830040	0.0384
15	1.375	4.956968	4.999306	0.0423
16	1.4	5.128158	5.174598	0.0464
17	1.425	5.305362	5.356098	0.0507
18	1.45	5.488761	5.543997	0.0552
19	1.475	5.678543	5.738489	0.0599
20	1.5	5.874897	5.939770	0.0649

Note that the absolute error increases as a function of k, the number of steps. Furthermore, the differences between successive errors are increasing slightly. For example, if you subtract the error for $k = 9$ from the error corresponding to $k = 10$, you get 0.0030, whereas subtracting the $k = 10$ error from the $k = 11$ error yields 0.0032.

If this error pattern seems vaguely familiar, it may be because you have seen error analysis applied to left- and right-hand Riemann sum approximations in calculus.   ◆

Let's try another problem. Practice makes perfect.

EXAMPLE 3.1.2   Euler's Method with Error Analysis

Consider the IVP $\dfrac{dy}{dt} = \dfrac{1}{t}$, $y(1) = 0$, and suppose we want to approximate $y(2)$.

You should recognize the solution of this IVP as $y = \ln t$, so we're really trying to approximate $\ln 2 = 0.69314718056.\ldots$ (*Hint:* If you check this "exact" answer on your calculator or CAS, realize that these devices are using very sophisticated approximation methods themselves!)

If we take $h = 0.05$, we'll need 20 steps to stretch from $t = 1$ to $t = 2$. In our example, Euler's method gives us the formula

$$y_{k+1} = y_k + \frac{0.05}{t_k}$$

for $t_k = 1 + 0.05\,k$ ($k = 0, \ldots, 20$). Table 3.3 gives the results.

The solution curve $y = \ln t$ is concave down, so the approximating tangent lines all lie *above* the solution curve, leading to an approximation of $\ln 2$ that's too large. Just as in Example 3.1.1, the errors increase with the value of k, but this time, if you subtract successive errors (corresponding to successive values of k), you'll see that the differences are *decreasing*. (To gain some insight into this phenomenon, compare $f(x, y)$ in Examples 3.1.1 and 3.1.2.)

Changing to $h = 0.025$ and $n = 40$ yields the approximate value 0.699436, whereas setting $h = 0.01$ and $n = 100$ gives us an approximation of 0.695653. Of course, technology (a spreadsheet) was used to obtain the last two approximations.   ◆

Next we'll see what happens when we are given an equation whose solution we don't know.

EXAMPLE 3.1.3   Euler's Method—Unknown Exact Solution

Suppose we're given the IVP $y' = \sqrt{x + y}$, $y(5) = 4$, and we want to find $y(4)$. The first thought that should occur to us is that the equation is neither separable nor linear. Are we in trouble here? *No,* not if we understand Euler's method.

In our problem, $f(x,y) = \sqrt{x + y}$, so formula (3.1.3) takes the form

$$y_{k+1} = y_k + h\sqrt{x_k + y_k},$$

TABLE 3.3 Euler's Method for $\dfrac{dy}{dt} = \dfrac{1}{t}$, $y(1) = 0$, with $h = 0.05$

k	t_k	y_k	True Value	Absolute Error
0	1	0.000000	0.000000	0.00000
1	1.05	0.050000	0.048790	0.00121
2	1.1	0.097619	0.095310	0.00231
3	1.15	0.143074	0.139762	0.00331
4	1.2	0.186552	0.182322	0.00423
5	1.25	0.228219	0.223144	0.00507
6	1.3	0.268219	0.262364	0.00585
7	1.35	0.306680	0.300105	0.00658
8	1.4	0.343717	0.336472	0.00724
9	1.45	0.379431	0.371564	0.00787
10	1.5	0.413914	0.405465	0.00845
11	1.55	0.447247	0.438255	0.00899
12	1.6	0.479506	0.470004	0.00950
13	1.65	0.510756	0.500775	0.00998
14	1.7	0.541059	0.530628	0.01043
15	1.75	0.570470	0.559616	0.01085
16	1.8	0.599042	0.587787	0.01126
17	1.85	0.626820	0.615186	0.01163
18	1.9	0.653847	0.641854	0.01199
19	1.95	0.680162	0.667829	0.01233
20	2	0.705803	0.693147	0.01266

where $x_k = 5 + kh$, $k = 0, 1, \ldots, n$. As usual, n denotes the number of steps we choose.

Let's start off by choosing five steps to get us from the initial point $x = 5$ to our destination $x = 4$. Each step has to have length 0.2, and because we are moving *backwards* from the initial point, we must take $h = -0.2$ in the formula. We'll carry out this first attempt at approximation by hand and then use a spreadsheet when the calculations become more numerous (and more tedious).

The formula gives us

$$y_1 = y_0 + h\sqrt{x_0 + y_0} = 4 + (-0.2)\sqrt{5 + 4} = 3.4$$

$$y_2 = y_1 + h\sqrt{x_1 + y_1} = 3.4 + (-0.2)\sqrt{4.8 + 3.4} = 2.82728716$$

$$y_3 = y_2 + h\sqrt{x_2 + y_2} = 2.82728716 + (-0.2)\sqrt{4.6 + 2.82728716} = 2.28222616$$

$$y_4 = y_3 + h\sqrt{x_3 + y_3} = 2.28222616 + (-0.2)\sqrt{4.4 + 2.28222616} = 1.76522612$$

$$y_5 = y_4 + h\sqrt{x_4 + y_4} = 1.76522612 + (-0.2)\sqrt{4.2 + 1.76522612} = 1.27674987.$$

These calculations tell us that $y(4) \approx 1.27674987$. The *true answer* is **1.34042895566892**. . . .[1] Therefore, when we round the "true" answer to eight places, the absolute error is $|1.34042896 - 1.27674987| = 0.06367909$. If we choose $h = -0.01$ and use 100 steps, a spreadsheet calculation gives us an approximation of 1.337296, with an absolute error of 0.0031.   ◆

So far we've been cheating a bit, discussing the numerical solutions of equations for which we could find an analytic solution (even if implicit). Knowing the exact solution enabled us to analyze the error—the gap between the true solution value and the approximate value of a solution at a point. However, it's time that we consider a more typical example.

EXAMPLE 3.1.4   Euler's Method—A Completely Unknown Solution

The initial-value problem $\dfrac{dy}{dt} = y^2 - t^2$, $y(0) = \frac{1}{2}$, cannot be solved by analytical methods. Nevertheless, we can approximate the solution at $t = 1$ (for example) so that it is accurate to, say, three decimal places.

"Without the exact answer as a guide, how do we know that these three decimal places are accurate?" you may be asking. Let's skip the detailed formula and see what happens for different step sizes (Table 3.4). We have rounded the approximations in the last column to six decimal places.

We have reached a stage at which the first three digits of the approximate values do not seem to be changing. The last approximate value agrees with the previous one to three decimal places after appropriate rounding, so we can

TABLE 3.4   The IVP $\dfrac{dy}{dt} = y^2 - t^2$, $y(0) = \dfrac{1}{2}$: Approximate

Values of $y(1)$ for Various Step Sizes

Step Size	Number of Steps	Approximate Value
1/100	100	0.512113
1/1000	1000	0.506106
1/2000	2000	0.505769
1/4000	4000	0.505600
1/8000	8000	0.505515
1/16000	16000	0.505473
1/20000	20000	0.505464

---

1. This answer is obtained by making a substitution to transform the given equation into a separable equation (see the explanation that precedes Problems 12–14 of Exercises 2.1), solving the equation to get an implicitly defined solution to the IVP, and then solving for the value of y when $x = 4$. Solving the implicit relation for y requires a calculator or CAS with a "solve" function for general equations. Even this "true" answer is only an approximation (although presumably a very accurate one), because the algebraic equation can't be solved exactly.

assume that the approximation is 0.505, accurate to three decimal places. The idea—a rule of thumb based on mathematical analysis—is to keep on using smaller step sizes until there are changes only *past* the decimal place in which we are interested. Then we can be sure of those decimal places that do *not* change. We'll say more about accuracy of numerical methods in the next section.   ◆

STIFF DIFFERENTIAL EQUATIONS

We can encounter difficulty in applying Euler's method (and some other methods) to approximate solutions when these solutions have components whose time scales differ widely. For instance, the solution of the circuit problem in Example 2.2.5 had two components: (1) a *transient term* of the form Ce^{-at}, with $C > 0$ and $a > 0$, that decreased rapidly to zero as $t \to \infty$ and (2) a *steady-state term* of the form $A\sin(\omega t) + B\cos(\omega t)$ that oscillated with time. Instead of approximating the steady-state part of the solution, Euler's method may allow the error associated with the transient part to dominate, producing meaningless results.

Equations exhibiting this characteristic behavior include many that arise in electrical circuit theory and in the study of chemical reactions. The term *stiff* is used because these numerical difficulties occur in analyzing the motion of spring-mass systems with large spring constants—that is, systems with "stiff" springs. (The stiffness of a spring depends on the materials of which it is made and on the specific manufacturing processes used.) In Chapters 4 and 5, we'll discuss such spring-mass problems in greater detail.

For now, let's give an example that highlights the difficulty.

EXAMPLE 3.1.5   **A Stiff Differential Equation**

Suppose we look at the IVP $\dfrac{dI}{dt} + 50I = \sin(\pi t)$, $I(0) = 0$, which is just the equation in Example 2.2.5 with $L = 1$, $R = 50$, $\nu_0 = 1$, and $\omega = \pi$. According to that example, the solution is

$$I(t) = \frac{1}{(2500 + \pi^2)} \{50\sin(\pi t) - \pi\cos(\pi t) + \pi e^{-50t}\}$$

(*Check this for yourself.*) Note both the transient component and the steady-state part.

Now say that we want to approximate the solution at $t = 2$. To understand the accuracy of the approximation, we can first use the solution formula to find the exact answer,

$$I(2) = \frac{-\pi}{2500 + \pi^2}\left(1 - \frac{1}{e^{100}}\right) = -0.001251695566.\ldots$$

For further comparison with approximations, here are some actual values of I at intermediate points between 0 and 2:

$$I(0.5) = 0.01992135365\ldots \qquad I(1.0) = 0.001251695566\ldots$$
$$I(1.5) = -0.1992135365.\ldots$$

Euler's method yields the formula $i_{k+1} = i_k + h(\sin(\pi t_k) - 50i_k)$. Table 3.5a displays the results of using Euler's method with $h = 0.1$, Table 3.5b shows the results when $h = 0.05$, and Table 3.5c shows what happens when $h = 0.01$. We omit the error column in each table because the discrepancies or agreements between actual values of I and approximate values are fairly obvious in each table.

You can see that the errors in approximating $I(2)$ are horribly large for $h = 0.1$ and $h = 0.05$, whereas there is very little error when we have reduced h to 0.01. Comparing the graph of the actual solution curve for $I(t)$ with the graphs of the approximation curves given by Euler's method for these three values of h is a real eye-opener (see Figures 3.7a, 3.7b, and 3.7c). In each graph, the solid line is the actual solution curve, and the dashed line is the approximation. Note that the scales are different from graph to graph.

TABLE 3.5a $h = 0.1$

k	t_k	i_k
0	0	0
5	0.5	-1.26575
10	1.0	1316.73504
15	1.5	-0.13483×10^7
20	2.0	0.13807×10^{10}

TABLE 3.5b $h = 0.05$

k	t_k	i_k
0	0	0
10	0.5	0.09262
20	1.0	4.18748
30	1.5	241.37834
40	2.0	13920.24471

TABLE 3.5c $h = 0.01$

k	t_k	i_k
0	0	0
50	0.5	0.01994
100	1.0	0.00125396
150	1.5	-0.01994
200	2.0	-0.00125396

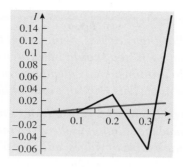

Figure 3.7a
$h = 0.1$

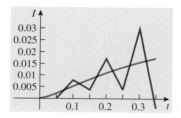

Figure 3.7b
$h = 0.05$

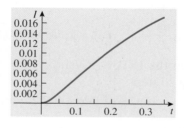

Figure 3.7c
$h = 0.01$

Figure 3.7c shows that you can hardly distinguish between the actual solution curve and its Euler method approximation when $h = 0.01$. The choice of the interval $[0, 0.35]$ for t was made after some experimentation. Using the technology available to you, you should look at the graphs of the approximations on larger intervals. For larger values of t, beginning around $t = 1$, you will find that the approximation curves are rather alarming distortions of the steady-state solution. ◆

In the next section, we will discuss approximation error in more detail, and then in subsequent sections of this chapter, we will investigate improved algorithms.

EXERCISES 3.1

In the following exercises, being asked to do a problem "by hand" or "manually" means that each step should be written out and that although calculators may be used to do arithmetic, no calculator routine or CAS program for Euler's method should be used. This is the opposite of being allowed to "use technology."

For each of Exercises 1–3, use Euler's method by hand with the given step sizes to approximate the solution to the given initial-value problem over the specified interval. Include a table of values, and give a sketch of the approximate solution by plotting the values you have calculated.

1. $\dfrac{dy}{dt} = t^2 - y^2$, $y(0) = 1$; $0 \le t \le 1$, $h = 0.25$

2. $\dfrac{dy}{dt} = e^{(2/y)}$, $y(0) = 2$; $0 \le t \le 2$, $h = 0.5$

3. $\dfrac{dy}{dt} = e^{(2/y)}$, $y(1) = 2$; $1 \le t \le 3$, $h = 0.5$

4. Compare your answers to Exercises 2 and 3, and explain what you see.

5. If y is the solution of the IVP $\dfrac{dy}{dt} = \cos t$, $y(0) = 0$, use Euler's method manually with $h = \pi/10$ to approximate $y(\pi/2)$. What is the absolute error?

6. Approximate $y(1.4)$ by hand if y is the solution to the IVP $\dfrac{dy}{dx} = x^3$, $y(1) = 1$. Use $h = 0.1$.

7. Given $\dfrac{dy}{dx} = \dfrac{x}{y}$, $y(0) = 1$, use $h = 0.1$ to approximate $y(1)$ manually.

8. Given the IVP $y' = y \sin 3t$, $y(0) = 1$, use technology to approximate $y(4)$ using 20 steps.

9. In Exercise 28 of Section 2.2, you were given the following model for the population of Botswana: $\dfrac{dP}{dt} = 0.0355P - 0.00160625t$, with $P(0) = 1.285$ (million).
   a. Use technology and Euler's method with $h = 0.01$ to approximate $P(1)$, the population in 1991.
   b. Using the approximation for $P(1)$ you found in part (a) as your starting point and $h = -0.01$, approximate $P(0)$.

10. In the area of pharmacokinetics, the *Michaelis-Menton equation* $\dfrac{dx}{dt} = \dfrac{-Kx}{A + x}$ describes the rate at which a body processes a drug. Here $x(t)$ is the concentration of the drug in the body at time t, and K and A are positive constants.
   a. For cocaine, let $A = 6$, $K = 1$, and $x(0) = 0.0025$. Use technology and Euler's method with $h = 0.1$ to evaluate x for $t = 1, 2, 3, 10$, and 20. Estimate how long it takes for the concentration to be half its initial value.

b. For alcohol, let $A = 0.005$, $K = 1$, and $x(0) = 0.025$. Use technology and Euler's method with $h = 0.01$ to evaluate x for $t = 0.01, 0.02, 0.03, 0.04$, and 0.05. Estimate how long it takes for the concentration to be half its initial value.

11. In modeling aircraft speed and altitude loss in a pull-up from a dive, basic laws of physics yield the differential equation

$$\frac{dV}{d\theta} = \frac{-gV \sin \theta}{kV^2 - g \cos \theta},$$

where θ denotes the dive angle (in radians), $V = V(\theta)$ is the speed of the plane, $g = 9.8$ m/s^2 is the acceleration constant, and k is a constant related to the wing surface area. For a particular plane, $k = 0.00145$, $\theta_0 = -0.786$, and $V(\theta_0) = V_0 = 150$ m/s. Use $h = 0.006$ (which divides θ_0 evenly) and $n = 131$ to estimate $V(0)$, the plane's speed at the completion of its pull-up—that is, when it levels out to $\theta = 0$. (Of course, use technology!)

12. Consider the IVP $y' = y^2$, $y(0) = 1$.

a. Using $h = 0.2$, approximate the solution y over the interval $[0, 1.2]$ by hand.

b. Show that the exact solution is given by $y = \dfrac{1}{1 - t}$.

c. Compare the values you found in part (a) with values given by the formula in part (b). Explain any strange numerical behavior. [*Hint:* A slope field or solution graph may help.]

13. Consider the IVP $y' = 1 - t + 4y$, $y(0) = 1$. Using technology and $h = 0.1$, approximate the solution on the interval $0 \le t \le 1$. What error is made at $t = \frac{1}{2}$ and $t = 1$?

14. Consider the IVP $y' = x^2 + y$, $y(0) = 1$. Approximate $y(0.1)$, $y(0.2)$, and $y(0.3)$ by hand, using both $h = 0.1$ and $h = 0.05$ for each approximation.

15. Use Euler's method manually with both $h = 0.5$ and $h = 0.25$ to approximate $x(2)$, where $x(t)$ is the solution of the IVP $\dfrac{dx}{dt} = \dfrac{3t^2}{2x}$, $x(0) = 1$. Solve the equation exactly, and compare the absolute errors you get with the different values of h.

16. Consider the IVP $y' = y(1 - y^2)$, $y(0) = 0.1$. Note that the equation has three equilibrium solutions.

a. Use a phase portrait analysis or a direction field to predict what *should* happen to the solution.

b. Use technology and Euler's method with $h = 0.1$ to step out to $x = 3$. What happens to the numerical solution?

17. Describe a class of differential equations for which Euler's method gives a *completely accurate* numerical solution—that is, for which y_k exactly equals the true solution $\varphi(x_k)$ for every k. [*Hint:* Try to think of differential equations for which all solution curves coincide with the tangent line segments.]

18. Suppose that $x' = x^3$.

    a. Find an expression for x'' in terms of x, assuming that x is a function of t.

    b. Suppose that $x(0) = 1$. Is the solution curve concave up or concave down? Use the result in part (a) to justify your answer.

    c. Does Euler's method overestimate or underestimate the true value of the solution at $t = 0.1$? Explain. (Don't actually carry out Euler's method.)

19. Consider the IVP $y' = y^\alpha$, $\alpha < 1$, $y(0) = 0$.

    a. Find the *exact* solution of the IVP.

    b. Show that Euler's method fails to determine an approximate solution to the IVP.

    c. Show what happens if the initial condition is changed to $y(0) = 0.01$.

20. Consider the stiff differential equation $\dfrac{dy}{dt} = -100y + 1$, with $y(0) = 1$.

    a. Solve this IVP and calculate the exact value of $y(1)$.

    b. Use technology and Euler's method to approximate $y(1)$ with $h = 0.1$, 0.05, and 0.01.

    c. Use technology to plot the exact solution and an approximate solution of the equation over the interval $[0, 0.03]$ on the same set of axes. Do this for each of the three values of h mentioned in this problem.

21. The equation $y' = -50(y - \cos x)$ is stiff.

    a. Use software to solve the equation with the initial condition $y(0) = 0$. Then calculate the exact value $y(0.2)$.

    b. Use Euler's method and technology to approximate $y(0.2)$ with step size $h = 1.974 / 50$. What is the absolute error?

    c. Use Euler's method and technology to approximate $y(0.2)$ with step size $h = 1.875 / 50$. What is the absolute error?

    d. Use Euler's method and technology to approximate $y(0.2)$ with step size $h = 2.1 / 50$. What is the absolute error now?

    e. Using technology, plot on the same axes the three approximation curves you found in parts (b), (c), and (d). Use the interval $[0, 1]$. Would you call the Euler method solution of the equation "sensitive to step size"?

3.2   A FEW MORE WORDS ABOUT ERROR

For example, when working with π, which has an infinite, nonrepeating decimal representation, we lose accuracy by using 3.14159 or even 3.14159265359 as its value. This, in turn, leads to what is called *propagated error,* the accumulated error resulting from many calculations with rounded values. If each item of data is inaccurate because of rounding of some sort, then the various steps in a calculation process can compound the error. A useful approximation method guarantees that the smaller the round-off error at each stage, the smaller the cumulative round-off error. Of course it turns out that sometimes round-off errors cancel each other out to a certain extent—approximate values that are too high may be balanced by values that are too low.

Truncation error occurs when we stop (or truncate) an approximation process after a certain number of steps. For example, when we approximate the values of sin x near $x = 0$ by using the first seven nonzero terms of its (infinite) Taylor series,

$$x - \frac{x^3}{3!} + \frac{x^5}{5!} - \frac{x^7}{7!} + \frac{x^9}{9!} - \frac{x^{11}}{11!} + \frac{x^{13}}{13!}$$

we are introducing truncation error. If we write sin $x = T_{13}(x) + R_{13}(x)$, where $T_{13}(x)$ is the 13th-degree polynomial just given, a formula we know from calculus gives an upper bound for the absolute truncation error:

$$|\sin x - T_{13}(x)| = |R_{13}(x)| = \frac{|\sin c|}{14!}|x|^{14} \le \frac{|x|^{14}}{14!},$$

where c is a positive number less than x. (See Appendix A.3.) Even if we use the 1001st-degree Taylor polynomial, we are still only approximating and will therefore have truncation error.

If y is the solution of the equation $y' = f(x, y)$, we can view Euler's method in this light by considering the Taylor expansion of $y(x)$ about $x = x_k$:

$$y(x_{k+1}) = y(x_k) + y'(x_k)h + y''(\xi_k)\frac{h^2}{2} = \underbrace{y(x_k) + f(x_k, y(x_k))h}_{y_{k+1}} + y''(\xi_k)\frac{h^2}{2}$$

with $x_k < \xi_k < x_{k+1}$. Assuming that $y(x)$ has a bounded second derivative and realizing that $h^2 < h$ for small values of h, we see that Euler's method is essentially using a first-degree Taylor polynomial to approximate the solution curve:

$$y(x_{k+1}) \approx y(x_k) + f(x_k, y(x_k))h.$$

Finally, we must be aware that there is usually a tradeoff in dealing with error. If we try to reduce the truncation error and increase the accuracy of our approximation by carrying out more steps (for example, by taking more terms of a Taylor series or more steps in Euler's method), we are increasing our calculation load and consequently running the risk of increasing propagated error. Figure 3.8 on page 114 shows the tradeoff in general terms.

Clearly, at each stage of Euler's method, we choose to round off entries in a certain way. Even if we assume for the sake of simplicity that round-off error is negligible, our use of local linearity—using straight lines to approximate curves—introduces truncation error. Now suppose that we are given the value, $y(x_0)$, of a solution at an initial point and want to approximate the value, $y(b)$, at some later point $b = x_0 + nh$. First there is **local truncation error** at each step, defined as $y(x_{k+1}) - y_{k+1}$ for each k ($k = 0, 1, 2, \ldots, n - 1$). This is the error introduced in computing the value y_{k+1} from the value y_k, *assuming that y_k is exact.* Then we have the **cumulative truncation error,** defined as $y(b) - y_n = y(x_n) - y_n$ which is

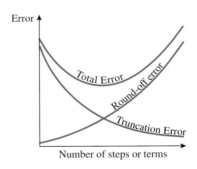

Figure 3.8
Total error = round-off error + truncation error

the (total) actual error in the value of $y(x_0 + nh)$, or $y(b)$, caused by all the previous approximations—that is, by the cumulative effect after n steps of the local errors from previous steps. (This is not just the sum of all the local truncation errors. *Life isn't that simple.*)

In any case, a mathematically rigorous analysis of the errors produced shows that *the local truncation error at any step of Euler's method behaves like a constant multiple of h^2,* which is smaller than h when h is small:

$$|\text{local truncation error at step } k| = |y(x_{k+1}) - y_{k+1}| \leq \frac{M}{2}h^2,$$

where $M = \max_{x_k < x < x_{k+1}} |y''(x)|$. This follows easily from the Taylor series expansion

$$y(x_{k+1}) = \underbrace{y(x_k) + f(x_k, y(x_k))h}_{y_{k+1}} + y''(\xi_k)\frac{h^2}{2}$$

that we had above.

It is also true that *for Euler's method, the cumulative truncation error is no greater than a constant multiple of the step size h:*

$$|\text{true value} - \text{approximation}| = |y(b) - y_n| \leq K \cdot h,$$

where K is independent of h but depends on $|y''(x)|$ and the interval $[x_0, b]$. Because the cumulative error is bounded by a constant multiple of the *first* power of the step size h, we say that Euler's method is a **first-order method.** (The number K is a *maximum* bound. In practice, the actual error incurred in a problem will usually be less than this bound.) Intuitively, we can reason as follows: There are $n = \dfrac{b - x_0}{h}$ steps in the Euler method approximation, each having a local error less than or equal to some multiple of h^2. If K^* is the largest of the multipliers, then the cumulative error is less than or equal to $\dfrac{b - x_0}{h} \cdot K^*h^2 = Kh$.

Therefore, if we ignore round-off error as essentially a statistical problem outside our range of interest right now, we can make the total error as small as we wish by making h "sufficiently small"—that is, by making the *number* of steps "sufficiently large." This is not very satisfactory because a larger number of steps *does* require more calculating time by hand or by computer, and in real-life problems, the larger number of steps often leads to a "snowballing" of round-off error. Take another look at Figure 3.8.

If you want to understand and improve the accuracy of your approximations, here are two rules of thumb you can use: (1) Start your calculations with many more decimal places than you need. (2) Keep on redoing your calculations with a step size h equal to one-half its previous value. If you reach a stage at which the new result agrees with the previous one to d decimal places after appropriate rounding, then you can assume that you have d decimal place accuracy. (Look back at Example 3.1.4 for a slight variation of this rule.)

Euler's method is not very accurate and is not used widely in practice. But the method is simple and displays the essential characteristics of more sophisticated methods. In the next section we discuss an improved method, one that uses Euler's basic idea in a more efficient way.

3.3   THE IMPROVED EULER METHOD

In Euler's original method. the slope $f(x, y)$ over any interval $x_k \leq x \leq x_{k+1}$ of length h is replaced by $f(x_k, y_k)$, so that x always takes the value of the left endpoint of the interval. (As noted just before Example 3.1.1, if $y' = f(x)$, a function of x alone, then Euler's method is equivalent to using a left-hand Riemann sum to approximate a definite integral.)

Now instead of always using the slope at the *left* endpoint of the interval $[x_k, x_{k+1}]$, we can think of using an *average* derivative value over the interval. The **improved Euler method** involves two stages that will be combined into one approximation formula. The first stage involves moving tentatively across the interval $[x_k, x_{k+1}]$ using Euler's original method, thereby producing a guess, or trial value, $\hat{y}_{k+1} = y_k + h \cdot f(x_k, y_k)$. Note that the values $f(x_k, y_k)$ and $f(x_{k+1}, \hat{y}_{k+1})$ approximate the slopes of the solution curve at $(x_k, y(x_k))$ and $(x_{k+1}, y(x_{k+1}))$, respectively. Now the second stage looks at the *average* of the derivative $f(x_k, y_k)$ and the guess $f(x_{k+1}, \hat{y}_{k+1}) = f(x_{k+1}, y_k + hf(x_k, y_k))$ and uses this average to take the *real* step across the interval.

Guess (tentative step): $\hat{y}_{k+1} = y_k + h \cdot f(x_k, y_k)$

Real step: $$y_{k+1} = y_k + h \left\{ \frac{f(x_k, y_k) + f(x_{k+1}, \hat{y}_{k+1})}{2} \right\}$$

$$= y_k + h \left\{ \frac{f(x_k, y_k) + f(x_{k+1}, y_k + h \cdot f(x_k, y_k))}{2} \right\}$$

$$= y_k + \frac{h}{2} \{ f(x_k, y_k) + f(x_{k+1}, y_k + h \cdot f(x_k, y_k)) \} \qquad (3.3.1)$$

Formula (3.3.1) describes the **improved Euler method** [or **Heun's method,** named for Karl Heun (1859–1929), a German applied mathematician who devised this scheme around 1900]. It is an example of a **predictor-corrector method:** We use $\hat{y}_{k+1}$ (via Euler's method) to *predict* a value of $y(x_{k+1})$ and then use y_{k+1} to *correct* this value by averaging.

Look carefully at equation (3.3.1). If $f(x, y)$ is really just $f(x)$, a function of x alone, then solving the IVP $y' = f(x, y)$, $y(x_0) = x_0$, amounts to solving the equation $y' = f(x)$, which is a matter of simple integration. In Section 1.2 (equation 1.2.1) we saw that we can write the solution as

$$y(x) = \int_{x_0}^{x} f(t)\,dt + y_0 = \sum_{k=0}^{n-1} \int_{x_k}^{x_{k+1}} f(t)\,dt + y_0,$$

where $x_n = x$. In this case, formula (3.3.1) reduces to

$$y_{k+1} = y_k + \frac{h}{2}\{f(x_k) + f(x_{k+1})\}$$

and the Fundamental Theorem of Calculus tells us that on the interval $[x_k, x_{k+1}]$,

$$\int_{x_k}^{x_{k+1}} \overbrace{f(t)}^{y'}\,dt = y(x_{k+1}) - y(x_k) \approx y_{k+1} - y_k = \frac{h}{2}\{f(x_k) + f(x_{k+1})\}.$$

In other words, in this situation we are using the *Trapezoid Rule* from calculus to approximate each integral on $[x_k, x_{k+1}]$.

Next we see some illustrations of why this method is called "improved."

EXAMPLE 3.3.1   The Improved Euler Method

Let's use the improved Euler formula, equation (3.3.1) to calculate an approximate value of the solution of the IVP $y' = y$, $y(0) = 1$, at $x = 1$. Of course, you realize that this is just a roundabout way of asking for an approximation of that important mathematical constant e. (*Right?*)

Let's start with $h = 0.1$, so we'll need 10 steps to reach $x = 1$ from the initial point $x = 0$. Thus in formula (3.3.1) we have $x_0 = 0$, $y_0 = 1$, $h = 0.1$, $x_k = 0 + kh$ $= kh$ ($k = 0, \ldots, 10$), and $f(x, y) = y$. When we put all this information together, we see that the formula takes a simplified form:

$$y_{k+1} = y_k + \frac{h}{2}\{y_k + (y_k + h\,y_k)\}$$

$$= y_k + \frac{h}{2}\{(2 + h)\,y_k\} = y_k + (0.05)(2.1)\,y_k = 1.105\,y_k.$$

Therefore, the calculations are

$$y_1 = 1.105\,y_0 = 1.105\,(1) = 1.105$$
$$y_2 = 1.105\,y_1 = (1.105)^2 = 1.221025$$
$$y_3 = 1.105\,y_2 = (1.105)^3 = 1.349232625$$
$$\vdots \qquad\qquad \vdots$$
$$y_{10} = 1.105\,y_9 = (1.105)^{10} = 2.71408084661$$

Comparing this approximation to the actual value 2.71828182846 (rounded to 11 decimal places), we find that the absolute error is 0.00420098185. (In Exercise 4, you'll be asked to try this with the original Euler method.)

Using 20 steps, a CAS gives the approximate value 2.71719105435, so the absolute error is now 0.00109077410. Note that when we doubled the number of steps from 10 to 20, the result was that the absolute error was roughly one-fourth what it was before.

$\blacklozenge$

Now let's revisit Example 3.1.1 to see how the improved method compares with the original process of approximation.

EXAMPLE 3.3.2   The Improved Euler Method—Example 3.1.1 Revisited

We want to approximate $x(1.5)$, given the IVP $\dfrac{dx}{dt} = t^2 + x$, $x(1) = 3$. The actual value is 5.939770. . . . We'll start with $h = 0.1$, so we'll need five steps to stretch between $t = 1$ and $t = 1.5$.

For this problem, the improved Euler formula is

$$x_{k+1} = x_k + \frac{h}{2}\{(t_k^2 + x_k) + t_{k+1}^2 + x_k + h(t_k^2 + x_k)\}$$

$$= x_k + \frac{h}{2}\{t_{k+1}^2 + (1 + h)t_k^2 + (2 + h)x_k\}$$

$$= x_k + (0.05)\{t_{k+1}^2 + 1.1t_k^2 + 2.1x_k\},$$

where $t_0 = 1$, $t_1 = 1.1$, $t_2 = 1.2$, $t_3 = 1.3$, $t_4 = 1.4$, and $t_5 = 1.5$. Therefore,

$x_1 = 3 + (0.05)\{(1.1)^2 + 1.1(1)^2 + 2.1(3)\} = 3.4305$
$x_2 = 3.4305 + (0.05)\{(1.2)^2 + 1.1(1.1)^2 + 2.1(3.4305)\} = 3.9292525$
$x_3 = 3.9292525 + (0.05)\{(1.3)^2 + 1.1(1.2)^2 + 2.1(3.9292525)\} = 4.5055240125$
$x_4 = 4.5055240125 + (0.05)\{(1.4)^2 + 1.1(1.3)^2 + 2.1(4.5055240125)\} = 5.16955403381$
$x_5 = 5.16955403381 + (0.05)\{(1.5)^2 + 1.1(1.4)^2 + 2.1(5.16955403381)\} = 5.93265720736.$

To five decimal places, we have $x(1.5) \approx 5.93266$. The absolute error is 0.00711. When we employed Euler's method in Example 3.1.1, the error was 0.244639.

If we use ten steps in the improved Euler method, then we get $x(1.5) \approx$ 5.943455, with absolute error 0.00369, compared to the Euler method's error of 0.12717.

$\blacklozenge$

Now let's go back and redo another earlier example with the new method.

EXAMPLE 3.3.3   Improved Euler Method—Example 3.1.3 Revisited

In Example 3.1.3 we discussed the IVP $y' = \sqrt{x + y}$, $y(5) = 4$, with the goal of approximating $y(4)$. If we apply the improved method to the problem, with five backward steps, each of length 0.2—that is, with $h = -0.2$—we get the values shown in Table 3.6.

Thus $y(4) \approx 1.341827$ by the improved method, compared to the "true" answer

TABLE 3.6   Improved Euler Method with $h = -0.2$

k	x_k	y_k	True Value	Absolute Error
0	5.0	4.000000	4.000000	0.000000
1	4.8	3.413644	3.413384	0.000260
2	4.6	2.854277	2.853750	0.000527
3	4.4	2.322249	2.321444	0.000805
4	4.2	1.817952	1.816857	0.001100
5	4.0	1.341827	1.34043	0.00140

1.34042895566892 and the original Euler method approximate value 1.27674987.   ◆

An analysis of error shows that the *local* truncation error at any stage of the improved Euler method behaves like a constant multiple of h^3 and that *the cumulative truncation error is no greater than a constant multiple of the square of the step size h:* $|$true value $-$ approximation$| \leq K \cdot h^2$, where K is a constant that depends on the function $f(x, y)$, on its partial derivatives, and on the interval involved but does not depend on h. We say that the improved Euler method is a **second-order method.**

In the next section, we'll look at a fourth-order method and a powerful combination of fourth- and fifth-order techniques.

EXERCISES 3.3

Use the table below to enter the data from Exercises 1 and 2.

	True Value	Euler's Method	Absolute Error	Improved Euler Method	Absolute Error
$h = 0.1$					
$h = 0.05$					
$h = 0.025$					

1. Use the improved Euler method to redo Example 3.1.1 with $h = 0.1, 0.05,$ and 0.025.

2. Use the improved Euler method to redo Example 3.1.2 with $h = 0.1, 0.05,$ and 0.025. (You'll also have to use Euler's method for $h = 0.1$.)

3. **a.** Find the exact solution to the IVP $\dfrac{dx}{dt} = t + x, x(0) = 1.$

b. Apply the improved Euler method with step size $h = 0.1$ to approximate the value $x(1)$.

c. Calculate the absolute error at each step of part (b).

4. Using technology, redo Example 3.3.1 with both Euler's method and the improved Euler method, using a step size of $h = 0.01$—that is, using 100 steps. For each method, calculate the absolute errors incurred in approximating $y(0.01)$, $y(0.02)$, ..., $y(0.99)$, $y(1.0)$. (A spreadsheet program can be particularly useful here.)

5. Redo Exercise 11 in Section 3.1 using the improved Euler method.

3.4 MORE SOPHISTICATED NUMERICAL METHODS: RUNGE-KUTTA AND OTHERS

Modern computers (and even hand-held calculators) have many algorithms for solving differential equations numerically. Some of these are highly specialized and are meant to handle very particular types of ODEs (such as stiff equations—see Section 3.1) and systems of ODEs. Euler's method and its improved version are useful for illustrating the idea behind numerical approximation, but they are not very efficient in terms of approximating a solution of an IVP very accurately and with a minimum number of steps.

A very good method, implemented in many computer algebra systems and in calculator firmware, is the **fourth-order Runge-Kutta method (RK4),** which was developed in an 1895 paper by Carl Runge (1856–1927), a German applied mathematician, and was generalized to *systems* of ODEs in 1901 by M. Wilhelm Kutta (1867–1944), a German mathematician and aerodynamicist. As the description indicates, in this method the total accumulated error is proportional to h^4, so reducing the step size by a factor of $\frac{1}{10}$ produces four more digits of accuracy—for example, reducing the step size from $h = 0.1$ to $h = 0.01$ generally decreases the total error by a factor of 0.0001. (The *local* truncation error behaves like h^5.) There are also second- and third-order Runge-Kutta methods. (Euler's method can be called a first-order Runge-Kutta method.)

Now suppose we have an IVP $y' = f(x, y)$, $y(x_0) = y_0$. The RK4 formula is a bit strange-looking, but not if we realize that it is approximating the value $y(x_{k+1})$ by a *weighted average*, y_{k+1}, of values of $f(x, y)$ calculated at different points in the interval $[x_k, x_{k+1}]$. For each interval $[x_k, x_{k+1}]$, we calculate the following (multiples of) slopes in the order given:

$$m_1 = hf(x_k, y_k)$$
$$m_2 = hf\left(x_k + \frac{h}{2}, y_k + \frac{m_1}{2}\right)$$
$$m_3 = hf\left(x_k + \frac{h}{2}, y_k + \frac{m_2}{2}\right)$$
$$m_4 = hf(x_k + h, y_k + m_3) = hf(x_{k+1}, y_k + m_3). \qquad (3.4.1)$$

Then the classical fourth-order Runge-Kutta formula is

$$y_{k+1} = y_k + \frac{1}{6}(m_1 + 2m_2 + 2m_3 + m_4). \qquad (3.4.2)$$

Perhaps this formula won't be so alarming if we look at the simplified situation when $f(x, y)$ is independent of y in the equation $y' = f(x, y)$. If $f(x, y) = g(x)$, then the formulas (3.4.1) for $m_1, m_2, m_3,$ and m_4 reduce to

$$m_1 = hg(x_k)$$
$$m_2 = hg\left(x_k + \frac{h}{2}\right)$$
$$m_3 = hg\left(x_k + \frac{h}{2}\right)$$
$$m_4 = hg(x_k + h) = hg(x_{k+1})$$

so formula (3.4.2) becomes

$$y_{k+1} = y_k + \frac{h}{6}\left\{g(x_k) + 2g\left(x_k + \frac{h}{2}\right) + 2g\left(x_k + \frac{h}{2}\right) + g(x_{k+1})\right\}$$

$$= y_k + \frac{h}{6}\left\{g(x_k) + 4g\left(x_k + \frac{h}{2}\right) + g(x_{k+1})\right\}$$

and you may recognize the expression $\frac{h}{6}\left\{g(x_k) + 4g\left(x_k + \frac{h}{2}\right) + g(x_{k+1})\right\}$ as

a form of *Simpson's Rule* for approximating $\displaystyle\int_{x_k}^{x_{k+1}} g(x)\,dx$. (Note that $x_k + \frac{h}{2}$ in

the expression is the *midpoint* of the interval $[x_k, x_{k+1}]$ because $h = x_{k+1} - x_k$.)

To get a feel for the calculations, let's choose an example that we've seen before.

EXAMPLE 3.4.1   RK4—Example 3.1.1 Yet Again

Let's approximate $x(1.5)$ by the Runge-Kutta method if we are given the IVP

$\dfrac{dx}{dt} = t^2 + x, x(1) = 3$. We'll use $h = 0.1$, so we need five steps.

Just to get the idea, let's focus on the interval $[t_0, t_1] = [1, 1.1]$. We calculate

$$m_1 = hf(t_0, x_0) = (0.1)f(1, 3) = 0.1(1^2 + 3) = 0.4$$
$$m_2 = hf\left(t_0 + \frac{h}{2}, x_0 + \frac{m_1}{2}\right) = (0.1)f(1 + 0.05, 3 + 0.2) = (0.1)(1.05^2 + 3.2)$$
$$= 0.43025$$
$$m_3 = hf\left(t_0 + \frac{h}{2}, x_0 + \frac{m_2}{2}\right) = (0.1)f(1 + 0.05, 3 + 0.215125)$$
$$= (0.1)(1.05^2 + 3.215125) = 0.4317625$$
$$m_4 = hf(t_0 + h, x_0 + m_3) = f(t_1, x_0 + m_3) = (0.1)f(1.1, 3 + 0.4317625)$$
$$= (0.1)(1.1^2 + 3.4317625) = 0.46417625$$

TABLE 3.7   Comparison of Methods with $h = 0.1$

t_k	True Value of $x(t_k)$	Euler's Method	Improved Euler Method	Runge-Kutta Method
1	3.00000	3.00000	3.00000	3.00000
1.1	3.43137	3.40000	3.43050	3.43137
1.2	3.93122	3.86100	3.92925	3.93122
1.3	4.50887	4.39110	4.50552	4.50887
1.4	5.17460	4.99921	5.16955	5.17460
1.5	5.93977	5.69513	5.93266	5.93977

so

$$x(1.1) \approx x_1 = 3 + \frac{1}{6}(0.4 + 2(0.43025) + 2(0.4317625) + 0.46417625)$$

$$= 3 + \frac{1}{6}(2.58820125) = 3.431366875.$$

The actual value of $x(1.1)$ (from the solution formula $x(t) = -t^2 - 2t - 2 + 8e^{t-1}$) is 3.4313673446. . . . Here the absolute error is 0.0000004696. *(This is an amazingly close approximation!)*

Table 3.7 shows, for the same value $h = 0.1$, the exact values and the approximate values given for this problem by Euler's method, the improved Euler method, and the Runge-Kutta method. We can see how accurate the Runge-Kutta method is at each step.   ◆

As accurate as the classical Runge-Kutta method is, there are improvements possible. For example, a very popular method, the **Runge-Kutta-Fehlberg algorithm,** combines fourth-order and fifth-order methods in a clever way announced by E. Fehlberg in 1969. The **rkf45** method, as its computer implementation is known, uses *variable* step sizes, choosing the step size at each stage to try to achieve a predetermined degree of accuracy. Such clever numerical techniques are called **adaptive methods.**

EXERCISES 3.4

In the exercises that follow, it is assumed that you have versions of the Runge-Kutta Fourth-Order method (RK4) and the Runge-Kutta-Fehlberg (rkf45) method available to you. Use the table that follows to enter the data from Exercises 1 and 2. You may go back to earlier examples to find needed values.

	True Value	Euler's Method	Improved Euler Method	RK4 Method
$h = 0.1$				
$h = 0.05$				
$h = 0.025$				

1. Use the RK4 method to redo Example 3.3.1 with $h = 0.1, 0.05$, and 0.025. (The case $h = 0.1$ has been done for you in the example.)
2. Use the RK4 method to redo Example 3.3.2 with $h = 0.1, 0.05$, and 0.025.
3. Use the rkf45 method to approximate the solution of $y' = y$, $y(0) = 1$, at $t = 1$. (That is, approximate the value of the constant e. See Example 3.3.1.)
4. **a.** Find the exact solution of the IVP $\dfrac{dx}{dt} = t + x$, $x(0) = 1$.

   b. Apply the rkf45 method to approximate $x(1)$, calculating the absolute error at each step.
5. **a.** Find the closed-form solution of the equation $\dfrac{dx}{dt} = -tx^2$.

   b. Using the rkf45 method, approximate the value $x(1)$ if x is the solution of the IVP $\dfrac{dx}{dt} = -tx^2$, $x(0) = 2$.
6. Approximate $y(0.8)$ using the rkf45 method if y is the solution of the IVP $\dfrac{dy}{dx} = \sin(xy)$, $y(0) = 0$.
7. A daredevil named Ayanna goes skydiving, jumping from a plane at an initial altitude of $10,000$ feet. At time t her velocity $v(t)$ satisfies the initial-value problem $\dfrac{dv}{dt} = f(v)$, $v(0) = 0$, where

$$f(v) = 32 - (0.000025) \cdot (100v + 10v^2 + v^3).$$

   If she does not open her parachute, she will reach a *terminal velocity* when the forces of gravity and air resistance balance.

   a. Use the rkf45 method to approximate her velocity at times $t = 5, 10, 15, 16, 17, 18, 19$, and 20, and so guess at her terminal velocity (accurate to three decimal places).

   b. Use technology to graph Ayanna's velocity over the interval $[0, 30]$.
8. Consider the IVP $\dfrac{dx}{dt} = x^2$, $x(0) = 2$.

   a. Use Euler's method with $h = 0.1$ to approximate $x(1)$. Does your answer seem strange?

b. Use the rkf45 method to approximate $x(1)$. Compare your answer to the answer in part (a)—if your calculator or CAS gives you a meaningful answer in both cases.

c. To help explain your difficulties in parts (a) and (b), find the closed-form solution of the IVP.

d. Use your answer to part (c) to explain why your answers to parts (a) and (b) are *both* wrong.

e. How do you think you may be able to avoid the difficulty uncovered in part (d)? Maybe by changing step size? Try to solve the problem again using the rkf45 method.

3.5 SUMMARY

Even if we can solve a first-order differential equation, we may not be able to find a closed-form solution. This difficulty has led to the development of numerical methods to *approximate* a solution to any degree of accuracy. Leaving aside *input error,* there are two main sources of error in numerical calculations done by hand, calculator, or computer: *round-off error* and *truncation error.* **Round-off error** is the kind of inaccuracy we get by taking a certain number of decimal places instead of taking the entire number. In particular, remember that our calculator or computer is limited in the number of decimal places it can handle. **Truncation error** occurs when we stop (or truncate) an approximation process after a certain number of steps. Finally, we must be aware that there is usually a tradeoff in dealing with error. If we try to reduce the truncation error and increase the accuracy of our approximation by carrying out more steps (for example, by taking more terms of a Taylor series), we are increasing your calculation load and consequently running the risk of increasing *propagated* (cumulative) *error.*

Euler's method uses the idea that values near a point on a curve can be approximated by values on the tangent line drawn to that point. If we want to approximate the solution of the IVP $y' = f(x, y)$, $y(x_0) = y_0$ on an interval $[a, b]$, we first partition $[a, b]$ by using $n + 1$ equally spaced points:

$$a = x_0 < x_1 < x_2 < \cdots < x_{n-1} < x_n = b,$$

where $x_{i+1} - x_i = \dfrac{b - a}{n} = h$ for $i = 0, 1, \ldots, n - 1$. Then, if y_i is an approximate value for $y(x_i)$, we can define the sequence of approximate solution values as follows: $y_{k+1} = y_k + hf(x_k, y_k)$. Generally speaking, you can increase the accuracy of the approximation (reduce the error) by making the step size h smaller—that is, by making the *number* of steps n larger. For Euler's method, a **first-order method,** the cumulative truncation error is bounded by a constant multiple of the step size: $|\text{true value} - \text{approximation}| \leq K \cdot h$, where K is independent of h but depends on $|y''(x)|$ and the interval $[x_0, b]$. In practice, the actual error incurred in a problem will usually be less than this bound.

An improvement of Euler's method called **Heun's method** guesses a value of $y(x_k)$ and then uses y_k to correct this guess by an averaging process. The algorithm can be expressed as follows:

Guess (tentative step): $\hat{y}_{k+1} = y_k + h \cdot f(x_k, y_k)$

Real Step: $y_{k+1} = y_k + h\left\{\dfrac{f(x_k, y_k) + f(x_{k+1}, \hat{y}_{k+1})}{2}\right\}$

$$= y_k + h\left\{\dfrac{f(x_k, y_k) + f(x_{k+1}, y_k + h \cdot f(x_k, y_k))}{2}\right\}$$

$$= y_k + \dfrac{h}{2}\{f(x_k, y_k) + f(x_{k+1}, y_k + h \cdot f(x_k, y_k))\}.$$

For the improved Euler method, the cumulative truncation error is no greater than a constant multiple of the square of the step size h: $|\text{true value} - \text{approximation}| \leq K \cdot h^2$, where K is a constant that depends on the function $f(x, y)$, on its partial derivatives, and on the interval involved but not on h. We say that the improved Euler method is a **second-order method.**

There are many more sophisticated algorithms for solving differential equations numerically. Two very effective methods implemented in many computer algebra systems and even some calculators are the **fourth-order Runge-Kutta method** and the **Runge-Kutta-Fehlberg algorithm.** The *rkf45* method, as the computer implementation of this last algorithm is known, uses *variable* step sizes, choosing the step size at each stage to try to achieve a predetermined degree of accuracy. Such clever numerical techniques are called **adaptive methods.**

PROJECT 3-1

Euler Backwards Is More Than reluE

A *stiff* differential equation, such as the one discussed in Example 3.1.5, does not respond well to Euler's method unless the step size is small, in which case the number of steps (and the accumulated round-off error) may be large. The solution of a problem that is stiff is impractical with numerical methods not designed specifically for such problems.

Suppose we have used the points $t_0, t_1, \ldots, t_n = b$ to divide the interval from t_0 to b into n equal subintervals of length h, as we would for Euler's method. Then the differential equation $\dfrac{dy}{dt} = f(t, y)$ at the point t_k can be written in the form $\dfrac{dy}{dt}(t_k) = f(t_k, y(t_k))$. Instead of approximating the derivative in the last equation by the *forward* difference quotient $\dfrac{y(t_{k+1}) - y(t_k)}{h}$, as we did for Euler's method, we use the *backward* difference quotient $\dfrac{y(t_k) - y(t_{k-1})}{h}$, so we get the formula

$$\frac{dy}{dt}(t_k) = f(t_k, y(t_k)) \approx \frac{y(t_k) - y(t_{k-1})}{h}$$

or $y(t_k) = y(t_{k-1}) + hf(t_k, y(t_k))$. Replacing k by $k + 1$, we get the **backward Euler formula:**

$$y_{k+1} = y_k + hf(t_{k+1}, y_{k+1}).$$

This is also called the *implicit* Euler method because the quantity y_{k+1} appears on *both* sides of the equation and has to be solved for.

a. By hand, use $h = 0.1$ in the backward Euler method on the stiff problem $y' = -2y$, $y(0) = 3$, to approximate $y(1)$. Compare your values to those given by the usual Euler method.

b. By hand, use $h = 0.1$ in the backward Euler method to approximate $y(0.5)$ if $y' = 25\cos(y)$, $y(0) = 1$. Use your calculator or CAS equation solver to find y_{k+1} at each step, keeping all digits shown for use in the next step.

c. Find out whether you have the backward Euler method available on a computer. If not, look for a numerical method described as being designed for stiff differential equations (maybe under the name LSODE) or perhaps a "multistep" algorithm. Use such a method to check your answers to part (b).

4 | Second- and Higher-Order Equations

4.0 INTRODUCTION

In Chapters 2 and 3, we analyzed first-order equations graphically, numerically, and analytically and introduced qualitative concepts that will be useful in later chapters.

In this chapter, we will make the jump from first-order equations to higher-order equations, especially second- and third-order equations. We'll start by investigating types of second-order equations that occur frequently in science and engineering applications. These equations have a fully developed theory that generalizes to higher-order equations of the same type.

Most of the chapter, however, will be devoted to a *systems* approach to higher-order equations. In particular, we will see how *any higher-order differential equation can be written as a system of first-order differential equations* and then learn how to handle such systems qualitatively and numerically. In fact, if we use a graphing calculator in our study of differential equations, this device will require us to input a higher-order equation as a system of first-order equations. As we'll see in Section 4.7, the numerical methods studied in Sections 3.1, 3.3, and 3.4 can be applied in a natural way to the first-order systems representation of any higher-order differential equation.

To illustrate the systems approach, we'll analyze some very interesting and important examples such as spring-mass problems, predator-prey relations, and arms races.

4.1 HOMOGENEOUS SECOND-ORDER LINEAR EQUATIONS WITH CONSTANT COEFFICIENTS

A very important application of differential equations is the analysis of an *RLC circuit* containing a resistance R, an inductance L, and a capacitance C. (We have already seen some first-order examples in Chapter 2.) In electrical circuit theory, if $I = I(t)$ represents the current, *Kirchhoff's Voltage Law* leads to the equation $L\dfrac{d^2I}{dt^2} + R\dfrac{dI}{dt} + \dfrac{1}{C}I = 0$ when the voltage applied to the circuit is constant. We describe any equation of the form $ay'' + by' + cy = 0$, where a, b, and c are constants, $a \neq 0$, as a **homogeneous linear second-order equation with constant coefficients.** In this section, we are going to develop a technique for solving any equation of this type.

Extending the way we considered first-order linear equation in Section 2.2, we see that a linear second-order equation with constant coefficients can be viewed in terms of an *operator L* transforming functions that have two derivatives: $L(y) = ay'' + by' + cy$. To solve a homogeneous equation, we must find a function y such that $L(y) = 0$. There is a natural extension of the *Superposition Principle* (see Section 2.2) for homogeneous equations: If y_1 and y_2 are solutions of the equation $ay'' + by' + cy = 0$, then so is any *linear combination* of these solutions—that is, $L(c_1y_1 + c_2y_2) = 0$ for arbitrary constants c_1 and c_2. (You'll be asked to show this in Exercise 11.)

If we consider a homogeneous *first*-order linear equation with constant coefficients, $ay' + by = 0$, where $a \neq 0$, we know that the general solution is $y = Ce^{-\frac{b}{a}t}$. In 1739, aware of this solution, Euler[1] thought of solving an *n*th-order homogeneous linear equation with constant coefficients by looking for solutions of the form $y = e^{\lambda t}$, where λ is a constant to be determined. Let's see how this works for the equation

$$ay'' + by' + cy = 0, \tag{4.1.1}$$

where a, b, and c are constants. But we should be aware that, although (for example) the combination of exponentials $y(t) = 3e^t - 2e^{-t}$ is a solution of the equation $y'' - y = 0$, the similar equation $y'' + y = 0$ has solutions that are combinations of $\sin t$ and $\cos t$. As we'll see, if we start by focusing on exponential solutions, the trigonometric possibilities will appear also. The exponential and trigonometric solutions are related in an important way, through the use of complex numbers.

If we assume that $y = e^{\lambda t}$ is a solution of equation (4.1.1), then $y' = \lambda e^{\lambda t}$ and $y'' = \lambda^2 e^{\lambda t}$. Substituting these derivatives into (4.1.1), we get $a(\lambda^2 e^{\lambda t}) + b(\lambda e^{\lambda t}) + c(e^{\lambda t}) = 0$, which simplifies to $(a\lambda^2 + b\lambda + c)e^{\lambda t} = 0$. Because the exponential factor is never zero, we must have $(a\lambda^2 + b\lambda + c) = 0$.

THE CHARACTERISTIC EQUATION AND EIGENVALUES

We have just concluded that if $y = e^{\lambda t}$ is a solution of equation (4.1.1), then λ must satisfy the equation $a\lambda^2 + b\lambda + c = 0$, which is called the **characteristic equation** (or **auxiliary equation**) of the differential equation (4.1.1). The roots of this

---

1. In a letter to John (Johannes) Bernoulli, who first solved the important type of differential equation devised by his brother Jakob. See Exercises 2.2, between Problems 18 and 19.

characteristic equation will reveal to us the nature of the solution(s) of (4.1.1). Note that we can go straight from the ODE to the characteristic equation as follows:

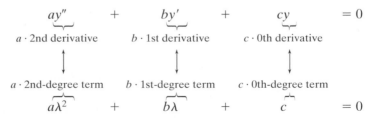

Because the characteristic equation of our second-order ODE is a quadratic equation, we know that there are two roots, called **characteristic values** or **eigenvalues**,[2] say λ_1 and λ_2. There are only three possibilities for these eigenvalues: (1) The eigenvalues are both real numbers with $\lambda_1 \neq \lambda_2$; (2) the eigenvalues are real numbers with $\lambda_1 = \lambda_2$; or (3) the eigenvalues are complex numbers: $\lambda_1 = p + qi$ and $\lambda_2 = p - qi$, where p and q are real numbers (called the *real part* and the *imaginary part*, respectively) and $i = \sqrt{-1}$. In possibility (3), we say that λ_1 and λ_2 are *complex conjugates* of each other. (Now would be a good time to review the quadratic formula and its implications. See Appendix C, especially C.3, for more information about complex numbers.)

Real but Unequal Eigenvalues

In possibility (1), where λ_1 and λ_2 are unequal real numbers, then both $y_1(t) = e^{\lambda_1 t}$ and $y_2(t) = e^{\lambda_2 t}$ are solutions of (4.1.1). By the extension of the Superposition Principle given above, any *linear combination* of the form $y(t) = c_1 e^{\lambda_1 t} + c_2 e^{\lambda_2 t}$ is also a solution, where c_1 and c_2 are arbitrary constants. It can be shown (see Section 4.2 for the details) that this is the *general* solution of (4.1.1)—that is, if the eigenvalues of (4.1.1) are real and distinct, then *any* solution of (4.1.1) must have the form $y(t) = c_1 e^{\lambda_1 t} + c_2 e^{\lambda_2 t}$ for some constants c_1 and c_2. The next example shows how to solve equations of the form (4.1.1) using eigenvalues.

EXAMPLE 4.1.1   The Characteristic Equation—Unequal Eigenvalues

Let's solve the homogeneous linear second-order equation with constant coefficients $6y'' + 13y' - 5y = 0$. We find that the characteristic equation of this ODE is $6\lambda^2 + 13\lambda - 5 = 0$:

$$\underbrace{6y''}_{6\,\cdot\,\text{2nd derivative}} + \underbrace{13y'}_{13\,\cdot\,\text{1st derivative}} + \underbrace{-5y}_{-5\,\cdot\,\text{0th derivative}} = 0$$

$$\downarrow \qquad\qquad \downarrow \qquad\qquad \downarrow$$

$$\underset{6\,\cdot\,\text{2nd-degree term}}{} \quad \underset{13\,\cdot\,\text{1st-degree term}}{} \quad \underset{-5\,\cdot\,\text{0th-degree term}}{}$$

$$\underbrace{6\lambda^2} + \underbrace{13\lambda} + \underbrace{(-5)} = 0$$

---

2. In German, the word *eigen* means "own, proper, inherent, special, characteristic," etc.

Using the quadratic formula, we find

$$\lambda = \frac{-13 \pm \sqrt{13^2 - 4(6)(-5)}}{2(6)} = \frac{-13 \pm \sqrt{289}}{12} = \frac{-13 \pm 17}{12} = \frac{1}{3} \text{ or } -\frac{5}{2},$$

so that we have two distinct real eigenvalues, $\lambda_1 = \frac{1}{3}$ and $\lambda_2 = -\frac{5}{2}$, and we can

write the general solution of our equation as $y(t) = c_1 e^{\frac{t}{3}} + c_2 e^{-\frac{5t}{2}}$. ◆

Real but Equal Eigenvalues

Next we consider possibility (2), that the eigenvalues are real numbers with $\lambda_1 = \lambda_2$. In this situation, we get only one solution, $y = e^{\lambda t}$, where λ is the value of the repeated eigenvalue. To obtain the general solution in this case, we have to find another solution that is not merely a constant multiple of $e^{\lambda t}$ (or else the "two" solutions can be merged into a single solution requiring only one arbitrary constant). Again Euler comes to the rescue (this time in 1743), suggesting that an independent[3] second solution might be found by considering functions of the form $y_2(t) = u(t)e^{\lambda t}$, where $u(t)$ is an unknown function that must be determined.

Rather than deriving the consequences of Euler's assumption in the general case (see Exercise 12), we'll illustrate his ingenious technique by an example.

EXAMPLE 4.1.2 The Characteristic Equation—Equal Eigenvalues

The equation $y'' - 4y' + 4y = 0$ has the characteristic equation $\lambda^2 - 4\lambda + 4 = (\lambda - 2)^2 = 0$, so $\lambda = 2$ is a repeated eigenvalue. We know that $y_1 = e^{2t}$ is one solution of the differential equation. Taking Euler's advice, we consider $y_2(t) = u(t)e^{2t}$.

Now, by the Product Rule (and the Chain Rule), $y_2' = 2ue^{2t} + u'e^{2t}$ and $y_2'' = 4ue^{2t} + 4u'e^{2t} + u''e^{2t}$. Substituting y_2 and its derivatives into our original differential equation, we obtain

$$\begin{aligned} y_2'' - 4y_2' + 4y_2 &= (4ue^{2t} + 4u'e^{2t} + u''e^{2t}) - 4(2ue^{2t} + u'e^{2t}) + 4(ue^{2t}) \\ &= u''e^{2t} = 0. \end{aligned}$$

Therefore, we must have $u''(t) = 0$, and two successive integrations give us $u'(t) = A$ and $u(t) = At + B$, where A and B are arbitrary constants. Our conclusion is that $y_2(t) = (At + B)e^{2t}$ is a solution of the original ODE that is *not* a constant multiple of $y_1 = e^{2t}$. The Superposition Principle tells us that the general solution is given by

$$y(t) = c_1 y_1(t) + c_2 y_2(t) = c_1 e^{2t} + c_2(At + B)e^{2t} = (C_1 t + C_2)e^{2t},$$

where $C_1 = c_2 A$ and $C_2 = c_1 + c_2 B$ are arbitrary constants. ◆

---

3. Two functions f_1 and f_2 are called (**linearly**) **independent** on an interval I if one is not a constant multiple of the other. Equivalently, if c_1 and c_2 are constants, then the only way for $c_1 f_1 + c_2 f_2$ to be the zero function on I is if $c_1 = c_2 = 0$. It can be shown that Euler's technique produces a new solution independent of the first.

Complex Conjugate Eigenvalues

When the eigenvalues are complex numbers—$\lambda_1 = p + qi$ and $\lambda_2 = p - qi$, where p and q are real numbers—the two corresponding solutions of the differential equation $ay'' + by' + cy = 0$ are $y_1(t) = e^{(p+qi)t}$ and $y_2(t) = e^{(p-qi)t}$. At this point, a crucial fact to know is **Euler's formula,**[4] which defines the exponential function for complex values of the argument (exponent):

$$e^{p+qi} = e^p(\cos(q) + i\sin(q)).$$

(If we let $p = 0$ and $q = \pi$, we get a particularly elegant formula connecting four of the most famous and useful constants in all of mathematics: $e^{\pi i} = -1$. Also see Appendix C.4.)

Using Euler's formula, we can write the solutions as

$$y_1(t) = e^{(p+qi)t} = e^{pt}e^{(qt)i} = e^{pt}(\cos(qt) + i\sin(qt))$$

and

$$y_2(t) = e^{(p-qi)t} = e^{pt}e^{-(qt)i} = e^{pt}(\cos(-qt) + i\sin(-qt))$$
$$= e^{pt}(\cos(qt) - i\sin(qt)),$$

where we have simplified $y_2(t)$ by recognizing that the cosine is an even function and the sine is an odd function. If we combine these complex-valued solutions carefully (see Exercise 13), we find that

$$y(t) = e^{pt}(C_1\cos(qt) + C_2\sin(qt)),$$

a real-valued function, is a solution of $ay'' + by' + cy = 0$ for all constants C_1 and C_2. **In fact, $y(t) = e^{pt}(C_1\cos(qt) + C_2\sin(qt))$ is the general solution of the homogeneous equation when the characteristic equation has complex conjugate roots $p \pm qi$.**

Now let's practice with complex eigenvalues.

EXAMPLE 4.1.3   Complex Conjugate Eigenvalues

The equation $\ddot{x} + 8\dot{x} + 25x = 0$ models the motion of a steel ball suspended from a spring, where $x(t)$ is the ball's distance (in feet) from its rest (equilibrium) position at time t seconds. Distance below the rest position is considered positive, and distance above is considered negative. We want to describe the motion of the ball by finding a formula for $x(t)$.

The characteristic equation is $\lambda^2 + 8\lambda + 25 = 0$. The quadratic formula gives us

$$\lambda = \frac{-8 \pm \sqrt{8^2 - 4(1)(25)}}{2} = \frac{-8 \pm \sqrt{-36}}{2} = \frac{-8 \pm 6i}{2} = -4 \pm 3i,$$

so the eigenvalues are $\lambda_1 = -4 + 3i$ and $\lambda_2 = -4 - 3i$. Using the solution formula derived above, with $p = -4$ and $q = 3$, we see that $x(t) = e^{-4t}(C_1\cos(3t) + C_2\sin(3t))$ for arbitrary constants C_1 and C_2.

---

4. Euler discovered this formula in 1740, while investigating solutions of the equation $y'' + y = 0$.

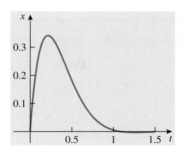

Figure 4.1

Graph of $\frac{4}{3}e^{-4t}\sin(3t)$, the solution of the IVP
$\ddot{x} + 8\dot{x} + 25x = 0; x(0) = 0, \dot{x}(0) = 4; 0 \le t \le 1.5$

Suppose that we specify initial conditions, say $x(0) = 0$ and $\dot{x}(0) = 4$. These conditions say that the ball is at its equilibrium position at the beginning of our investigation and that the ball is started in motion from its equilibrium position with an initial velocity of 4 ft/sec in the downward direction. Applying these conditions, we have

$$x(0) = e^{-4(0)}(C_1 \cos(0) + C_2 \sin(0)) = C_1 = 0$$

and

$$\dot{x}(0) = e^{-4(0)}(-3C_1 \sin(0) + 3C_2 \cos(0)) - 4e^{-4(0)}(C_1 \cos(0) + C_2 \sin(0))$$
$$= 3C_2 - 4C_1 = 4.$$

Therefore, $C_1 = 0, C_2 = \frac{4}{3}$, and the solution of our IVP is $x(t) = \frac{4}{3}e^{-4t}\sin(3t)$. The graph of this solution (Figure 4.1) shows that the motion is dying out as time passes—that is, $x \to 0$ as $t \to \infty$.

As we'll see later in this chapter, the differential equation has a term in it that represents air resistance, and this results in what is called *damped* motion. ◆

Summary

We can summarize the situation for *homogeneous* linear second-order equations with constant coefficients as follows:

> Suppose that we have the equation $ax'' + bx' + cx = 0$, where a, b, and c are constants, $a \ne 0$, and λ_1 and λ_2 are the zeros of the characteristic equation $a\lambda^2 + b\lambda + c = 0$. Then
>
> 1. If there are two distinct real eigenvalues—λ_1, λ_2, with $\lambda_1 \ne \lambda_2$—corresponding to our equation, the general solution is $x(t) = c_1 e^{\lambda_1 t} + c_2 e^{\lambda_2 t}$.
> 2. If there is a repeated real eigenvalue λ, the general solution has the form $x(t) = c_1 e^{\lambda t} + c_2 t e^{\lambda t} = (c_1 + c_2 t)e^{\lambda t}$.
> 3. If the eigenvalues form a complex conjugate pair $p \pm qi$, then Euler's formula can be used to show that the (real-valued) general solution has the form $x(t) = e^{pt}(c_1 \cos(qt) + c_2 \sin(qt))$.

EXERCISES 4.1

Find the general solution of each of the equations in Exercises 1–10.

1. $y'' - 4y' + 4y = 0$
2. $\ddot{x} + 4\dot{x} - 5x = 0$
3. $x'' - 2x' + 2x = 0$
4. $x'' + 5x' + 6x = 0$
5. $\ddot{x} + 2\dot{x} = 0$
6. $\ddot{x} - x = 0$
7. $y'' + 4y = 0$
8. $6\ddot{x} - 11\dot{x} + 4x = 0$
9. $\ddot{r} - 4\dot{r} + 20r = 0$
10. $y'' + 4ky' - 12k^2 y = 0$ (k is a parameter.)

11. **a.** Show that if a, b, and c are constants and y is any function having at least two derivatives, then the *differential operator L* defined by the relation $L(y) = ay'' + by' + cy$ is *linear:* $L(c_1 y_1 + c_2 y_2) = c_1 L(y_1) + c_2 L(y_2)$ for any twice-differentiable functions y_1 and y_2 and any constants c_1 and c_2.

    b. Show that if y_1 and y_2 are two solutions of $L(y) = 0$, then the function $c_1 y_1 + c_2 y_2$ is also a solution of $L(y) = 0$.

12. Suppose that we have a constant-coefficient equation $ay'' + by' + cy = 0$ whose characteristic equation has a repeated root r. Then we know that $y_1(t) = e^{rt}$ is a solution of the equation. If we form the new function $y_2(t) = u(t)e^{rt}$, where $u(t)$ is unknown, we want to determine $u(t)$ such that y_2 is a solution of the differential equation but is not a constant multiple of y_1.

    a. Show that any constant-coefficient equation $ay'' + by' + cy = 0$ whose characteristic equation has a double root r must have the form

$$y'' - 2ry' + r^2 y = 0$$

    b. Find y_2' and y_2'' and then substitute y_2 and these derivatives into the equation $y'' - 2ry' + r^2 y = 0$. Simplify the result.

    c. Solve the equation you get in part (b) for $u(t)$.

13. We know that $y_1(t) = e^{pt}(\cos(qt) + i\sin(qt))$ and $y_2(t) = e^{pt}(\cos(qt) - i\sin(qt))$ are complex-valued solutions of the homogeneous equation (4.1.1) when the eigenvalues are complex conjugate numbers $p \pm qi$. In what follows, you may assume that complex constants are valid in the Superposition Principle.

    a. Calculate $Y_1 = \dfrac{y_1 + y_2}{2}$ and show that Y_1 is a real-valued solution of (4.1.1).

    b. Calculate $Y_2 = \dfrac{y_1 - y_2}{2i}$ and show that Y_2 is a real-valued solution of (4.1.1).

    c. Calculate $Y = c_1 Y_1 + c_2 Y_2$ and conclude that Y is a real-valued solution of (4.1.1) for arbitrary real constants c_1 and c_2.

14. Consider the equation $ay'' + by' + cy = 0$. Another approach to the situation where the characteristic equation has a double real root λ was devised by the French mathematician d'Alembert (1717–1783) in about 1748. He proposed splitting this root into two "neighboring" roots λ and $\lambda + \varepsilon$, where ε is small.

a. Show that $e^{\lambda t}$, $e^{(\lambda + \varepsilon)t}$, and also the combination $y(t) = \dfrac{e^{(\lambda + \varepsilon)t} - e^{\lambda t}}{\varepsilon}$ are solutions of the equation $ay'' + by' + cy = 0$.

b. Show that as $\varepsilon \to 0$, $y(t)$ becomes the solution $te^{\lambda t}$.

As we noted at the beginning of Section 4.1, if $I = I(t)$ represents the current in an electrical circuit, then Kirchhoff's Voltage Law *gives us the equation*

$$L\frac{d^2I}{dt^2} + R\frac{dI}{dt} + \frac{1}{C}I = 0 \text{ when the voltage applied to the circuit is constant. In this}$$

equation, L is the inductance, R is the resistance, and C is the capacitance. Use this equation in Exercises 15–16.

15. An RLC circuit with $R = 6$ ohms, $L = 0.1$ henry, and $C = 0.02$ farad has a constant voltage of 6 volts. Assume that there is no initial current and that $\dfrac{dI}{dt} = 60$ when the voltage is first applied.

    a. Find an expression for the current in the circuit at time $t > 0$.

    b. Use technology to graph the answer found in part (a) for $0 \le t \le 0.5$.

    c. From the graph in part (b), estimate the maximum value of I and find the exact value by calculus techniques applied to the expression found in part (a).

    d. At what time is the maximum value found in part (c) achieved? (You can use a calculator or a CAS for this.)

16. An RLC circuit has $R = 10$ ohms, $C = 0.01$ farad, $L = 0.5$ henry, and an applied voltage of 12 volts. The charge, Q, on the capacitor is defined in terms of the current, I, by $I = \dfrac{dQ}{dt}$. Assuming no initial current and no initial charge on the capacitor, find the charge on the capacitor at time $t > 0$.

17. Solve the IVP $\ddot{x} - 3\dot{x} + 2x = 0$, $x(0) = 1$, $\dot{x}(0) = 0$.

18. Solve the IVP $y'' - 2y' + y = 0$, $y(0) = 1$, $y'(0) = 0$.

19. Solve the IVP $y'' - 4y' + 20y = 0$, $y(\pi/2) = 0$, $y'(\pi/2) = 1$.

According to Newton's Second Law of Motion (see Section 4.5 for a further discussion), if an object with mass m is suspended from a spring attached to the ceiling, then the motion of the object is governed by the equation $m\ddot{x} + a\dot{x} + kx = 0$. In this equation, $x(t)$ is the object's distance from its rest (equilibrium) position at time t seconds. Distance below the rest position is considered positive, and distance above is considered negative. Also, a is a constant representing the air resistance and/or friction present in the system, and k is the spring constant, *describing the "give" in the spring. (Recall that* mass = weight/g, *where g is the gravitational constant—32 ft/sec² or 9.8 m/sec².) Use this equation to do Exercises 20–23.*

20. An object of mass 4 slugs ($= 128$ lb / 32 ft/sec²) is suspended from a spring having a spring constant of 64 lb/ft. The object is started in motion, with no initial velocity, by pulling it 6 inches (*Watch the units!*) below the equilibrium position and then releasing it. If there is no air resistance, find a formula for the position of the object at any time $t > 0$. (Note that the problem statement contains two initial conditions.)

21. A 20-g mass hangs from the end of a spring having a spring constant of 2880 dynes/cm and is allowed to come to rest. It is then set in motion by stretching the spring 3 cm from its equilibrium position and releasing the mass with an initial velocity of 10 cm/sec in the downward (positive) direction. Find the position of the mass at time $t > 0$ if there is no air resistance.

22. A $\frac{1}{2}$-kg mass is attached to a spring having a spring constant of 6 lb/ft. The mass is set in motion by displacing it 6 inches below its equilibrium position with no initial velocity. Find the subsequent motion of the mass if a, the constant representing air resistance, is 4 lb/ft.

23. A $\frac{1}{2}$-kg mass is attached to a spring having a spring constant of 8 N/m, where N stands for newtons. The mass is set in motion by displacing it 10 cm above its equilibrium position with an initial velocity of 2 m/sec in the upward direction.

    a. Find the subsequent motion of the mass if the constant representing air resistance is 2 N.sec/m.

    b. Graph the function $x(t)$ found in part (a) for $0 \le t \le 3$, $2 \le t \le 3$, and $3 \le t \le 4$. Describe the motion of the mass in your own words.

    c. Estimate the greatest distance of the mass above its equilibrium position.

24. The equation $\theta'' = -4\theta - 5\theta'$ represents the angle $\theta(t)$ made by a swinging door, where θ is measured from the equilibrium position of the door, which is the closed position. The initial conditions are $\theta(0) = \dfrac{\pi}{3}$ and $\theta'(0) = 0$.

    a. Determine the angle $\theta(t)$ as a function of time $(t > 0)$.

    b. What does your solution tell you is going to happen as t becomes large?

    c. Use technology to graph the solution $\theta(t)$ on the interval $[0, 5]$.

4.2 NONHOMOGENEOUS SECOND-ORDER LINEAR EQUATIONS WITH CONSTANT COEFFICIENTS

THE STRUCTURE OF SOLUTIONS

If we take the same RLC circuit that we considered at the beginning of the last section and hook up a generator supplying alternating current to it, Kirchhoff's Voltage Law will now take the form $L\dfrac{d^2I}{dt^2} + R\dfrac{dI}{dt} + \dfrac{1}{C}I = \dfrac{dE}{dt}$, where E is the applied nonconstant voltage. Such an equation is called a **nonhomogeneous linear second-order equation with constant coefficients.** (The nonzero right-hand side of such an equation is often called the **forcing function** or the **input.** The solution of the equation is the **output.** See Section 2.2.)

To get a handle on solving a nonhomogeneous linear equation, let's think a bit about the difference between a nonhomogeneous equation $ay'' + by' + cy = f(t)$ and its **associated homogeneous equation** $ay'' + by' + cy = 0$. If y is the general solution of the *homogeneous* system, then y doesn't quite "reach" all the way to $f(t)$ under the transformation L. It stops short at 0. Perhaps we could enhance the solution y in some way so that operating on this new function *does* give us all of f. We have to be able to capture the "leftover" term $f(t)$.

For nonhomogeneous equations, the proper form of the Superposition Principle is the following: If y_1 is a solution of $ay'' + by' + cy = f_1(t)$ and y_2 is a solution of $ay'' + by' + cy = f_2(t)$, then $y = c_1y_1 + c_2y_2$ is a solution of $ay'' + by' + cy = c_1f_1(t) + c_2f_2(t)$ for any constants c_1 and c_2.

Here's a fundamental fact about linear equations:

> The general solution, y_{GNH}, of a linear nonhomogeneous equation $ay'' + by' + cy = f(t)$ is obtained by finding a *particular* solution, y_{PNH}, of the nonhomogeneous equation and adding it to the *general* solution, y_{GH}, of the associated homogeneous equation.

We can prove this easily using operator notation, where $L(y) = ay'' + by' + cy$:

1. First note that $L(y_{GH}) = 0$ and $L(y_{PNH}) = f(t)$ by definition.
2. Then if $y = y_{GH} + y_{PNH}$, we have $L(y) = L(y_{GH} + y_{PNH}) = L(y_{GH}) + L(y_{PNH}) = 0 + f(t) = f(t)$, so y is a solution of the nonhomogeneous equation.
3. Now we must show that *every* solution of the nonhomogeneous equation has the form $y = y_{GH} + y_{PNH}$. To do this, we assume that y^* is any solution of $L(y) = f(t)$ and let $z = y^* - y_{PNH}$. Then

$$L(z) = L(y^* - y_{PNH}) = L(y^*) - L(y_{PNH}) = f(t) - f(t) = 0$$

which shows that z is a solution to the homogeneous equation $L(y) = 0$. Because $z = y^* - y_{PNH}$, it follows that $y^* = z + y_{PNH}$, where z is a solution of $L(y) = 0$.

(See Problem 39 of Exercises 1.2 and Problem 37 of Exercises 2.2 for related results.)

Let's go through a simple example to get a feel for the solutions of nonhomogeneous equations.

EXAMPLE 4.2.1 Solving a Nonhomogeneous Equation

Suppose we want to find the general solution of $y'' + 3y' + 2y = 12e^t$. Because the characteristic equation of the associated homogeneous equation is $\lambda^2 + 3\lambda + 2 = 0$, with roots -1 and -2, we know that the general solution of the homogeneous equation is $y_{GH} = c_1e^{-t} + c_2e^{-2t}$.

Now we look carefully at the form of the nonhomogeneous equation. In looking for a particular solution y_{PNH}, we can ignore any terms of the form e^{-t} *or* e^{-2t} because these are part of the homogeneous solution and won't contribute anything new. But somehow, after differentiations and additions, we have to wind up with the term $12e^t$. We guess that $y = ce^t$ for some undetermined constant c. Substituting this expression into the left-hand side of the nonhomogeneous equation, we get $(ce^t) + 3(ce^t) + 2(ce^t) = 6ce^t$. If we choose $c = 2$, then $y_{PNH} = 2e^t$ is a particular solution of the nonhomogeneous equation.

Putting these two components together, we can write the general solution of the nonhomogeneous equation as $y_{GNH} = y_{GH} + y_{PNH} = c_1e^{-t} + c_2e^{-2t} + 2e^t$.   ◆

To show the importance of nonhomogeneous second-order differential equations with constant coefficients, let's look at an interesting application.

EXAMPLE 4.2.2

A 2560-lb car supported by a MacPherson strut (a particular type of shock-absorbing system) is traveling over a bumpy road at a constant velocity v. The equation modeling the motion is

$$80\ddot{x} + 10000x = 2500\cos\left(\frac{\pi vt}{6}\right),$$

where x represents the vertical position of the car's axle relative to its equilibrium position, and the basic units of measurement are feet and feet per second where appropriate. (Note that the coefficient of $\ddot{x}$ is $2560/g = 2560/32 = 80$, the *mass* of the car.) We want to determine the velocity that induces **resonance** in the car, vibrations of unbounded magnitude. (We'll encounter resonance again in Example 4.5.8 and discuss it further in the section that follows that example.)

This is a nonhomogeneous equation, and we know that the general solution can be expressed as the sum of the general solution of the associated homogeneous equation and a particular solution of the nonhomogeneous equation: $x_{GNH} = x_{GH} + x_{PNH}$.

The characteristic equation corresponding to the homogeneous equation is $80\lambda^2 + 10000 = 0$, or $\lambda^2 + 125 = 0$, with eigenvalues $\pm 5\sqrt{5}i = 0 \pm 5\sqrt{5}i$. Case (3) of the summary at the end of Section 4.1 tells us that $x_{GH} = c_1\sin(5\sqrt{5}t) + c_2\cos(5\sqrt{5}t)$. Now we can guess (see Exercise 11) that a particular solution of the original nonhomogeneous equation is given by

$$x_{PNH} = \frac{2500}{10000 - 80(\frac{\pi v}{6})^2}\cos\left(\frac{\pi vt}{6}\right).$$

The general solution of our equation of motion is

$$x(t) = c_1\sin(5\sqrt{5}t) + c_2\cos(5\sqrt{5}t) + \frac{2500}{10000 - 80(\frac{\pi v}{6})^2}\cos\left(\frac{\pi vt}{6}\right).$$

Clearly the first two trigonometric terms help describe the bumpy ride, but they have fixed amplitudes c_1 and c_2, so there isn't any possibility of unbounded oscillations. However, the amplitude of the last term is given by $\dfrac{2500}{10000 - 80(\frac{\pi v}{6})^2}$,

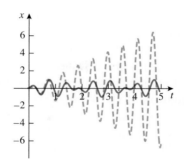

Figure 4.2

$$x(t) = c_1 \sin(5\sqrt{5}t) + c_2 \cos(5\sqrt{5}t) + \frac{2500}{10000 - 80(\frac{\pi v}{6})^2}\cos\left(\frac{\pi vt}{6}\right);$$

$$x(0) = 0, \dot{x}(0) = 0; 0 \le t \le 5$$

$$v = 15(\text{solid curve}); v = 21(\text{dashed curve})$$

which grows larger and larger as the denominator expression $10000 - 80\left(\dfrac{\pi v}{6}\right)^2$

gets closer and closer to zero. Thus we get resonance when $10000 - 80\left(\dfrac{\pi v}{6}\right)^2 = 0$

—that is, when $v = \sqrt{\dfrac{4500}{\pi^2}} \approx 21.35$ ft/sec. The dimensional equation $\dfrac{\text{miles}}{\text{hr}} =$

$\dfrac{\text{miles}}{\text{ft}} \cdot \dfrac{\text{ft}}{\text{sec}} \cdot \dfrac{\text{sec}}{\text{hr}}$ allows us to express our answer as $\dfrac{1}{5280} \cdot \dfrac{21.35}{1} \cdot \dfrac{3600}{1} \approx 14.56$

miles per hour.

Assuming the initial conditions $x(0) = 0$ and $\dot{x}(0) = 0$, Figure 4.2 shows two graphs of $x(t)$ against t, the solid graph using $v = 15$ ft/sec and the dashed-line graph using $v = 21$ ft/sec. You can see that the car's vibrations become wilder over time for a speed close to 21.35 feet per second. ◆

The intelligent guessing used in the last two examples can be formalized into the **method of undetermined coefficients.** But this method is effective only when the forcing function $f(t)$ in the equation $ay'' + by' + cy = f(t)$ is of a special type. (Exercise 12 asks for a report on this method.) Next, we'll focus our attention on a more generally applicable method.

VARIATION OF PARAMETERS

There are various techniques for finding a particular solution of the nonhomogeneous equation. The **method of variation of parameters** (or **variation of constants**) was developed by the French-Italian mathematician Lagrange (1736–1818) in 1775.

Let's look at the nonhomogeneous equation $ay'' + by' + cy = f(t)$ and assume that $y_1(x)$ and $y_2(x)$ are two known solutions of the homogeneous equation $ay'' + by' + cy = 0$ that are independent of each other. Then we know that $c_1 y_1(t) + c_2 y_2(t)$ is also a solution of the homogeneous equation for any constants c_1 and c_2. Lagrange's idea was to look for a particular solution of the *nonhomogeneous* equation of the form $c_1(t) y_1(t) + c_2(t) y_2(t)$, where $c_1(t)$ and $c_2(t)$ are unknown *functions* that must be determined.

Rather than go through this method in complete generality, we'll illustrate the technique in specific examples.

EXAMPLE 4.2.3   Using Variation of Parameters

Suppose we want to solve $y'' + 3y' + 2y = 3e^{-2x} + x$. The characteristic equation of the associated homogeneous equation is $\lambda^2 + 3\lambda + 2 = 0$, with roots -2 and -1. Then $y_1(x) = e^{-2x}$ and $y_2(x) = e^{-x}$ are two independent solutions of the homogeneous equation, and the general solution of the homogeneous equation is $y_{GH} = C_1 e^{-2x} + C_2 e^{-x}$, where C_1 and C_2 are arbitrary constants.

Now assume that $y = c_1 y_1 + c_2 y_2 = c_1 e^{-2x} + c_2 e^{-x}$ is a particular solution of the nonhomogeneous equation, where $c_1 = c_1(x)$ and $c_2 = c_2(x)$ are unknown functions. Differentiating, we obtain

$$y' = -2c_1 e^{-2x} + c_1' e^{-2x} - c_2 e^{-x} + c_2' e^{-x}$$
$$= (-2c_1 e^{-2x} - c_2 e^{-x}) + (c_1' e^{-2x} + c_2' e^{-x}).$$

To avoid messy higher derivatives, Lagrange's method requires that

$$c_1' e^{-2x} + c_2' e^{-x} = 0 \qquad (*)$$

Accepting this condition, we have $y' = -2c_1 e^{-2x} - c_2 e^{-x}$, from which we calculate $y'' = 4c_1 e^{-2x} - 2c_1' e^{-2x} + c_2 e^{-x} - c_2' e^{-x}$.

Substituting these expressions for y, y', and y'' into the equation $y'' + 3y' + 2y = 3e^{-2x} + x$, we find that

$$(4c_1 e^{-2x} - 2c_1' e^{-2x} + c_2 e^{-x} - c_2' e^{-x})$$
$$+ 3(-2c_1 e^{-2x} - c_2 e^{-x}) + 2(c_1 e^{-2x} + c_2 e^{-x}) = 3e^{-2x} + x,$$

or

$$-2c_1' e^{-2x} - c_2' e^{-x} = 3e^{-2x} + x. \qquad (**)$$

Equations $(*)$ and $(**)$ form a system of equations that we must solve for c_1' and c_2':

$$c_1' e^{-2x} + c_2' e^{-x} = 0 \qquad (*)$$
$$-2c_1' e^{-2x} - c_2' e^{-x} = 3e^{-2x} + x \qquad (**)$$

Adding $(*)$ and $(**)$ gives us $-c_1' e^{-2x} = 3e^{-2x} + x$, so $c_1' = -3 - xe^{2x}$. Integration (by parts, manually, or by CAS) yields $c_1(x) = -3x - \dfrac{x}{2}e^{2x} + \dfrac{1}{4}e^{2x}$. In using variation of parameters, we make all constants of integration 0 because we want only a *particular* solution.

Next, we use (*) to find that $c_2' = e^x(-c_1'e^{-2x}) = e^x(3e^{-2x} + x) = 3e^{-x} + xe^x$. Integration gives us $c_2(x) = -3e^{-x} + xe^x - e^x$.

Finally,

$$y_{\text{PNH}} = c_1 y_1 + c_2 y_2 = \left(-3x - \frac{x}{2}e^{2x} + \frac{1}{4}e^{2x}\right)(e^{-2x}) + (-3e^{-x} + xe^x - e^x)(e^{-x})$$

$$= -3xe^{-2x} - \frac{x}{2} + \frac{1}{4} - 3e^{-2x} + x - 1 = -3xe^{-2x} - 3e^{-2x} + \frac{x}{2} - \frac{3}{4},$$

so that $y_{\text{GNH}} = C_1 e^{-2x} + C_2 e^{-x} + \dfrac{x}{2} - \dfrac{3}{4} - 3xe^{-2x}$ is the general solution of the original nonhomogeneous equation. (Note that the term $-3e^{-2x}$ in y_{PNH} has been absorbed by the term $C_1 e^{-2x}$ in y_{GH}.) ◆

This last example involved quite a bit of algebra and calculus, but the method of variation of parameters is guaranteed to work. Even if the integrations of $c_1'(x)$ and $c_2'(x)$ can't be done in closed form, we can still use numerical methods such as Simpson's Rule to approximate the solution. The next example illustrates this kind of integration difficulty.

EXAMPLE 4.2.4 **Variation of Parameters—No Closed-Form Solution**
Let's solve the equation $y'' + y' - 2y = \ln x$. The characteristic equation of the associated homogeneous equation is $\lambda^2 + \lambda - 2 = 0$, with roots -2 and 1, so we know that $y_{\text{GH}} = C_1 e^{-2x} + C_2 e^x$, where C_1 and C_2 are constants.

Next we consider $y = c_1 e^{-2x} + c_2 e^x$, where c_1 and c_2 are unknown functions of x. Differentiating, we get

$$y' = (-2c_1 e^{-2x} + c_2 e^x) + (c_1' e^{-2x} + c_2' e^x) = -2c_1 e^{-2x} + c_2 e^x$$

because we must assume that

$$c_1' e^{-2x} + c_2' e^x = 0. \tag{#}$$

Then $y'' = 4c_1 e^{-2x} - 2c_1' e^{-2x} + c_2 e^x + c_2' e^x$.

After substituting these expressions for y, y', and y'' into the equation $y'' + y' - 2y = \ln x$, we get

$$-2c_1' e^{-2x} + c_2' e^x = \ln x. \tag{# #}$$

Now we must solve the following system for c_1' and c_2':

$$c_1' e^{-2x} + c_2' e^x = 0 \tag{#}$$
$$-2c_1' e^{-2x} + c_2' e^x = \ln x. \tag{# #}$$

Subtracting (# #) from (#) gives us $3c_1' e^{-2x} = -\ln x$, so $c_1' = -\dfrac{1}{3}e^{2x}\ln x$ and

$c_1(x) = -\dfrac{1}{3}\displaystyle\int e^{2x}\ln x\,dx = -\dfrac{1}{6}e^{2x}\ln x + \dfrac{1}{6}\displaystyle\int \dfrac{e^{2x}}{x}dx$. This integration was done manually (integration by parts: $u = \ln x$, $dv = e^{2x}dx$, etc.). A CAS might give an answer in terms of the "exponential integral,"[5] which you may not recognize. In any case, the integral $\int \dfrac{e^{2x}}{x}dx$ cannot be expressed in closed form.

Equation (#) tells us that $c_2' = e^{-x}(-c_1'e^{-2x}) = -e^{-3x}\left(-\dfrac{1}{3}e^{2x}\ln x\right) = \dfrac{1}{3}e^{-x}\ln x$, and an integration by parts leads to the conclusion that $c_2(x) = $

$$-\frac{1}{3}e^{-x}\ln x + \frac{1}{3}\int\frac{e^{-x}}{x}dx.$$

The next to the last step is to calculate

$$y_{\text{PNH}} = c_1 y_1 + c_2 y_2$$

$$= \left(-\frac{1}{6}e^{2x}\ln x + \frac{1}{6}\int\frac{e^{2x}}{x}dx\right)(e^{-2x}) + \left(-\frac{1}{3}e^{-x}\ln x + \frac{1}{3}\int\frac{e^{-x}}{x}dx\right)(e^{x})$$

$$= -\frac{\ln x}{2} + \frac{e^{-2x}}{6}\int\frac{e^{2x}}{x}dx + \frac{e^{x}}{3}\int\frac{e^{-x}}{x}dx.$$

Finally, the general solution is given by the formula

$$y_{\text{GNH}} = y_{\text{GH}} + y_{\text{PNH}} = C_1 e^{-2x} + C_2 e^{x} - \frac{\ln x}{2} + \frac{e^{-2x}}{6}\int\frac{e^{2x}}{x}dx + \frac{e^{x}}{3}\int\frac{e^{-x}}{x}dx. \qquad \blacklozenge$$

EXERCISES 4.2

Find the general solution of each of the equations in Exercises 1–10 by using the method of variation of parameters.

1. $y'' - 2y' - 3y = e^{4t}$ **2.** $\ddot{x} - 3\dot{x} + 2x = \sin t$

3. $x'' - 2x' + 2x = e^{t} + t\cos t$ **4.** $x'' - 2x' + x = \dfrac{e^{t}}{t}$

5. $\ddot{x} + \dot{x} = 4\sin t$ **6.** $\ddot{x} - x = 2e^{t} - t^2$

7. $y'' + 4y = 2\tan x$ **8.** $6\ddot{x} - 11\dot{x} + 4x = t$

9. $\ddot{r} + r = \dfrac{1}{\sin t}$ **10.** $y'' + 2y' + y = \dfrac{e^{-x}}{x}$

11. Consider the equation $80\ddot{x} + 10000x = 2500\cos\left(\dfrac{\pi vt}{6}\right)$ in Example 4.2.2.

Guess that a particular solution x_{PNH} has the form $A\sin\left(\dfrac{\pi vt}{6}\right) + B\cos\left(\dfrac{\pi vt}{6}\right)$

and show that $A = 0$ and $B = \dfrac{2500}{10000 - 80\left(\frac{\pi v}{6}\right)^2}$.

---

5. For a positive real number x and a nonnegative integer n, the *exponential integral* is defined as
$$Ei(n, x) = \int_{1}^{\infty} \exp(-xt)/t^{n}dt.$$

12. Look up the method of **undetermined coefficients** in another book and write a report on it or perhaps make an oral presentation (details to be determined by your instructor). Be sure to indicate when the method works well and when it doesn't apply. Also, show by example that when it is applicable, the method of undetermined coefficients is generally easier to use than the method of variation of parameters.

As we noted at the beginning of Section 4.2, if I $= I(t)$ *represents the current in an electrical circuit, then* Kirchhoff's Voltage Law *gives us the nonhomogeneous equation* $L\dfrac{d^2I}{dt^2} + R\dfrac{dI}{dt} + \dfrac{1}{C}I = \dfrac{dE}{dt}$, *where E is the applied nonconstant voltage. In this equation, L is the inductance, R is the resistance, and C is the capacitance. Use this equation in Exercises 13–14.*

13. An *RLC* circuit has a resistance of 5 ohms, an inductance of 0.05 henry, a capacitance of 0.0004 farad, and an applied alternating voltage of 200 cos(100 *t*) volts.

    a. Without using technology, find an expression for the current flowing through this circuit if the initial current is zero and $\dfrac{dI}{dt}(0)$ is 4000.

    b. Check your answer to part (a) by using technology.

14. An *RLC* circuit has $R = 10$ ohms, $C = 0.01$ farad, $L = 0.5$ henry, and an applied voltage given by $E(t) = 16 \cos(2 t)$. The charge, Q, on the capacitor is defined in terms of the current, I, by $I = \dfrac{dQ}{dt}$. Assuming no initial current and no initial charge on the capacitor, find the charge on the capacitor at time $t > 0$.

15. Look for a particular solution of $y'' + 0.2y' + y = \sin(\omega x)$ and investigate its amplitude as a function of ω. Use technology to graph the particular solution for values of ω that seem significant to you and describe the behavior of this solution.

16. In her dorm room, a student attaches a weight to a spring hanging from the ceiling. She starts the mass in motion from the equilibrium position with an initial velocity in the upward direction. But during this experiment there is rhythmic stomping (dancing or pest control?) from the student upstairs that causes the ceiling and the entire spring-mass system to vibrate. Taking into account air resistance and this "external force," she determines that the equation of motion is $\ddot{x} + 9\dot{x} + 14x = \frac{1}{2}\sin t$, with $x(0) = 0$ and $\dot{x}(0) = -1$.

    a. Solve this equation for $x(t)$, the position of the weight relative to its rest position.

    b. Use technology to graph $x(t)$ for $0 \le t \le 10$.

17. Solve the IVP $y'' - 3y' - 4y = 3e^{4x}$; $y(0) = 0$, $y'(0) = 0$.

18. Using a CAS, solve the IVP $y'' + y' + y = t^2 e^{-t} \cos t$, $y(0) = 1$, $y'(0) = 0$. [*Warning:* Serious mental injury may result from attempting to do this manually.]

4.3 HIGHER-ORDER LINEAR EQUATIONS WITH CONSTANT COEFFICIENTS

The good news is that linearity is such a marvelous property that we can generalize our work in the last two sections in a very natural way. The details may get a bit complicated, but the theory is crisp and clear.

If y is a function that is n-times differentiable and $a_0, a_1, a_2, \ldots, a_n$ are constants, then we can define the nth-order linear operator L as follows:

$$L(y) = a_n y^{(n)} + a_{n-1} y^{(n-1)} + \cdots + a_2 y'' + a_1 y' + a_0 y.$$

Any nth-order linear differential equation with constant coefficients can be expressed concisely as $L(y) = f(t)$. If $f(t) \equiv 0$, then the equation is called a **homogeneous nth-order linear equation with constant coefficients.** If $f(t)$ is not the zero function, then we have a **nonhomogeneous nth-order linear equation with constant coefficients.**

An important property of such nth-order equations is the (extended) Superposition Principle:

Superposition Principle

If y_j is a solution of $L(y) = f_j$ for $j = 1, 2, \ldots, n$, and $c_1, c_2, \ldots, c_n$ are arbitrary constants, then $c_1 y_1 + c_2 y_2 + \cdots + c_n y_n$ is a solution of $L(y) = c_1 f_1 + c_2 f_2 + \cdots + c_n f_n$—that is,

$$L(c_1 y_1 + c_2 y_2 + \cdots + c_n y_n) = c_1 L(y_1) + c_2 L(y_2) + \cdots + c_n L(y_n)$$
$$= c_1 f_1 + c_2 f_2 + \cdots + c_n f_n.$$

First, let's look at an nth-order *homogeneous* linear equation with constant coefficients

$$a_n y^{(n)} + a_{n-1} y^{(n-1)} + \cdots + a_2 y'' + a_1 y' + a_0 y = 0.$$

For such an equation, there's a neat algorithm for finding the general solution, a generalization of the procedure we've already seen:

First find the roots of the characteristic equation

$$a_n \lambda^n + a_{n-1} \lambda^{n-1} + \cdots + a_1 \lambda + a_0 = 0.$$

You should see how to form this equation. Focus on the fact that the characteristic equation of an nth-order linear equation is an nth-degree polynomial. But once a polynomial has degree greater than or equal to 5, there is no longer a general formula that gives the zeros. (Even the formulas that exist for the zeros of third- and fourth-order polynomials are very unwieldy.) In general, the only practical way to tackle such equations is to use *approximation* methods. A CAS or a graphing calculator should have various algorithms implemented to solve polynomial equations.

Next, group these roots as follows:

1. Distinct real roots
2. Distinct complex conjugate pairs $p \pm qi$
3. Multiple real roots
4. Multiple complex roots

Then the general solution is a sum of n terms of the forms

1. $c_i e^{\lambda_i t}$ for each distinct real root λ_i
2. $e^{pt}(c_1 \cos qt + c_2 \sin qt)$ for each distinct complex pair $p \pm qi$
3. $(c_1 + c_2 t + \cdots + c_k t^{k-1})e^{\lambda_i t}$ for each multiple real root λ_i, where k is the multiplicity of that root
4. $e^{pt}(c_1 \cos qt + c_2 \sin qt) + te^{pt}(c_3 \cos qt + c_4 \sin qt) + \cdots + t^{k-1}e^{pt}(c_{2k-1} \cos qt + c_{2k} \sin qt)$ for each multiple complex pair of roots $p \pm qi$, where k is the multiplicity of the pair $p \pm qi$

Now let's see how to use this procedure to solve some higher-order homogeneous linear equations with constant coefficients.

EXAMPLE 4.3.1 Solving a Fourth-Order Homogeneous Linear Equation
Let's find the general solution of the fourth-order equation

$$x^{(4)} - 3x'' + 2x' = 0.$$

The characteristic equation is $\lambda^4 - 3\lambda^2 + 2\lambda = \lambda(\lambda^3 - 3\lambda + 2) = 0$, whose roots are 0, 1, 1, and −2. (*Verify this.*) Thus we have two distinct real roots and another real root of multiplicity 2.

According to the process described above, the general solution is

$$x = c_1 e^{0 \cdot t} + c_2 e^{-2t} + (c_3 + c_4 t)e^{1 \cdot t} = c_1 + c_2 e^{-2t} + (c_3 + c_4 t)e^{t}.$$

You should check that this is a solution, manually or by using a CAS. ◆

EXAMPLE 4.3.2 Solving an Eighth-Order Homogeneous Linear Equation
The equation $64y^{(8)} + 48y^{(6)} + 12y^{(4)} + y'' = 0$ should be interesting to tackle. The characteristic equation is $64\lambda^8 + 48\lambda^6 + 12\lambda^4 + \lambda^2 = 0$. A CAS gives the roots 0, 0, $i/2$, $-i/2$, $i/2$, $-i/2$, $i/2$, and $-i/2$. Grouping these, we see that 0 is a real root of multiplicity 2, whereas the complex conjugate pair $\pm i/2 (= 0 \pm i/2)$ has multiplicity 3. Therefore, the form of the general solution of this eighth-order equation is $y(t) =$

$$(c_1 + c_2 t)e^{0 \cdot t} + e^{0 \cdot t}\left(c_3 \cos\left(\frac{t}{2}\right) + c_4 \sin\left(\frac{t}{2}\right)\right) + te^{0 \cdot t}$$

$$\left(c_5 \cos\left(\frac{t}{2}\right) + c_6 \sin\left(\frac{t}{2}\right)\right) + t^2 e^{0 \cdot t}\left(c_7 \cos\left(\frac{t}{2}\right) + c_8 \sin\left(\frac{t}{2}\right)\right)$$

$$= c_1 + c_2 t + (c_3 + c_5 t + c_7 t^2)\cos\left(\frac{t}{2}\right) + (c_4 + c_6 t + c_8 t^2)\sin\left(\frac{t}{2}\right).$$ ◆

For the nonhomogeneous case, once again the theory is simple:

The general solution, y_{GNH}, of an nth-order linear nonhomogeneous equation $a_n y^{(n)} + a_{n-1} y^{(n-1)} + \cdots + a_2 y'' + a_1 y' + a_0 y = f(t)$ is obtained by finding a particular solution, y_{PNH}, of the nonhomogeneous equation and adding it to the general solution, y_{GH}, of the associated homogeneous equation: $y_{\text{GNH}} = y_{\text{GH}} + y_{\text{PNH}}$.

As before, the challenge is to find a particular solution of the nonhomogeneous equation. But once again we can use the method of variation of parameters or the method of undetermined coefficients ("educated guessing").

If we look back at Examples 4.2.3 and 4.2.4 to see the number of calculations required to implement variation of parameters, we realize that the work can be formidable for equations of order 3 and above. But there is no need to do problems of higher order manually because any CAS will use the appropriate method efficiently to give us a general solution or solve an IVP. (One of the best methods for handling single linear equations and systems of linear equations is the *Laplace transform*, which we'll study in Chapter 6.) For now we'll just give an example of solving a higher-order linear equation, with some of the gory details left out.

EXAMPLE 4.3.3   Solving a Nonhomogeneous Third-Order Equation

Suppose we want to find the general solution of $y''' - y'' - 6y' = 3t^2 + 2$. The first thing to do is to find the general solution of the associated homogeneous equation $y''' - y'' - 6y' = 0$. The characteristic equation is $\lambda^3 - \lambda^2 - 6\lambda = \lambda(\lambda^2 - \lambda - 6) = \lambda(\lambda - 3)(\lambda + 2) = 0$, with roots 0, 3, and -2, so the general solution of the homogeneous equation is $c_1 e^{0 \cdot t} + c_2 e^{3t} + c_3 e^{-2t}$, or $c_1 + c_2 e^{3t} + c_3 e^{-2t}$.

Next, we look for a particular solution of the original nonhomogeneous equation. Examining the right-hand side of the equation, we can guess that a particular solution will be a polynomial in t. If the degree of this guessed-at polynomial is n, then the three individual derivative terms making up the differential equation will leave behind polynomials of degrees $n - 3$, $n - 2$, and $n - 1$. In order for the combination $y''' - y'' - 6y'$ to produce the second-degree polynomial $3t^2 + 2$, we must have $n - 1 = 2$—that is, the polynomial we're looking for must be a third-degree polynomial, say $y(t) = At^3 + Bt^2 + Ct + D$, where A, B, C, and D are *undetermined coefficients*. (Think about the reasoning that led to this form for y.)

Substituting this guess into the nonhomogeneous equation, we find that

$$-18At^2 - (12B + 6A)t + (6A - 2B - 6C) = 3t^2 + 2.$$

Equating coefficients of terms of equal degree on both sides, we get the algebraic equations

$$
\begin{aligned}
-18A &= 3 && \text{[Second-degree terms must match.]} \\
-(12B + 6A) &= 0 && \text{[First-degree terms must match.]} \\
6A - 2B - 6C &= 2 && \text{[Constant terms must match.]}
\end{aligned}
$$

Starting from the top, we can solve the equations successively to obtain $A = -\frac{1}{6}$, $B = \frac{1}{12}$, and $C = -\frac{19}{36}$.

Therefore, $y_{\text{PNH}} = -\frac{1}{6}t^3 + \frac{1}{12}t^2 - \frac{19}{36}t$ and the general solution of the non-homogeneous equation is given by

$$y = y_{\text{GNH}} = y_{\text{GH}} + y_{\text{PNH}} = c_1 + c_2 e^{3t} + c_3 e^{-2t} - \frac{1}{6}t^3 + \frac{1}{12}t^2 - \frac{19}{36}t. \qquad \blacklozenge$$

EXERCISES 4.3

Find the general solution of each of the higher-order equations in Exercises 1–8, using a graphing calculator or CAS only to solve each characteristic equation.

1. $y^{(4)} - 13y'' + 36y = 0$

2. $y''' - 3y'' + 3y' - y = 0$

3. $y^{(5)} + 2y''' + y' = 0$

4. $y^{(4)} + 13y'' + 36y = 0$

5. $y^{(4)} - 3y'' + 2y' = 0$

6. $y^{(4)} + 2y'' + y = 0$

7. $y''' - 12y'' + 22y' - 20y = 0$

8. $y^{(7)} - 14y^{(6)} + 80y^{(5)} - 242y^{(4)} + 419y^{(3)} - 416y'' + 220y' - 48y = 0$

9. Apply your CAS solver to find the general solution of the equation in Exercise 8.

10. The author of a classic differential equations text[6] once wrote

In preparing problems and examinations . . . teachers (including the author) must use some restraint. It is not reasonable to expect students in this course to have computing skill and equipment necessary for efficient solving of equations such as

$$4.317\frac{d^4y}{dx^4} + 2.179\frac{d^3y}{dx^3} + 1.416\frac{d^2y}{dx^2} + 1.295\frac{dy}{dx} + 3.169y = 0.$$

Demonstrate that technology has advanced in the last 40 years by feeding this equation into your CAS and obtaining the general solution. (You may have to use some "simplify" command to get a neat answer.)

11. Solve the IVP $3y''' + 5y'' + y' - y = 0$; $y(0) = 0$, $y'(0) = 1$, $y''(0) = -1$.

12. A uniform horizontal beam sags by an amount $y = y(x)$ at a distance x from one end. For a fairly rigid beam with uniform loading, $y(x)$ typically satisfies an equation of the form $d^4y/dx^4 = R$, where R is a constant depending on the load being carried and on the characteristics of the beam itself. If the ends of the beam are supported at $x = 0$ and $x = L$, then $y(0) = y(L) = 0$. The extended beam also behaves as though its profile had an inflection point at each support so that $y''(0) = y''(L) = 0$.

a. Use the multiple eigenvalue of the associated homogeneous equation to find the general solution of the homogeneous equation.

b. Show that the sag (vertical deflection) at point x is

$$\tfrac{1}{24}R(x^4 - 2Lx^3 + L^3x), \ 0 \le x \le L.$$

---

6. Ralph P. Agnew, *Differential Equations*, 2nd ed. (New York: McGraw-Hill, 1960): 176.

13. Solve the IVP $y^{(5)} = y'$; $y(0) = 0$, $y'(0) = 1$, $y''(0) = 0$, $y'''(0) = 1$, $y^{(4)}(0) = 2$.

14. For all positive integers $n \geq 2$, find the general solution of the equation $x^{(n)} = x^{(n-2)}$.

15. Find the general solution of the equation $y'' - 6y' + 13y = 15\cos 2x$.

16. Find the general solution of the equation $\ddot{y} + 5\dot{y} - 6\dot{y} = 9e^{3t}$.

17. Find the general solution of the equation $y''' + 6y'' + 11y' + 6y = 6x - 7$.

18. Solve the IVP $y'' - y' - 2y = e^{-x} + x$; $y(0) = 0$, $y'(0) = 1$.

19. Solve the IVP $y''' + 5y'' - 6y' = 3e^x$; $y(0) = 1$, $y'(0) = \frac{3}{7}$, $y''(0) = \frac{6}{7}$.

20. If $y_1(x) = x$ and $y_2(x) = 1/x$ are two linearly independent solutions of the differential equation $x^2 y'' + xy' - y = 0$, use the method of variation of parameters to find the general solution of $x^2 y'' + xy' - y = x$, $x \neq 0$. (Although we have discussed variation of parameters only for a linear equation with constant coefficients, the method remains valid for a linear equation whose coefficients are continuous functions of the independent variable.)

4.4   HIGHER-ORDER EQUATIONS AND THEIR EQUIVALENT SYSTEMS

To see where we're headed, think back to the first time you had to solve the following kind of word problem:

> Lenston has 21 coins, all nickels and dimes, in his pockets. They amount to $1.75. How many dimes does he have?

The first time you saw this problem, you were probably shown a solution like this one:

> Let x be the number of dimes. Then the total *amount* corresponding to dimes is $10x$ cents. The *number* of nickels must be $21 - x$, so the *amount* corresponding to nickels is $5(21 - x)$ cents. Because the total amount of money in Lenston's pockets is $1.75—or 175 cents—we have the equation $10x + 5(21 - x) = 175$, equivalent to $5x + 105 = 175$, which has the solution $x = 14$. Thus Lenston has 14 dimes (and $21 - 14 = 7$ nickels).

A bit later in your algebra course, you could have seen the same problem again, but this time you were probably shown how to turn this problem into a *system* problem:

> Let x be the number of dimes and let y be the number of nickels. Then the words of the problem tell us two things, one fact about the *number* of coins and one fact about the *amount* of money: (1) $x + y = 21$ and (2) $10x + 5y = 175$. In other words, viewed this way, the problem gives us the *system* of equations

$$x + y = 21$$
$$10x + 5y = 175.$$

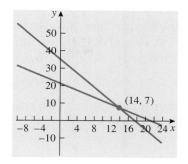

Figure 4.3
Graphs of $x + y = 21$ and $10x + 5y = 175$

This system can be solved by elimination (multiply the first equation by -5 and then add the result to the second equation) or by substitution (solve the first equation for x, for example, and then substitute for x in the second equation).

The most important consequence of looking at our problem as a system problem is that the system has a very nice geometrical interpretation as a set of two straight lines (Figure 4.3). The solution of the system (and of our original problem) is given by the coordinates of the point where the lines intersect: $x = 14$, $y = 7$.

The important thing here is that a systems approach has certain advantages, especially the graphical interpretation of a problem and its solution. Also, certain problems may naturally occur in system form. For example, we may want to compute the trajectory of a baseball. In this case, it is natural to consider the components, u and v, of the ball's velocity in both its horizontal (x) and vertical (y) directions, respectively. A system[7] arising from this problem is

$$mu\frac{du}{dx} = -F_L \sin\theta - F_D \cos\theta$$

$$mv\frac{dv}{dy} = F_L \cos\theta - F_D \sin\theta - mg.$$

Similarly, in an ecological study, we may want to analyze the interaction of two or more biological species, each of which needs its own equation to represent its growth rate and its relationship to the other species.

CONVERSION TECHNIQUE I: CONVERTING A HIGHER-ORDER EQUATION INTO A SYSTEM

Now that our previous discussion has prepared us to see even simple problems as systems, we can tackle some higher-order differential equations. The key here is the following result:

---

7. Robert B. Banks, *Towing Icebergs, Falling Dominoes, and Other Adventures in Applied Mathematics* (Princeton, N.J.: Princeton University Press, 1998).

Any single *n*th-order differential equation can be converted into an equivalent system of first-order equations. More precisely, any *n*th-order differential equation of the form

$$x^{(n)} = F(t, x, x', x'', \ldots, x^{(n-1)})$$

can be converted into an equivalent system of *n* first-order equations by letting $x_1 = x$, $x_2 = x'$, $x_3 = x''$, $\ldots$, $x_n = x^{(n-1)}$.

After looking at some examples of how this conversion technique works, we'll introduce the geometric/graphical significance of this method.

EXAMPLE 4.4.1   Converting a Second-Order Linear Equation

As we saw in Section 4.1 and will see again in Section 4.5 and its exercises, the second-order linear equation $2\dfrac{d^2x}{dt^2} + 3\dfrac{dx}{dt} + x = 0$ could represent the motion of a weight attached to a spring, the flow of electricity through a circuit, or other important phenomena.

Using the substitutions described above, we introduce new variables x_1 and x_2: Let $x_1 = x$ and $x_2 = \dfrac{dx}{dt}$. Now isolate the highest derivative (the second) in the original equation, and then substitute the new variables on the right-hand side:

$$(1) \quad \frac{d^2x}{dt^2} + \frac{3}{2}\frac{dx}{dt} + \frac{1}{2}x = 0$$

$$(2) \quad \frac{d^2x}{dt^2} = -\frac{3}{2}\frac{dx}{dt} - \frac{1}{2}x$$

$$(3) \quad \frac{d^2x}{dt^2} = -\frac{3}{2}x_2 - \frac{1}{2}x_1$$

In terms of the new variables, we see that $\dfrac{dx_1}{dt} = \dfrac{dx}{dt} = x_2$ and $\dfrac{dx_2}{dt} = \dfrac{d}{dt}\left(\dfrac{dx}{dt}\right) = \dfrac{d^2x}{dt^2} = $ [from step (3) above] $-\dfrac{3}{2}x_2 - \dfrac{1}{2}x_1$. From this we see that our original second-order equation leads to the following system of linear first-order equations in two unknown functions x_1 and x_2:

$$(A) \quad \frac{dx_1}{dt} = x_2$$

$$(B) \quad \frac{dx_2}{dt} = -\frac{3}{2}x_2 - \frac{1}{2}x_1$$

This system is *equivalent* to the original single differential equation in the sense that any solution $x(t)$ of the original equation yields solutions $x_1(t) = x(t)$ and $x_2(t) = \dfrac{d}{dt}x(t)$ of the system, and any solution $(x_1(t), x_2(t))$ of the system gives us a solution $x(t) = x_1(t)$ of the original equation.

Let's follow up on the first part of that statement. From our work in Section 4.1, we know that $x(t) = e^{-t/2} + 2e^{-t}$ is a solution of the original second-order equation. Then the pair $x_1(t) = x(t) = e^{-t/2} + 2e^{-t}$ and $x_2(t) = \dfrac{d}{dt}x(t) = -\dfrac{1}{2}e^{-t/2} - 2e^{-t}$ constitutes a solution of the system. (*Verify this!*) ◆

Let's look at a few more examples of this technique of converting a higher-order equation into a system of first-order equations.

EXAMPLE 4.4.2 **Converting a Second-Order Nonlinear Equation**

Suppose we have the second-order nonlinear equation $y'' = y^3 + (y')^3$. Let $x_1 = y$ and $x_2 = y'$. Then $x_1' = y' = x_2$, $y'' = x_2'$, $y^3 = x_1^3$, and $(y')^3 = x_2^3$, so we can rewrite $y'' = y^3 + (y')^3$ as $x_2' = x_1^3 + x_2^3$.

Finally, putting these pieces together, we can write the original equation as the following equivalent nonlinear system in x_1 and x_2:

$$\begin{aligned} x_1' &= x_2 \\ x_2' &= x_1^3 + x_2^3 \end{aligned}$$ ◆

EXAMPLE 4.4.3 **Converting a Third-Order Equation**

The nonautonomous third-order linear equation

$$\frac{d^3x}{dt^3} - \frac{d^2x}{dt^2} + 2t\frac{dx}{dt} - 3x + 6 = 0$$

can be changed into a system of first-order equations as follows: Let $x_1 = x$, $x_2 = \dfrac{dx}{dt}$, and $x_3 = \dfrac{d^2x}{dt^2}$. Then $\dfrac{dx_1}{dt} = \dfrac{dx}{dt} = x_2$, $\dfrac{dx_2}{dt} = \dfrac{d}{dt}\left(\dfrac{dx}{dt}\right) = \dfrac{d^2x}{dt^2} = x_3$, and $\dfrac{dx_3}{dt} = \dfrac{d}{dt}\left(\dfrac{d^2x}{dt^2}\right) = \dfrac{d^3x}{dt^3}$.

Solving the original equation for $\dfrac{d^3x}{dt^3}$ and then substituting the new variables x_1, x_2, and x_3, we have

$$\frac{d^3x}{dt^3} = \frac{d^2x}{dt^2} - 2t\frac{dx}{dt} + 3x - 6 = x_3 - 2tx_2 + 3x_1 - 6.$$

Putting all the information together, we see that the original third-order equation is equivalent to the system of three first-order equations

$$\frac{dx_1}{dt} = x_2$$

$$\frac{dx_2}{dt} = x_3$$

$$\frac{dx_3}{dt} = x_3 - 2tx_2 + 3x_1 - 6.$$

To be mathematically precise, we can describe this system as a *three-dimensional nonautonomous linear system with independent variable t and dependent variables* x_1, x_2, *and* x_3.   ◆

As we'll see later in this chapter, an autonomous system has a nice graphical interpretation that gives us a neat qualitative analysis. We lose some of this power when we are dealing with a nonautonomous system. But even when we are confronted with a nonautonomous equation, a simple variation of the conversion technique we've been illustrating will allow us to transform the equation into an autonomous system. To convert a single *nonautonomous* nth-order equation into an equivalent *autonomous* system (one whose equations do not explicitly contain the independent variable t), we need $n + 1$ first-order equations: $x_1 = x$, $x_2 = x'$, $x_3 = x''$, ..., $x_n = x^{(n-1)}$, $x_{n+1} = t$. We see this in the next example.

EXAMPLE 4.4.4   Converting a Nonautonomous Equation Into an Autonomous System

The nonautonomous second-order linear equation $2\dfrac{d^2x}{dt^2} + 3\dfrac{dx}{dt} + x = 50 \sin t$ could be handled in the same way as the equation in Example 4.4.3, but instead we'll demonstrate the extension of the conversion technique.

Start by letting $x_1 = x$ and $x_2 = \dfrac{dx}{dt}$ as before, but also introduce $x_3 = t$. Then

$$\frac{dx_1}{dt} = \frac{dx}{dt} = x_2,$$

$$\frac{dx_2}{dt} = \frac{d^2x}{dt^2} = \frac{1}{2}\left(-3\frac{dx}{dt} - x + 50 \sin t\right) = \frac{1}{2}(-3x_2 - x_1 + 50 \sin x_3),$$

and

$$\frac{dx_3}{dt} = 1,$$

so the equivalent system is

$$\frac{dx_1}{dt} = x_2$$

$$\frac{dx_2}{dt} = \frac{1}{2}(-x_1 - 3x_2 + 50 \sin x_3)$$

$$\frac{dx_3}{dt} = 1.$$

Our second-order equation has been replaced by an equivalent *autonomous three-dimensional* system. If we had not used the third variable x_3 and had written our equation as a system of *two* equations, the second equation would have been non-autonomous. We would have had $\dfrac{dx_1}{dt} = x_2$ and $\dfrac{dx_2}{dt} = \dfrac{1}{2}(-x_1 - 3x_2 + 50 \sin t)$ as the system, with the explicit presence of t in the second equation making this equation (and therefore the system) nonautonomous.   ◆

Of course, we should be able to convert an initial-value problem into a system IVP as well. If you think about this, we would expect that the original initial conditions would have to expand to cover each first-order equation in the system. The next example shows how this works.

EXAMPLE 4.4.5   **Converting a Second-Order Initial-Value Problem**
The nonautonomous second-order linear IVP

$$y'' - xy' - x^2y = 0; \quad y(0) = 1, y'(0) = 2$$

can be transformed into a system IVP as follows. Let $u_1 = y$ and $u_2 = y'$. (We're using a different letter for the new variables to avoid confusion with the original independent variable x.) We see that $u_1' = y' = u_2$ and $u_2' = y'' = xy' + x^2y = xu_2 + x^2u_1$. Then, because $u_1 = y$, $y(0) = 1$, implies that $u_1(0) = 1$; and $y'(0) = 2$ implies that $u_2(0) = 2$ because $u_2 = y'$. Therefore, the original IVP becomes the system IVP

$$\begin{aligned} u_1' &= u_2 \\ u_2' &= xu_2 + x^2u_1; \quad u_1(0) = 1, u_2(0) = 2. \end{aligned}$$

Note that because each equation in the system is first-order, we need only one initial condition for each new variable. *What would the equivalent autonomous system look like?*   ◆

CONVERSION TECHNIQUE II: CONVERTING A SYSTEM INTO A HIGHER-ORDER EQUATION

We indicated in Example 4.4.1 that a higher-order equation and its related system are equivalent, so it seems reasonable that we should be able to convert a system into a single higher-order equation. We accomplish this in the next example.

EXAMPLE 4.4.6   **Converting a System into a Single Equation**
Can you convert the system

$$\begin{aligned} (1) \quad & y' = z \\ (2) \quad & z' = w \\ (3) \quad & w' = x - 3y - 6z - 3w, \end{aligned}$$

where y, z, and w are functions of x, into an equivalent single higher-order equation?

Sure you can. Just look back at what we did in our earlier examples, but start with the *last* equation and work backwards: Differentiating equation (2) gives us $z'' = w'$. But equation (1) says that $z'' = (y')'' = y'''$, so that $w' = y'''$. Now we use this last fact to rewrite equation (3) as

$$\begin{aligned} y''' &= x - 3y - 6z - 3w \\ &= x - 3y - 6y' - 3z' \quad \text{[from (1) and (2)]} \\ &= x - 3y - 6y' - 3y'' \quad \text{[from (1)]} \end{aligned}$$

or $y''' + 3y'' + 6y' + 3y = x$, a third-order linear nonautonomous differential equation.   ◆

LOOKING AHEAD

Now that we've seen how to transform any differential equation of order greater than 1 into a system of first-order equations and vice versa, how can we use this information to gain insight into the behavior of solutions of higher-order equations?

The next section will exploit the geometric (graphical) aspects of an autonomous system of equations and give us qualitative tools for analysis. We'll see that the qualitative approach will give us useful information not easily obtained otherwise. We'll also discuss important applied examples, both linear and nonlinear. Later in Chapter 4, we'll deal with numerical approximations to solutions of systems of equations. Chapter 5 will explore linear autonomous systems thoroughly, and Chapter 7 will introduce valuable methods for analyzing nonlinear systems of equations.

EXERCISES 4.4

Write each of the higher-order ODEs or systems of ODEs in Exercises 1–9 as a system of first-order equations. If initial conditions are given, rewrite them in terms of the first-order system.

1. $\dfrac{d^2x}{dt^2} - x = 1$

2. $\ddot{y} + y = t$; $y(0) = 1$, $y'(0) = 0$

3. $x'' + 3x' + 2x = 1$; $x(0) = 1$, $x'(0) = 0$

4. $y^{(4)} + y = 0$

5. $w^{(4)} - 2w''' + 5w'' + 3w' - 8w = 6\sin(4t)$

6. $(x'')^2 - (\sin t)x' = x\cos t$

7. $x^2y'' - 3xy' + 4y = 5\ln x$

8. $\ddot{x} + (\dot{x})^2 + x(x - 1) = 0$

9. $\dfrac{d^2x}{dt^2} = -x, \dfrac{d^2y}{dt^2} = y$ [*Hint*: Write each second-order equation as two first-order equations.]

10. The equation $\dfrac{d^2x}{dt^2} + 4\dfrac{dx}{dt} + 4x = 0$ describes the position, $x(t)$, of a particular mass attached to a spring and set in motion by pulling it down 2 ft below its equilibrium position ($x = 0$) and giving it an initial velocity of 2 ft/sec in the upward direction. Some air resistance is assumed. Express this equation as a system of first-order equations and describe what each equation of the system represents.

11. An object placed in water, pushed down a certain distance below the water, and then released has its bobbing motion described by the equation

$$\frac{d^2y}{dt^2} + \left(\frac{g}{s_0}\right)y = 0,$$

where y is the vertical displacement from its equilibrium position, g is the acceleration due to gravity, and s_0 is the initial depth. Express this equation as a system of first-order equations.

12. The second-order nonlinear equation $\dfrac{d^2x}{dt^2} + \dfrac{g}{L}\sin x = 0$ describes the swinging of a pendulum, where x is the angle the pendulum makes with the vertical, g is the acceleration due to gravity, and L is the pendulum's length. Convert this equation into a nonlinear system of first-order equations.

13. The equation $y''' + y' - \cos y = 0$ describes a geometrical model of crystal growth. Express this third-order equation as a system of three first-order equations.

14. The equation $y^{(4)} + \lambda(yy''' - y'y'') - y' = 0$, where λ is a positive parameter, arises in a nonlinear "boundary layer" problem in physical oceanography. Write this equation as a system of four first-order equations.

15. Rewrite the system IVP given in Example 4.4.5 as an equivalent *autonomous* system.

16. Write each of the following systems of equations as a single second-order equation, rewriting initial conditions as necessary.

  a. $\dfrac{dy}{dt} = x, \dfrac{dx}{dt} = -y; \quad y(0) = 0, x(0) = 1$

  b. $\dfrac{du}{dx} = 2v - 1, \dfrac{dv}{dx} = 1 + 2u$

  c. $x' = x + y,\ y' = x - y$

  d. $\dfrac{dx}{dt} = 7y - 4x - 13, \dfrac{dy}{dt} = 2x - 5y + 11; \quad x(0) = 2, y(0) = 3$

17. Write the following system as a system of four first-order equations:

$$x\frac{d^2y}{dt^2} - y = 4t, \quad 2\frac{d^2x}{dt^2} + \left(\frac{dy}{dt}\right)^2 = x.$$

(Convert each second-order equation into two first-order equations.)

18. Write the following system of equations as a single fourth-order equation, with appropriate initial conditions:

$$\frac{d^2x}{dt^2} + 2\frac{dy}{dt} + 8x = 32t$$

$$\frac{d^2y}{dt^2} + 3\frac{dx}{dt} - 2y = 60e^{-t}; \quad x(0) = 6, x'(0) = 8,$$

$$y(0) = -24, \text{ and } y'(0) = 0.$$

19. Suppose you are given the linear system of first-order equations

$$t\frac{dx}{dt} = -3x + 4y$$

$$t\frac{dy}{dt} = -2x + 3y.$$

Introduce a new independent variable w by the substitution $w = \ln t$ (or $t = e^w$) and show that this substitution allows you to write the system as a new system with constant coefficients.

4.5 QUALITATIVE ANALYSIS OF AUTONOMOUS SYSTEMS

In this section, we investigate the graphical representation of a system of first-order equations. Because many systems—especially nonlinear systems—cannot be solved in closed form, the ability to analyze systems graphically is very important. The first thing we have to realize is that the very useful graphical tool of *slope fields* can't be applied directly to higher-order equations; this technique depends on a knowledge of the first derivative alone. However, there's a clever way of using our knowledge of first-order qualitative methods in the analysis of higher-order differential equations.

For convenience, we'll spend most of our time analyzing autonomous two-dimensional systems, although we will also tackle some nonautonomous systems and some three-dimensional systems toward the end of this section.

PHASE PORTRAITS FOR SYSTEMS OF EQUATIONS

Suppose we have an autonomous system of the form

$$\frac{dx}{dt} = f(x, y)$$
$$\frac{dy}{dt} = g(x, y). \tag{4.5.1}$$

For example, let's take the system

$$\frac{dx}{dt} = y$$
$$\frac{dy}{dt} = -17x - 2y$$

and work with it throughout our initial discussions.

First we can eliminate the variable t by dividing the second equation by the first equation:

$$\frac{dy}{dx} = \frac{dy}{dt} \cdot \frac{dt}{dx} = \frac{\dfrac{dy}{dt}}{\dfrac{dx}{dt}} = \frac{g(x, y)}{f(x, y)}. \tag{4.5.2}$$

(See Appendix A.2 for a reminder of the Chain Rule used in this process.) For our example, $g(x, y) = -17x - 2y$ and $f(x, y) = y$ in (4.5.2), and we get $\dfrac{dy}{dx} = \dfrac{-17x - 2y}{y}$.

Now we have a single first-order differential equation $\dfrac{dy}{dx} = \dfrac{g(x, y)}{f(x, y)}$ in the variables y and x. If we could solve equation (4.5.2) for y in terms of x, or even implicitly, we would have a solution curve in the x-y plane. The plane of the variables x and y (with x- and y-axes) is called the **phase plane** of the original system of differential equations. As we saw in Section 1.2, each individual solution curve in the phase plane, $x = x(t)$, $y = y(t)$, is called a **trajectory** (or **orbit**) of the system of equations. Although the independent variable t is not present explicitly, the passage of time is represented by the *direction* that a point $(x(t), y(t))$ takes on a particular trajectory. The way the curve is followed as the values of t increase (offstage) is called the **positive direction** on the trajectory. The collection of plots of the trajectories is called the system's **phase portrait** or **phase-plane diagram.** (You may want to review the qualitative analysis for first-order equations in Section 2.4.)

Even if we can't solve the system, we can look at the slope field of the single equation (4.5.2), the outline of the phase portrait of the system. If we give some initial points $(x_0^i, y_0^i) = (x^i(t_0), y^i(t_0))$, $i = 1, 2, \ldots, n$, through which we want the trajectories to pass, we can plot a few specific trajectories and get a less complicated view of the phase plane. Let's do this for the system we've been discussing.

EXAMPLE 4.5.1 **Phase Portrait—One Trajectory**
Our system is

$$\frac{dx}{dt} = y$$

$$\frac{dy}{dt} = -17x - 2y,$$

which gives us the first-order equation $\dfrac{dy}{dx} = \dfrac{-17x - 2y}{y}$ when we eliminate the variable t. You should use your calculator or CAS to draw a piece of the slope field for this first-order equation. Figure 4.4a shows the slope field, and Figure 4.4b shows a single trajectory satisfying the initial condition $x(0) = 4$, $y(0) = 0$—that is, a trajectory passing through the point $(4, 0)$ in the x-y (phase) plane—superimposed on the slope field.

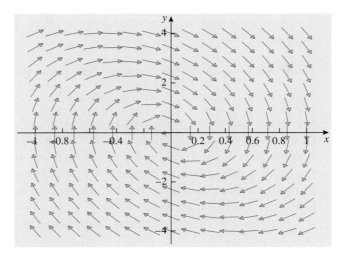

Figure 4.4a

Slope field for $\dfrac{dy}{dx} = \dfrac{-17x - 2y}{y}$

$0 \le t \le 5; \quad -1 \le x \le 1, -4 \le y \le 4$

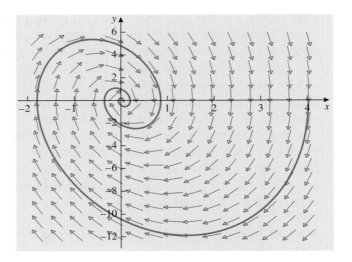

Figure 4.4b
Trajectory for

$$\left\{ \frac{dx}{dt} = y, \frac{dy}{dt} = -17x - 2y; \quad x(0) = 4, y(0) = 0 \right\}$$

$$0 \le t \le 5; \quad -2 \le x \le 4, -12 \le y \le 7$$

Because the trajectory starts at $(4, 0)$, you can see that the positive direction on the trajectory is clockwise, and the curve seems to spiral into the origin. (Try using technology to draw the trajectory for $0 \le t \le b$, letting b get larger and larger.) To get an accurate phase portrait, you may want to use the slope field to suggest good initial points to use. Each dynamical system has its own appropriate range for t. ◆

Now let's look at a more elaborate phase portrait, one showing several trajectories.

EXAMPLE 4.5.2   Phase Portrait—Several Trajectories

The system consists of the two equations (1) $\dfrac{dx}{dt} = x + y$ and (2) $\dfrac{dy}{dt} = -x + y$.

Whatever quantities these equations describe, certain facts should be obvious from the very nature of the equations. First of all, from equation (1), the growth of quantity x depends on itself and on the other quantity y in a positive way. On the other hand, equation (2) indicates that quantity y depends on itself positively, but its growth is hampered by the presence of quantity x—a larger value of x leads to a slowdown in the growth of y.

Let's look at the phase portrait corresponding to this problem. For our system, equation (4.5.2) looks like

$$\frac{dy}{dx} = \frac{\dfrac{dy}{dt}}{\dfrac{dx}{dt}} = \frac{-x + y}{x + y}.$$

This first-order equation is neither separable nor linear, but it is *homogeneous* and can be solved implicitly. (See the explanation for Problems 15–18 of Exercises 2.1.) Figure 4.5a shows several trajectories, obtained by specifying nine initial points $(x(0), y(0))$, superimposed on the slope field for $\dfrac{dy}{dx} = \dfrac{-x + y}{x + y}$.

Because points on a trajectory are calculated by numerical methods, your CAS may allow you (or require you) to specify a step size and the actual numerical approximation method to be used. Numerical methods for systems of differential equations will be discussed in Section 4.7.

Each point on a particular curve in Figure 4.5a represents a *state* of the system: For each value of t, the point $(x(t), y(t))$ on the curve provides a snapshot of this dynamical system. If the variables x and y are supposed to represent animal or human populations, for example, then the proper place to view the trajectories is the first quadrant. Figure 4.5b describes the first quadrant of the phase plane for our problem, with four trajectories determined by four initial points.

These trajectories tell us that for the initial points chosen, the quantity y increases to a maximum value and then decreases to zero, while the quantity x also increases until it reaches its maximum level after quantity y has disappeared.

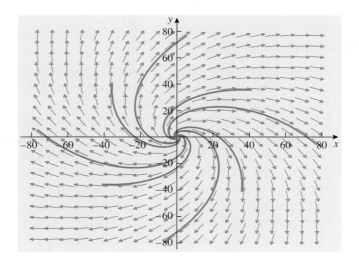

Figure 4.5a

Trajectories for $\left\{ \dfrac{dx}{dt} = x + y, \dfrac{dy}{dt} = -x + y \right\}$

$(x(0), y(0)) = (-1, -1), (-1, 0), (-1, 1), (0, -1), (0, 0), (0, 1), (1, -1), (1, 0),$ and $(1, 1)$

$0 \le t \le 4$

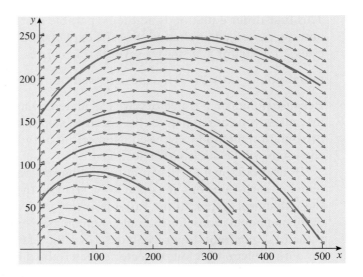

Figure 4.5b

Trajectories for $\left\{\dfrac{dx}{dt} = x + y, \dfrac{dy}{dt} = -x + y\right\}$

$(x(0), y(0)) = (0, 60), (25, 100), (50, 140),$ and $(0, 160)$

$0 \le t \le 1.2$ ◆

Other Graphical Representations

Using technology again in our last example, we can graph $x(t)$ and $y(t)$ in the t-x and t-y planes, respectively. Figures 4.6a and 4.6b show solution curves with $x(0) = 50$ and $y(0) = 140$, respectively.

These graphs show clearly that the quantity y reaches a maximum of about 164 when $t \approx 0.4$ and that the x quantity hits a peak of about 800 when $t \approx 2$. Note that the horizontal and vertical scales are different for Figures 4.6a and 4.6b.

Viewed another way, the system in the last example had *three* variables—the independent variable t and the dependent variables x and y. To be precise about all this, we state that *a solution of our system is a pair of functions $x = x(t)$, $y = y(t)$, and the graphical representation of such a solution is a curve in three-dimensional t-x-y space—a set of points of the form $(t, x(t), y(t))$.* Figure 4.7 shows what the solution with initial point $(0, 50, 140)$ looks like for our problem.

Your CAS will probably allow you to manipulate the axes and get different views of this space curve. Figure 4.5a represents the *projection* of several such space curves onto the x-y plane, a much less confusing way of viewing the behavior of the system. These projections can be thought of as the shadows that would be cast by the space curves if a very bright light were shining on them from the front (the x-y face) of Figure 4.7.

Note that because the system we started with in the last example is autonomous, the solution curves are *independent of the starting time*. This means that if you pick a starting point (x^*, y^*) at time t^*, then the path of a population starting at this point is the same as the path of a population starting at the same

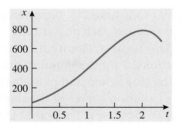

Figure 4.6a
$x(t); x(0) = 50$
$0 \le t \le 2.355$

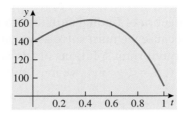

Figure 4.6b
$y(t); y(0) = 140$
$0 \le t \le 1$

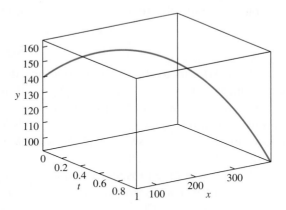

Figure 4.7
Solution of $\left\{ \dfrac{dx}{dt} = x + y, \dfrac{dy}{dt} = -x + y; \quad x(0) = 50, y(0) = 140 \right\}$
$0 \le t \le 1$

point at any other time t^{**}. Geometrically, this says that there is only one path (trajectory) through each point of the *x-y* plane. This is a consequence of an Existence and Uniqueness Theorem for systems that we'll see in Section 4.6. (Look back at Section 2.7 for the theorem that applies to first-order ODEs.)

From the slope field and phase portrait in Figure 4.5a, it seems clear that all trajectories (solution curves of the single differential equation) are escaping from the origin as *t* increases. The variable *t* is behind the scenes in a phase portrait, but you should experiment with different ranges of *t* in your CAS or graphing calculator to verify the last statement. The **critical point** or **equilibrium point** $(0, 0)$—where both $dx/dt = 0$ and $dy/dt = 0$—is called a **source** in this case. We have used this terminology before (for the one-dimensional case, in Section 2.5), we will use it in this chapter, and we'll see it again as part of the discussion of systems in Chapter 5.

The algebra of finding equilibrium points (or equilibrium *solutions*) is trickier now because we must solve a *system* of equations. For example, if we want to find the equilibrium solutions of the nonlinear system of differential equations $\{\dot{x} = x - y, \dot{y} = 1 - xy\}$, we must solve the algebraic system

$$(1) \quad x - y = 0$$
$$(2) \quad 1 - xy = 0.$$

We can solve equation (1) for *y*, finding that $y = x$, and then substitute for *y* in the second equation. We get $1 - x^2 = 0$, which implies that $x = \pm 1$. Because $y = x$, the only equilibrium points are $(-1, -1)$ and $(1, 1)$.

Before we move on, let's look at the system $\{\dot{x} = 4 - 4x^2 - y^2, \dot{y} = 3xy\}$. Any equilibrium solution has to satisfy the equations

$$(A) \quad 4 - 4x^2 - y^2 = 0$$
$$(B) \quad \qquad 3xy = 0.$$

Equation (B) tells us that we have two possibilities: (i) $x = 0$ or (ii) $y = 0$. (We can eliminate $x = y = 0$ because (A) wouldn't be satisfied with this choice.) If $x = 0$, substituting in (A) gives us $4 - y^2 = 0$, so $y = \pm 2$. Then we have two equilibrium solutions, $(0, 2)$ and $(0, -2)$. Alternatively, if $y = 0$, substituting in (A) yields $4 - 4x^2 = 0$, so $x = \pm 1$. Now we have the remaining two equilibrium solutions, $(1, 0)$ and $(-1, 0)$.

The next example presents a simple system model of an *arms race*. Models of this general form were proposed by the English scientist Lewis F. Richardson (1881–1953) in the 1930s. As a Quaker, he was greatly interested in the causes and avoidance of war. We'll see how a qualitative analysis helps us to understand the situation being modeled.

EXAMPLE 4.5.3   An Arms Race Model
Let's look at an autonomous linear system:

$$\frac{dx}{dt} = 7y - 4x - 13$$
$$\frac{dy}{dt} = 2x - 5y + 11.$$

The functions $x(t)$ and $y(t)$ could represent the readiness for war of two nations, X and Y, respectively. This readiness might be measured, for example, in terms of the level of expenditures for weapons for each country at time t. To get the first equation, this model assumes that the rate of increase of x is a linear function of both x and y. In particular, if y increases, then so does the rate at which x increases. *This makes sense, doesn't it?* But the cost of building up and maintaining a supply of weapons also puts the brakes on *too* much expansion. The term $-4x$ in the first equation suggests a sense of restraint proportional to the arms level of country X. Finally, the constant term -13 can represent some basic, constant relationship of country X to country Y—probably some underlying feelings of good will that diminish the threat and therefore decrease dependence on weaponry. The second equation can be interpreted in a similar way, but here the positive constant 11 probably signifies a grievance by Y against X that results in an accumulation of arms. Now what does this model tell us about the situation? We don't know how to solve such a system, but we can still learn a lot about the arms race between the two countries.

As in Example 4.5.1, we can start constructing the phase portrait of the system by eliminating the variable t:

$$\frac{dy}{dx} = \frac{\dfrac{dy}{dx}}{\dfrac{dx}{dt}} = \frac{2x - 5y + 11}{7y - 4x - 13}.$$

Now we can look at the slope field and some trajectories corresponding to this single equation (Figure 4.8). Several initial points were chosen. (Try a smaller set

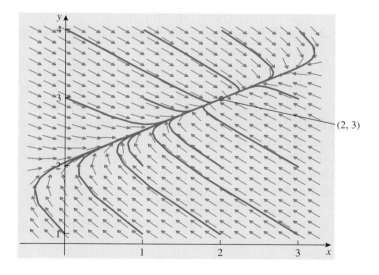

Figure 4.8

Trajectories for $\left\{\dfrac{dx}{dt} = 7y - 4x - 13, \dfrac{dy}{dt} = 2x - 5y + 11\right\}$

Initial points (i, j), $i = 0, 1, 2, 3$; $j = 1, 2, 3, 4$

$0 \leq t \leq 5$

of initial points yourself.) For this to be a realistic model of an arms race, the values of *x* and *y* should be positive; hence our focus on the first quadrant.

First of all, note that one solution of the system is the pair of functions $x(t) \equiv 2$, $y(t) \equiv 3$. In this phase portrait, if we look hard enough, we may notice that the points $(x(t), y(t))$ on every trajectory seem to be moving toward the point (2, 3) as *t* increases. (To verify this last statement, you should plot the phase portrait for $0 \leq t \leq b$ and let *b* increase.) The point (2, 3) is an equilibrium point—as we've seen above, a point (x, y) at which both *dx/dt* and *dy/dt* equal 0. The behavior of trajectories near this point entitles it to be called a **sink.** In real-life terms, this means that the arms race represented by this system would *stabilize* as time passes, approaching a state in which the level of military expenditures for nation X would be 2 and the level for nation Y would be 3, where the units could be millions or billions.   ◆

A PREDATOR-PREY MODEL: THE LOTKA-VOLTERRA EQUATIONS

An important type of real-life problem that can be modeled by a system of differential equations is a **predator-prey problem,** in which we assume that there are two species of animals, X and Y, in a small geographical region such as an island. One species (the **predator**) thinks of the other species (the **prey**) as food and is very dependent on this food supply for survival.

Let $x(t)$ and $y(t)$ represent the populations of the two species at time *t*. We can make the following reasonable assumptions:

1. If there are no predators, the prey species will grow at a rate proportional to its own population (assuming an unlimited food supply).
2. If there are no prey, the predator species will decline at a rate proportional to the predator population.
3. The presence of both predators and prey is beneficial to the growth of the predator species and is harmful to the growth of the prey species.

The third assumption says that interactions (or close encounters of the hungry kind) between the predator and prey lead to a decrease in the prey population and to a resulting increase in the predator population. As we will see, these contacts are indicated mathematically by a *multiplication* of the variables that represent predator and prey. These assumptions lead to a system of *nonlinear* first-order differential equations such as the following:

$$\frac{dx}{dt} = 0.2\,x - 0.002\,xy, \frac{dy}{dt} = -0.1\,y + 0.001\,xy. \tag{4.5.3}$$

For this system, how can we see that $x(t)$ is the size of the *prey* population at any time *t* and $y(t)$ is the number of *predators* at time *t*?

First of all, note that if there are *no* predators—that is, if *y* is always 0—the system reduces to $dx/dt = 0.2\,x$, $dy/dt = 0$. This says that the prey population would increase at a rate that is proportional to the actual prey population at any time. Also, the predator population is constant—at zero. This is realistic and consistent with assumption 1. Furthermore, if there are no *prey*—that is, if $x \equiv 0$—the system becomes $dx/dt = 0$, $dy/dt = -0.1\,y$, which means that the number of predators would decrease at a rate proportional to the predator population, where 0.1 is the constant of proportionality, the predator's *intrinsic death rate*. Again, this is realistic because in the absence of a crucial food supply, the bottom line would be starvation and a net decline in the predator population.

The intriguing terms in (4.5.3) are the terms involving the product xy. We've already suggested that these terms represent *the number of possible interactions* between the two species. To illustrate this point, suppose there were four foxes and three rabbits on an island. If we label the foxes F_1, F_2, F_3, and F_4 and the rabbits R_1, R_2, and R_3, then we have the following possible one-on-one encounters between foxes and rabbits: (F_1, R_1), (F_1, R_2), (F_1, R_3), (F_2, R_1), (F_2, R_2), (F_2, R_3), (F_3, R_1), (F_3, R_2), (F_3, R_3), (F_4, R_1), (F_4, R_2), and (F_4, R_3). Note that there are $4 \times 3 = 12$, or x times y, possible interactions. Of course, we can have two foxes meeting up with one rabbit or one fox coming upon three rabbits, and so on, but the idea is that *the number of interactions is proportional to the product of the two populations*. The coefficient of xy in the first equation, -0.002, is a measure of the predator's effectiveness in terms of prey capture, whereas the coefficient 0.001 in the second equation is an indicator of the predator's efficiency in terms of prey consumption.

This nonlinear system above is a particular example of a system called the **Lotka-Volterra equations:**

$$\frac{dx}{dt} = a_1 x - a_2 xy$$

$$\frac{dy}{dt} = -b_1 y + b_2 xy,$$

where a_2 and b_2 are positive constants. [Alfred Lotka (1880–1949) was a chemist and demographer and Vito Volterra (1860–1940) was a mathematical physicist. In the 1920's they derived these equations independently—Lotka from a chemical reaction problem and Volterra from a problem concerned with fish catches in the Adriatic Sea.] In general, there is no explicit solution of the Lotka-Volterra equations in terms of elementary functions. We'll discuss numerical solutions of systems in Section 4.7.

However, as the next example shows, we can understand the relationship between the two species by using a *qualitative* analysis.

EXAMPLE 4.5.4   Qualitative Analysis of a Predator-Prey Model
Figure 4.9 shows the trajectory corresponding to our system

$$\frac{dx}{dt} = 0.2\,x - 0.002xy$$

$$\frac{dy}{dt} = -0.1\,y + 0.001xy,$$

with $x(0) = 100$, $y(0) = 25$, and $0 \le t \le 52$. What does this picture tell us? First realize that the horizontal axis (x) represents the prey, the vertical axis (y) the predators. Our starting point, corresponding to $t = 0$, is (100, 25), and the direction of the trajectory is counterclockwise. To see the direction, use technology

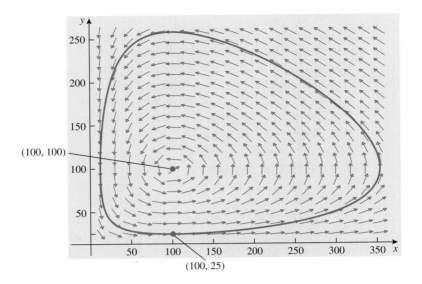

Figure 4.9

Trajectory for $\left\{ \dfrac{dx}{dt} = 0.2\,x - 0.002\,xy, \dfrac{dy}{dt} = -0.1\,y + 0.001\,xy; \quad x(0) = 100, y(0) = 25 \right\}$

$$0 \le t \le 52$$

to look at partial trajectories such as those given by $0 \le t \le 10$, $0 \le t \le 15$, or $0 \le t \le 25$. Figure 4.9 illustrates a cyclic behavior that seems a bit too neat to be found in the wild. However, regular population cycles do seem to occur in nature.[8] In our graph, both prey and predator populations increase as the number of prey increases, but when the prey population exceeds about 350, the predators seem to overwhelm their prey to the extent that there are more and more predators but a declining prey population. The predators continue to increase until their number is about 260, at which time the effect of a dwindling food supply catches up to the predators and their population begins to decline. The predators may starve or start killing each other as competition for diminishing resources grows fierce. Finally, the predator population is low enough for the prey population to recover, and the cycle begins again.

Figure 4.9 highlights the point $(100, 100)$ because $x = 100$, $y = 100$ is an equilibrium solution of the system, called a **center** in this case. (Verify the last statement.) If this system were to have initial point $(100, 100)$, neither population would move from this state. The origin is also an equilibrium point.

Other Graphical Representations

With the aid of technology, we can look at plots of $x(t)$ against t and $y(t)$ against t separately (Figures 4.10a and 4.10b). Compare these graphs, noting the way in

---

8. Examination of the records of the Hudson's Bay Company, which trapped fur-bearing animals in Canada for almost 200 years, suggests a periodic pattern in the number of lynx pelts harvested in the period 1821 to 1934. The lynx, a weasel-like predator, has the snowshoe hare as its main prey. For an analysis of the data, see J. D. Murray, *Mathematical Biology*, 2nd rev. ed. (New York: Springer-Verlag, 1993): 66–68.

which one population lags behind the other over time. The trajectory (Figure 4.9) gives the big picture, the state $(x(t), y(t))$ of the ecological system as time marches on, whereas Figures 4.10a and 4.10b show the individual population fluctuations. Figure 4.11 exhibits the cyclic nature of the predator fluctuation and that of the prey fluctuation on the same set of axes. Each graph in this example was done by a CAS using a numerical approximation to the actual system solution.

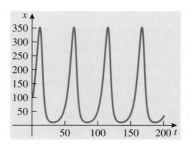

Figure 4.10a
$x(t)$, Prey Population
$x(0) = 100; 0 \leq t \leq 200$

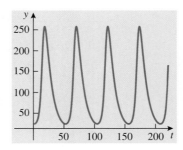

Figure 4.10b
$y(t)$, Predator Population
$y(0) = 25; 0 \leq t \leq 220$

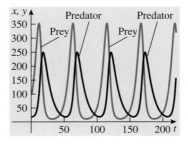

Figure 4.11
Predator and prey populations vs. t

SPRING-MASS PROBLEMS

Simple Harmonic Motion

To start with, suppose we have a spring attached to the ceiling and a weight (mass) hanging from the bottom of the spring, as in Figure 4.12a.

If we set the mass in motion by giving it an upward or downward push, we can use Newtonian mechanics and the qualitative analysis of systems of ODEs to investigate the forces acting on the mass during its motion. We want to describe the state of this system, giving the mass's position and velocity at any time *t*. First we'll assume that there's no air resistance, friction, or other impeding force. The resulting situation is called **simple harmonic motion** or **free undamped motion.**

Fundamental to understanding the mass's movement is **Newton's Second Law of Motion,** which can be stated as $F = m \cdot a$, where F is a force (or sum of forces) acting on a body (such as the weight hanging from the spring), m is the body's mass, and a is the acceleration of the body. If x denotes the displacement (distance) of the mass from its equilibrium (rest) position, where a move downward is considered a *positive* displacement (Figure 4.12b), we can write this expression for the force as $m \cdot \dfrac{d^2x}{dt^2}$.

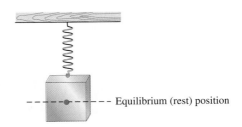

Figure 4.12a

Spring-mass system, mass in the equilibrium position

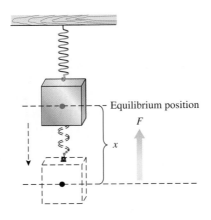

Figure 4.12b

Spring-mass system, mass displaced from the equilibrium position

Now note that if you pull *down* on the weight (stretching the spring in the process), you can feel a certain tension—a tendency for the spring to pull the weight back *up*. Similarly, if you push *up* on the weight, thereby compressing the spring, you feel a force that tends to push the weight *down*. This behavior is described by **Hooke's Law:** The force *F* (called the restoring force) exerted by a spring, tending to restore the weight to the equilibrium position, is proportional to the distance *x* of the weight from the equilibrium position. Stated simply, *force is proportional to stretch.* Mathematically, we write $F = -kx$, where *k* is a positive constant called the *spring constant*. Note that if *x* is *positive*, then the restoring force is *negative*, whereas if *x* is *negative*, then *F* is *positive*.

Because we are ignoring any other kind of force acting on the weight, we can equate the two expressions for the force to get

$$m \cdot \frac{d^2x}{dt^2} = -kx,$$

which we can write in the form

$$\frac{d^2x}{dt^2} + \beta x = 0, \quad \text{where } \beta = \frac{k}{m}. \tag{4.5.4}$$

We saw this kind of homogeneous second-order linear equation in Section 4.1; and from our work in Section 4.4 we know how to convert this equation into an equivalent system of first-order equations. Earlier in this section, we learned how to understand what a phase portrait is telling us. Now let's analyze this problem qualitatively.

EXAMPLE 4.5.5   A Spring-Mass System—Simple Harmonic Motion

Given the equation $\frac{d^2x}{dt^2} + \beta x = 0$, where $\beta = \frac{k}{m}$, we let $x_1 = x$ and $x_2 = \dot{x}$. We see that $\dot{x}_1 = \dot{x} = x_2$ and $\dot{x}_2 = \ddot{x} = -\beta x$ (by solving the second-order equation for the second derivative) $= -\beta x_1$, so we have the two-dimensional system

$$\begin{aligned} \dot{x}_1 &= x_2 \\ \dot{x}_2 &= -\beta x_1. \end{aligned} \tag{4.5.5}$$

First of all, note that x_2 represents the *velocity* of the mass: $x_2 = \dot{x}$, the rate of change of the position, or displacement of the mass. Using the language developed in Example 4.5.1, we say that if we could solve the system (4.5.5) for x_1 and x_2, then the ordered pair $(x_1(t), x_2(t))$, consisting of the mass's current position and velocity, would give the *state* of the system at time *t*.

Now we can look at some trajectories in the phase plane of (4.5.5)—that is, some solution curves in the x_1-x_2 plane. Using initial points $(x_1(0), x_2(0)) = (1,0),(0,1)$, and $(2, 0)$, Figure 4.13 shows what these curves look like when $\beta = \frac{2}{5}$ and we take the interval $0 \le t \le 10$. You should use technology to plot your own trajectories, with different initial points and smaller ranges for *t*.

Analysis

Is this the behavior you would expect from a bouncing mass? First of all, note that the origin is a special point, an equilibrium solution, because both equations of our system vanish at $(x_1, x_2) = (0, 0)$. Physically, this means that a mass-spring

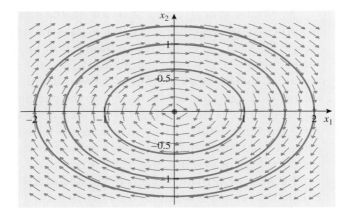

Figure 4.13
Trajectories for $\{\dot{x}_1 = x_2, \dot{x}_2 = -\frac{2}{5}x_1\}$, initial points $(1, 0)$, $(0, 1)$, $(2, 0)$; $0 \le t \le 10$

system that starts at its equilibrium position ($x(0) = 0$) and has no initial push or pull ($\dot{x}(0) = x_2(0) = 0$) will remain at rest forever, which makes sense.

Now look closely at a typical **closed orbit,** as one of these elliptical trajectories is called. Assume that $x_1 = 0$ and x_2 is positive—that is, the mass is at its equilibrium position and is given an initial tug downward. When the mass is at rest ($x_1 = 0$) and it is pushed or pulled in a downward direction ($dx_1/dt = x_2 > 0$), the flow moves in a clockwise direction (note the direction of the slope field arrows), with x_2 decreasing and x_1 increasing until the trajectory is at the x_1-axis. Physically, this means that the mass moves downward until the spring reaches its maximum extension (x_1 is at its most positive value), depending on how much force was applied initially to pull the mass downward, at which time the mass has lost all its initial velocity (that is, $x_2 = 0$). Then the energy stored in the spring serves to pull the mass back up toward its equilibrium position, so that x_1 is decreasing at the same time that the velocity x_2 is increasing—*but in a negative direction* (*upward*). Graphically, this is taking place in the fourth quadrant of the phase plane. When the flow has reached the state $(0, x_2)$, where x_2 is negative, the mass has reached its original position and has attained its maximum velocity upward.

As the trajectory takes us into the third quadrant, the mass is overshooting its original position but is slowing down: $x_1 < 0$ and $x_2 < 0$. When the trajectory has reached the point $(x_1, 0)$, where x_1 is negative, the spring is most compressed and the mass is (for an instant) not moving.

As the trajectory moves through the second quadrant, the mass is headed back toward its initial position with increasing velocity in a downward (positive) direction: $x_1 < 0$ and $x_2 > 0$. Finally, the mass reaches its initial position with its initial velocity in the positive (downward) direction—$x_1 = 0$, $x_2 > 0$—and the cycle begins all over again.

This analysis seems to say that the mass will never stop, bobbing up and down forever. This apparently nonsensical conclusion is perfectly reasonable when you realize that a real mass-spring system is always subject to some air resistance and some sort of friction that slows the system down and eventually forces the mass to stop moving. Our analysis assumes no such impeding force, so the conclusion is rational, even though the assumption is unrealistic.

Another View—Solution Curves

As we did in Examples 4.5.1 and 4.5.3, we can use technology to plot each solution of our system against t. Figures 4.14a and 4.14b show the solution with $\beta = \dfrac{2}{5}$, $x_1(0) = 1$, and $x_2(0) = 1$, corresponding to a spring-mass system that starts 1 unit below its equilibrium position and has been given an initial velocity of 1 in a downward direction. We should not be surprised at the appearance of these solution curves. The closed orbits in Figure 4.13 reflect the periodic nature of the motion of the mass. Such motions are called **oscillations.** Using methods that we saw in Section 4.1, we can determine that when $\beta = \frac{2}{5}$, the general solution of system (4.5.5) is

$$x_1(t) = \frac{C_1}{2}\sqrt{10}\sin\left(\frac{1}{5}\sqrt{10}t\right) + C_2\cos\left(\frac{1}{5}\sqrt{10}t\right)$$
$$x_2(t) = C_1\cos\left(\frac{1}{5}\sqrt{10}t\right) - \frac{C_2}{5}\sqrt{10}\sin\left(\frac{1}{5}\sqrt{10}t\right),$$

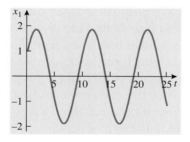

Figure 4.14a
$x_1(t)$, displacement
$x_1(0) = 1, 0 \le t \le 25$

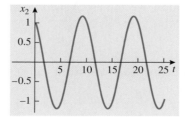

Figure 4.14b
$x_2(t)$, velocity
$x_2(0) = 1, 0 \le t \le 25$

and we can see that the explicit source of the oscillations is the trigonometric terms. The particular system solution shown in Figures 4.14a and 4.14b corresponds to the initial conditions $x_1(0) = 1$, $x_2(0) = 1$, so $C_1 = C_2 = 1$. (*Verify this.*)

Remembering the discussion of the *equivalence* of a second-order equation and a system in Example 4.4.1, we realize that

$$x_1(t) = \frac{C_1}{2}\sqrt{10}\,\sin\!\left(\frac{1}{5}\sqrt{10}\,t\right) + C_2 \cos\!\left(\frac{1}{5}\sqrt{10}\,t\right)$$

is the general solution of the original single differential equation $\dfrac{d^2x}{dt^2} + \beta x = 0$,

where $\beta = \dfrac{2}{5}$. (Review Section 4.1. It happens that $x_2(t) = dx_1/dt$ is also a solution,

but this is true only because the equation is *homogeneous*.)   ◆

Free Damped Motion

Now let's look at a more realistic version of a spring-mass system. This time we'll assume the existence of a combination of air resistance and some friction in the spring-mass system, called a **damping force,** to slow the mass down. To dramatize the situation, you may think of the mass as being immersed in a bucket of water, oil, or maple syrup, so that any initial force imparted to the mass is opposed by a force in the opposite direction as the mass meets resistance. The motion that results is called **free damped motion.** For instance, the damping produced by automobile shock absorbers provides a more comfortable ride.

The damping force works *against* the motion of the mass, so when the mass is moving *down* (the positive direction), the damping force acts in an *upward* direction, and when the mass is moving *up* (the negative direction), the damping force acts in a *downward* direction. In algebraic terms, this damping force's sign must be opposite to the sign of the direction of the velocity. For small velocities, experiments have shown that the damping force is proportional to the velocity of the

mass. We can express the last two sentences mathematically as $F = -\alpha\dfrac{dx}{dt}$, where

α is a positive constant of proportionality called the **damping constant.** Realizing that both the spring's restoring force and this damping force are opposed to the mass's motion, we can use *Newton's Second Law of Motion* to derive the equation

$$m \cdot \frac{d^2x}{dt^2} = -\alpha\frac{dx}{dt} - kx,$$

which we can write in the form

$$\frac{d^2x}{dt^2} + b\frac{dx}{dt} + cx = 0, \quad \text{where } b = \frac{\alpha}{m} \text{ and } c = \frac{k}{m}. \tag{4.5.6}$$

Now we can convert this second-order differential equation into a system and analyze our problem qualitatively.

EXAMPLE 4.5.6 **A Spring-Mass System—Free Damped Motion**

The second-order linear equation $\dfrac{d^2x}{dt^2} + b\dfrac{dx}{dt} + cx = 0$ is equivalent to the two-dimensional system

$$\frac{dx_1}{dt} = x_2$$

$$\frac{dx_2}{dt} = -bx_2 - cx_1. \tag{4.5.7}$$

Phase Portrait Analysis

To understand the motion of the mass, we'll look first at the trajectory we get when we take $b = \frac{1}{4}$, $c = 2$, $x_1(0) = 1$, and $x_2(0) = 0$ (Figure 4.15). In particular, you should see that the mass starts off 1 unit *below* its equilibrium position with *no* initial velocity in any direction.

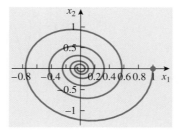

Figure 4.15 Damped Free Motion

Trajectory for the system

$$\left\{ \frac{dx_1}{dt} = x_2, \frac{dx_2}{dt} = -\tfrac{1}{4}x_2 - 2x_1; x_1(0) = 1, x_2(0) = 0 \right\}$$

$$0 \le t \le 25$$

The direction of the trajectory in Figure 4.15 indicates very dramatically that the state of the system is spiraling in to the origin—that is, $x_1(t) \to 0$ and $x_2(t) \to 0$ as $t \to \infty$. Every time the spiral trajectory in Figure 4.15 crosses the x_2-axis (so that $x_1 = 0$), the mass is at its equilibrium position—on its way up when the velocity x_2 is negative and on its way down when x_2 is positive. (Remember our agreement on which direction is positive and which direction is negative.) This type of spiral clearly indicates why we can say that the origin is a **sink** for the system.

Another View

We can also look at the graphs of $x_1(t)$ against t (Figure 4.16a) and $x_2(t)$ against t (Figure 4.16b) for the same system. The oscillations shown in Figures 4.16a and 4.16b reflect the behavior of the system in a different way.

The mass reaches its equilibrium position when the $x_1(t)$ curve crosses the t-axis. If $x_1(t^*) = 0$, then look at Figure 4.16b to see what the value of $x_2(t^*)$ is. If $x_2(t^*) > 0$, for example, the mass is on its way *down*. Also note how Figures 4.15, 4.16a, and 4.16b show that the successive rises and falls get progressively smaller.

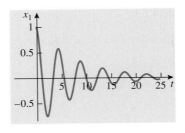

Figure 4.16a

$x_1(t)$, displacement

$x_1(0) = 1$

$0 \le t \le 25$

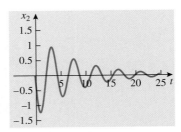

Figure 4.16b

$x_2(t)$, velocity

$x_2(0) = 0$

$0 \le t \le 25$

The figures all reflect the initial conditions and seem to say that the mass eventually comes to rest at its equilibrium position. If you were to hit a brass gong with a special ceremonial hammer, the vibrations would be loud at the beginning but would gradually fade to nothing. This is roughly what we are seeing here.

A Look at the Actual Solution

The curves in Figures 4.16a and 4.16b are *not* periodic, despite their resemblance to familiar trigonometric curves that are. In Section 4.1 we saw how to determine that the solution of our IVP $\dfrac{d^2x}{dt^2} + \dfrac{1}{4}\dfrac{dx}{dt} + 2x = 0$, with $x(0) = 1$ and $\dfrac{dx}{dt}(0) = 0$, is given by

$$x(t) = e^{(-\frac{1}{8}t)}\left(\cos\left(\frac{1}{8}\sqrt{127}\,t\right) + \frac{\sqrt{127}}{127}\,\sin\left(\frac{1}{8}\sqrt{127}\,t\right)\right),$$

which is not a pure trigonometric function because of the exponential factor. You should verify that this is a solution. In terms of our system (4.5.7), we have $x_1(t) = x(t)$ and $x_2(t) = \dfrac{dx}{dt}$. (Do the differentiation to see what $x_2(t)$ looks like.)

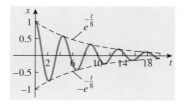

Figure 4.17

The graphs of $x(t)$, $e^{(-\frac{1}{8}t)}$, and $-e^{(-\frac{1}{8}t)}$

The exponential factor $e^{(-\frac{1}{8}t)}$, called the **time-varying amplitude,** forces the decay of the oscillations indicated by the trigonometric terms. Figure 4.17 shows the graph of the solution $x(t)$, together with the graphs of $e^{(-\frac{1}{8}t)}$ and $-e^{(-\frac{1}{8}t)}$. ◆

Different Kinds of Damping

You should be aware that there are different kinds of damped motion. The behavior of a damped system described by the equation $m\dfrac{d^2x}{dt^2} + \alpha\dfrac{dx}{dt} + kx = 0$ depends on the relationship among the three constants m, α, and k—the mass, the damping coefficient, and the spring constant, respectively. The example we've just analyzed is a case of **underdamped motion,** occurring when the damping coefficient is relatively small compared to the other constants: $\alpha^2 < 4mk$, technically. The other two possibilities, **overdamped motion** ($\alpha^2 > 4mk$) and **critically damped motion** ($\alpha^2 = 4mk$), are explored in Problems 9 and 10 in Exercises 4.5. We'll give a detailed explanation of the significance of the relationship among m, α, and k in Chapter 5.

Forced Motion

Sometimes a physical system is subject to external forces, which must appear in its mathematical representation. For example, the motion of an automobile (whose body-suspension combination can be considered a spring-mass system) is influenced by irregularities in the road surface. Similarly, a tall building may be subjected to strong winds that will cause it to sway in an uncharacteristic way.

We're going to look at an initial-value problem related to Example 2.2.5 and to Problems 25–27 in Exercises 2.2. This discussion will involve an important type of second-order linear equation with a forcing term.

EXAMPLE 4.5.7 Forced Damped Motion

Suppose we have an electrical circuit with an inductance of 0.5 henry, a resistance of 6 ohms, a capacitance of 0.02 farad, and a generator providing alternating voltage given by 24 sin (10 t) for $t \geq 0$. The alternating voltage is the external force applied to the circuit, and the resistance is a damping coefficient. Then, letting Q

denote the instantaneous charge on the capacitor, *Kirchhoff's Law* gives us the equation

$$0.5\frac{d^2Q}{dt^2} + 6\frac{dQ}{dt} + 50Q = 24\sin 10t,$$

or

$$\frac{d^2Q}{dt^2} + 12\frac{dQ}{dt} + 100Q = 48\sin 10t.$$

Let's assume that $Q(0) = 0$ and $\frac{dQ}{dt}(0) = 0$.

This second-order nonhomogeneous equation is equivalent to the non-autonomous system

$$\frac{dx_1}{dt} = x_2$$

$$\frac{dx_2}{dt} = 48\sin 10t - 12x_2 - 100x_1,$$

with initial conditions $x_1(0) = 0$ and $x_2(0) = 0$. (*You should work this out for yourself.*)

Phase Portrait
The phase portrait (Figure 4.18a) corresponding to this system, for $0 \le t \le 0.94$, is interesting. At first, we suspect that we may get a spiral opening outward. But with an expanded range for t—say, from 0 to 5—the phase portrait resembles a closed orbit around the origin (Figure 4.18b). We can understand the initial "blip" by using the explicit solution found by the techniques discussed in Section 4.2:

$$Q(t) = \frac{1}{10}e^{-6t}(4\cos 8t + 3\sin 8t) - \frac{2}{5}\cos 10t.$$

As in Example 2.2.5, we see that there is a *transient term*, $\frac{1}{10}e^{-6t}(4\cos 8t + 3\sin 8t)$, that becomes negligible as t grows large (*Why?*), and a

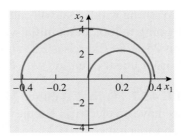

Figure 4.18a

Trajectory for the system

$$\left\{\frac{dx_1}{dt} = x_2, \frac{dx_2}{dt} = 48\sin t - 12x_2 - 100x_1; \quad x_1(0) = 0 = x_2(0)\right\}$$

$$0 \le t \le 0.94$$

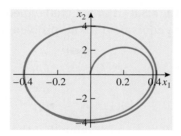

Figure 4.18b
Trajectory for the system

$$\left\{\frac{dx_1}{dt} = x_2, \frac{dx_2}{dt} = 48\sin t - 12x_2 - 100x_1; \quad x_1(0) = 0 = x_2\right\}$$

$$0 \le t \le 5$$

steady-state term, $\frac{2}{5}\cos 10t$, that controls the behavior of $Q(t)$ $(= x_1)$ eventually. This steady-state term is periodic with the same period $\left(\dfrac{2\pi}{10} = \dfrac{\pi}{5}\right)$ as the forcing term and has the amplitude $\frac{2}{5}$. The *current* in the circuit is given by $I = \dfrac{dQ}{dt} = x_2$. ◆

Let's look at one other example of a spring-mass system. First we suppose that there is no air resistance or friction. Next we assume that the spring to which the mass is attached is supported by a board. Now we set the mass into motion by moving the supporting board up and down in a periodic manner. This situation is described as **driven undamped motion** or **forced undamped motion.** As in the last example, a force external to the spring-mass system itself is being applied to the system, and we want to understand the behavior of the system.

When we apply Newton's Second Law of Motion, an analysis similar to that provided in Example 4.5.5 gives us the equation

$$m \cdot \frac{d^2x}{dt^2} = -kx + f(t),$$

which we can write as

$$\frac{d^2x}{dt^2} + \beta x = F(t) \text{ where } \beta = \frac{k}{m} \text{ and } F(t) = \frac{f(t)}{m}. \tag{4.5.8}$$

The **forcing function** $f(t)$ (or $F(t)$) describes the external force that jiggles the supporting board up and down rhythmically. Remember that we are assuming that this force is *periodic*, so $f(t)$ is sometimes positive and sometimes negative— that is, sometimes the board is moved downward and sometimes it is moved upward. (Did you ever see the toy consisting of a paddle with a rubber ball attached to it by a rubber cord?)

The next example gives us the qualitative analysis of this problem.

EXAMPLE 4.5.8   **Forced Undamped Motion**
The system equivalent to our problem is

$$\frac{dx_1}{dt} = x_2$$

$$\frac{dx_2}{dt} = F(t) - \beta x_1. \tag{4.5.9}$$

Let's take $\beta = 4$ and assume that the forcing function is $F(t) = \cos(2t)$. Furthermore, let's assume that the mass starts from its equilibrium position, $x_1(0) = x(0) = 0$, and that it has no initial motion before the external force is applied—that is, $x_2(0) = \dfrac{dx}{dt}(0) = 0$. Figure 4.19 shows the phase portrait corresponding to this IVP for $0 \le t \le 20$.

Analysis
Note that because the initial point is the origin, it is obvious that the spiral trajectory is moving *outward*—that is, in a clockwise direction. (You should contrast this with Figure 4.15 in Example 4.5.6.) Figure 4.19 indicates that both the displacement of the mass and its velocity are growing without bound. The graphs of $x_1(t)$ and $x_2(t)$ against t (Figures 4.20a–b) confirm this.

The Actual Solution
The solution of the system we have chosen as an example is $x_1(t) = x(t) = \dfrac{t}{4}\sin(2t)$. (Check that this is a solution of the IVP.) The sine term contributes an oscillation between -1 and 1, but the factor $\dfrac{t}{4}$ affects the *amplitude* of the oscillations: $|x(t)| = \left|\dfrac{t}{4}\right||\sin(2t)|$, so that $-\dfrac{t}{4} \le x(t) \le \dfrac{t}{4}$ for $t \ge 0$, and $x(t)$ gets larger and larger in both the positive and negative directions as t gets larger.

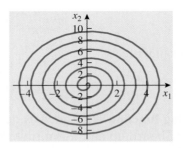

Figure 4.19
Trajectory for the system
$$\left\{\frac{dx_1}{dt} = x_2, \frac{dx_2}{dt} = \cos 2t - 4x_1; \quad x_1(0) = 0 = x_2(0)\right\}$$
$$0 \le t \le 20$$

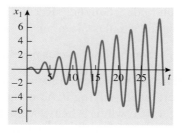

Figure 4.20a

$x_1(t)$, displacement

$x_1(t)$ in Example 4.5.8, $0 \le t \le 30$

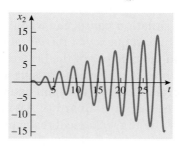

Figure 4.20b

$x_2(t)$, velocity

$x_2(t)$ in Example 4.5.8, $0 \le t \le 30$

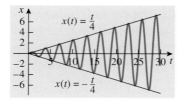

Figure 4.21

$x_1(t) = x(t) = \dfrac{t}{4} \sin(2t), 0 \le t \le 30$

Figure 4.21 shows how the linear factor $\dfrac{t}{4}$ magnifies the oscillation caused by the trigonometric factor. ◆

Resonance

A situation in which we have unbounded oscillation, as shown in the last example, is called **resonance**. This is particularly important because all mechanical systems have **natural** or **characteristic frequencies**—that is, each atom making up the system is vibrating at a particular frequency, and the composite system has its own characteristic frequency. Recall that if a function g is periodic with period T (so that T is the smallest number for which $g(t + T) = g(t)$ for all t), then its **frequency** f is the number of cycles per unit of time: $f = \dfrac{1}{T}$. Resonance occurs when the frequency of an external force coincides with the natural frequency of the system, thereby amplifying it. You may have experienced having the windows in your home rattle when a heavy vehicle drives by. Going faster than a certain speed in a car may cause a disturbing rattling. In our last example, the natural frequency of the system is $\dfrac{1}{\pi}$ cycles per unit of time, which is equal to the frequency of the forcing function $F(t) = \cos(2t)$.

An unfortunately frequent physical consequence of such amplified vibration is the destruction of the system. In a spring-mass system, the spring could break. A serious situation can occur when numbers of people march in step over a bridge and the frequency of the vibrations set up by the marching feet causes resonance and the collapse of the bridge. (This is why military columns and parade marchers "break step" when crossing a bridge.) As another example, in 1959 and 1960, several models of the same plane crashed, seeming to explode in midair. The Civil Aeronautics Board (CAB) determined that the disintegration of the planes was due to mechanical resonance: A component within the planes, when not fastened securely, generated oscillations that acted as an excessive external force on the wings, breaking them within 30 seconds.[9] Similarly, resonance occurs when the ocean's waves hit a human-made barrier or when wind swirls around a bridge or tower.

A less disastrous example of resonance is the shattering of a glass by a powerful singer hitting a very high note. The external force here is the sound wave that amplifies the natural frequency of the glass.

It should be pointed out, however, that resonance can also be our friend. The great scientist Galileo (1564–1642) made the following observation about resonance used in the ringing of heavy, free-swinging bells in a tower:

> Even as a boy, I observed that one man alone by giving these impulses at the right instant was able to ring a bell so large that when four, or even six, men seized the rope and tried to stop it they were lifted from the ground, all of them together being unable to counterbalance the momentum which a single man, by properly timed pulls, had given it.[10]

A parent pushing a child's swing, timing the pushes to coincide with the swing's motion, is using resonance to increase the amplitude of each swing. A motorist rocking his or her car to get it out of a muddy rut or a snow bank is applying an external force to amplify the car's natural frequency.

---

9. For examples of resonance, see Alice B. Dickinson, *Differential Equations: Theory and Use in Time and Motion* (Reading, Mass.: Addison-Wesley, 1972): 100 ff.
10. Galileo Galilei, *Dialogues Concerning Two New Sciences*, translated by H. Crew and A. DeSalvio (New York: Macmillan, 1914), "First Day," 98.

THREE-DIMENSIONAL SYSTEMS

We have been focusing on second-order equations and their equivalent systems, but the techniques we have discussed apply to any differential equation of order greater than 1. The main difficulty with equations of order 3 and higher is that we lose some aspects of the graphical interpretation of the solution. Let's look at the next example, which presents us with a three-dimensional system.

EXAMPLE 4.5.9   A System of Three First-Order Equations
We want to examine the behavior of the three-dimensional system

$$\dot{x} = -0.1x - y$$
$$\dot{y} = x - 0.1y$$
$$\dot{z} = -0.2z.$$

A Three-Dimensional Trajectory
The complete picture of this linear system is given by the set of points $(t, x(t), y(t), z(t))$, a *four-dimensional* situation. Assuming that x, y, and z are functions of the parameter t and that we have the initial condition $x(0) = 5$, $y(0) = 5$, and $z(0) = 10$, we get the three-dimensional trajectory shown in Figure 4.22. This is a *projection* of the four-dimensional picture onto three-dimensional space, just as the two-dimensional phase portraits we've seen previously are projections of three-dimensional curves onto two-dimensional planes.

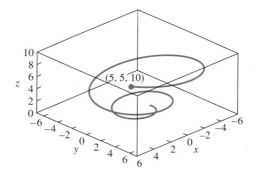

Figure 4.22
An x-y-z plane trajectory for the system
$\{\dot{x} = -0.1x - y, \dot{y} = x - 0.1y, \dot{z} = -0.2z; x(0) = 5, y(0) = 5, z(0) = 10\}$
$0 \leq t \leq 15$

By plotting this with your CAS and rotating the axes (if possible), you should be able to see that the solution spirals into the origin in the x-y plane, while it moves toward the origin in the variable z as well.

A Two-Dimensional Trajectory
Now we can, for example, project the three-dimensional spiral (in x-y-z space) onto the x-y plane (Figure 4.23).

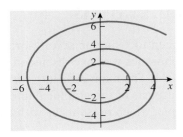

Figure 4.23
An x-y plane trajectory for the system
$$\{\dot{x} = -0.1x - y, \dot{y} = x - 0.1y, \dot{z} = -0.2z; \quad x(0) = 5, y(0) = 5, z(0) = 10\}$$
$$0 \leq t \leq 15$$

If you increase the range of t, you will get a tighter spiral and see that the origin is a *sink* for this system.   ◆

In this section, we have seen how any differential equation of order greater than 1 can be turned into an equivalent system of first-order equations. We've looked at different ways to view such systems graphically. In the next sections, we will discuss other aspects of first-order equations that can be extended to systems. In particular, we'll investigate questions of the existence and uniqueness of solutions and the numerical approximation of solutions of systems.

EXERCISES 4.5

Assume that each function in Exercises 1–6 is a function of time, t. For each of these initial-value problems, (a) convert to a system, (b) use technology to find the graph of the solution in the phase plane, and (c) show a graph of the two components of the solution relative to the t-axis.

1. $x'' + x' = 0; \quad x(0) = 1, x'(0) = 2$ **2.** $\ddot{r} - r = 0; \quad r(0) = 0, \dot{r}(0) = 1$

3. $\ddot{y} + y = 0; \quad y(0) = 2, \dot{y}(0) = 0$ **4.** $y'' = -4; \quad y(0) = y'(0) = 0$

5. $\ddot{x} - \dot{x} = 0; \quad x(0) = 1, \dot{x}(0) = 1$

6. $x'' - 2x' + x = 0; \quad x(0) = -1, x'(0) = -1$

7. Read the explanation before Problems 15–18 of Exercises 2.1 and solve the equation $\dfrac{dy}{dx} = \dfrac{-x + y}{x + y}$ that arises in Example 4.5.2.

8. Consider the specific Lotka-Volterra equations (4.5.3) in Example 4.5.4.

   a. Find the *first-order* differential equation that defines the *trajectories* of this system in the phase plane.

   b. Solve this separable equation to find the implicit algebraic equation of the trajectories.

9. The IVP $\ddot{x} + 20\dot{x} + 64x = 0$, with $x(0) = \frac{1}{3}$ and $\dot{x}(0) = 0$, models the motion of a spring-mass system with a damping force. The initial conditions indicate that the mass has been pulled below its equilibrium position and released.

   a. Express this IVP as a system of two first-order equations, with the appropriate initial conditions.

   b. Use technology to graph the solution of the system in the phase plane.

   c. Use technology to graph the solution of the original second-order equation relative to the *t*-axis.

   d. Comparing the results of parts (b) and (c) to the appropriate graphs in Examples 4.5.5 and 4.5.6, why do you think that the motion shown in this exercise should be called *overdamped*?

10. Consider the spring-mass system modeled by the IVP $\ddot{x} + c\dot{x} + 0.25x = 0$, with $x(0) = \dfrac{1}{2}$ and $\dot{x}(0) = \dfrac{7}{4}$. Here c is a positive parameter.

    a. Express the IVP in terms of a system of first-order equations, including initial conditions.

    b. For each of the values $c = 0.5$, 1, and 1.5, use technology to graph the solution of the system in the phase plane, $0 \leq t \leq 20$.

    c. For each of the values $c = 0.5$, 1, and 1.5, use technology to graph the solution of the original equation with respect to t on the interval $0 \leq t \leq 20$.

    d. On the basis of your answers to parts (b) and (c), describe how the nature of the solution changes as the value of c passes through the value 1. (When $c = 1$, the system is *critically damped*.)

11. Consider the following model of a spring-mass system: $\ddot{x} + 64x = 16\cos 8t$, with $x(0) = 0$ and $\dot{x}(0) = 0$.

    a. Express the IVP in terms of a system of first-order equations, including initial conditions.

    b. Use technology to graph the solution of the system in the phase plane.

    c. Use technology to graph the solution of the original second-order equation relative to the *t*-axis.

    d. What is the relationship of the graph in part (c) to the two half-lines $x = t$ and $x = -t$ for $t \geq 0$?

12. The equation $\ddot{Q} + 9\dot{Q} + 14Q = \frac{1}{2}\sin t$ models an electrical circuit with resistance 180 ohms, capacitance $\frac{1}{280}$ farad, inductance 20 henrys, and an applied voltage given by $E(t) = 10\sin t$. $Q = Q(t)$ denotes the *capacitance*, the charge on the capacitor at time t, and $\dot{Q}(t)$ denotes the current in the circuit. Assume that $Q(0) = 0$ and $\dot{Q}(0) = 0.1$.

    a. Express this IVP as a system of two first-order equations, with the appropriate initial conditions.

    b. Use technology to graph the solution of the system in the phase plane, with $0 \leq t \leq 8$.

    c. Use technology to graph the solution of the original second-order equation relative to the *t*-axis, considering first the interval $0 \leq t \leq 2$ and then $0 \leq t \leq 8$.

    d. Describe the behavior of the capacitance as $t \to \infty$.

13. Convert each of the following systems to a single second-order equation. Then interpret each equation to determine which (if any) *cannot* represent a spring-mass system. Explain your reasoning.

    a. $Q' = -6Q + 3R$
        $R' = -Q - 2R$

    b. $\dot{x} = 3x - y$
        $\dot{y} = x + 3y$

14. Find all equilibrium solutions of each of the following systems.

    a. $\dot{x} = x - 3y, \dot{y} = 3x + y$
    b. $x' = 2x + 4y, y' = 3x + 6y$

    c. $\dot{r} = -2rs + 1, \dot{s} = 2rs - 3s$
    d. $x' = \cos y, y' = \sin x - 1$

    e. $\dot{x} = x - y^2, \dot{y} = x^2 - y$
    f. $r' = 1 - s, s' = r^3 + s$

15. Convert the equation $x'' + x' - x + x^3 = 0$ to a system and find all equilibrium points.

16. The equation $\dfrac{d^2\theta}{dt^2} + k^2 \sin \theta = 0$ describes the motion of an *undamped pendulum*, where θ is the angle the pendulum makes with the vertical. Convert this equation to a system and describe all its equilibrium points.

4.6   EXISTENCE AND UNIQUENESS

Now that we've learned how to convert higher-order equations to equivalent systems of first-order equations and we've seen some qualitative analyses of these systems, it's time to ask that important question we first considered in Section 2.7 in the context of first-order equations: How do we know that a given higher-order equation or equivalent system *has* a solution—and do we know that any such solution is *unique*?

We don't want to waste human and computer resources searching for a solution that may not exist or that may merely be one of many solutions. For now we'll focus on second-order equations and their corresponding systems. In Chapter 5 we'll look at generalizations to higher-order equations and larger systems.

The first example shows that when there is one solution of a system, there may be many.

EXAMPLE 4.6.1   A System IVP with Many Solutions

Let's look at the initial-value problem

$$t^2 x'' - 2tx' + 2x = 0, \quad \text{with } x(0) = 0 \text{ and } x'(0) = 0.$$

This is equivalent to the system IVP

$$x_1' = x_2$$
$$x_2' = \frac{2}{t} x_2 - \frac{2}{t^2} x_1, \quad \text{with } x_1(0) = 0 = x_2(0).$$

Then $x(t) \equiv 0$ and any function of the form $x(t) = Kt^2$ (where K is any constant) are solutions of the original IVP. (*Verify this.*) With respect to the equivalent system of equations, $x_1(t) \equiv 0$, $x_2(t) \equiv 0$ is a solution, and any pair of functions $x_1(t) = Kt^2$, $x_2(t) = 2Kt$ is a solution. What we are saying here is that *our IVP has infinitely many solutions.* ◆

In contrast to the IVP in the last example, we can have a system of differential equations with *no* solution.

EXAMPLE 4.6.2 **A System IVP with No Solution**
Let's look at the IVP

$$x_1' = \frac{1}{x_1^2}, \ x_2' = 2x_1 - x_2, \quad \text{with } x_1(0) = 0 \text{ and } x_2(0) = 1.$$

When we examine this situation carefully, we see that if $x_1(t)$ is part of a solution pair for this IVP, then x_1' doesn't exist for $t = 0$ because $x_1'(0) = \dfrac{1}{[x_1(0)]^2}$ and $x_1(0) = 0$. This says that there is no solution to this IVP. ◆

What we want in most real-life situations is one and only one solution to an initial-value problem. The next example shows such a case.

EXAMPLE 4.6.3 **A System IVP with a Unique Solution**
The IVP $\left\{ \dfrac{dx}{dt} = y, \dfrac{dy}{dt} = x; \quad x(0) = 1, y(0) = 0 \right\}$ has the *unique* solution $x(t) = \dfrac{1}{2}(e^t + e^{-t})$, $y(t) = \dfrac{1}{2}(e^t - e^{-t})$. You may recognize x and y as the *hyperbolic cosine* (cosh) and *hyperbolic sine* (sinh), respectively.

This system is equivalent to the single equation $\ddot{x} - x = 0$, or $\ddot{x} = x$, with $x(0) = 1$ and $\dot{x}(0) = 0$, and it isn't too difficult to guess what kind of function is equal to its own second derivative. Exercise 1 following this section will ask you to explore this further. ◆

AN EXISTENCE AND UNIQUENESS THEOREM

At this point we have seen that the possibilities for second-order IVPs are similar to those we saw in Section 2.7 for first-order IVPs. We can have *no* solution, *infinitely many solutions*, or *exactly one solution*. Once again we would like to determine when there is one and only one solution of an initial-value problem.

The simplest Existence and Uniqueness Theorem for second-order differential equations or two-dimensional systems of first-order equations is one that is a natural extension of the result we saw in Section 2.7. We'll state two forms of this.

Existence and Uniqueness

Suppose we have a second-order IVP $\dfrac{d^2y}{dt^2} = f(t, y, \dot{y})$, with $y(t_0) = y_0$ and $\dot{y}(t_0) = \dot{y}_0$. If $f, \dfrac{\partial f}{\partial y}$, and $\dfrac{\partial f}{\partial \dot{y}}$ are continuous in a closed box B in three-dimensional space (t-y-$\dot{y}$ space) and the point $(t_0, y_0, \dot{y}_0)$ lies inside B, then the IVP has a unique solution $y(t)$ on some t-interval I containing t_0.

Equivalently,

Existence and Uniqueness

Suppose we have a two-dimensional system of first-order equations

$$\frac{dx_1}{dt} = f(t, x_1, x_2)$$

$$\frac{dx_2}{dt} = g(t, x_1, x_2),$$

where $x_1(t_0) = x_1^0$ and $x_2(t_0) = x_2^0$. If $f, g, \dfrac{\partial f}{\partial x_1}, \dfrac{\partial g}{\partial x_1}, \dfrac{\partial f}{\partial x_2}$, and $\dfrac{\partial g}{\partial x_2}$ are all continuous in a box B in t-x_1-x_2 space containing the point (t_0, x_1^0, x_2^0), then there is an interval I containing t_0 in which there exists a unique solution $x_1 = y_1(t), x_2 = y_2(t)$ of the IVP.

Many Solutions

We can write the equation in Example 4.6.1 in the form $\ddot{x} = f(t, x, \dot{x}) = \dfrac{2t\dot{x} - 2x}{t^2}$, so we see that f doesn't exist in any box in which $t = 0$. Therefore, we should not expect exactly one solution, and, in fact, although there *is* a solution to the IVP with initial conditions $x(0) = 0$ and $\dot{x}(0) = 0$, any such solution is not unique.

No Solution

In Example 4.6.2, we can use the system form of our Existence and Uniqueness Theorem to see that the function $f(t, x_1, x_2) = 1/x_1^2$ does not exist at the point $(t, x_1^0, x_2^0) = (0, 0, 1)$, so once again we are not guaranteed exactly one solution— and, in fact, there is *no* solution of the IVP.

Exactly One Solution

Finally, if we examine the IVP in Example 4.6.3 from either the single-equation or the systems point of view, we should see that in this situation we are guaranteed the existence of one and only one solution of the initial-value problem. (*Check this.*)

The nice thing about these questions is that in most common applied problems, the functions and their derivatives are well-behaved (continuous, etc.), so that we *do* have both existence and uniqueness.

EXERCISES 4.6

1. In Example 4.6.3 you saw an IVP for a system of equations that was equivalent to the single equation IVP $x'' - x = 0$, or $x'' = x$, with $x(0) = 1$ and $x'(0) = 0$. Using the technique of Section 4.1, show that $x(t) = \dfrac{1}{2}(e^t + e^{-t})$ is the solution of the IVP $x'' - x = 0$ with $x(0) = 1$ and $x'(0) = 0$.

2. Verify that each of the following initial-value problems has a solution that is guaranteed unique *everywhere* in three-dimensional space.
   a. $x_1' = x_2, \, x_2' = 3x_1 - 5x_2; \quad x_1(0) = 1, x_2(0) = 0$
   b. $x_1' = x_1^2, \, x_2' = \sin x_1 - x_2^2; \quad x_1(0) = 0, x_2(0) = 0$
   c. $x_1' = x_2^3, \, x_2' = tx_1 - x_2; \quad x_1(0) = 0, x_2(0) = 1$

3. Two students are looking at the equation $x'' + f(t)x = 0$, where $f(t)$ is a given continuous function and $x(0) = 0$. The first student claims that the trivial function $x(t) \equiv 0$ satisfies these conditions and uses the Existence and Uniqueness Theorem to conclude that the zero function is the only solution of the problem for any function $f(t)$. The second student sees that if $f(t) \equiv 1$, then the function $x(t) = \sin t$ satisfies the conditions of the problem. Explain this contradiction.

4. Show that the initial-value problem
$$\{yx' = y - 4t, \, (x - 3)y' = -4x + \sin t; \quad x(0) = 3, y(0) = 0\}$$
has no solution. Does this contradict the "existence" part of the result we have given in this section? Explain.

5. **a.** Show that
$$\{x_1(t) = e^{-t}\sin(3t), \, y_1(t) = e^{-t}\cos(3t)\}$$
and
$$\{x_2(t) = e^{-(t-1)}\sin(3(t-1)), \, y_2(t) = e^{-(t-1)}\cos(3(t-1))\}$$
are solutions of the system
$$\frac{dx}{dt} = -x + 3y$$
$$\frac{dy}{dt} = -3x - y.$$

   b. Use technology to draw the graphs of each of the solutions in part (a) in the *x-y* phase plane.

   c. Explain why the solutions in part (a) don't contradict the "uniqueness" part of the result in this section.

6. Consider the equation

$$5x^2y^{(5)} - (6\sin x)y''' + 2xy'' + \pi x^3 y' + (3x - 5)y = 0.$$

Suppose that $Y(x)$ is a solution of this equation such that $Y(1) = 0$, $Y'(1) = 0$, $Y''(1) = 0$, $Y'''(1) = 0$, $Y^{(4)}(1) = 0$, and $Y^{(5)}(1) = 0$. Why must $Y(x)$ be equal to 0 for *all* values of x?

7. Use technology to plot some trajectories of the nonautonomous system

$$\frac{dx}{dt} = (1 - t)x - ty$$

$$\frac{dy}{dt} = tx + (1 - t)y.$$

Your graph should show some intersecting curves. Does the graph contradict the Existence and Uniqueness Theorem? *Explain.*

4.7   NUMERICAL SOLUTIONS

The difficulty of finding closed-form solutions of single differential equations is compounded when it comes to systems of equations. We've already seen some useful ways in which systems are analyzed qualitatively. However, you should realize that a graphing calculator or computer produces phase portraits by using numerical methods. As for any computer graph, individual points are calculated and then connected by a series of small line segments that give the impression of a continuous curve.

Now it is time to see that any of the numerical techniques introduced for first-order equations in Sections 3.1, 3.3, and 3.4 can be extended to *systems* of first-order equations in a natural way. In this section, we'll work with two-dimensional systems, leaving the obvious generalizations to Chapters 5 and 7. Even though it is important to be able to solve simple numerical problems by hand, most systems of differential equations are solved using numerical methods implemented on computers.

EULER'S METHOD APPLIED TO SYSTEMS

Let's start by recalling *Euler's method* for solving the first-order initial-value problem $y' = f(x, y)$, $y'(x_0) = y_0$. This algorithm was originally given as formula (3.1.3):

$$y_{k+1} = y_k + h \cdot f(x_k, y_k).$$

Here h is the step size and y_k denotes the approximate value of the solution at the point $x_k = x_0 + kh$.

Now suppose we have a system of two first-order differential equations

$$\frac{dx}{dt} = f(t, x, y)$$

$$\frac{dy}{dt} = g(t, x, y),$$

with $x(t_0) = x_0$ and $y(t_0) = y_0$. If we let $t_k = t_0 + kh$, $x_k \approx x(t_k)$, and $y_k \approx y(t_k)$, we can apply Euler's algorithm to each equation separately to get the result

$$x_{k+1} = x_k + h \cdot f(t_k, x_k, y_k)$$
$$y_{k+1} = y_k + h \cdot f(t_k, x_k, y_k). \tag{4.7.1}$$

Let's see how this method works on a system we've already seen.

EXAMPLE 4.7.1   Euler's Method for a System—by Hand

As a simple illustration of Euler's method applied to a system, let's approximate the solution of the IVP of Example 4.6.3 at $t = 0.5$. The system, which we know has a unique solution, is $\dfrac{dx}{dt} = y$, $\dfrac{dy}{dt} = x$; $x(0) = 1$, $y(0) = 0$.

Equations

Using a step size $h = 0.1$, the algorithm given by equations (4.7.1) looks like

$$x_{k+1} = x_k + (0.1)y_k$$
$$y_{k+1} = y_k + (0.1)x_k,$$

where $x_0 = x(0) = 1$ and $y_0 = y(0) = 0$.

Calculation

We approximate the solution at $t = 0.5$ by taking five steps:

$$x_1 = x_0 + (0.1)y_0 = 1 + (0.1)(0) = 1$$
$$y_1 = y_0 + (0.1)x_0 = 0 + (0.1)(1) = 0.1$$

$$x_2 = x_1 + (0.1)y_1 = 1 + (0.1)(0.1) = 1.01$$
$$y_2 = y_1 + (0.1)x_1 = 0.1 + (0.1)(1) = 0.2$$

$$x_3 = x_2 + (0.1)y_2 = 1.01 + (0.1)(0.2) = 1.03$$
$$y_3 = y_2 + (0.1)x_2 = 0.2 + (0.1)(1.01) = 0.301$$

$$x_4 = x_3 + (0.1)y_3 = 1.03 + (0.1)(0.301) = 1.0601$$
$$y_4 = y_3 + (0.1)x_3 = 0.301 + (0.1)(1.03) = 0.404$$

$$x_5 = x_4 + (0.1)y_4 = 1.0601 + (0.1)(0.404) = 1.1005$$
$$y_5 = y_4 + (0.1)x_4 = 0.404 + (0.1)(1.0601) = 0.51001$$

Result

These calculations indicate that $x(0.5) \approx 1.1005$ and $y(0.5) \approx 0.5100$. But to four decimal places, the *exact* solution is $x(0.5) = \cosh(0.5) = (\tfrac{1}{2})(\exp(0.5) + \exp(-0.5)) = 1.1276$ and $y(0.5) = \sinh(0.5) = (\tfrac{1}{2})(\exp(0.5) - \exp(-0.5)) = 0.5211$. Thus the absolute error is 0.0271 for x and 0.0111 for y.

    If we cut our step size in half, letting $h = 0.05$ and using technology, we need 10 steps and find that our approximations are $x(0.5) \approx 1.1138$ and $y(0.5) \approx 0.5151$, to four decimal places. Now the error—0.0130 for x and 0.006 for y—is roughly half of what these errors were when $h = 0.1$. Having computer resources at our command, it's hard to resist another run, this time with $h = 0.01$. Taking 50 steps, we have $x(0.5) \approx 1.1248$ and $y(0.5) \approx 0.5198$, with errors 0.0028 and 0.0013 for x

and y, respectively. You should experiment with a few other values of h on your own CAS or graphing calculator.   ◆

Exercise 1 following this section asks you to write the system form of the *improved Euler method* (*Heun's method*).

THE FOURTH-ORDER RUNGE-KUTTA
METHOD FOR SYSTEMS

As an additional example, let's look at the system form of the Runge-Kutta algorithm introduced in Section 3.4. (As we mentioned in that discussion, it was Kutta who generalized the basic method to *systems* of ODEs in 1901.)

We start with the same general first-order system we considered before:

$$\frac{dx}{dt} = f(t, x, y)$$

$$\frac{dy}{dt} = g(t, x, y),$$

with $x(t_0) = x_0$ and $y(t_0) = y_0$. Again let $t_k = t_0 + kh$, $x_k \approx x(t_k)$, and $y_k \approx y(t_k)$. Then the system version of the classic Runge-Kutta formula (3.4.2) is

$$x_{k+1} = x_k + \frac{1}{6}(m_1 + 2m_2 + 2m_3 + m_4)$$

$$y_{k+1} = y_k + \frac{1}{6}(M_1 + 2M_2 + 2M_3 + M_4),$$

where

$$m_1 = hf(t_k, x_k, y_k)$$

$$m_2 = hf\left(t_k + \frac{h}{2}, x_k + \frac{m_1}{2}, y_k + \frac{M_1}{2}\right)$$

$$m_3 = hf\left(t_k + h, x_k + \frac{m_2}{2}, y_k + \frac{M_2}{2}\right)$$

$$m_4 = hf(t_k + h, x_k + m_3, y_k + M_3) = hf(t_{k+1}, x_k + m_3, y_k + M_3)$$

and

$$M_1 = hg(t_k, x_k, y_k)$$

$$M_2 = hg\left(t_k + \frac{h}{2}, x_k + \frac{m_1}{2}, y_k + \frac{M_1}{2}\right)$$

$$M_3 = hg\left(t_k + \frac{h}{2}, x_k + \frac{m_2}{2}, y_k + \frac{M_2}{2}\right)$$

$$M_4 = hg(t_k + h, x_k + m_3, y_k + M_3) = hg(t_{k+1}, x_k + m_3, y_k + M_3).$$

Now let's put this algorithm to use—with the aid of technology, of course.

EXAMPLE 4.7.2   Using Runge-Kutta (RK4) and a CAS

Let's look again at the initial-value problem analyzed in Example 4.7.1. The system IVP is $\dfrac{dx}{dt} = y, \dfrac{dy}{dt} = x; x(0) = 1, y(0) = 0$, and we want to approximate $x(0.5)$ and $y(0.5)$. Rather than wearing ourselves out trying to implement the fourth-order Runge-Kutta method by hand, we can enter the equations and initial conditions into our CAS, specify the method (in whatever way you must describe the RK4 method), and choose a step size $h = 0.1$.

What we get is an approximation for $x(0.5)$ of 1.1276 and an approximation for $y(0.5)$ of 0.5211, both rounded to four decimal places. To four decimal places, the absolute error for each approximation is 0!   ◆

Our final example shows how the Runge-Kutta-Fehlberg fourth- and fifth-order algorithm works on an interesting system application.

EXAMPLE 4.7.3   Using Runge-Kutta-Fehlberg (rkf45) and a CAS

The British mathematician E. C. Zeeman (1925–) developed a simple nonlinear model of the human heartbeat:

$$\varepsilon \frac{dx}{dt} = -(x^3 - Ax + c)$$

$$\frac{dc}{dt} = x,$$

where $x(t)$ is the displacement from equilibrium of the heart's muscle fiber, $c = c(t)$ is the concentration of a chemical control at time t, and ε and A are positive constants. Because the levels of c determine the contraction and expansion (relaxation) of the muscle fibers, we can think of c as a *stimulus* and of x as a *response*.

We want to investigate the nature of the model's solution, and for convenience we'll assume that $\varepsilon = 1.0$ and $A = 3$. Also, let $x(0) = 1.7$ and $c(0) = 0.3$. (The initial conditions were determined after experimenting with various values on a CAS.) The calculations producing the graphs and the values discussed below were carried out using the rkf45 method for the system.

Because one important feature of a heartbeat is that it is periodic (lub-*dub*, lub-*dub*, . . .), the solution should reveal this in the *x-c* phase plane—and in fact it does (Figure 4.24). Both the *systole*, corresponding to a fully relaxed heart muscle, and the *diastole*, indicating a state of full contraction, are labeled on Figure 4.24.

We see that the heart muscle starts at (1.7, 0.3) and, under the influence of increasing c, contracts until it is fully contracted at D. Then the muscle begins to relax until it attains systole at S, returns to the initial point, and (we hope) begins the cycle again. Superimposing the trajectory on the slope field makes it easy to see the direction of the trajectory, but the numerical values in Table 4.1 also tell the story.

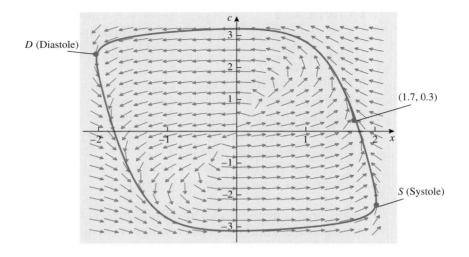

Figure 4.24

Slope field and trajectory for the system IVP

$$\left\{ \frac{dx}{dt} = -(x^3 - 3x + c), \frac{dc}{dt} = x; \quad x(0) = 1.7, c(0) = 0.3 \right\}$$

$$0 \le t \le 30$$

TABLE 4.1 Solution values of

$$\left\{ \frac{dx}{dt} = -(x^3 - 3x + c), \frac{dc}{dt} = x; \right.$$

$$\left. x(0) = 1.7, c(0) = 0.3 \right\}$$

t	$x(t)$	$c(t)$
0	1.7000	0.3000
2	0.7499	2.9990
4	−1.8417	0.4728
6	−1.1132	−2.5911
8	1.9436	−1.2862
10	1.3384	2.0618

If we examine the signs of x and c as t increases, we see that the points (x, c) are moving *counterclockwise* through the quadrants of the x-c plane. Looking carefully at the data in the table, we can see that the trajectory returns to its initial point $(1.7, 0.3)$ sometime between 8 and 10. In fact, a more detailed analysis reveals that the solution of our IVP has period approximately equal to 8.88. (See Exercise 10.)

Solving the system with rkf45 and then plotting x against t (Figure 4.25a), we see the periodic nature of the heart muscle's expansions and contractions. Figure 4.25b shows how the electrochemical activity represented by the variable c also varies periodically.

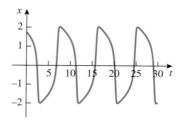

Figure 4.25a

Graph of $x(t)$ vs. t for the system IVP

$$\left\{ \frac{dx}{dt} = -(x^3 - 3x + c), \frac{dc}{dt} = x; \quad x(0) = 1.7, c(0) = 0.3 \right\}$$

$$0 \le t \le 30, -2.5 \le x \le 2.5$$

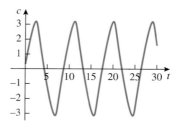

Figure 4.25b

Graph of $c(t)$ vs. t for the system IVP

$$\left\{ \frac{dx}{dt} = -(x^3 - 3x + c), \frac{dc}{dt} = x; \quad x(0) = 1.7, c(0) = 0.3 \right\}$$

$$0 \le t \le 30, -4 \le c \le 4$$

We will investigate interesting nonlinear systems again in Chapter 7. ◆

Just as for a single first-order equation, we can use spreadsheet commands to carry out the calculations needed to approximate the solutions of systems. Systems versions of the standard numerical techniques may be a bit more difficult to program, may require more intermediate storage, and may take a little more time, but they work well. Graphing calculators also handle systems of differential equations. In fact, as we remarked in the Introduction for this chapter, they usually deal with a single higher-order equation by requiring the user to write it in terms of a system and then solving the system numerically.

Whatever technology you use, try to understand what methods have been implemented by reading your documentation or checking out your software's "Help" features.

EXERCISES 4.7

All problems are to be done using technology, unless otherwise indicated.

1. a. Extend the improved Euler method given by formula (3.3.1) to a system of two first-order equations.

    b. *By hand*, redo Example 4.7.1, using $h = 0.1$ to find approximations to $x(0.5)$ and $y(0.5)$.

    c. Calculate the absolute error in part (b).

    d. Use technology and the improved Euler method with $h = 0.1$ to check your answers to part (b).

2. Consider the system $x' = x - 4y$, $y' = -x + y$, with $x(0) = 1$ and $y(0) = 0$. The exact solution is $x(t) = (e^{-t} + e^{3t})$, $y(t) = (e^{-t} - e^{3t})$.

    a. Verify that the exact solution of the IVP is the solution given above.

    b. Approximate the value of the solution at the point $t = 0.2$ using Euler's method with $h = 0.1$. Compare your result with the values of the exact solution, calculating the absolute error.

    c. Approximate the value of the solution at the point $t = 0.2$ using the fourth-order Runge-Kutta method with $h = 0.2$. Calculate the absolute error.

3. Consider the initial-value problem $y'' + y' - 2y = 2x$, with $y(0) = 1$ and $y'(0) = 1$.

    a. Convert this problem into a system of two first-order equations. (*Choose your new variables carefully.*)

    b. Determine approximate values of the solution at $x = 0.5$ and $x = 1.0$ by using Euler's method with $h = 0.1$.

    c. Determine approximate values of the solution at $x = 0.5$ and $x = 1.0$ by using the fourth-order Runge-Kutta method with $h = 0.1$.

4. In Example 4.5.6 you were told that the solution to the IVP

$$\frac{d^2x}{dt^2} + \frac{1}{4}\frac{dx}{dt} + 2x = 0, \text{ with } x(0) = 1, \dot{x}(0) = 0,$$

is $x(t) = \frac{1}{127}e^{(-\frac{1}{8}t)}\left(127\cos\left(\frac{1}{8}\sqrt{127}t\right) + \sqrt{127}\sin\left(\frac{1}{8}\sqrt{127}t\right)\right)$.

    a. Convert this IVP into a system of first-order equations.

    b. Determine the approximate value of the solution at $t = 0.6$ by using the Runge-Kutta-Fehlberg method (rkf45), if available. Otherwise, use the highest-order Runge-Kutta method available to you, with $h = 0.01$. Compare your values with the exact solution above.

5. A particle moves in three-dimensional space according to the equations

$$\frac{dx}{dt} = yz, \qquad \frac{dy}{dt} = zx, \qquad \frac{dz}{dt} = xy.$$

a. Assuming that $x(0) = 0$, $y(0) = 5$, and $z(0) = 0$, use the Runge-Kutta-Fehlberg method, if available, to approximate the solution at $t = 0.1, 0.2,$ 0.3, 0.4, 0.5, 0.6, 0.7, 0.8, 0.9, 1.0, 5.0, and 37. (Otherwise, use the highest-order Runge-Kutta method available to you, with $h = 0.01$.) Describe what these values seem to be telling you about the motion of the particle.

b. Now assume that $x(0) = y(0) = 1$ and $z(0) = 0$. Approximate the solution at $t = 0.1, 0.2, 0.3, 1.5, 1.6, 1.7, 1.8,$ and 1.9 using the same procedure you used in part (a). What seems to be happening?

6. The system

$$\frac{dx}{dt} = 7y - 4x - 13$$

$$\frac{dy}{dt} = 2x - 5y + 11$$

appeared in Example 4.5.3, where it was described as a possible arms race model.

a. Suppose that $x(0) = 1$ and $y(0) = 1$. Use technology and the Runge-Kutta-Fehlberg fourth- and fifth-order method (or a reasonable substitute) to estimate x and y for $t = 1, 2, 3, 4, 5, 10, 15,$ and 20.

b. On the basis of the values found in part (a), guess at $\lim_{t \to \infty} x(t)$ and $\lim_{t \to \infty} y(t)$.

7. The Lotka-Volterra system (Section 4.5)

$$\dot{x} = 3x - 2xy$$
$$\dot{y} = 0.5xy - y$$

has solutions $(x(t), y(t))$ that are periodic because a given trajectory always returns to its initial point in some finite time t^*: $x(t + t^*) = x(t)$ and $y(t + t^*) = y(t)$. By using technology and the rkf45 method, estimate the smallest value of t^* to two decimal places if $x(0) = 3$ and $y(0) = 2$. (Try different values of $t \neq 0$ until you get $x(t) \approx 3$ and $y(t) \approx 2.0$.)

8. The equations

$$\dot{x} = y$$
$$\dot{y} = -0.25\,y - 2\,x; \quad x(0) = 1, \, y(0) = 0$$

represent a certain spring-mass system with damping. As usual, assume that the positive direction for $x(t)$ and $y(t)$ is downward and that time is measured in seconds.

a. Using technology, approximate $x(t)$ and $y(t)$ for $t = 1, 2, 3,$ and 4, and interpret the position and velocity in each case.

b. Estimate (to the nearest hundredth of a second) the time when the mass *first* reaches its equilibrium position, $x = 0$.

9. A famous model for the spread of a disease is the *S-I-R* model. At a given time t, S represents the population of *susceptibles*, those who have never had the disease and can get it; I stands for the *infected*, those who have the disease now and can give it to others; and R denotes the *recovered*, people who

have already had the disease and are immune. Suppose that these populations are related by the system

$$\frac{dS}{dt} = (-0.00001)SI$$

$$\frac{dI}{dt} = (0.00001)SI - \frac{I}{14}$$

$$\frac{dR}{dt} = \frac{I}{14},$$

with $S(0) = 45400$, $I(0) = 2100$, and $R(0) = 2500$.

  a. Add the three differential equations and interpret the result in terms of a
  population.
  b. Use your CAS to plot S, I, and R as functions of t on separate graphs.
  (*Warning*: Some mathematical software (such as Maple) may reserve the
  letter I for the imaginary number $\sqrt{-1}$. If this is your situation, use IN to
  denote the infected population.)
  c. Use your CAS to plot phase portraits in the S-I, S-R, and I-R planes.
  d. Use a powerful numerical method with $h = 0.1$ to approximate the values
  of S, I, and R at $t = 1, 2, 3, 10, 15, 16$, and 17 days. What do you see?
  e. Approximate the value of t at which $I = 0$.

10. Use the rkf45 method to show why the period of the trajectory in Figure 4.24
is approximately 8.88. (Use the method suggested in Exercise 7.)

11. Investigate the Zeeman heartbeat model in Example 4.7.3 with $\varepsilon = 0.025$,
$A = 0.1575$, and $(x_0, c_0) = (0.45, -0.02025)$.

  a. Use the rkf45 method to approximate $x(t)$ and $c(t)$ for $t = 0.01, 0.02, \ldots,$
  0.10 seconds. What do your calculations tell you about the direction of the
  solution curve in the x-c plane?
  b. Draw the trajectory corresponding to the initial conditions given above.
  c. Approximate the period of the trajectory found in part (b).
  d. Estimate the coordinates of the diastole and the systole.

4.8   SUMMARY

For second-order homogeneous linear equations with constant coefficients—
equations of the form

$$ax'' + bx' + cx = 0,$$

where a, b, and c are constants, $a \neq 0$—we can describe the solutions explicitly in
terms of the roots of the associated **characteristic equation** $a\lambda^2 + b\lambda + c = 0$ as
follows:

  1. If there are two distinct real roots—λ_1, λ_2, with $\lambda_1 \neq \lambda_2$—then the general solution is $x(t) = c_1 e^{\lambda_1 t} + c_2 e^{\lambda_2 t}$.

2. If there is a repeated real root λ, then the general solution has the form
$x(t) = c_1 e^{\lambda t} + c_2 t e^{\lambda t} = (c_1 + c_2 t) e^{\lambda t}$.

3. If the roots form a complex conjugate pair $p \pm qi$, then the (real) general solution has the form $x(t) = e^{pt}(c_1 \cos(qt) + c_2 \sin(qt))$. Here we need Euler's formula to deal with complex exponentials.

The general solution, y_{GNH}, of a linear nonhomogeneous system is obtained by finding a particular solution, y_{PNH}, of the nonhomogeneous system and adding it to the general solution, y_{GH}, of the homogeneous system: $y_{\text{GNH}} = y_{\text{GH}} + y_{\text{PNH}}$. A particular solution can be found using the method of **variation of parameters.**

For a linear equation of any order, we have the **Superposition Principle:** If y_j is a solution of $L(y) = f_j$ for $j = 1, 2, \ldots, n$, and $c_1, c_2, \ldots, c_n$ are arbitrary constants, then $c_1 y_1 + c_2 y_2 + \cdots + c_n y_n$ is a solution of $L(y) = c_1 f_1 + c_2 f_2 + \cdots + c_n f_n$. That is, $L(c_1 y_1 + c_2 y_2 + \cdots + c_n y_n) = c_1 L(y_1) + c_2 L(y_2) + \cdots + c_n L(y_n) = c_1 f_1 + c_2 f_2 + \cdots + c_n f_n$.

As a consequence of the Superposition Principle, the formula $y_{\text{GNH}} = y_{\text{GH}} + y_{\text{PNH}}$ is valid for a linear equation of any order n. We have an algorithm to find the general solution y_{GH} of the associated nth-order homogeneous equation $a_n y^{(n)} + a_{n-1} y^{(n-1)} + \cdots + a_2 y'' + a_1 y' + a_0 y = 0$, where y is a function of the independent variable t and $a_n, a_{n-1}, \ldots, a_1, a_0$ are constants:

First find the roots of the characteristic equation

$$a_n \lambda^n + a_{n-1} \lambda^{n-1} + \cdots + a_1 \lambda + a_0 = 0.$$

Use a CAS to solve the equation if n is greater than or equal to 3. Next group these roots as follows: (a) distinct real roots; (b) distinct complex conjugate pairs $p \pm qi$; (c) multiple real roots; (d) multiple complex roots. Then the general solution is a sum of n terms of the forms

1. $c_i e^{\lambda_i t}$ for each distinct real root λ_i
2. $e^{pt}(c_1 \cos qt + c_2 \sin qt)$ for each distinct complex pair $p \pm q i$
3. $(c_1 + c_2 t + \cdots + c_k t^{k-1}) e^{\lambda_i t}$ for each multiple real root λ_i, where k is the multiplicity of that root
4. $e^{pt}(c_1 \cos qt + c_2 \sin qt) + t e^{pt}(c_3 \cos qt + c_4 \sin qt) + \cdots + t^{k-1} e^{pt}(c_{2k-1} \cos qt + c_{2k} \sin qt)$ for each multiple complex pair of roots $p \pm qi$, where k is the multiplicity of the pair $p \pm qi$

To find a particular solution of the nth-order nonhomogeneous equation, we can use the method of **variation of parameters** as we did in the second-order case (although more work is involved).

The most important fact in this chapter is that **any single nth-order differential equation can be converted into an equivalent system of first-order equations.** More precisely, **any nth-order differential equation**

$$x^{(n)} = F(t, x, x', x'', \ldots, x^{(n-1)})$$

can be converted into an equivalent system of first-order equations by letting $x_1 = x$, $x_2 = x'$, $x_3 = x''$, $\ldots$, $x_n = x^{(n-1)}$. However, to convert a single *nonautonomous* nth-order equation into an equivalent *autonomous* system (one whose equations do not explicitly contain the independent variable t), we need $n + 1$ first-order equations: $x_1 = x$, $x_2 = x'$, $x_3 = x''$, ..., $x_n = x^{(n-1)}$, $x_{n+1} = t$. The system is linear or nonlinear, autonomous or nonautonomous, according to the nature of the individual equations in the system. Linear systems are easier to calculate with and understand than nonlinear systems. Similarly, autonomous systems are nicer than nonautonomous systems.

Converting a single higher-order equation to a system often provides graphical insights that cannot be obtained from the one equation. This conversion also allows us to use the first-order methods in Chapters 2 and 3 to understand higher-order equations.

A two-dimensional system has the form

$$x' = F(t, x, y)$$
$$y' = G(t, x, y).$$

A particular **solution** of such a system consists of a *pair* of functions $x(t)$, $y(t)$ that, when substituted into the equations of the system, give true statements. The proper graphical representation of a solution is a curve in three-dimensional t-x-y space, the set of points of the form $(t, x(t), y(t))$; but often it is useful to think of the points $(x(t), y(t))$ as tracing out a path (also called an **orbit** or a **trajectory**) in the x-y plane (called the **phase plane**) as the parameter t varies "offstage." The *positive* direction of the path is the direction that corresponds to increasing values of t. The collection of all trajectories is the **phase portrait** of the system. Technology also enables us to study the graphs of x vs. t and y vs. t.

For *autonomous* systems $x' = f(x, y)$, $y' = g(x, y)$, we can eliminate any explicit reliance on the parameter t by using the Chain Rule to form the first-order differential equation

$$\frac{dy}{dx} = \frac{dy}{dt} \cdot \frac{dt}{dx} = \frac{\dfrac{dy}{dt}}{\dfrac{dx}{dt}} = \frac{g(x, y)}{f(x, y)}.$$

This gives the slope of the tangent line at points of the phase plane. The slope field of this first-order equation outlines the phase portrait of the system.

Given any two-dimensional autonomous system $x' = f(x, y)$, $y' = g(x, y)$, an **equilibrium point** is a point (x, y) such that $f(x, y) = 0 = g(x, y)$. This means, for example, that a particle at this point in the phase plane is not moving. The language of **sinks** and **sources** used in Section 2.5 can be extended to apply to equilibrium solutions of systems. The behavior of trajectories near equilibrium points of linear systems will be discussed systematically in Chapter 5. Trajectories for nonlinear systems are treated in Chapter 7.

As examples of these ideas, we discussed **predator-prey systems,** in particular the **Lotka-Volterra equations.** Several examples of **spring-mass problems** were also analyzed, including one exhibiting the phenomenon of **resonance.**

Before getting too immersed in trying to solve higher-order equations or their equivalent systems, we have to determine when solutions *exist*—and whether existing solutions are *unique*. A useful result applies to a second-order

IVP, $\dfrac{d^2y}{dt^2} = f\left(t, y, \dfrac{dy}{dt}\right)$ with $y(t_0) = y_0$ and $\dfrac{dy}{dt}(t_0) = \dot{y}_0$. If $f, \dfrac{\partial f}{\partial y}$, and $\dfrac{\partial f}{\partial \dot{y}}$ are continuous in a closed box B in three-dimensional space (t-y-ẏ space) and the point $(t_0, y_0, \dot{y}_0)$ lies inside B, then the IVP has a unique solution $y(t)$ on some t-interval I containing t_0.

Equivalently, suppose we have a two-dimensional system of first-order equations

$$\frac{dx_1}{dt} = f(t, x_1, x_2)$$

$$\frac{dx_2}{dt} = g(t, x_1, x_2),$$

where $x_1(t_0) = x_1^0$ and $x_2(t_0) = x_2^0$. Then if $f, g, \dfrac{\partial f}{\partial x_1}, \dfrac{\partial g}{\partial x_1}, \dfrac{\partial f}{\partial x_2}$, and $\dfrac{\partial g}{\partial x_2}$ are all continuous in a box B in t-x_1-x_2 space containing the point (t_0, x_1^0, x_2^0), there is an interval I containing t_0 in which there exists a unique solution $x_1 = y_1(t)$, $x_2 = y_2(t)$ of the IVP.

Once we are confident that an IVP involving a higher-order equation or its system equivalent *has* a unique solution, we can apply natural two-dimensional generalizations of the numerical solution methods introduced in Sections 3.1, 3.3, and 3.4: Euler's method; the improved Euler method; and higher-order techniques such as the fourth-order Runge-Kutta and Runge-Kutta-Fehlberg methods. Technology is indispensable in the numerical solution of both single equations and systems.

PROJECT 4-1

Get the Lead Out

In analyzing the flow of lead pollution in a human body among the three compartments bone, blood, and tissue, the following system was developed:[11]

$$\dot{x}_1 = -\frac{65}{1800}x_1 + \frac{1088}{87500}x_2 + \frac{7}{200000}x_3 + \frac{6162}{125}$$

$$\dot{x}_2 = \frac{20}{1800}x_1 - \frac{20}{700}x_2$$

$$\dot{x}_3 = \frac{7}{1800}x_1 - \frac{7}{200000}x_3.$$

Here $x_1(t)$ is the amount of lead in the blood at time t (in years), $x_2(t)$ is the amount of lead in tissue, and $x_3(t)$ is the amount of lead in bone. Assume that $x_1(0) = x_2(0) = x_3(0) = 0$.

---

11. E. Batschelet, L. Brand, and A. Steiner, "On the Kinetics of Lead in the Human Body," *Journal of Mathematical Biology* 8 (1979): 15–23.

a. Use technology to graph the three-dimensional trajectory in x_1-x_2-x_3 space with $0 \le t \le 250$. (Move the axes around to get a good view.)

b. Use technology to graph the solution in the t-x_1 plane, $0 \le t \le 150$. What seems to be the equilibrium level of lead in the blood?

c. Use technology to graph the solution in the t-x_2 plane, $0 \le t \le 250$. What seems to be the equilibrium level of lead in tissue?

d. Use technology to graph the solution in the t-x_3 plane, $0 \le t \le 70,000$ (Yes!). In your CAS, specify a step size of 50 if you can. (*Warning: It may take a long time for your CAS to produce the graph.*) What seems to be the equilibrium level of lead in bone?

e. What do the graphs in parts (b), (c), and (d) say about the comparative times it takes blood, tissue, and bone to reach their equilibrium levels of lead?

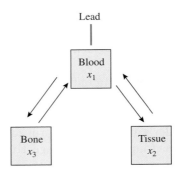

5 | Systems of Linear Differential Equations

5.0 INTRODUCTION

In Chapter 4, we saw how any higher-order differential equation can be written as an equivalent system of first-order differential equations. The examples we discussed introduced some algebraic manipulations and some geometric aspects of second- and third-order systems such as the *phase plane*, but there was no attempt to give a systematic approach.

In this chapter, we will explore (for the most part) autonomous systems of first-order *linear* differential equations, for which the theory is neat and complete. An important component of this theory is the *Superposition Principle,* which we discussed in Chapters 2 and 4 and which is the distinguishing characteristic of linear systems, as we shall see in the sections to come. This fundamental principle will help us to determine the general solution of linear systems in essentially the same way in which we solved single second-order linear equations in Sections 4.1 and 4.2.

To understand the important ideas underlying the theory and application of linear systems, we'll introduce some of the language and concepts from the area of mathematics called *linear algebra* without probing too deeply into the intricacies of this valuable and useful subject. For the most part, we'll stick to two-dimensional systems (two equations in two unknown functions) for the sake of geometric intuition, but we will also look at some higher-order systems. Chapter 6 will enhance our ability to handle linear systems, and in Chapter 7 we'll see how *nonlinear* systems can be analyzed in terms of certain related linear systems.

5.1   SYSTEMS AND MATRICES

MATRICES AND VECTORS

Suppose we look at the linear system

$$\dot{x} = ax + by$$
$$\dot{y} = cx + dy \, , \tag{5.1.1}$$

where x and y are functions of t, and a, b, c, and d are constants.

There is a useful notation for linear systems that was invented by the English mathematician Arthur Cayley and named by his fellow countryman James Sylvester around 1850. This notation allows us to pick out the coefficients a, b, c, and d in system (5.1.1) and write them in a square array $A = \begin{bmatrix} a & b \\ c & d \end{bmatrix}$ called a **matrix**—in this case, the **matrix of coefficients** of the linear system. (The plural of *matrix* is *matrices*.) In general, a matrix is just a rectangular array of mathematical objects (numbers or functions in this book) and can describe linear systems of all sizes. The size of a matrix is given in terms of the number of its **rows** and **columns.** For example, $A = \begin{bmatrix} a & b \\ c & d \end{bmatrix}$ is described as a 2×2 matrix because it has two rows, $(a \ b)$ and $(c \ d)$, and two columns, $\begin{bmatrix} a \\ c \end{bmatrix}$ and $\begin{bmatrix} b \\ d \end{bmatrix}$. The matrix $B = \begin{bmatrix} -4 & 0 & 1 & 5 \\ 2 & 6 & 7 & -\pi \\ 0 & \sqrt{5} & 3 & 5/9 \end{bmatrix}$ is a 3×4 matrix because it has three rows and four columns.

In describing the linear system (5.1.1), we can also introduce a **column matrix** or **vector** $X = \begin{bmatrix} x \\ y \end{bmatrix}$. (This is a 2×1 matrix.) If $x(t)$ and $y(t)$ are solutions of the system (5.1.1), we call $X = \begin{bmatrix} x \\ y \end{bmatrix}$ a **solution vector** of the system. We can view X as a point in the x-y plane, or phase plane, whose coordinates are written vertically instead of in the usual horizontal ordered-pair configuration. If a vector is made up of constants, then the *direction* of the vector is taken as the direction of an arrow from the origin to the point (x, y) in the x-y plane. (See Appendix B.1 for more information.)

If $A = \begin{bmatrix} a & b \\ c & d \end{bmatrix}$ and $B = \begin{bmatrix} e & f \\ g & h \end{bmatrix}$, then we say the matrices are **equal** and write $A = B$ if and only if $a = e$, $b = f$, $c = g$, and $d = h$. We say that "corresponding elements must be equal." Similarly, if $V = \begin{bmatrix} x_1 \\ y_1 \end{bmatrix}$ and $W = \begin{bmatrix} x_2 \\ y_2 \end{bmatrix}$, we can say that $V = W$ if and only if $x_1 = x_2$ and $y_1 = y_2$.

If a vector (or, more generally, a matrix) is made up of objects (**elements** or **entries**) that are *functions*, we can define the *derivative* of such a vector as the vector whose elements are the derivatives of the original elements, provided that all these individual derivatives exist. For example, if $X = \begin{bmatrix} -t^2 \\ \sin t \end{bmatrix}$, then

$$\frac{d}{dt}X = \begin{bmatrix} \dfrac{d}{dt}(-t^2) \\ \dfrac{d}{dt}(\sin t) \end{bmatrix} = \begin{bmatrix} -2t \\ \cos t \end{bmatrix}.$$

The Matrix Representation of a Linear System

We can write the system

$$\dot{x} = ax + by$$
$$\dot{y} = cx + dy$$

compactly and symbolically as

$$\begin{bmatrix} \dot{x} \\ \dot{y} \end{bmatrix} = \begin{bmatrix} a & b \\ c & d \end{bmatrix}\begin{bmatrix} x \\ y \end{bmatrix},$$

or $\dot{X} = AX$, where $X = \begin{bmatrix} x \\ y \end{bmatrix}$ and $A = \begin{bmatrix} a & b \\ c & d \end{bmatrix}$. The juxtaposition ("product") $\begin{bmatrix} a & b \\ c & d \end{bmatrix}\begin{bmatrix} x \\ y \end{bmatrix}$ represents the vector $\begin{bmatrix} ax + by \\ cx + dy \end{bmatrix}$. For example, $\begin{bmatrix} 3 & -2 \\ 1 & 4 \end{bmatrix}\begin{bmatrix} x \\ y \end{bmatrix} = \begin{bmatrix} 3x - 2y \\ x + 4y \end{bmatrix}$. There is a way to define and interpret this product of a matrix and a vector in the context of linear algebra (see Appendix B2 for details), but **we will take this product as a symbolic representation of the system, highlighting the matrix of coefficients and the solution vector.** Soon we will see how a linear system's solutions—its behavior in the phase plane—are determined by the matrix of coefficients. For now, let's look at some examples of the use of matrix notation.

EXAMPLE 5.1.1   **Matrix Form of a Two-Dimensional Linear System**
We can write the linear system of ODEs

$$\dot{x} = -3x + 5y$$
$$\dot{y} = x - 4y$$

in matrix terms as $\begin{bmatrix} \dot{x} \\ \dot{y} \end{bmatrix} = \begin{bmatrix} -3 & 5 \\ 1 & -4 \end{bmatrix}\begin{bmatrix} x \\ y \end{bmatrix}.$ ◆

The next example demonstrates how we have to be careful in extracting the right matrix of coefficients from a linear-system problem.

EXAMPLE 5.1.2   Matrix Form of a Two-Dimensional Linear System

The linear system $\dfrac{dx}{dt} = y, \dfrac{dy}{dt} = -x$ should be written as

$$\frac{dx}{dt} = 0 \cdot x + 1 \cdot y$$
$$\frac{dy}{dt} = -1 \cdot x + 0 \cdot y$$

first. Then it is clear that the matrix representation of the system is

$$\frac{d}{dt}\begin{bmatrix} x \\ y \end{bmatrix} = \begin{bmatrix} \dfrac{dx}{dt} \\ \dfrac{dy}{dt} \end{bmatrix} = \begin{bmatrix} 0 & 1 \\ -1 & 0 \end{bmatrix}\begin{bmatrix} x \\ y \end{bmatrix}. \qquad \blacklozenge$$

Some Matrix Algebra

Before we go on, we should discuss some properties of matrix algebra that we'll be using in the rest of this chapter. For example, if we're given the system

$$x' = -4x + 6y = 2(-2x + 3y)$$
$$y' = 2x - 8y = 2(x - 4y)$$

it is natural to write $\begin{bmatrix} x' \\ y' \end{bmatrix} = \begin{bmatrix} -4 & 6 \\ 2 & -8 \end{bmatrix} = 2\begin{bmatrix} -2 & 3 \\ 1 & -4 \end{bmatrix}$, or $X' = 2AX$, where

$A = \begin{bmatrix} -2 & 3 \\ 1 & -4 \end{bmatrix}$. More generally, if $A = \begin{bmatrix} a & b \\ c & d \end{bmatrix}$ and k is a constant (called a

scalar to distinguish it from a matrix), then $kA = k\begin{bmatrix} a & b \\ c & d \end{bmatrix} = \begin{bmatrix} ka & kb \\ kc & kd \end{bmatrix}$. In

particular, for vectors, we have $k\begin{bmatrix} u \\ v \end{bmatrix} = \begin{bmatrix} ku \\ kv \end{bmatrix}$. Put simply, *multiplying a matrix by a number requires multiplying each element of that matrix by the number.*

For example, if $A = \begin{bmatrix} 2 & -3 \\ 5 & 0 \end{bmatrix}$ and $k = -2$, then

$$kA = -2A = \begin{bmatrix} -2(2) & -2(-3) \\ -2(5) & -2(0) \end{bmatrix} = \begin{bmatrix} -4 & 6 \\ -10 & 0 \end{bmatrix}.$$

Two matrices, A and B, of the same size (that is, having the same number of rows and the same number of columns) can be added in an element-by-element way. For example, if $A = \begin{bmatrix} -2 & 3 \\ 4 & -1 \end{bmatrix}$ and $B = \begin{bmatrix} 1 & 2 \\ 3 & 4 \end{bmatrix}$, then $A + B = $

$\begin{bmatrix} -2+1 & 3+2 \\ 4+3 & (-1)+4 \end{bmatrix} = \begin{bmatrix} -1 & 5 \\ 7 & 3 \end{bmatrix}$ and

$$A - B = A + (-1)B = \begin{bmatrix} -2 & 3 \\ 4 & -1 \end{bmatrix} + \begin{bmatrix} -1 & -2 \\ -3 & -4 \end{bmatrix} = \begin{bmatrix} -3 & 1 \\ 1 & -5 \end{bmatrix}.$$

Similarly, if $V = \begin{bmatrix} x_1 \\ y_1 \end{bmatrix}$ and $W = \begin{bmatrix} x_2 \\ y_2 \end{bmatrix}$, then $V + W = \begin{bmatrix} x_1 + x_2 \\ y_1 + y_2 \end{bmatrix}$. The vector defined as $\mathbf{0} = \begin{bmatrix} 0 \\ 0 \end{bmatrix}$, which is called the **zero vector,** behaves in the world of vectors the way the *number* 0 acts in arithmetic: $V + \mathbf{0} = V = \mathbf{0} + V$ for any vector V. Similarly, we can define the **zero matrix,** $Z = \begin{bmatrix} 0 & 0 \\ 0 & 0 \end{bmatrix}$, having the same property for matrix addition. Note that $X = \begin{bmatrix} x \\ y \end{bmatrix}$ is an *equilibrium point* for the system $\begin{bmatrix} \dot{x} \\ \dot{y} \end{bmatrix} = \begin{bmatrix} a & b \\ c & d \end{bmatrix} \begin{bmatrix} x \\ y \end{bmatrix}$ if and only if $\begin{bmatrix} a & b \\ c & d \end{bmatrix} \begin{bmatrix} x \\ y \end{bmatrix} = \mathbf{0}$—that is, if and only if X is a solution of the matrix equation $AX = \mathbf{0}$.

A particularly useful idea for our future work is a *linear combination* of vectors. Given two vectors $V = \begin{bmatrix} x_1 \\ y_1 \end{bmatrix}$ and $W = \begin{bmatrix} x_2 \\ y_2 \end{bmatrix}$, any vector of the form $k_1 V + k_2 W$, where k_1 and k_2 are scalars, is called a **linear combination** of V and W. In terms of our given vectors, a linear combination of V and W is any vector of the form $k_1 V + k_2 W = \begin{bmatrix} k_1 x_1 \\ k_1 y_1 \end{bmatrix} + \begin{bmatrix} k_2 x_2 \\ k_2 y_2 \end{bmatrix} = \begin{bmatrix} k_1 x_1 + k_2 x_2 \\ k_1 y_1 + k_2 y_2 \end{bmatrix}$. As an example, for the specific vectors $V = \begin{bmatrix} \sin t \\ 2 \end{bmatrix}$ and $W = \begin{bmatrix} \cos t \\ e^t \end{bmatrix}$, a linear combination has the form $\begin{bmatrix} k_1 \sin t + k_2 \cos t \\ 2k_1 + k_2 e^t \end{bmatrix}$.

It is important to know that the *associative* and *distributive rules* of algebra hold for matrix addition and the product of a matrix and a vector. For example, if A and B are matrices, V and W are vectors, and k, k_1, and k_2 are scalars, then

$$A(kV) = k(AV)$$
$$A(V + W) = AV + AW,$$

and

$$A(k_1 V + k_2 W) = A(k_1 V) + A(k_2 W) = k_1(AV) + k_2(AW).$$

These results are proved in Appendix B.3.

Finally, note that the matrix $\begin{bmatrix} 1 & 0 \\ 0 & 1 \end{bmatrix}$ acts as an *identity* for multiplication: $\begin{bmatrix} 1 & 0 \\ 0 & 1 \end{bmatrix} \begin{bmatrix} x \\ y \end{bmatrix} = \begin{bmatrix} x \\ y \end{bmatrix}$ for any vector $\begin{bmatrix} x \\ y \end{bmatrix}$. In the context of two-dimensional systems, the matrix $\begin{bmatrix} 1 & 0 \\ 0 & 1 \end{bmatrix}$ is called the **identity matrix** and is denoted by I.

In the next section we will see how matrix notation gives us insight into the nature of a system's solutions. To understand the solutions more fully, we will introduce some additional concepts from linear algebra.

EXERCISES 5.1

1. Express each of the following systems of linear equations in matrix terms—that is, in the form $AX = B$, where A, X, and B are matrices.

   a. $3x + 4y = -7$
   $-x - 2y = 5$

   b. $\pi a - 3b = 4$
   $5a + 2b = -3$

   c. $x - y + z = 7$
   $-x + 2y - 3z = 9$
   $2x - 3y + 5z = 11$

   [*Think about what would make sense in part (c).*]

   Convert each system of differential equations in Exercises 2–6 to the matrix form $\dot{X} = AX$.

2. $\dot{x} = 2x + y$
   $\dot{y} = 3x + 4y$

3. $\dot{x} = x - y$
   $\dot{y} = y - 4x$

4. $\dot{x} = 2x + y$
   $\dot{y} = 4y - x$

5. $\dot{x} = x$
   $\dot{y} = y$

6. $\dot{x} = -2x + y$
   $\dot{y} = -2y$

7. Using the technique shown in Section 4.4, write each of the following second-order equations as a system of first-order equations and then express the system in matrix form.

   a. $y'' - 3y' + 2y = 0$

   b. $5y'' + 3y' - y = 0$

8. Find the derivative of each of the following vectors:

   a. $X(t) = \begin{bmatrix} t^3 - 2t^2 + t \\ e^t \sin t \end{bmatrix}$

   b. $V(x) = \begin{bmatrix} 2\cos x \\ -3e^{-2x} \end{bmatrix}$

   c. $B(u) = \begin{bmatrix} e^{-u} + e^u \\ 2\cos u - 5\sin u \end{bmatrix}$

9. If $A(t) = \begin{bmatrix} a_{11}(t) & a_{12}(t) \\ a_{21}(t) & a_{22}(t) \end{bmatrix}$ and $B(t) = \begin{bmatrix} b_{11}(t) \\ b_{21}(t) \end{bmatrix}$, both matrices having entries that are differentiable functions of t, show that

$$\frac{d}{dt}[A(t)B(t)] = \frac{dA(t)}{dt}B(t) + A(t)\frac{dB(t)}{dt}.$$

10. If $A = \begin{bmatrix} 1 & 2 \\ 3 & 4 \end{bmatrix}$, $B = \begin{bmatrix} 0 & -2 \\ 3 & 1 \end{bmatrix}$, $I = \begin{bmatrix} 1 & 0 \\ 0 & 1 \end{bmatrix}$, $V = \begin{bmatrix} 2 \\ -1 \end{bmatrix}$, and $W = \begin{bmatrix} 3 \\ 2 \end{bmatrix}$, calculate each of the following:

    a. $2A - 3B$

    b. AV

    c. BW

    d. $-2V + 5W$

    e. $A(3V - 2W)$

    f. $(A - 5I)W$

11. If $A = \begin{bmatrix} 1 & 2 \\ 3 & 4 \end{bmatrix}$ and $V = \begin{bmatrix} x \\ y \end{bmatrix}$, solve the equation $AV = \begin{bmatrix} 1 \\ 3 \end{bmatrix}$ for V (that is, find values for x and y).

12. Show that the origin is the only equilibrium point of the system

$$\dot{x} = ax + by$$
$$\dot{y} = cx + dy,$$

    where a, b, c, and d are constants, with $ad - bc \neq 0$.

5.2   TWO-DIMENSIONAL SYSTEMS OF FIRST-ORDER LINEAR EQUATIONS

EIGENVALUES AND EIGENVECTORS

To get a handle on linear systems of ordinary differential equations, including their qualitative behavior and their possible closed-form solutions, we will focus on linear two-dimensional systems of the form

$$\dot{x} = ax + by$$
$$\dot{y} = cx + dy, \tag{5.2.1}$$

where x and y both depend on the variable t, and a, b, c, and d are constants. Our analysis of such simple (but important) systems will prepare us to understand the treatment of higher-order linear systems in Section 5.7.

First let's write the system (5.2.1) in matrix form:

$$\begin{bmatrix} \dot{x} \\ \dot{y} \end{bmatrix} = \begin{bmatrix} a & b \\ c & d \end{bmatrix} \begin{bmatrix} x \\ y \end{bmatrix},$$

or

$$\frac{d}{dt} \begin{bmatrix} x \\ y \end{bmatrix} = \begin{bmatrix} a & b \\ c & d \end{bmatrix} \begin{bmatrix} x \\ y \end{bmatrix},$$

or

$$\dot{X} = AX, \tag{5.2.2}$$

where $X = \begin{bmatrix} x \\ y \end{bmatrix}$ and $A = \begin{bmatrix} a & b \\ c & d \end{bmatrix}$. Ignoring the fact that the capital letters represent matrices, what does the form of equation (5.2.2) remind you of? *Think about this.* Have you seen a differential equation of this form before? If we use lower-case letters and write the equation as $\dot{x} = ax$, we get a familiar separable equation representing exponential growth or decay. (See Section 2.1, especially Example 2.1.1.) This observation suggests that the solution of system (5.2.1) or the matrix equation (5.2.2) may have something to do with exponentials.

Let's make a shrewd guess and then examine the consequences of our guess. (This was Euler's strategy, described in Section 4.1.) In particular, let us assume that $x(t) = c_1 e^{\lambda t}$ and $y(t) = c_2 e^{\lambda t}$ for constants λ, c_1, and c_2. (Stating that λ, the coefficient of t, is the same for both x and y is a simplifying assumption.) Substituting $\begin{bmatrix} c_1 e^{\lambda t} \\ c_2 e^{\lambda t} \end{bmatrix}$ for X in (5.2.2), we get

$$\begin{bmatrix} c_1 \lambda e^{\lambda t} \\ c_2 \lambda e^{\lambda t} \end{bmatrix} = \begin{bmatrix} a & b \\ c & d \end{bmatrix} \begin{bmatrix} c_1 e^{\lambda t} \\ c_2 e^{\lambda t} \end{bmatrix} = \begin{bmatrix} a & b \\ c & d \end{bmatrix} e^{\lambda t} \begin{bmatrix} c_1 \\ c_2 \end{bmatrix},$$

or

$$\lambda e^{\lambda t} \begin{bmatrix} c_1 \\ c_2 \end{bmatrix} = \begin{bmatrix} a & b \\ c & d \end{bmatrix} e^{\lambda t} \begin{bmatrix} c_1 \\ c_2 \end{bmatrix},$$

or (dividing out the exponential factor and switching right and left sides)

$$A\tilde{X} = \lambda\tilde{X},$$ (5.2.3)

where $\tilde{X} = \begin{bmatrix} c_1 \\ c_2 \end{bmatrix}$. Note that our reasonable guess about x and y has allowed us to replace our original differential equation problem with a pure algebra problem. Equation (5.2.3) is in matrix terms and has nothing (apparently) to do with differential equations. Given a 2×2 matrix A and a 2×1 column matrix $\tilde{X}$, we can try to solve (5.2.3) for the value or values of λ, each called a **characteristic value** or **eigenvalue** of the matrix A. (Remember how we used this term in Sections 4.1–4.3. The connection between the earlier use of the term *eigenvalue* and the current use will be established shortly.) Eigenvalues will play an important role in solving linear systems and in understanding the qualitative behavior of solutions.

Furthermore, if we have solved equation (5.2.3) for its eigenvalues λ, then for each value of λ we can solve (5.2.3) for the corresponding vector or vectors $\tilde{X}$. Each such *nonzero* vector $\tilde{X}$ is called an **eigenvector** (or **characteristic vector**) of the system. We see that if both entries of $\tilde{X}$ are zero, then $\tilde{X}$ satisfies (5.2.3) for any value of λ, but this is the trivial case. **In all the discussion that follows, we will assume that c_1 and c_2 are not both zero**—that is, at least one of the two constants is not zero.

Before getting bogged down in symbolism, terminology, and the general problem of solving the matrix equation $A\tilde{X} = \lambda\tilde{X}$, let's look at a specific example in detail.

EXAMPLE 5.2.1   Solving a Linear System with Eigenvalues and Eigenvectors
Suppose we have the system

$$\dot{x} = -2x + y$$
$$\dot{y} = -4x + 3y,$$ (*)

which we can write as $\dot{X} = \begin{bmatrix} \dot{x} \\ \dot{y} \end{bmatrix} = \begin{bmatrix} -2 & 1 \\ -4 & 3 \end{bmatrix} \begin{bmatrix} x \\ y \end{bmatrix}$. We want to find the general solution of this system.

Substitution
Assuming, say, that $c_1 \neq 0$, we substitute $x = c_1 e^{\lambda t}$ and $y = c_2 e^{\lambda t}$ into (*) and get $\lambda c_1 e^{\lambda t} = -2c_1 e^{\lambda t} + c_2 e^{\lambda t} = e^{\lambda t}(-2c_1 + c_2)$ and $\lambda c_2 e^{\lambda t} = -4c_1 e^{\lambda t} + 3c_2 e^{\lambda t} = e^{\lambda t}(-4c_1 + 3c_2)$. If we simplify each equation by dividing out the exponential term and moving all terms to the left-hand side, we get

(A) $(\lambda + 2)c_1 - c_2 = 0$
(B) $4c_1 + (\lambda - 3)c_2 = 0.$ (* *)

We want to solve (* *) for λ.

Solving for λ
If we multiply equation (A) by $(\lambda - 3)$ and then add the resulting equation to (B), we get $(\lambda - 3)(\lambda + 2)c_1 + 4c_1 = 0$, or $(\lambda^2 - \lambda - 2)c_1 = 0$. Because we have

assumed that c_1 is not zero, we must have $\lambda^2 - \lambda - 2 = 0$. This means that the eigenvalues of A are $\lambda = 2$ and $\lambda = -1$. (Go through all the algebra carefully.) Note that we didn't have to know c_1 to find λ. We just had to know that c_1 was not zero. It is important that we could have assumed that c_2 was not zero and come to the same conclusion. (*Check this.*)

Solving the System of ODEs

If we take the eigenvalue $\lambda = 2$, we have $x(t) = c_1 e^{2t}$ and $y(t) = c_2 e^{2t}$, so that $X_1 = \begin{bmatrix} x \\ y \end{bmatrix} = \begin{bmatrix} c_1 e^{2t} \\ c_2 e^{2t} \end{bmatrix} = e^{2t} \begin{bmatrix} c_1 \\ c_2 \end{bmatrix}$. But when $\lambda = 2$, the equations in (* *) both represent the single equation $4c_1 - c_2 = 0$, so that we have the relation $c_2 = 4c_1$. Then we can write $X_1 = e^{2t} \begin{bmatrix} c_1 \\ c_2 \end{bmatrix} = e^{2t} \begin{bmatrix} c_1 \\ 4c_1 \end{bmatrix} = c_1 \begin{bmatrix} 1 \\ 4 \end{bmatrix} e^{2t}$, which is a one-parameter family of solutions of the system (*). We're saying that the pair of functions $x(t) = c_1 e^{2t}$ and $y(t) = 4c_1 e^{2t}$ is a nontrivial solution of our system for any nonzero constant c_1.

Similarly, if we take the eigenvalue $\lambda = -1$, then the system (* *) reduces to the single equation $c_1 - c_2 = 0$ and we can define $X_2 = e^{-t} \begin{bmatrix} c_1 \\ c_2 \end{bmatrix} = e^{-t} \begin{bmatrix} c_1 \\ c_1 \end{bmatrix} = c_1 \begin{bmatrix} 1 \\ 1 \end{bmatrix} e^{-t}$, which is also a one-parameter family of solutions of the system. In other words, the pair of functions $x(t) = c_1 e^{-t}$ and $y(t) = c_1 e^{-t}$ is also a nontrivial solution of our system for any nonzero constant c_1.

It is easy to see that the Superposition Principle we have been using since Chapter 2 allows us to conclude that

$$X = k_1 X_1 + k_2 X_2 = k_1 \left[c_1 \begin{bmatrix} 1 \\ 4 \end{bmatrix} e^{2t} \right] + k_2 \left[c_1 \begin{bmatrix} 1 \\ 1 \end{bmatrix} e^{-t} \right] = C_1 \begin{bmatrix} 1 \\ 4 \end{bmatrix} e^{2t} + C_2 \begin{bmatrix} 1 \\ 1 \end{bmatrix} e^{-t}$$

is the general solution of the system $\dot{x} = -2x + y$, $\dot{y} = -4x + 3y$. The constants C_1 and C_2 can be chosen to match arbitrary initial data.   ◆

Geometric Interpretation of Eigenvectors

The vector $\begin{bmatrix} 1 \\ 4 \end{bmatrix}$ that appears in the last example is called an *eigenvector* (or *characteristic vector*) corresponding to the eigenvalue (or characteristic value) $\lambda = 2$. This vector is a nonzero solution, $\tilde{X}$, of $A\tilde{X} = \lambda\tilde{X}$ when $\lambda = 2$. This means that there are infinitely many eigenvectors corresponding to the eigenvalue $\lambda = 2$— all the vectors $\begin{bmatrix} c_1 \\ c_2 \end{bmatrix}$ such that $4c_1 - c_2 = 0$, or all the nonzero vectors of the form $\begin{bmatrix} c_1 \\ 4c_1 \end{bmatrix}$ are eigenvectors associated with $\lambda = 2$. Choosing $c_1 = 1$ gives us the simple particular vector $V_1 = \begin{bmatrix} 1 \\ 4 \end{bmatrix}$, which can be called the *representative* eigenvector.

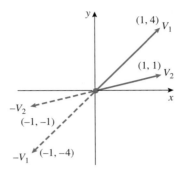

Figure 5.1

Representative eigenvectors

$$V_1 = \begin{bmatrix} 1 \\ 4 \end{bmatrix} \text{ and } V_2 = \begin{bmatrix} 1 \\ 1 \end{bmatrix}$$

Graphically, this eigenvector represents a straight line from the origin to the point $(1, 4)$ in the c_1-c_2 plane. Similarly, the vector $V_2 = \begin{bmatrix} 1 \\ 1 \end{bmatrix}$ is the representative eigenvector corresponding to the eigenvalue $\lambda = -1$ and can be interpreted as a straight line from $(0, 0)$ to $(1, 1)$ in the c_1-c_2 plane. (See the description of vectors in Appendix B.1.) Figure 5.1 shows V_1 and V_2 in the c_1-c_2 plane.

The General Problem

Now let's consider the equation $A\tilde{X} = \lambda\tilde{X}$, where $A = \begin{bmatrix} a & b \\ c & d \end{bmatrix}$ and $\tilde{X} = \begin{bmatrix} c_1 \\ c_2 \end{bmatrix}$, and at least one of the numbers c_1 and c_2 is nonzero. In the discussion that follows, we'll assume that $c_1 \neq 0$.

Written out as individual equations, $A\tilde{X} = \lambda\tilde{X}$ has the form

$$ac_1 + bc_2 = \lambda c_1$$
$$cc_1 + dc_2 = \lambda c_2$$

or

(A) $(a - \lambda)c_1 + bc_2 = 0$
(B) $cc_1 + (d - \lambda)c_2 = 0$

and we want to determine λ.

We can solve this algebraic system by the method of elimination as follows:

1. Multiply equation (A) by $d - \lambda$ to obtain

$$(d - \lambda)(a - \lambda)c_1 + b(d - \lambda)c_2 = 0.$$

2. Multiply equation (B) by $-b$ to get

$$-bcc_1 - b(d - \lambda)c_2 = 0.$$

3. Add the equations found in steps 1 and 2 to get

$$(d - \lambda)(a - \lambda)c_1 - bcc_1 = 0,$$

or $[\lambda^2 - (a + d)\lambda + (ad - bc)]c_1 = 0$. (*Check the algebra.*)

4. Because we assumed that $c_1 \neq 0$ at the beginning of this discussion, we must have

$$\lambda^2 - (a + d)\lambda + (ad - bc) = 0. \qquad (5.2.4)$$

This equation is called the **characteristic equation** of the matrix A, and its roots are the eigenvalues of A. We'll see the connection between this equation and the characteristic equation we introduced in Section 4.1 shortly.

Using the quadratic formula, we find that

$$\lambda = \frac{(a + d) \pm \sqrt{(a + d)^2 - 4(ad - bc)}}{2}.$$

If we had assumed that $c_2 \neq 0$ at the beginning, we would have found the same solution for λ.

Then for each distinct value of λ that we find, we can substitute that value into the system

$$(a - \lambda)c_1 + bc_2 = 0$$
$$cc_1 + (d - \lambda)c_2 = 0$$

and solve for $\tilde{X} = \begin{bmatrix} c_1 \\ c_2 \end{bmatrix}$, the corresponding eigenvector.

There are two things to notice about the characteristic equation of A,

$$\lambda^2 - (a + d)\lambda + (ad - bc) = 0$$

and the resulting formula for λ:

1. The expression $a + d$ is just the sum of the *main diagonal* (upper left, lower right) elements of the matrix $A = \begin{bmatrix} a & b \\ c & d \end{bmatrix}$. In linear algebra, this is called the **trace** of A. For example, if $A = \begin{bmatrix} -7 & 2 \\ 0 & 4 \end{bmatrix}$, then the trace of A is $(-7) + 4 = -3$.

2. The expression $ad - bc$ is formed from the matrix of coefficients $A = \begin{bmatrix} a & b \\ c & d \end{bmatrix}$ as follows: Multiply the main diagonal elements and then subtract the product of the other diagonal elements (upper right, lower left). The number calculated this way is called the **determinant** of the coefficient matrix. Symbolically, $\det(A) = \det \begin{bmatrix} a & b \\ c & d \end{bmatrix} = ad - bc$. For example, if

$A = \begin{bmatrix} -7 & -3 \\ 2 & 4 \end{bmatrix}$, then $\det(A) = (-7)(4) - (-3)(2) = -28 - (-6) = -28 + 6 = -22$. For any 2×2 matrix A, the determinant of A is often denoted by the symbol $|A|$, so the rule for calculation can be given as

$$|A| = \begin{vmatrix} a & b \\ c & d \end{vmatrix} = ad - bc.$$

Observations 1 and 2 provide us with an alternative way of viewing the characteristic equation:

$$\lambda^2 - (\text{trace of } A)\lambda + \det(A) = 0. \tag{5.2.5}$$

The roots of the characteristic equation (5.2.5)—the eigenvalues—lead to eigenvectors and ultimately to the general solution of a linear system. Let's look at an example using this shortcut.

EXAMPLE 5.2.2   Solving a Linear System with Eigenvalues and Eigenvectors
The following equations constitute a simple model for detecting diabetes:

$$\frac{dg}{dt} = -2.92g - 4.34h$$

$$\frac{dh}{dt} = 0.208g - 0.780h,$$

where $g(t)$ denotes excess glucose concentration in the bloodstream and $h(t)$ represents excess insulin concentration. "Excess" refers to concentrations above the equilibrium values. We want to determine the solution at any time t.

Eigenvalues

The matrix form of our equations is $\frac{d}{dt}X = \begin{bmatrix} dg/dt \\ dh/dt \end{bmatrix} = \begin{bmatrix} -2.92 & -4.34 \\ 0.208 & -0.780 \end{bmatrix} \begin{bmatrix} g \\ h \end{bmatrix}$, so

that the matrix of coefficients is $A = \begin{bmatrix} -2.92 & -4.34 \\ 0.208 & -0.780 \end{bmatrix}$. We see that the *trace* of A is $-2.92 + (-0.780) = -3.7$ and the *determinant* of A is $-2.92(-0.780) - (-4.34)(0.208) = 3.18032$. We see that the characteristic equation is

$$\lambda^2 - (\text{trace of } A)\lambda + \det(A) = \lambda^2 + 3.7\lambda + 3.18032 = 0$$

Solving this by calculator or CAS, we find that the eigenvalues are $\lambda_1 = -2.34212$ and $\lambda_2 = -1.35788$, rounded to five decimal places.

Eigenvectors
Now we substitute each eigenvalue in the equations

$$(a - \lambda)c_1 + bc_2 = 0$$
$$cc_1 + (d - \lambda)c_2 = 0$$

and solve for the corresponding eigenvector $\tilde{X} = \begin{bmatrix} c_1 \\ c_2 \end{bmatrix}$. In our problem, we must

substitute in the equations

$$(-2.92 - \lambda)c_1 - 4.34c_2 = 0$$
$$0.208c_1 + (-0.780 - \lambda)c_2 = 0.$$

If $\lambda = -2.34212$, then the equations are

$$-0.57788c_1 - 4.34c_2 = 0$$
$$0.208c_1 + 1.56212c_2 = 0.$$

But these two equations are really only one distinct equation, $c_2 = -0.13315c_1$. (Solve each equation for c_2 and see for yourself.) Therefore, to ensure that at least one element of the eigenvector is an integer, we can take $c_1 = 1$ and $c_2 = -0.13315$, so that an eigenvector corresponding to the eigenvalue $\lambda = -2.34212$ is $\tilde{X}_1 = \begin{bmatrix} 1 \\ -0.13315 \end{bmatrix}$.

Similarly, if we use the other eigenvalue, $\lambda = -1.35788$, we can take the single equation $(-2.92 - \lambda)c_1 - 4.34c_2 = 0$ and substitute the eigenvalue to get $(-2.92 + 1.35788)c_1 - 4.34c_2 = 0$, so that $c_2 = -0.35994c_1$. If we take $c_1 = 1$, we must have $c_2 = -0.35994$, and an eigenvector corresponding to the eigenvalue $\lambda = -2.34212$ is $\tilde{X}_2 = \begin{bmatrix} 1 \\ -0.35994 \end{bmatrix}$.

The Solution
The Superposition Principle gives the general solution as

$$\tilde{X} = C_1\tilde{X}_1 + C_2\tilde{X}_2 = C_1 \begin{bmatrix} 1 \\ -0.13315 \end{bmatrix} e^{-2.34212t} + C_2 \begin{bmatrix} 1 \\ -0.35994 \end{bmatrix} e^{-1.35788t}.$$

If we were given initial concentrations of glucose and insulin, we could determine the constants C_1 and C_2. (See Exercise 8.)   ◆

THE GEOMETRIC BEHAVIOR OF SOLUTIONS

In the next few examples, we will get a preview of how the behavior of a two-dimensional system of linear differential equations with constant coefficients depends on the eigenvalues and eigenvectors of its matrix of coefficients. We'll illustrate some typical phase portraits. Then, in Sections 5.3–5.5, we'll give a systematic description of all possible behaviors of such linear systems, using the nature of their eigenvalues and eigenvectors.

EXAMPLE 5.2.3   **Example 5.2.1 Revisited—A Saddle Point**
Let's look again at the system from Example 5.2.1:

$$\dot{x} = -2x + y$$
$$\dot{y} = -4x + 3y.$$

As we saw earlier, the eigenvalues of this system are $\lambda_1 = 2$ and $\lambda_2 = -1$, with corresponding representative eigenvectors $V_1 = \begin{bmatrix} 1 \\ 4 \end{bmatrix}$ and $V_2 = \begin{bmatrix} 1 \\ 1 \end{bmatrix}$. The general solution was given by $X = C_1 \begin{bmatrix} 1 \\ 4 \end{bmatrix} e^{2t} + C_2 \begin{bmatrix} 1 \\ 1 \end{bmatrix} e^{-t} = \begin{bmatrix} C_1 e^{2t} + C_2 e^{-t} \\ 4C_1 e^{2t} + C_2 e^{-t} \end{bmatrix}.$

Figure 5.2 shows some trajectories for this system of linear equations. These are particular solutions of $\dfrac{dy}{dx} = \dfrac{dy/dt}{dx/dt} = \dfrac{-4x + 3y}{-2x + y}$. Note in particular that the lines $y = 4x$ and $y = x$ appear as trajectories. These trajectories are actually four *half-lines*: $y = 4x$ for $x > 0$, $y = 4x$ for $x < 0$, $y = x$ for $x > 0$, and $y = x$ for $x < 0$.

A little basic algebra shows us that the origin is the only equilibrium point, and it is called a **saddle point** in this situation. A saddle point is the two-dimensional version of the *node* we discussed in Section 2.5. What characterizes a saddle point is that solutions can approach the equilibrium point along one direction (as though it were a *sink*), yet move away from the point in another direction (as though it were a *source*).[1] In particular, it turns out that one trajectory is the half-line $y = 4x$ in the first quadrant, along which the motion is *away* from the origin, and another trajectory is the line $y = x$ also in the first quadrant, along which the movement is *toward* the origin. The straight lines $y = 4x$ and $y = x$ are *asymptotes* for the other trajectories (as $t \to \pm\infty$). You may not be able to see this clearly from the phase portrait that your graphing utility generates unless you play with the range of t and choose initial values carefully, but you can see this and other behavior *analytically* (see Exercise 9). ◆

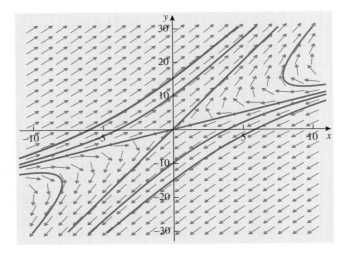

Figure 5.2
Phase portrait of the system $\dot{x} = -2x + y$, $\dot{y} = -4x + 3y$

---

1. This terminology is usually seen in a multivariable calculus course: If you look at a horse's saddle in the tail-to-head direction, it appears that the center of the saddle is lower than the front or back, so that the center seems to be a *minimum* point on the saddle's surface. However, if you look *across* the saddle from one side of the horse, it appears that the center is at the peak of a stirrup-to-stirrup curve, so the center seems like a *maximum* point. In fact, a *saddle point* is neither a minimum nor a maximum.

EXAMPLE 5.2.4   A Source

Now let's look at the system of differential equations

$$\dot{x} = 2x + y$$
$$\dot{y} = 3x + 4y.$$

First of all, note that the system's only equilibrium point—where $\dot{x} = 0$ and $\dot{y} = 0$—is the origin of the phase plane, $(x, y) = (0, 0)$. (You should verify this using the ordinary algebra of simultaneous equations.)

Using the formula given by equation (5.2.5), we see that the characteristic equation of our system is $\lambda^2 - (2 + 4)\lambda + ((2)(4) - (1)(3)) = 0$, or $\lambda^2 - 6\lambda + 5 = 0$, which has the roots $\lambda_1 = 5$ and $\lambda_2 = 1$. To find the eigenvectors corresponding to these eigenvalues, we must solve the matrix equation $A\tilde{X} = \lambda\tilde{X}$, where $A = \begin{bmatrix} 2 & 1 \\ 3 & 4 \end{bmatrix}$, $\lambda = 5$ or 1, and $\tilde{X} = \begin{bmatrix} c_1 \\ c_2 \end{bmatrix}$. This matrix equation is equivalent to the system

$$\begin{align}
(1) \quad & (2 - \lambda)c_1 + \quad\quad\quad c_2 = 0 \\
(2) \quad & \quad\quad\quad 3c_1 + (4 - \lambda)c_2 = 0.
\end{align} \tag{5.2.6}$$

Substituting the first eigenvalue, $\lambda = 5$, in (5.2.6) gives us

$$\begin{align}
(1) \quad & -3c_1 + c_2 = 0 \\
(2) \quad & \quad 3c_1 - c_2 = 0.
\end{align}$$

There is really only one equation here, and its solution is given by $c_2 = 3c_1$. Thus the eigenvectors corresponding to the eigenvalue $\lambda = 2$ have the form $\begin{bmatrix} c_1 \\ 3c_1 \end{bmatrix}$, or $c_1 \begin{bmatrix} 1 \\ 3 \end{bmatrix}$. If we let $c_1 = 1$, we get the "neat" representative eigenvector $V_1 = \begin{bmatrix} 1 \\ 3 \end{bmatrix}$.

When we use the other eigenvalue, $\lambda = 1$, in the system (5.2.6), we find that

$$\begin{align}
(1) \quad & c_1 + \quad c_2 = 0 \\
(2) \quad & 3c_1 + 3c_2 = 0,
\end{align}$$

which has the solution $c_2 = -c_1$. Therefore, the eigenvectors in this case have the form $\begin{bmatrix} c_1 \\ -c_1 \end{bmatrix}$, or $c_1 \begin{bmatrix} 1 \\ -1 \end{bmatrix}$. Thus our representative eigenvector can be $V_2 = \begin{bmatrix} 1 \\ -1 \end{bmatrix}$.

Now let's look at the phase portrait corresponding to the original system, a family of trajectories corresponding to the first-order equation

$$\frac{dy}{dx} = \frac{\dfrac{dy}{dt}}{\dfrac{dx}{dt}} = \frac{3x + 4y}{2x + y}.$$

This phase portrait is shown in Figure 5.3. The curves $y = 3x$ and $y = -x$, which are straight-line trajectories, are labeled so that we can see the significance of the eigenvectors.

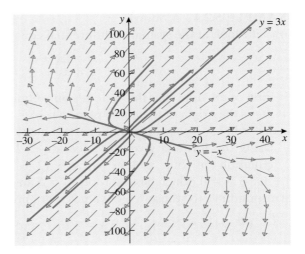

Figure 5.3
Phase portrait of the system
$\dot{x} = 2x + y, \dot{y} = 3x + 4y$

If you look carefully (or find your own phase portrait), you may notice that the trajectories shown are fleeing the origin (as $t \to \infty$) in such a way that any trajectory is tangent to the line $y = -x$ at the origin—that is, as $t \to -\infty$. In this situation, the origin is called an **unstable node** (specifically, a **source** or **repeller**). We'll come to a better understanding of this behavior in Section 5.3.   ◆

The next example reveals another type of source for a two-dimensional system.

EXAMPLE 5.2.5   A Spiral Source

Look at the system

$$\frac{dx}{dt} = x + y$$

$$\frac{dy}{dt} = -4x + y.$$

Note that once again the origin is this system's only equilibrium point. (Check this for yourself.)

Because the matrix of coefficients has $a = 1, b = 1, c = -4$, and $d = 1$, we use formula (5.2.4) to determine that the characteristic equation of this system is $\lambda^2 - 2\lambda + 5 = 0$, so the quadratic formula gives us the eigenvalues $\lambda_1 = 1 + 2i$ and $\lambda_2 = 1 - 2i$. When we get complex eigenvalues such as this complex conjugate pair, the eigenvectors will turn out to have complex numbers as entries and to have no useful direct geometric significance. We'll deal with this situation in more detail in Section 5.5. The phase portrait for this system is shown in Figure 5.4.

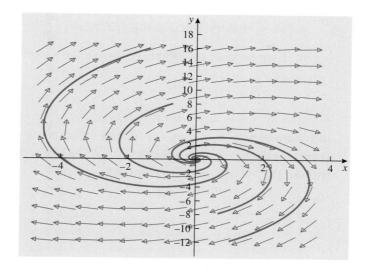

Figure 5.4

Phase portrait of the system $\dfrac{dx}{dt} = x + y, \dfrac{dy}{dt} = -4x + y$

$(x(0), y(0)) = (-4, 0), (-2, 0), (2, 0), (3, 0); -5 \le t \le 0.7$

We can see that the trajectories are spirals that move outward, away from the equilibrium point, in a clockwise direction. In this case, as in the last example, the equilibrium point is called a *source* (or a *repeller*). Other systems with complex eigenvalues may correspond to spirals that move in a *counterclockwise* direction or to spirals that move in *toward* the equilibrium point (clockwise or counterclockwise). ◆

These examples should convince you that trajectories can behave quite differently near equilibrium points. In the next section, we will examine how the trajectories of a two-dimensional system can be classified.

EXERCISES 5.2

1. Calculate the determinant of each of the following matrices by hand.

a. $\begin{bmatrix} -3 & 5 \\ -4 & 1 \end{bmatrix}$ **b.** $\begin{bmatrix} 4 & 2 \\ 10 & 5 \end{bmatrix}$

c. $\begin{bmatrix} 6t & -4 \\ \sin t & t^3 \end{bmatrix}$ **d.** $\begin{bmatrix} \cos\theta & \sin\theta \\ -\sin\theta & \cos\theta \end{bmatrix}$

2. Consider the system

$$ax + by = e$$
$$cx + dy = f,$$

where $a, b, c, d, e,$ and f are constants, with $ad - bc \ne 0$.

a. Show that the solution is given by $x = \dfrac{de - bf}{ad - bc}$, $y = \dfrac{af - ce}{ad - bc}$.

b. Express your solution in part (a) in terms of the determinants
$$\begin{vmatrix} a & e \\ c & f \end{vmatrix}, \begin{vmatrix} a & b \\ c & d \end{vmatrix}, \text{ and } \begin{vmatrix} e & b \\ f & d \end{vmatrix}.$$

For each system in Exercises 3–7, (a) convert to the matrix form $\dot{X} = AX$; (b) find the characteristic equation; (c) find all eigenvalues; (d) describe all eigenvectors corresponding to each eigenvalue found in part (c). Parts (a)–(d) should be done without the aid of a calculator or CAS.

3. $\dot{x} = 2x + y$
   $\dot{y} = 3x + 4y$

4. $\dot{x} = x - y$
   $\dot{y} = y - 4x$

5. $\dot{x} = -4x + 2y$
   $\dot{y} = 2x - y$

6. $\dot{x} = x$
   $\dot{y} = y$

7. $\dot{x} = -6x + 4y$
   $\dot{y} = -3x + y$

8. In Example 5.2.2, find the solution of the system satisfying the initial conditions $g(0) = g_0$ and $h(0) = 0$. (You may use technology to solve the resulting system of algebraic equations.)

9. In Example 5.2.3, the system $\dot{x} = -2x + y$, $\dot{y} = -4x + 3y$ was shown to have the solution $X = \begin{bmatrix} C_1 e^{2t} + C_2 e^{-t} \\ 4C_1 e^{2t} + C_2 e^{-t} \end{bmatrix}$.

a. Substitute for $x(t)$ and $y(t)$ in the right-hand side of the expression $\dfrac{dy}{dx} = \dfrac{-4x + 3y}{-2x + y}$.

b. Use the result of part (a) to show that as $t \to \infty$ the slope of any trajectory not on either of the lines determined by the eigenvectors approaches 4, the slope of the eigenvector corresponding to the larger of the two distinct eigenvalues. [*Hint*: Factor out e^{2t}, the dominant term for large positive values of t.]

c. Use the result of part (a) to show that as $t \to -\infty$ the slope of any trajectory not on either of the lines determined by the eigenvectors approaches 1, the slope of the eigenvector corresponding to the smaller of the two eigenvalues. [*Hint*: Factor out e^{-t}, the dominant term for large negative values of t.]

10. Use technology to sketch the phase portrait of the system in Exercise 3. Then sketch in the eigenvectors (getting them from the answers in the back of the book if necessary) and comment on the behavior of the trajectories with respect to the origin. (Use both positive and negative values of t.)

11. Use technology to sketch the phase portrait of the system in Exercise 4. Then sketch in the eigenvectors (using your CAS if necessary) and comment on the behavior of the trajectories with respect to the origin. (Use both positive and negative values of t.)

12. A substance X decays into substance Y at rate $k_1 > 0$, and Y in turn decays into another substance at rate $k_2 > 0$. The system

$$\frac{dx}{dt} = -k_1 x$$

$$\frac{dy}{dt} = k_1 x - k_2 y$$

describes the process, where $x(t)$ and $y(t)$ represent the amount of X and Y, respectively. Assume that $k_1 \neq k_2$.

a. Find the eigenvalues of the system.

b. Find the eigenvectors corresponding to each of the eigenvalues found in part (a).

c. Solve for $x(t)$ and $y(t)$ and then find $\lim_{t \to \infty} x(t)$ and $\lim_{t \to \infty} y(t)$, interpreting your answers in physical terms.

13. The following system models the exchange of nutrients between mother and fetus in the placenta.

$$\frac{dc_1}{dx} = -\alpha_1(c_1 - c_2)$$

$$\frac{dc_2}{dx} = -\alpha_2(c_1 - c_2),$$

where $c_1(x)$ is the concentration of nutrient in the maternal bloodstream at a distance x along the placental membrane and $c_2(x)$ is the concentration of nutrient in the fetal bloodstream at a distance x. Here α_1 and α_2 are constants, $\alpha_1 \neq \alpha_2$.

a. If $c_1(0) = c_0$ and $c_2(0) = C_0$, use eigenvalues and eigenvectors to solve the system for $c_1(x)$ and $c_2(x)$.

b. Solve for $c_1(x)$ and $c_2(x)$ by converting the system into a single second-order differential equation and using the techniques of Section 4.1.

5.3 STABILITY OF LINEAR SYSTEMS: UNEQUAL REAL EIGENVALUES

First of all, we should have guessed by now that a linear system $\dot{X} = AX$ of ordinary differential equations, where $\det(A) \neq 0$, has exactly one equilibrium point, the origin $(0, 0)$. (See Problem 12 of Exercises 5.1.) If $\det(A) = 0$, however, the system may have many other equilibrium solutions. As promised in the last section, the *stability* of a system—the behavior of trajectories with respect to the equilibrium point(s)—will be explained completely in terms of the eigenvalues and eigenvectors of the matrix A.

Because the characteristic equation of a two-dimensional system is a quadratic equation, we know that there are two eigenvalues, λ_1 and λ_2. There are only three possibilities for these eigenvalues: (1) The eigenvalues are both real numbers with $\lambda_1 \neq \lambda_2$; (2) the eigenvalues are real numbers with $\lambda_1 = \lambda_2$; or

(3) the eigenvalues are complex numbers: $\lambda_1 = p + qi$ and $\lambda_2 = p - qi$, where p and q are real numbers (called the *real part* and the *imaginary part*, respectively) and $i = \sqrt{-1}$. In situation 3, we say that λ_1 and λ_2 are *complex conjugates* of each other. (You may want to review Appendix C, especially C.3, for more information about complex numbers.) The nature of the eigenvalues will play an important role in the analysis of systems of linear equations, just as it did for second- and higher-order linear equations with constant coefficients in Sections 4.1–4.3. In this section we will deal with possibility 1 in the foregoing list, leaving situations 2 and 3 for the next two sections.

UNEQUAL REAL EIGENVALUES

First suppose that the matrix A in the system $\dot{X} = AX$ has two real eigenvalues λ_1 and λ_2, with $\lambda_1 \neq \lambda_2$. Let V_1 and V_2 be the corresponding representative eigenvectors. Then we'll show that the general solution of the system is given by

$$X(t) = c_1 e^{\lambda_1 t} V_1 + c_2 e^{\lambda_2 t} V_2, \tag{5.3.1}$$

where c_1 and c_2 are arbitrary constants.

Geometrically, the first term on the right-hand side of (5.3.1) represents a straight-line trajectory parallel[2] to V_1, and the second term describes a line parallel to V_2 (see Figure 5.5). Note that these trajectories lie in the phase plane (the *x-y* plane).

If both c_1 and c_2 are nonzero, then the solution $X(t)$ is a linear combination of the two basic terms whose relative contributions change with time. In this situation, the trajectories curve in a way that will be described below.

To see why (5.3.1) is the general solution, first note that each term is itself a solution of the system. If, for example, we consider $X_1(t) = c_1 e^{\lambda_1 t} V_1$, then $\dot{X}_1(t) =$

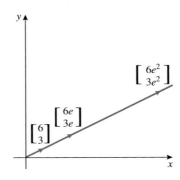

Figure 5.5

$$V = 3e^t \begin{bmatrix} 2 \\ 1 \end{bmatrix} \text{ for } t = 0, 1, \text{ and } 2$$

---

2. Two vectors V and W are **parallel** if $W = cV$ for some nonzero constant c. In other words, parallel vectors lie on the same straight line through the origin, pointing in the same direction (if $c > 0$) or in opposite directions (if $c < 0$). See Appendix B.1.

$c_1\lambda_1 e^{\lambda_1 t}V_1$ and $AX_1 = A(c_1 e^{\lambda_1 t}V_1) = c_1 e^{\lambda_1 t}(AV_1) = c_1 e^{\lambda_1 t}(\lambda_1 V_1) = \lambda_1 c_1 e^{\lambda_1 t}V_1$ because V_1 is an eigenvector corresponding to λ_1. (See Section 5.1 for properties of matrix multiplication.) Therefore, $\dot{X}_1(t) = AX_1$. Now let's see that if X_1 and X_2 are any solutions of the system, then the linear combination $X = k_1 X_1 + k_2 X_2$ is also a solution for any constants k_1 and k_2:

$$\dot{X} = \overbrace{(k_1 X_1 + k_2 X_2)}^{\bullet} = k_1 \dot{X}_1 + k_2 \dot{X}_2 = k_1(AX_1) + k_2(AX_2)$$
$$= A(k_1 X_1) + A(k_2 X_2)$$
$$= A(k_1 X_1 + k_2 X_2) = AX.$$

These steps follow from the algebraic properties of matrices (Section 5.1) and of derivatives, and this property of solutions of linear systems is another version of the Superposition Principle that we have encountered several times before. (For example, see Section 4.1.)

We can argue (somewhat loosely) that (5.3.1) represents a solution of a two-dimensional system (or its equivalent second-order equation) and has two arbitrary constants and hence is the *general* solution of the system $\dot{X} = AX$. To be rigorous, we can use the fact that any initial condition $X_0 = X(t_0) = \begin{bmatrix} x(t_0) \\ y(t_0) \end{bmatrix} = \begin{bmatrix} x_0 \\ y_0 \end{bmatrix}$ for the system can be written as a linear combination of the eigenvectors—$X_0 = k_1 V_1 + k_2 V_2$ for some constants k_1 and k_2—so a solution (5.3.1) can be found to satisfy any initial condition $X(t_0) = X_0$. (You'll be asked to prove these assertions in Exercises 17 and 18 at the end of this section.) Finally, the Existence and Uniqueness Theorem of Section 4.6 allows us to say that (5.3.1) is the *only* solution.

The Impossibility of Dependent Eigenvectors

If one of the eigenvectors is a scalar multiple of the other—say V_2 is a multiple of V_1—then the expression in (5.3.1) collapses to a scalar multiple of V_1 and there is only one arbitrary constant. This expression can't represent the general solution of a second-order equation.

Fortunately, this collapse can't happen with our current assumption. It is easy to prove that *if a 2×2 matrix A has distinct eigenvalues λ_1 and λ_2 with corresponding eigenvectors V_1 and V_2, then neither eigenvector is a scalar multiple of the other*. Suppose that $V_2 = cV_1$, where c is a nonzero scalar. Then $V_2 - cV_1 = \mathbf{0}$, the zero vector, and we must have

$$\mathbf{0} = A(V_2 - cV_1) = AV_2 - c(AV_1) = \lambda_2 V_2 - c(\lambda_1 V_1)$$
$$= \lambda_2(cV_1) - c(\lambda_1 V_1) = c(\lambda_2 - \lambda_1)V_1.$$

But then, because $c \neq 0$ and V_1 (as an eigenvector) is nonzero, we must conclude that $(\lambda_2 - \lambda_1) = 0$—contradicting the assumption that we have distinct eigenvalues.

Unequal Positive Eigenvalues

In the expression for the general solution, $c_1 e^{\lambda_1 t}V_1 + c_2 e^{\lambda_2 t}V_2$, suppose that $\lambda_1 > \lambda_2 > 0$. First note that as t increases, both eigenvector multiples point *away* from the origin so all solutions *grow* with time. (The algebraic signs of the constants c_1 and c_2 influence the quadrants in which the solutions grow.) To understand the *relative* rates at which the individual terms grow, we can factor out the exponential corresponding to the larger eigenvalue and write $X(t) = e^{\lambda_1 t}(c_1 V_1 + c_2 e^{(\lambda_2 - \lambda_1)t}V_2)$.

Note that $e^{(\lambda_2 - \lambda_1)t} \to 0$ as $t \to +\infty$ because $\lambda_2 - \lambda_1 < 0$. Therefore $X(t) \approx e^{\lambda_1 t}c_1 V_1$ as t gets larger and larger. Noting that $e^{\lambda_1 t}c_1 V_1$ is parallel to V_1, we see that the slope of any trajectory $X(t)$ approaches the slope of the line determined by V_1. This says that trajectories will curve *away* from the origin and their slopes will approach the slope of the line determined by the eigenvector V_1, corresponding to the larger eigenvalue. In this situation, the equilibrium point $(0, 0)$ is called a *source* (*unstable node, repeller*). (Recall our discussions in Section 2.5.) In "backward time," as $t \to -\infty$, the trajectories will be asymptotic to the line determined by the eigenvector V_2 because then the first term in the linear combination $c_1 e^{\lambda_1 t}V_1 + c_2 e^{\lambda_2 t}V_2$ is approaching zero faster than the second term. This says that if we move *backwards*, the trajectories *enter* the origin tangent to the line determined by V_2.

We are now ready to re-examine an earlier example in light of the last two paragraphs.

EXAMPLE 5.3.1   Unequal Positive Eigenvalues: A Source

First of all, the system

$$\dot{x} = 2x + y$$
$$\dot{y} = 3x + 4y$$

that we saw in Example 5.2.4 has two positive unequal eigenvalues, $\lambda_1 = 5$ and $\lambda_2 = 1$, with corresponding eigenvectors $V_1 = \begin{bmatrix} 1 \\ 3 \end{bmatrix}$ and $V_2 = \begin{bmatrix} 1 \\ -1 \end{bmatrix}$. Therefore, the general solution is

$$X(t) = c_1 e^{5t}\begin{bmatrix} 1 \\ 3 \end{bmatrix} + c_2 e^t\begin{bmatrix} 1 \\ -1 \end{bmatrix} = \begin{bmatrix} c_1 e^{5t} + c_2 e^t \\ 3c_1 e^{5t} - c_2 e^t \end{bmatrix} = \begin{bmatrix} x(t) \\ y(t) \end{bmatrix}.$$

Figure 5.6 is a more detailed version of Figure 5.3, the phase portrait of our system. The new graph shows several trajectories and the way in which they curve away from the origin, their slopes approaching the slope of the line determined by the eigenvector $V_1 = \begin{bmatrix} 1 \\ 3 \end{bmatrix}$ corresponding to the larger eigenvalue $\lambda = 5$.

Analytically, we can examine the equation $\dfrac{dy}{dx} = \dfrac{\dfrac{dy}{dt}}{\dfrac{dx}{dt}} = \dfrac{3x + 4y}{2x + y}$, whose solutions make up the phase portrait—that is, the equation giving the slopes of trajectories in the *x-y* plane. Substituting $x(t) = c_1 e^{5t} + c_2 e^t$ and $y(t) = 3c_1 e^{5t} - c_2 e^t$

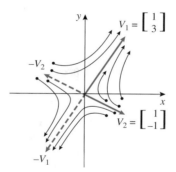

Figure 5.6

Trajectories of the system $\dot{x} = 2x + y$, $\dot{y} = 3x + 4y$

Bold points • indicate initial positions ($t = 0$) for trajectories.

from the general solution given above, we get $\dfrac{dy}{dx} = \dfrac{15c_1e^{5t} - c_2e^t}{5c_1e^{5t} + c_2e^t}$. For large

values of t, the expression for $\dfrac{dy}{dx}$ is dominated by the e^{5t} terms, which we can

factor out:

$$\frac{dy}{dx} = \frac{e^{5t}(15c_1 - c_2e^{-4t})}{e^{5t}(5c_1 + c_2e^{-4t})} = \frac{15c_1 - c_2e^{-4t}}{5c_1 + c_2e^{-4t}}.$$

The condition $c_1 = 0$ would mean that we are dealing with the straight-line trajec-

tory determined by the eigenvector $V_2 = \begin{bmatrix} 1 \\ -1 \end{bmatrix}$. But if $c_1 \neq 0$, as $t \to \infty$, we see

that the slope of any trajectory tends to $\dfrac{15c_1 - 0}{5c_1 + 0} = 3$, the slope of the line deter-

mined by the eigenvector $V_1 = \begin{bmatrix} 1 \\ 3 \end{bmatrix}$.

If we consider large *negative* values of t—that is, if we run the trajectories

backwards in time—then e^t is the dominant term in the expression for $\dfrac{dy}{dx}$ and we

can factor it out:

$$\frac{dy}{dx} = \frac{15c_1e^{5t} - c_2e^t}{5c_1e^{5t} + c_2e^t} = \frac{e^t(15c_1e^{4t} - c_2)}{e^t(5c_1e^{4t} + c_2)} = \frac{15c_1e^{4t} - c_2}{5c_1e^{4t} + c_2}.$$

This last expression tells us that if $c_2 \neq 0$, then as $t \to -\infty$, the slope of any trajec-

tory tends to $\dfrac{0 - c_2}{0 + c_2} = -1$, the slope of the line determined by the eigenvector

$V_2 = \begin{bmatrix} 1 \\ -1 \end{bmatrix}$. If we have $c_2 = 0$, we will be on the straight-line trajectory deter-

mined by the eigenvector $V_1 = \begin{bmatrix} 1 \\ 3 \end{bmatrix}$. We conclude that if $c_2 \neq 0$, then any trajectory

is tangent to the line $y = -x$ at the origin—that is, as $t \to -\infty$.   ◆

Unequal Negative Eigenvalues

If both eigenvalues are *negative* (say $\lambda_1 < \lambda_2 < 0$), then both eigenvector multiples point *toward* the origin, and all solutions *decrease* or *decay* with time. To see this, write (5.3.1) in the form

$$X(t) = \left[\begin{array}{c} c_1 \\ e^{-\lambda_1 t} \end{array}\right] V_1 + \left[\begin{array}{c} c_2 \\ e^{-\lambda_2 t} \end{array}\right] V_2 = \left[\begin{array}{c} c_1 \\ e^{Kt} \end{array}\right] V_1 + \left[\begin{array}{c} c_2 \\ e^{Mt} \end{array}\right] V_2,$$

where $K = -\lambda_1$ and $M = -\lambda_2$ are *positive* constants. Then clearly, both terms of $X(t)$ approach the origin as $t \to +\infty$. Because $\lambda_1 < \lambda_2$ we have $-\lambda_1 > -\lambda_2$, or $K > M$, so the first term in the expression for $X(t)$ approaches the origin faster than the second term. We will see in the next example that as t increases, trajectories curve *toward* the origin, closer to the eigenvector V_2 (or its negative if $c_2 < 0$), corresponding to the larger eigenvalue. Under these circumstances, we say that $(0, 0)$ is a *stable node*, or *sink*.

EXAMPLE 5.3.2   Unequal Negative Eigenvalues: A Sink

Suppose we look at the system

$$\dot{x} = -4x + y$$
$$\dot{y} = 3x - 2y.$$

The characteristic equation is $\lambda^2 + 6\lambda + 5 = 0$ and the eigenvalues are negative and unequal: $\lambda_1 = -5$ and $\lambda_2 = -1$. Using the linear algebra capabilities of a CAS, we find that the corresponding representative eigenvectors are $V_1 = \left[\begin{array}{c} -1 \\ 1 \end{array}\right]$ and $V_2 = \left[\begin{array}{c} 1 \\ 3 \end{array}\right]$. (Don't be disturbed if your CAS produces eigenvectors that are different from the book's—yours should lie on the same line as the ones given here. Your slopes y / x should be -1 and 3.)

The general solution of our system is $X(t) = c_1 e^{-5t} \left[\begin{array}{c} -1 \\ 1 \end{array}\right] + c_2 e^{-t} \left[\begin{array}{c} 1 \\ 3 \end{array}\right] = \left[\begin{array}{c} -c_1 e^{-5t} + c_2 e^{-t} \\ c_1 e^{-5t} + 3c_2 e^{-t} \end{array}\right].$

It is clear from the negative exponents in the expression for $X(t)$ that $X(t) \to \left[\begin{array}{c} 0 \\ 0 \end{array}\right]$ as $t \to \infty$, so the origin is a *sink*. Figure 5.7 shows some typical trajectories and seems to indicate that the trajectories are tangent to the line determined by the eigenvector $V_2 = \left[\begin{array}{c} 1 \\ 3 \end{array}\right]$.

Recognizing that e^{-t} is larger than e^{-5t} for large values of t, we look at

$$\frac{dy}{dx} = \frac{3x - 2y}{-4x + y} = \frac{-5c_1 e^{-5t} - 3c_2 e^{-t}}{5c_1 e^{-5t} - c_2 e^{-t}}$$
$$= \frac{e^{-t}(-5c_1 e^{-4t} - 3c_2)}{e^{-t}(5c_1 e^{-4t} - c_2)} = \frac{-5c_1 e^{-4t} - 3c_2}{5c_1 e^{-4t} - c_2}.$$

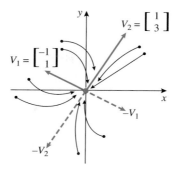

Figure 5.7

Trajectories for the system $\dot{x} = -4x + y$, $\dot{y} = 3x - 2y$
Bold points • indicate initial positions ($t = 0$) for trajectories.

If $c_2 \neq 0$, then $\dfrac{dy}{dx}$ approaches $\dfrac{-3c_2}{-c_2} = 3$, the slope of the eigenvector $V_2 = \begin{bmatrix} 1 \\ 3 \end{bmatrix}$, as $t \to \infty$. If $c_2 = 0$, then the trajectory is on the straight line determined by the eigenvector $\begin{bmatrix} -1 \\ 1 \end{bmatrix}$.   ◆

Unequal Eigenvalues with Opposite Signs

If the eigenvalues have *opposite* signs (say $\lambda_1 < 0 < \lambda_2$), then look at the general solution $X(t) = c_1 e^{\lambda_1 t} V_1 + c_2 e^{\lambda_2 t} V_2$ to see that the term $c_1 e^{\lambda_1 t} V_1$ (corresponding to the negative eigenvalue λ_1) points *toward* the origin, whereas $c_2 e^{\lambda_2 t} V$ points *away* from the origin (Figure 5.8).

In this case, trajectories *approach* the origin along one direction and veer *away* from the origin along another. In this situation we describe $(0, 0)$ as a *saddle point*. Look back at Example 5.2.3, especially Figure 5.2.

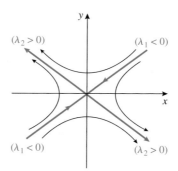

Figure 5.8

Typical eigenvectors for the case $\lambda_1 < 0 < \lambda_2$

Let's consider a new example of what happens when the eigenvalues of a system have opposite signs.

EXAMPLE 5.3.3   Unequal Eigenvalues with Opposite Signs: A Saddle Point

Let's investigate the system $\dfrac{dx}{dt} = x + 5y$, $\dfrac{dy}{dt} = x - 3y$. The characteristic equation is $\lambda^2 + 2\lambda - 8 = 0$. The eigenvalues and their corresponding eigenvectors are $\lambda_1 = -4$, $V_1 = \begin{bmatrix} 1 \\ -1 \end{bmatrix}$; $\lambda_2 = 2$, $V_2 = \begin{bmatrix} 5 \\ 1 \end{bmatrix}$. The general solution is

$$X(t) = c_1 e^{-4t} \begin{bmatrix} 1 \\ -1 \end{bmatrix} + c_2 e^{2t} \begin{bmatrix} 5 \\ 1 \end{bmatrix} = \begin{bmatrix} c_1 e^{-4t} + 5c_2 e^{2t} \\ -c_1 e^{-4t} + c_2 e^{2t} \end{bmatrix}.$$

We can see that the straight-line trajectory $c_1 e^{-4t} V_1 = c_1 e^{-4t} \begin{bmatrix} 1 \\ -1 \end{bmatrix} = \begin{bmatrix} c_1 e^{-4t} \\ -c_1 e^{-4t} \end{bmatrix}$ approaches the origin as $t \to \infty$. (There are actually *two* half-line trajectories, one for positive c_1 and one for negative c_1. See Figure 5.9.) But the half-line trajectories corresponding to $c_2 e^{2t} V_2 = c_2 e^{2t} \begin{bmatrix} 5 \\ 1 \end{bmatrix} = \begin{bmatrix} 5c_2 e^{2t} \\ c_2 e^{2t} \end{bmatrix}$ for positive and negative values of c_2 are clearly growing *away* from the origin with increasing t.

Substituting the expressions for $x(t)$ and $y(t)$ in the formula for $\dfrac{dy}{dx}$ and factoring out the dominant term for large t, we get

$$\frac{dy}{dx} = \frac{x - 3y}{x + 5y} = \frac{4c_1 e^{-4t} + 2c_2 e^{2t}}{-4c_1 e^{-4t} + 10c_2 e^{2t}}$$
$$= \frac{e^{2t}(4c_1 e^{-6t} + 2c_2)}{e^{2t}(-4c_1 e^{-6t} + 10c_2)} = \frac{4c_1 e^{-6t} + 2c_2}{-4c_1 e^{-6t} + 10c_2}.$$

If $c_2 \neq 0$, we see that as $t \to \infty$, $\dfrac{dy}{dx}$ tends to $\dfrac{2c_2}{10c_2} = \dfrac{1}{5}$, the slope of the eigenvector V_2. What this says is that the slopes of trajectories not on the straight lines deter-

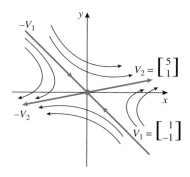

Figure 5.9

Trajectories for the system $\dfrac{dx}{dt} = x + 5y$, $\dfrac{dy}{dt} = x - 3y$

mined by V_1 and V_2 approach the slope of V_2, the eigenvalue associated with the positive eigenvalue. As $t \to -\infty$, the slopes of these trajectories tend to the slope of V_1. Figure 5.9 shows this partial-source/partial-sink behavior with respect to the origin, which is a *saddle point*. ◆

Unequal Eigenvalues, One Eigenvalue Equal to Zero

Finally, we consider the situation in which we have two unequal eigenvalues, but one of them is 0. Suppose that $\lambda_1 = 0$ and $\lambda_2 \neq 0$. This means that the characteristic equation can be written in the form $0 = (\lambda - 0)(\lambda - \lambda_2) = \lambda^2 - \lambda_2\lambda$. We know from Section 5.2 that the constant term of the characteristic equation equals $\det(A)$. Clearly in this case we have $\det(A) = 0$. Therefore, we should not expect the origin to be the only equilibrium point (see Problem 12 of Exercises 5.1). In fact, *every point $(x, 0)$ of the horizontal axis may be an equilibrium point for such a system.* (Exercise 20 at the end of this section asks for a proof of this assertion.) If V_1 is the eigenvector associated with $\lambda_1 = 0$, we know that $A(c_1V_1) = c_1A(V_1) = c_1\lambda_1V_1 = \mathbf{0}$—that is, *each point on the line determined by V_1 is an equilibrium point.*

The general solution in this situation has the form $X(t) = c_1e^{(0)t}V_1 + c_2e^{\lambda_2t}V_2 = c_1V_1 + c_2e^{\lambda_2t}V_2$. Note that if $\lambda_2 > 0$ and $t \to \infty$, then $X(t)$ grows without bound. But if $t \to -\infty$, so that we are traveling backward along a trajectory, then the trajectory approaches c_1V_1, the line determined by V_1. Similarly, if $\lambda_2 < 0$ and $t \to \infty$, then $X(t)$ approaches the line determined by V_1, whereas if $t \to -\infty$, then $X(t)$ grows without bound. In any case, each trajectory will be a half-line parallel (in the usual plane-geometry sense) to the eigenvector V_2, with one endpoint on the line determined by V_1. (The constant vector c_1V_1 just shifts $c_2e^{\lambda_2t}V_2$ horizontally and vertically.)

The next example should explain the geometry of the trajectories when we have one eigenvalue equal to 0.

EXAMPLE 5.3.4   Unequal Eigenvalues, One Eigenvalue Equal to Zero
Figure 5.10 shows the phase portrait for the system $\dot{x} = y, \dot{y} = y$, whose eigenvalues are 0 and 1 and whose corresponding eigenvectors are $\begin{bmatrix} 1 \\ 0 \end{bmatrix}$ and $\begin{bmatrix} 1 \\ 1 \end{bmatrix}$, respectively. Therefore, the equations of the trajectories are $X(t) = c_1\begin{bmatrix} 1 \\ 0 \end{bmatrix} + c_2e^t\begin{bmatrix} 1 \\ 1 \end{bmatrix} = \begin{bmatrix} c_1 + c_2e^t \\ c_2e^t \end{bmatrix}$. This says (Exercise 19) that any trajectory not on the line determined by $V = \begin{bmatrix} 1 \\ 0 \end{bmatrix}$ has the equation $y(t) = x(t) + k$, so these trajectories form an infinite family of straight lines parallel to $y = x$. Note that the eigenvector $\begin{bmatrix} 1 \\ 0 \end{bmatrix}$ corresponding to the zero eigenvalue determines two half-line trajectories, the positive x-axis and the negative x-axis. In our example, it is easy to see that every point $(x, 0)$ of the horizontal axis is an equilibrium point: $\dot{x} = y = 0$ and $\dot{y} = y = 0$ imply that $y = 0$ and the x-coordinate is completely

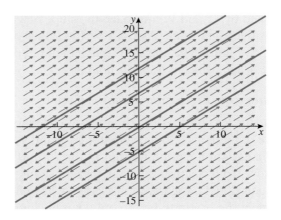

Figure 5.10
Phase portrait for the system $\dot{x} = y, \dot{y} = y$

arbitrary. The fact that the nonzero eigenvalue is positive makes the points on the x-axis *sources*. (If necessary, review the last full paragraph before this example.) ◆

By looking at Examples 5.2.3–5.2.5 and the examples in this section, we notice that a solution starting in a direction different from those of the eigenvectors is curved, representing [as we know from (5.3.1)] a linear combination, $c_1 e^{\lambda_1 t} V_1 + c_2 e^{\lambda_2 t} V_2$, of two exponential solutions that have different rates of change (indicated by the eigenvalues). If we look at enough phase portraits, we may also realize that there is a tendency for the "fast" eigenvector (associated with the larger of two unequal eigenvalues) to have the stronger influence on the solutions. Trajectories curve toward the direction of this eigenvector as $t \to \infty$.

In the next section, we'll investigate what happens when there is a repeated real eigenvalue and when there seems to be only one eigenvector corresponding to two real eigenvalues.

EXERCISES 5.3

For each of the systems in Exercises 1–10, (a) find the eigenvalues and their corresponding eigenvectors and (b) sketch or plot a few trajectories and show the position(s) of the eigenvector(s). Do part (a) manually, but if the eigenvalues are irrational numbers, you may use technology to find the corresponding eigenvectors.

1. $\dot{x} = 3x, \dot{y} = 2y$

2. $\dot{x} = -x, \dot{y} = -2y$

3. $x' = -3x - y, y' = 4x + 2y$

4. $\dot{r} = 5r + 4s, \dot{s} = -2r - s$

5. $\dot{x} = x + 5y, \dot{y} = x - 3y$

6. $\dot{x} = 2x + 3y, \dot{y} = x + y$

7. $\dot{x} = -3x + y, \dot{y} = 4x - 2y$

8. $x' = -4x + 2y, y' = -3x + y$

9. $x' = -2x - y, y' = -x + 2y$

10. $\dot{x} = 2x + y, \dot{y} = 2x + 3y$

11. Consider the system $\dot{x} = 4x - 3y$, $\dot{y} = 8x - 6y$.

    a. Find the eigenvalues of this system.

    b. Find the eigenvectors corresponding to the eigenvalues in part (a).

    c. Sketch or plot some trajectories and explain what you see.

    d. Write the general solution of the system in the form $X(t) = c_1 e^{\lambda_1 t} V_1 + c_2 e^{\lambda_2 t} V_2$, and then re-examine your explanation in part (c).

12. Show that if X is an eigenvector of A corresponding to eigenvalue λ, then any nonzero multiple of X is also an eigenvector of A corresponding to λ.

13. Write a system of first-order linear equations whose trajectories show the following behaviors.

    a. $(0, 0)$ is a sink with eigenvalues $\lambda_1 = -3$ and $\lambda_2 = -5$.

    b. $(0, 0)$ is a saddle point with eigenvalues $\lambda_1 = -1$ and $\lambda_2 = 4$.

    c. $(0, 0)$ is a source with eigenvalues $\lambda_1 = 2$ and $\lambda_2 = 3$.

14. Consider the system $\dot{x} = -x + \alpha y$, $\dot{y} = -2y$, where α is a constant.

    a. Show that the origin is a *sink* regardless of the value of α.

    b. Assume that $X(t)$ is the solution vector of the system satisfying the initial condition $X(0) = \begin{bmatrix} 0 \\ 0.5 \end{bmatrix}$. Sketch the trajectory for different values of α and describe how the trajectory $X(t)$, for $t \geq 0$, depends on the value of α.

15. Two quantities of a chemical solution are separated by a membrane. If $x(t)$ and $y(t)$ represent the amounts of the chemical at time t on each side of the membrane, and if V_1 and V_2 represent the (constant) volume of each solution, respectively, then the *diffusion problem* can be modeled by the system

$$\dot{x} = P\left[\frac{y}{V_2} - \frac{x}{V_1}\right]$$

$$\dot{y} = P\left[\frac{x}{V_1} - \frac{y}{V_2}\right],$$

where P is a positive constant called the *permeability* of the membrane. Note that $\dfrac{x(t)}{V_1}$ and $\dfrac{y(t)}{V_2}$ represent the *concentrations* of solution on each side.

    a. Assuming that $x(0) = x_0$ and $y(0) = y_0$, find the solution of the system IVP *without using technology.*

    b. Calculate $\lim\limits_{t \to \infty} x(t)$ and $\lim\limits_{t \to \infty} y(t)$.

    c. Using part (b), interpret the result $\lim\limits_{t \to \infty}[x(t) + y(t)]$ physically.

    d. Note that if $\dfrac{y}{V_2} > \dfrac{x}{V_1}$, then $\dot{x} > 0$. Does this say that the chemical moves across the membrane from the side with a lower concentration to the side with a higher concentration, or vice versa? Confirm your answer by considering what happens if $\dfrac{x}{V_1} > \dfrac{y}{V_2}$ in the second equation.

16. Consider the system

$$\dot{r} = -r - s$$
$$\dot{s} = -\beta r - s$$

where β is a parameter.

   a. Find the general solution of the system when $\beta = 0.5$. Use the eigenvalues of the coefficient matrix to determine what kind of equilibrium the system has at the origin.

   b. Find the general solution of the system when $\beta = 2$. Use the eigenvalues of the coefficient matrix to determine what kind of equilibrium the system has at the origin.

   c. The solutions of the system show two rather different types of behavior for the two values of β considered in parts (a) and (b). Find a formula for the eigenvalues in terms of β, and determine the value of β between 0.5 and 2 where the transition from one type of behavior to the other occurs. (This critical value of the parameter is called a *bifurcation point*. See Section 2.6.)

17. Suppose that we have the system $\dot{X} = AX$ and that V_1 and V_2 are eigenvectors of A such that neither V_1 nor V_2 is a scalar multiple of the other. Show that any initial condition $X_0 = X(0) = \begin{bmatrix} x_0 \\ y_0 \end{bmatrix}$ can be written as a linear combination of V_1 and V_2. In other words, show that you can always find scalars c_1 and c_2 so that $X_0 = c_1 V_1 + c_2 V_2$.

[*Hint:* Let $V_1 = \begin{bmatrix} x_1 \\ y_1 \end{bmatrix}$ and $V_2 = \begin{bmatrix} x_2 \\ y_2 \end{bmatrix}$ be the eigenvectors, where you assume that x_1, x_2, y_1, and y_2 are known. Now convert the equation $X_0 = c_1 V_1 + c_2 V_2$ into a system of algebraic linear equations and go from there.]

18. If the system $\dot{X} = AX$ has two real eigenvalues λ_1 and λ_2, with $\lambda_1 \neq \lambda_2$, and if V_1 and V_2 are the corresponding (distinct) eigenvectors, show that $X(t) = c_1 e^{\lambda_1 t} V_1 + c_2 e^{\lambda_2 t} V_2$ satisfies the initial condition $X(0) = X_0 = \begin{bmatrix} x_0 \\ y_0 \end{bmatrix} = c_1 V_1 + c_2 V_2$. (See the previous problem for the justification of this representation of X_0 for some scalars c_1 and c_2.)

19. As indicated in Example 5.3.4, the system $\dot{x} = y, \dot{y} = y$ has the solution $X(t) = \begin{bmatrix} c_1 + c_2 e^t \\ c_2 e^t \end{bmatrix}$. Show that any trajectory not on the line determined by $V = \begin{bmatrix} 1 \\ 0 \end{bmatrix}$ satisfies the equation $y(t) = x(t) + k$ (in the phase plane) for some constant k. (This says that the trajectories form an infinite family of lines parallel to $y = x$.)

20. Consider the system

$$\dot{x} = ax + by$$
$$\dot{y} = cx + dy,$$

where a, b, c, and d are constants. Show that if $ad - bc = 0$, then every point $(x, 0)$ of the horizontal axis may be an equilibrium point for the system. [*Hint:* Solve the system $ax + by = 0$, $cx + dy = 0$ for y.]

5.4   STABILITY OF LINEAR SYSTEMS: EQUAL REAL EIGENVALUES

Now let's see what happens if both eigenvalues are real and equal. In other words, the characteristic equation has a *repeated root*, or *double root*. (See Section 4.1 for the second-order homogeneous linear equation case.) A full understanding of this situation requires more linear algebra than we want to pursue right now. However, the following discussions and examples should give us a good idea of what's going on.

EQUAL NONZERO EIGENVALUES, TWO INDEPENDENT EIGENVECTORS

First suppose that $\lambda_1 = \lambda_2 \neq 0$. If we can find distinct representative eigenvectors V_1 and V_2 that are not scalar multiples of each other, then we can still write the general solution of the system using (5.3.1): $X(t) = c_1 e^{\lambda_1 t} V_1 + c_2 e^{\lambda_2 t} V_2 = c_1 e^{\lambda_1 t} V_1 + c_2 e^{\lambda_1 t} V_2 = e^{\lambda_1 t}(c_1 V_1 + c_2 V_2)$. If we let $t = 0$, we see that $X(0) = e^{\lambda_1(0)}(c_1 V_1 + c_2 V_2) = c_1 V_1 + c_2 V_2$, so we can write $X(t) = e^{\lambda_1 t} X_0$, where $X_0 = X(0)$. (See Problem 17 of Exercises 5.3.) Under these conditions, all trajectories are straight lines through the origin because they are constant multiples of the constant vector $X_0 = c_1 V_1 + c_2 V_2$. The origin is called a **star node** in this case and

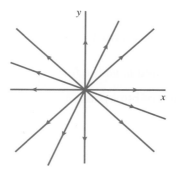

Figure 5.11a
Source: $\lambda > 0$

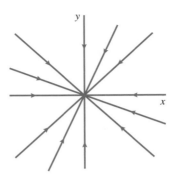

Figure 5.11b
Sink: $\lambda < 0$

will be a *source* if $\lambda_1 > 0$ and a *sink* if $\lambda_1 < 0$. Figures 5.11a and 5.11b show possible trajectories for various initial vectors X_0.

Let's examine a system for which the origin is a star node.

EXAMPLE 5.4.1   The Origin as a Star Node (a Source)

Look at the system $\dfrac{dx}{dt} = x, \dfrac{dy}{dt} = y$. We can write this in matrix form as $\dot{X} = AX$, where $A = \begin{bmatrix} 1 & 0 \\ 0 & 1 \end{bmatrix}$. It is easy to see that A has eigenvalues $\lambda_1 = 1 = \lambda_2$. (Check this.) By the way we defined the product of a matrix and a vector in Section 5.1, we see that our matrix of coefficients A is such that $AV = V = 1 \cdot V = \lambda_1 V$ for *every* vector V. In particular, any *nonzero* vector V is an eigenvector corresponding to the eigenvalue 1. *Be sure you understand this last statement.* A particularly simple eigenvector to work with is $V_1 = \begin{bmatrix} 1 \\ 0 \end{bmatrix}$. It is easy to see that the vector $V_2 = \begin{bmatrix} 0 \\ 1 \end{bmatrix}$ is not a multiple of V_1 because any scalar multiple of V_1 would have the form $\begin{bmatrix} c \\ 0 \end{bmatrix}$, where c is a constant. Therefore, we can write the solution of our system as

$$X(t) = c_1 e^t V_1 + c_2 e^t V_2 = c_1 e^t \begin{bmatrix} 1 \\ 0 \end{bmatrix} + c_2 e^t \begin{bmatrix} 0 \\ 1 \end{bmatrix} = \begin{bmatrix} c_1 e^t \\ c_2 e^t \end{bmatrix}.$$

Of course, because each of our original (separable) differential equations contains only one variable, we could solve each one separately to get the same result in the form $x(t) = c_1 e^t, y(t) = c_2 e^t$. As we indicated in the discussion right before this example, the trajectories are straight lines through the origin, and Figure 5.12 shows that the origin, a star node, is a *source*.

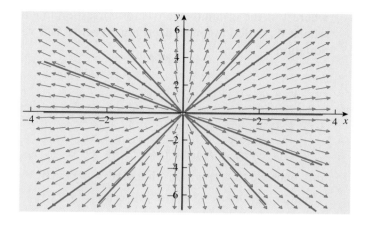

Figure 5.12

Phase portrait of the system $\dfrac{dx}{dt} = x, \dfrac{dy}{dt} = y$ ◆

EQUAL NONZERO EIGENVALUES, ONLY ONE INDEPENDENT EIGENVECTOR

Now suppose that $\lambda_1 = \lambda_2 \neq 0$, but our single eigenvalue has *only one distinct representative eigenvector*. What we mean is that all eigenvectors corresponding to the single distinct eigenvalue are scalar multiples of each other. Geometrically, this says that all eigenvectors lie on the same straight line through the origin. Then if we tried to use the solution form (5.3.1), we would get

$$X(t) = c_1 e^{\lambda_1 t}V + c_2 e^{\lambda_1 t}V = (c_1 + c_2)e^{\lambda_1 t}V = ke^{\lambda_1 t}V.$$

But how can the general solution of a two-dimensional system or second-order equation have only one arbitrary constant?

What we have to do here is find another solution of the system that is *independent* of the one solution we found using the single eigenvalue and its representative eigenvector. This is similar to the technique we used in solving a second-order linear equation with a repeated eigenvalue (see Section 4.1). In our situation, an independent solution is one that is not a scalar multiple of the first solution. If we *do* find another eigenvector corresponding to the single eigenvalue, but one that is independent of the original eigenvector, then the solution can still be written in the form $X(t) = c_1 e^{\lambda_1 t}V_1 + c_2 e^{\lambda_2 t}V_2$.

It turns out that we *can* find a substitute for an independent eigenvector. Although we won't go into all the linear algebraic details here, we can at least try to explain the end result. Another (independent) solution of the system must have the form

$$X_2(t) = te^{\lambda t}V + e^{\lambda t}W, \tag{5.4.1}$$

where V is the original eigenvector corresponding to the single eigenvalue λ, and where W, called a **generalized eigenvector,** is a vector that satisfies the matrix equation

$$(A - \lambda I)W = V. \tag{5.4.2}$$

(See Exercise 8.)

It's easy to see that the vector defined by (5.4.1) is a solution of the system. If $X(t) = te^{\lambda t}V + e^{\lambda t}W$, then

$$\dot{X}(t) = t(\lambda e^{\lambda t}V) + e^{\lambda t}V + \lambda e^{\lambda t}W = (\lambda t + 1)e^{\lambda t}V + \lambda e^{\lambda t}W$$

and, because (5.4.2) implies that $AW = V + \lambda W$,

$$AX = A(te^{\lambda t}V + e^{\lambda t}W) = te^{\lambda t}(AV) + e^{\lambda t}(AW) = te^{\lambda t}(\lambda V) + e^{\lambda t}(V + \lambda W)$$
$$= (\lambda t + 1)e^{\lambda t}V + \lambda e^{\lambda t}W.$$

Thus $\dot{X} = AX$—that is, (5.4.1) defines a solution of the system.

Next we must solve equation (5.4.2) for W, and then we can write the general solution of the system as

$$X(t) = c_1 e^{\lambda t}V + c_2[te^{\lambda t}V + e^{\lambda t}W]. \tag{5.4.3}$$

(The theory of linear algebra shows that we can always solve for W in equation (5.4.2) if V is an eigenvector of A corresponding to eigenvalue λ.)

Now let's look at an example in which we have equal nonzero eigenvalues but only one distinct representative eigenvector.

EXAMPLE 5.4.2   Equal Nonzero Eigenvalues, Only One Distinct Eigenvector
Consider the system $\dot{x} = -2x + y$, $\dot{y} = -2y$. We can write this in matrix form as $\dot{X} = AX$, where $A = \begin{bmatrix} -2 & 1 \\ 0 & -2 \end{bmatrix}$. The characteristic polynomial of A is $\lambda^2 + 4\lambda + 4 = (\lambda + 2)^2$, so $\lambda = -2$ is a repeated root. Then the matrix equation $AV = \lambda V = -2V$ is equivalent to the system

$$\begin{array}{rl} -2x + y = & -2x \\ -2y = & -2y, \end{array}$$

or

$$\begin{array}{l} y = 0 \\ -2y = -2y. \end{array}$$

From this we see that any eigenvector $\begin{bmatrix} x \\ y \end{bmatrix}$ must have the form $\begin{bmatrix} x \\ 0 \end{bmatrix} = x\begin{bmatrix} 1 \\ 0 \end{bmatrix}$ for arbitrary values of x. Therefore, we can take $V = \begin{bmatrix} 1 \\ 0 \end{bmatrix}$ as the only independent eigenvector that corresponds to the eigenvalue -2. Now we must find a vector $W = \begin{bmatrix} r \\ s \end{bmatrix}$ satisfying $(A - \lambda I)W = V$.

In our problem, $(A - \lambda I)W = V$ becomes

$$\left[\begin{bmatrix} -2 & 1 \\ 0 & -2 \end{bmatrix} - (-2)\begin{bmatrix} 1 & 0 \\ 0 & 1 \end{bmatrix}\right]\begin{bmatrix} r \\ s \end{bmatrix} = \begin{bmatrix} 1 \\ 0 \end{bmatrix},$$

$$\left[\begin{bmatrix} -2 & 1 \\ 0 & -2 \end{bmatrix} + \begin{bmatrix} 2 & 0 \\ 0 & 2 \end{bmatrix}\right]\begin{bmatrix} r \\ s \end{bmatrix} = \begin{bmatrix} 1 \\ 0 \end{bmatrix},$$

or

$$\begin{bmatrix} 0 & 1 \\ 0 & 0 \end{bmatrix}\begin{bmatrix} r \\ s \end{bmatrix} = \begin{bmatrix} 1 \\ 0 \end{bmatrix},$$

which is equivalent to the algebraic system

$$0 \cdot r + 1 \cdot s = 1$$
$$0 \cdot r + 0 \cdot s = 0.$$

This tells us that $s = 1$ and r is a "free variable"—that is, r is completely arbitrary. For convenience, let $r = 0$ so that our generalized eigenvector is $W = \begin{bmatrix} 0 \\ 1 \end{bmatrix}$. Finally, we can write the general solution of our system in the form (5.4.3):

$$X(t) = c_1 e^{-2t}\begin{bmatrix} 1 \\ 0 \end{bmatrix} + c_2\left[te^{-2t}\begin{bmatrix} 1 \\ 0 \end{bmatrix} + e^{-2t}\begin{bmatrix} 0 \\ 1 \end{bmatrix}\right]$$

$$= \begin{bmatrix} c_1 e^{-2t} + c_2 t e^{-2t} \\ c_2 e^{-2t} \end{bmatrix}.$$

Figure 5.13, generated by a CAS, shows that the trajectories spiral in toward the origin, in such a way that they are tangent to the eigenvector $V = \begin{bmatrix} 1 \\ 0 \end{bmatrix}$ or its negative at the origin. (Note that the vector V is part of the positive x-axis.)

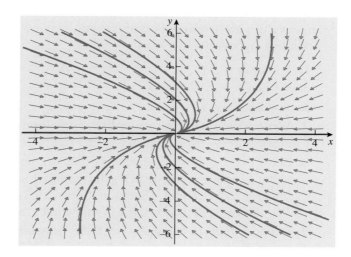

Figure 5.13
Trajectories for the system $\dot{x} = -2x + y$, $\dot{y} = -2y$

Whenever we have a system with equal nonzero eigenvalues but only one distinct eigenvector, the phase portrait will consist of spirals *approaching* the origin when the repeated eigenvalue is *negative*, and the phase portrait will consist of spirals moving *outward* if the eigenvalue is *positive*. A negative eigenvalue makes the origin a **spiral sink;** a positive eigenvalue makes the origin a **spiral source.** Furthermore, if the eigenvalue is negative, the slopes of all trajectories not on the line determined by the one eigenvector approach the slope of this line as $t \to \infty$. A positive eigenvalue indicates that the slopes of all trajectories not on the line determined by the one eigenvector approach the slope of this line as $t \to -\infty$. (Exercise 9 asks for a proof of the last two assertions.)   ◆

BOTH EIGENVALUES ZERO

Finally, let's assume that $\lambda_1 = \lambda_2 = 0$. If there are two linearly independent eigenvectors V_1 and V_2, then the general solution is $X(t) = c_1 e^{0 \cdot t} V_1 + c_2 e^{0 \cdot t} V_2 = c_1 V_1 + c_2 V_2$, a single vector of constants. If there is only one linearly independent eigenvector V corresponding to the eigenvalue 0, then we can find a generalized eigenvector and use formula (5.4.3):

$$X(t) = c_1 e^{\lambda t} V + c_2 [t e^{\lambda t} V + e^{\lambda t} W].$$

For $\lambda = 0$, we get $X(t) = c_1 V + c_2 [tV + W] = (c_1 + c_2 t)V + c_2 W$. In Exercise 10 you will investigate a system that has both eigenvalues zero.

EXERCISES 5.4

For each of the systems in Exercises 1–6, (a) find the eigenvalues and their corresponding linearly independent eigenvectors and (b) sketch or plot a few trajectories and show the position(s) of the eigenvector(s). Do part (a) manually, but if the eigenvalues are irrational numbers, you may use technology to find the corresponding eigenvectors.

1. $\dot{x} = 3x, \dot{y} = 3y$ **2.** $\dot{x} = -4x, \dot{y} = x - 4y$

3. $\dot{x} = 2x + y, \dot{y} = 4y - x$ **4.** $\dot{x} = 3x - y, \dot{y} = 4x - y$

5. $\dot{x} = 2y - 3x, \dot{y} = y - 2x$ **6.** $\dot{x} = 5x + 3y, \dot{y} = -3x - y$

7. Write a system of first-order linear equations for which $(0, 0)$ is a sink with eigenvalues $\lambda_1 = -2$ and $\lambda_2 = -2$.

8. Show that if V is an eigenvector of a 2×2 matrix A corresponding to eigenvalue λ, and if vector W is a solution of $(A - \lambda I)W = V$, then V and W are linearly independent. (See equations (5.4.2)–(5.4.3).) [*Hint:* Suppose that $W = cV$ for some scalar c. Then show that V must be the zero vector.]

9. Suppose that a system $\dot{X} = AX$ has only one eigenvalue λ and that every eigenvector is a scalar multiple of one fixed eigenvector, V. Then equation (5.4.3) tells us that any trajectory has the form

$$X(t) = c_1 e^{\lambda t} V + c_2[t e^{\lambda t} V + e^{\lambda t} W] = t e^{\lambda t}\left[\frac{1}{t}(c_1 V + W) + c_2 V\right].$$

a. If $\lambda < 0$, show that the slope of $X(t)$ approaches the slope of the line determined by V as $t \to \infty$. [*Hint:* $\dfrac{e^{-\lambda t}}{t}X(t)$, as a scalar multiple of $X(t)$, is parallel to $X(t)$.]

b. If $\lambda < 0$, show that the slope of $X(t)$ approaches the slope of the line determined by V as $t \to -\infty$.

10. Consider the system $\dot{x} = 6x + 4y$, $\dot{y} = -9x - 6y$.

a. Show that the only eigenvalue of the system is 0.

b. Find the single independent eigenvector V corresponding to $\lambda = 0$.

c. Show that every trajectory of this system is a straight line parallel to V, with trajectories on opposite sides of V moving in opposite directions.

5.5 STABILITY OF LINEAR SYSTEMS: COMPLEX EIGENVALUES

COMPLEX EIGENVALUES AND COMPLEX EIGENVECTORS

Now let's examine what occurs when the matrix A in the system $\dot{X} = AX$ has *complex* eigenvalues. As we've already stated, any complex root λ of the quadratic characteristic equation $\lambda^2 - (a + d)\lambda + (ad - bc) = 0$ must occur as part of a *complex conjugate pair*: $\lambda = p \pm qi$. As we'll see, the behavior of trajectories in the case of complex eigenvalues depends on the *real part*, p, of the complex eigenvalues. When the eigenvalues of a matrix are complex numbers, the eigenvectors will also have complex entries (see Appendix C), and therefore the algebra of the situation will be slightly more complicated.

The most important thing to realize is that when A has complex eigenvalues, the general solution of $\dot{X} = AX$ has the same form as (5.3.1), $X(t) = c_1 e^{\lambda_1 t} V_1 + c_2 e^{\lambda_2 t} V_2$. In other words, the Superposition Principle holds, but we have to deal with the fact that this formula will produce vectors whose elements are complex functions or numbers. For example, in the context of the general solution formula, the phrase *multiplying by a scalar* refers to multiplying vectors (whose entries may be complex numbers) by complex numbers.

Fortunately, there are some useful results that aid us in our work with complex eigenvalues and eigenvectors:

1. A crucial fact to recall is *Euler's formula*, which we saw in Section 4.1:

$$e^{p+qi} = e^p(\cos(q) + i\sin(q)).$$

This result will be useful in simplifying complex-valued expressions and will show us how to obtain real-valued solutions of $\dot{X} = AX$.

2. Another important fact is that *eigenvectors corresponding to complex conjugate eigenvalues are conjugate to each other.* If the eigenvalue $\lambda_1 = p + qi$ has a corresponding eigenvector $V_1 = \begin{bmatrix} a_1 + b_1 i \\ a_2 + b_2 i \end{bmatrix} = \begin{bmatrix} a_1 \\ a_2 \end{bmatrix} + i \begin{bmatrix} b_1 \\ b_2 \end{bmatrix} = U + iW$, then $\lambda_2 = \overline{\lambda}_1 = p - qi$ has a corresponding eigenvector $V_2 = \overline{V}_1 = \overline{\begin{bmatrix} a_1 + b_1 i \\ a_2 + b_2 i \end{bmatrix}} = \begin{bmatrix} a_1 - b_1 i \\ a_2 - b_2 i \end{bmatrix} = \begin{bmatrix} a_1 \\ a_2 \end{bmatrix} - i \begin{bmatrix} b_1 \\ b_2 \end{bmatrix} = U - iW$. The proof of this result follows from the properties of the conjugate. Suppose that $AV_1 = \lambda_1 V_1$. Then $\overline{(AV_1)} = \overline{(\lambda_1 V_1)}$, so $\overline{A} \, \overline{V}_1 = \overline{\lambda}_1 \overline{V}_1$, or (because all elements of A are real) $A\overline{V}_1 = \overline{\lambda}_1 \overline{V}_1 = \lambda_2 \overline{V}_1$. That is, $\overline{V}_1$ is an eigenvector corresponding to $\lambda_2 = \overline{\lambda}_1$.

To see how valuable results 1 and 2 are, let's suppose that $\lambda = p + qi$ is an eigenvalue of the matrix A and that $V = U + iW$ is a corresponding eigenvector. If we define $X(t) = e^{\lambda t} V$, then $AX = A(e^{\lambda t}V) = e^{\lambda t}(AV) = e^{\lambda t}(\lambda V) = \lambda e^{\lambda t}V = \dot{X}$, so $X(t)$ is a solution of $\dot{X} = AX$. Using Euler's formula and the properties of complex multiplication (see Appendix C), we have

$$X(t) = e^{\lambda t}V = e^{(p+qi)t}V = e^{pt}(\cos qt + i \sin qt)(U + iW)$$
$$= e^{pt}\{(\cos qt)U - (\sin qt)W\} + ie^{pt}\{(\cos qt)W + (\sin qt)U\}.$$

Then the *real part* and the *imaginary part* of $X(t)$ can be considered separately.

$$X_1(t) = \text{Re}\{X(t)\} = e^{pt}\{(\cos qt)U - (\sin qt)W\}$$
$$X_2(t) = \text{Im}\{X(t)\} = e^{pt}\{(\cos qt)W + (\sin qt)U\}.$$

The important observation here is that $X_1(t)$ and $X_2(t)$ *are real-valued linearly independent solutions of the system* $\dot{X} = AX$. (Exercise 10 asks for a proof that the same two solutions result from taking the real and imaginary parts of $e^{\overline{\lambda}t}\overline{V}$.) We will justify this observation for the real part of $X(t)$, leaving the proof for the imaginary part as Exercise 11. First we write $X_R = \text{Re}\{X(t)\} = \dfrac{X + \overline{X}}{2}$ (see Appendix C.1 if necessary). Then

$$AX_R = A\left(\frac{X + \overline{X}}{2}\right) = \tfrac{1}{2}A(X + \overline{X}) = \tfrac{1}{2}(AX + A\overline{X})$$

$$= \tfrac{1}{2}(\dot{X} + \overline{(AX)}) = \tfrac{1}{2}(\dot{X} + \overline{\dot{X}}) = \text{Re}(\dot{X}) = (\dot{X})_R = \overset{\displaystyle\cdot}{\overbrace{(X_R)}}.$$

Now the Superposition Principle tells us that $c_1 X_1(t) + c_2 X_2(t)$ is also a solution—in fact, it is the *general solution* of the system. The proofs of this last fact in Section 5.3 are valid here. We can take the scalars c_1 and c_2 to be real numbers.

As a first example of working with complex eigenvalues and eigenvectors, let's look at the equation $\dfrac{d^2\theta}{dt^2} + k^2 \sin\theta = 0$, which describes the motion of an *undamped pendulum*. Here θ is the angle the pendulum makes with the vertical, and $k^2 = \dfrac{g}{L}$, where g is the acceleration due to gravity and L is the length of the pendulum. This famous equation is nonlinear and will be treated fully in Chapter 7, but for small angles θ, $\sin\theta \approx \theta$, so we can consider the *linearized* equation $\frac{d^2\theta}{dt^2} + k^2\theta = 0$. The system form of the linear pendulum equation has complex eigenvalues.

Let's see how to work with the complexities (pun intended) of this situation.

EXAMPLE 5.5.1   A System with Complex Eigenvalues

First we convert the linearized pendulum equation to a system (see Problem 16 of Exercises 4.5 for the nonlinear case). Letting $x = \theta$ and $y = \frac{d\theta}{dt} = \frac{dx}{dt}$, we convert our linear second-order homogeneous equation into the system $\frac{dx}{dt} = y$, $\frac{dy}{dt} = -k^2x$. (Be sure that you remember how to carry out this conversion.)

In matrix form, we have the system $\frac{d}{dt}\begin{bmatrix} x \\ y \end{bmatrix} = \begin{bmatrix} 0 & 1 \\ -k^2 & 0 \end{bmatrix}\begin{bmatrix} x \\ y \end{bmatrix}$, with characteristic equation $\lambda^2 + k^2 = 0$ and complex conjugate eigenvalues $\lambda_1 = ki$ and $\lambda_2 = -ki$. (Verify all the statements in the last sentence.) The equation $AV = \lambda_1 V$ has the form $\begin{bmatrix} 0 & 1 \\ -k^2 & 0 \end{bmatrix}\begin{bmatrix} x \\ y \end{bmatrix} = ki\begin{bmatrix} x \\ y \end{bmatrix} = \begin{bmatrix} kix \\ kiy \end{bmatrix}$, which is equivalent to the algebraic system

$$y = kix$$
$$-k^2x = kiy.$$

Because the second equation is just ki times the first, we see that we can take x as arbitrary and $y = kix$, which gives us the eigenvector $V = \begin{bmatrix} x \\ kix \end{bmatrix} = x\begin{bmatrix} 1 \\ ki \end{bmatrix}$. Letting $x = 1$, we get the representative eigenvector $V_1 = \begin{bmatrix} 1 \\ ki \end{bmatrix} = \begin{bmatrix} 1 \\ 0 \end{bmatrix} + i\begin{bmatrix} 0 \\ k \end{bmatrix}$.

From the discussion preceding this example, we realize that we don't have to worry about the second (conjugate) eigenvalue and its associated eigenvector. The general solution of our original equation and its system version can be obtained from the information we already have. We start with the solution

$$\hat{X}(t) = e^{kit}\left(\begin{bmatrix} 1 \\ 0 \end{bmatrix} + i\begin{bmatrix} 0 \\ k \end{bmatrix}\right) = (\cos kt + i\sin kt)\left(\begin{bmatrix} 1 \\ 0 \end{bmatrix} + i\begin{bmatrix} 0 \\ k \end{bmatrix}\right)$$

$$= \left((\cos kt)\begin{bmatrix} 1 \\ 0 \end{bmatrix} - (\sin kt)\begin{bmatrix} 0 \\ k \end{bmatrix}\right) + i\left((\cos kt)\begin{bmatrix} 0 \\ k \end{bmatrix} + (\sin kt)\begin{bmatrix} 1 \\ 0 \end{bmatrix}\right).$$

Because the real and imaginary parts of the last expression are linearly independent solutions of the system, the general solution is given by

$$X(t) = c_1\left((\cos kt)\begin{bmatrix} 1 \\ 0 \end{bmatrix} - (\sin kt)\begin{bmatrix} 0 \\ k \end{bmatrix}\right) + c_2\left((\cos kt)\begin{bmatrix} 0 \\ k \end{bmatrix} + (\sin kt)\begin{bmatrix} 1 \\ 0 \end{bmatrix}\right)$$

$$= c_1\begin{bmatrix} \cos kt \\ -k\sin kt \end{bmatrix} + c_2\begin{bmatrix} \sin kt \\ k\cos kt \end{bmatrix} = \begin{bmatrix} c_1\cos kt + c_2\sin kt \\ -kc_1\sin kt + kc_2\cos kt \end{bmatrix}.$$

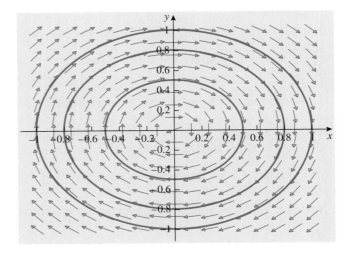

Figure 5.14

Trajectories for the system $\dfrac{dx}{dt} = y, \dfrac{dy}{dt} = -x, 0 \le t \le 7$

Initial points: $(x(0), y(0)) = (1, 0), (0.5, 0), (0, 0.8)$

Figure 5.14 shows some trajectories for this system when $k = 1$. These curves are circles centered at the origin. We say that the origin is a **center** for the system. You should try to generate your own phase portrait by choosing different values of k and various initial points for each value of k. ◆

The next example provides a more challenging problem algebraically.

EXAMPLE 5.5.2   A System with Complex Eigenvalues
According to *Kirchhoff's Second Law*, an electric circuit with resistance of 2 ohms, capacitance of 0.5 farad, inductance of 1 henry, and no driving electromotive force can be modeled by the second-order linear equation $\ddot{Q} + 2\dot{Q} + 2Q = 0$, where $Q = Q(t)$ is the charge on the capacitor at time t. If $Q(0) = 1$ and $\dot{Q}(0) = 0$, we want to determine the charge on the capacitor at time $t \ge 0$.

We write our second-order equation as a system of first-order equations by introducing new variables: Let $x = Q$ and $y = \dot{x} = \dot{Q}$, so $\dot{y} = \ddot{Q} = -2Q - 2\dot{Q} = -2x - 2y$. Then the original second-order equation is equivalent to the system

$$\dot{x} = y$$
$$\dot{y} = -2x - 2y,$$

which can be written in matrix form as $\dot{X} = \begin{bmatrix} 0 & 1 \\ -2 & -2 \end{bmatrix} \begin{bmatrix} x \\ y \end{bmatrix}$. The matrix of coefficients has characteristic equation $\lambda^2 + 2\lambda + 2 = 0$, with roots $-1 + i$ and $-1 - i$. Working with the first of these eigenvalues, we see that any eigenvector must satisfy the matrix equation

$$\begin{bmatrix} 0 & 1 \\ -2 & -2 \end{bmatrix} \begin{bmatrix} x \\ y \end{bmatrix} = (-1 + i) \begin{bmatrix} x \\ y \end{bmatrix},$$

which is equivalent to the equations

$$y = -x + ix$$
$$-2x - 2y = -y + iy.$$

Substituting the first equation in the second equation, we get

$$-2x - 2[-x + ix] = -[-x + ix] + i[-x + ix]$$
$$-2x + 2x - 2ix = x - ix - ix - x \qquad \text{(remembering that } i^2 = -1\text{)}$$
$$-2ix = -2ix.$$

This last equation, an identity, says that *any* value of x will be a solution. If we choose $x = 1$ for convenience, then the first equation gives us $y = -1 + i$, so the representative eigenvector is

$$V_1 = \begin{bmatrix} 1 \\ i - 1 \end{bmatrix} = \begin{bmatrix} 1 \\ -1 \end{bmatrix} + i \cdot \begin{bmatrix} 0 \\ 1 \end{bmatrix} = U + iW.$$

As in the last example, we work with the solution provided by one of the complex conjugate eigenvalues and its representative eigenvector:

$$\hat{X}(t) = e^{(-1+i)t}\left(\begin{bmatrix} 1 \\ -1 \end{bmatrix} + i\begin{bmatrix} 0 \\ 1 \end{bmatrix} \right) = e^{-t}(\cos t + i \sin t)\left(\begin{bmatrix} 1 \\ -1 \end{bmatrix} + i\begin{bmatrix} 0 \\ 1 \end{bmatrix} \right)$$
$$= e^{-t}\left((\cos t)\begin{bmatrix} 1 \\ -1 \end{bmatrix} - (\sin t)\begin{bmatrix} 0 \\ 1 \end{bmatrix} \right) + ie^{-t}\left((\cos t)\begin{bmatrix} 0 \\ 1 \end{bmatrix} + (\sin t)\begin{bmatrix} 1 \\ -1 \end{bmatrix} \right).$$

Extracting the real and imaginary parts of this last complex-valued expression, we express the general solution as

$$X(t) = c_1 e^{-t}\left((\cos t)\begin{bmatrix} 1 \\ -1 \end{bmatrix} - (\sin t)\begin{bmatrix} 0 \\ 1 \end{bmatrix} \right) + c_2 e^{-t}\left((\cos t)\begin{bmatrix} 0 \\ 1 \end{bmatrix} + (\sin t)\begin{bmatrix} 1 \\ -1 \end{bmatrix} \right)$$
$$= c_1\begin{bmatrix} e^{-t}\cos t \\ -e^{-t}\cos t - e^{-t}\sin t \end{bmatrix} + c_2\begin{bmatrix} e^{-t}\sin t \\ e^{-t}\cos t - e^{-t}\sin t \end{bmatrix}$$
$$= e^{-t}\begin{bmatrix} c_1 \cos t + c_2 \sin t \\ (c_2 - c_1)\cos t - (c_2 + c_1)\sin t \end{bmatrix} = \begin{bmatrix} x(t) \\ y(t) \end{bmatrix}.$$

Now, using the initial conditions $x(0) = Q(0) = 1$ and $y(0) = \dot{Q}(0) = 0$ in the general solution just given, we get the condition $\begin{bmatrix} c_1 \\ c_2 - c_1 \end{bmatrix} = \begin{bmatrix} 1 \\ 0 \end{bmatrix}$, which implies that $c_1 = 1$ and $c_2 = 1$. Thus the solution of our original initial-value problem is $Q(t) = x(t) = e^{-t}(\cos t + \sin t)$. Because the current, I, is defined as the rate of change of Q, we get a bonus: $I(t) = \dot{Q}(t) = y(t) = -2e^{-t}\sin t$.

As satisfying as this analytical solution may be, a natural question is what the trajectories for this system look like. Figure 5.15 shows five trajectories, corresponding to different initial conditions. The trajectory for the IVP we started with is second from the bottom.

Note that these trajectories are *spirals* moving *toward* the equilibrium solution, the origin. We say that the origin is a **spiral sink.** If we examine the general solution, we can see why the trajectories behave this way. First of all, there is no

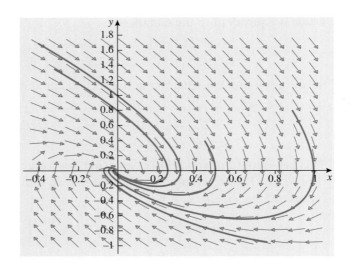

Figure 5.15

Trajectories for the system $\dot{x} = y$, $\dot{y} = -2x - 2y$, $-0.3 \leq t \leq 4$

Initial conditions: $(x(0), y(0)) = (1, 0), (0.5, 0), (0, 0.8), (0, 1), (0.5, -0.8)$

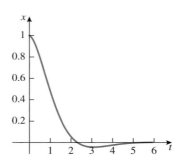

Figure 5.16

Graph of $x(t) = e^{-t}(\cos t + \sin t)$, $0 \leq t \leq 6$

straight-line direction along which the trajectories approach the origin. The expressions for both $x(t)$ and $y(t)$ have trigonometric terms that contribute oscillations, movements back and forth across the x-axis. But in addition, each entry of the general solution has a factor of e^{-t}, which *dampens* these oscillations for positive values of t. Thus as t increases in a positive direction, the amplitudes of these oscillations tend to 0. A look at Euler's formula explains the existence of this decaying exponential: *The real part, p, of the eigenvalue pair is negative.* Figure 5.16 shows a plot of x against t for the particular solution with $x(0) = 1$ and $y(0) = 0$.

The graph of y against t is similar. In terms of the spring-mass problems we analyzed in various examples of Section 4.5, we can interpret our problem as representing a system with *damped* oscillations. (See Example 4.5.6, especially Figure 4.16a.) ◆

TABLE 5.1 Summary of Stability Criteria for Two-Dimensional Linear Systems

Eigenvalues	Stability	References
REAL		
Unequal		
Both > 0	Unstable node (source, repeller)	Examples 5.2.4 and 5.3.1
Both < 0	Stable node (sink, attractor)	Examples 5.2.2 and 5.3.2
Different signs	Saddle point	Examples 5.2.1, 5.2.3, and 5.3.3
One $= 0$, the other $\neq 0$	Whole line of equilibrium points	Example 5.3.4 and Problem 20 of Exercises 5.3
Equal		
Both > 0	Unstable node (source, repeller)	Example 5.4.1
Both < 0	Stable node (sink, attractor)	Example 5.4.2
Both $= 0$	"Algebraically unstable"	Problem 10 of Exercises 5.4
COMPLEX		
Real part > 0	Spiral source (unstable spiral, repeller)	Example 5.2.5
Real part < 0	Spiral sink (stable spiral)	Example 5.5.2
Real part $= 0$	Center (neutral center, stable center)	Example 5.5.1

As we'll see in some of the exercises following this section, if the eigenvalues are $p \pm qi$ and $p > 0$, then we get spirals that wind *away* from $(0, 0)$ as t increases. Here we say that the origin is a *spiral source*. This corresponds to oscillatory solutions with increasing amplitudes and describes *resonance*. (See Example 4.5.8, especially Figure 4.19.)

The case where $p = 0$, so that we have *pure imaginary eigenvalues*, is interesting. Now the trajectories are *closed, nonintersecting curves that encircle the origin*. This corresponds to the situation where we have *undamped oscillations*. (See Example 5.5.1 and Example 4.5.5, especially Figure 4.13.)

Now let us stand back and summarize all these cases. Table 5.1 categorizes the stability of two-dimensional autonomous systems, referring to relevant examples or exercises.

EXERCISES 5.5

For each of the systems in Exercises 1–7, (a) find the eigenvalues and their corresponding eigenvectors. Then (b) sketch or plot a few trajectories and show the position(s) of the eigenvector(s) if they do not have complex entries.

1. $\dot{r} = -r - 2s, \dot{s} = 2r - s$

2. $\dot{x} = 3x - 2y, \dot{y} = 2x + 3y$

3. $\dot{x} = -0.5x - y, \dot{y} = x - 0.5y$

4. $\dot{x} = x + y, \dot{y} = -3x - y$

5. $\dot{x} = 2x + y, \dot{y} = -3x - y$

6. $\dot{x} = -0.5x - y, \dot{y} = x - 0.5y$

7. $\dot{x} = y - 7x, \dot{y} = -2x - 5y$

8. Write systems of first-order linear equations whose trajectories show the following behaviors:

   a. $(0, 0)$ is a spiral source with eigenvalues $\lambda_1 = 2 + 2i$ and $\lambda_2 = 2 - 2i$.

b. $(0, 0)$ is a stable center with eigenvalues $\lambda_1 = -3i$ and $\lambda_2 = 3i$.

c. $(0, 0)$ is a spiral sink with eigenvalues $\lambda_1 = -1 + 2i$ and $\lambda_2 = -1 - 2i$.

9. Consider the system

$$\dot{x} = y$$
$$\dot{y} = -x - \beta y,$$

where β is a parameter.

a. By using technology to draw trajectories, examine the stability of the equilibrium solution for $\beta = -1, -0.1, 0, 0.1$, and 1.

b. Does there seem to be a *bifurcation point*—that is, a critical value of β at which the stability changes its nature? (You may want to look at Section 2.6.)

c. Find a formula for the eigenvalues of the system, showing their dependence on β.

d. Relate the information found in part (c) to the stability summary in Table 5.1 and answer the question in part (b) with increased authority.

10. If λ is a complex eigenvalue of matrix A, $V = U + iW$ is a corresponding eigenvector, and $X(t) = e^{\lambda t}V$, then we have seen that

$$X_1(t) = \mathrm{Re}\{X(t)\} = e^{pt}\{(\cos qt)U - (\sin qt)W\}$$
$$X_2(t) = \mathrm{Im}\{X(t)\} = e^{pt}\{(\cos qt)W + (\sin qt)U\}$$

are real-valued linearly independent solutions of the system $\dot{X} = AX$. Show that the same two solutions can be obtained by taking the real and imaginary parts of $e^{\bar{\lambda} t}\bar{V}$. (Thus the second term of the familiar solution formula $c_1 e^{\lambda_1 t}V_1 + c_2 e^{\lambda_2 t}V_2 = c_1 e^{\lambda_1 t}V_1 + c_2 e^{\bar{\lambda}_1 t}\bar{V}_1$ is unnecessary.)

11. Show that if $X(t)$ is a complex-valued solution of the system $\dot{X} = AX$, then so is $X_1 = \mathrm{Im}(X) = \dfrac{X - \bar{X}}{2i}$, the imaginary part of $X(t)$.

12. The two-loop electrical circuit in the accompanying illustration can be modeled by the system

$$\frac{di_1}{dt} = -\left(\frac{R_1 + R_2}{L}\right)i_1 + \frac{R_2}{L}i_2$$
$$\frac{di_2}{dt} = -\left(\frac{R_1 + R_2}{L}\right)i_1 + \left(\frac{R_2}{L} - \frac{1}{R_2 C}\right)i_2.$$

Using eigenvalues and eigenvectors, solve the initial-value problem $i_1(0) = 1$, $i_2(0) = 0$, when $R_1 = R_2 = 1$, $L = 1$, and $C = 3$. (Use technology to find the eigenvectors.)

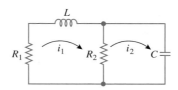

5.6 NONHOMOGENEOUS SYSTEMS

THE GENERAL SOLUTION

The linear systems we have been dealing with so far are called **homogeneous** systems. Basically, this means that they can be expressed in the form $\dot{X} = AX$ with no "leftover" terms. If a linear system has to be written as $\dot{X} = AX + B(t)$, where $B(t)$ is a vector of the form $\begin{bmatrix} b_1(t) \\ b_2(t) \end{bmatrix}$, then we say that the system is **nonhomogeneous.** For example, in matrix terms, the system $\dfrac{dx}{dt} = x + \sin t, \dfrac{dy}{dt} = t - y$ must be written as $\begin{bmatrix} \dot{x}(t) \\ \dot{y}(t) \end{bmatrix} = \begin{bmatrix} 1 & 0 \\ 0 & -1 \end{bmatrix} \begin{bmatrix} x \\ y \end{bmatrix} + \begin{bmatrix} \sin t \\ t \end{bmatrix}$ and so is nonhomogeneous.

Don't confuse the distinction between *autonomous* and *nonautonomous* with that between *homogeneous* and *nonhomogeneous*. For example, if both $b_1(t)$ and $b_2(t)$ are constant functions, then we have a system that is both autonomous and nonhomogeneous. (See, for instance, Example 4.5.3.)

The techniques that were introduced in Section 4.2 for second-order nonhomogeneous equations generalize to systems, but the calculations are more complicated. To get a handle on solving a nonhomogeneous linear system, we need a fundamental fact about linear systems:

The general solution, X_{GNH}, of a linear nonhomogeneous system is obtained by finding a *particular* solution, X_{PNH}, of the *nonhomogeneous* system and adding it to the *general* solution, X_{GH}, of the associated *homogeneous* system.

You should see this as an application of the Superposition Principle and as an extension of the result we saw for single linear differential equations (Section 4.2). Symbolically, we can write $X_{\text{GNH}} = X_{\text{GH}} + X_{\text{PNH}}$. Using the definitions of these terms, we can see that this sum of vectors is a solution of the nonhomogeneous system:

$$\dot{X}_{\text{GNH}} = \dot{X}_{\text{GH}} + \dot{X}_{\text{PNH}} = AX_{\text{GH}} + \{AX_{\text{PNH}} + B(t)\}$$
$$= A(X_{\text{GH}} + X_{\text{PNH}}) + B(t) = AX_{\text{GNH}} + B(t).$$

(*Be sure you follow this.*) You should see that X_{GH}, as a general solution, must contain two arbitrary constants, so the expression for X_{GNH} contains two arbitrary constants.

Let's look at a simple example showing the structure of a nonhomogeneous system's solution.

EXAMPLE 5.6.1   The Solution of a Nonhomogeneous System
The system

$$\dot{x} = x + y + 2e^{-t}$$
$$\dot{y} = 4x + y + 4e^{-t}$$

can be written in the form $\dot{X}(t) = \begin{bmatrix} 1 & 1 \\ 4 & 1 \end{bmatrix} X + \begin{bmatrix} 2e^{-t} \\ 4e^{-t} \end{bmatrix} = \begin{bmatrix} 1 & 1 \\ 4 & 1 \end{bmatrix} X + 2e^{-t} \begin{bmatrix} 1 \\ 2 \end{bmatrix}$.
The system has eigenvalues $\lambda_1 = 3$ and $\lambda_2 = -1$, with corresponding eigen-
vectors $V_1 = \begin{bmatrix} 1 \\ 2 \end{bmatrix}$ and $V_2 = \begin{bmatrix} 1 \\ -2 \end{bmatrix}$. (*Check this.*) Then the general solution of the
associated homogeneous system $\dot{X}(t) = \begin{bmatrix} 1 & 1 \\ 4 & 1 \end{bmatrix} X$ is

$$X_{\text{GH}} = c_1 e^{3t} \begin{bmatrix} 1 \\ 2 \end{bmatrix} + c_2 e^{-t} \begin{bmatrix} 1 \\ -2 \end{bmatrix}.$$

You should verify that a particular solution of the original nonhomogeneous
system is given by $X_{\text{PNH}} = e^{-t} \begin{bmatrix} 0 \\ -2 \end{bmatrix} = \begin{bmatrix} 0 \\ -2e^{-t} \end{bmatrix}$. Therefore, the general solution
of the nonhomogeneous system is

$$\begin{aligned} X_{\text{GNH}} = X_{\text{GH}} + X_{\text{PNH}} &= c_1 e^{3t} \begin{bmatrix} 1 \\ 2 \end{bmatrix} + c_2 e^{-t} \begin{bmatrix} 1 \\ -2 \end{bmatrix} + \begin{bmatrix} 0 \\ -2e^{-t} \end{bmatrix} \\ &= \begin{bmatrix} c_1 e^{3t} + c_2 e^{-t} \\ 2c_1 e^{3t} - 2c_2 e^{-t} - 2e^{-t} \end{bmatrix}. \end{aligned}$$

You should check that this is the general solution of the original nonhomogeneous system.   ◆

THE METHOD OF UNDETERMINED COEFFICIENTS

The challenge in working with a nonhomogeneous system is to find a particular
solution of the nonhomogeneous system. There are various techniques for finding a particular solution. We can use the *variation of parameters* technique of Section 4.2, but for systems the calculations involved are very tedious. Therefore,
we'll restrict our attention to the method of **undetermined coefficients,** which is
not so powerful but is easier to use. As we saw in Examples 4.2.1 and 4.2.2, this
method requires intelligent guessing. We have to ask ourselves what terms are
contained in $B(t)$ but not in X_{GH}—and then guess at the form of X_{PNH} on the basis of this information.

We should note that this method of undetermined coefficients can be used
only when the vector $B(t) = \begin{bmatrix} b_1(t) \\ b_2(t) \end{bmatrix}$ contains terms that are constants, exponential functions, sines, cosines, polynomials, or any sum or product of such terms.
For other kinds of functions making up $B(t)$, X_{PNH} must be found using some
other technique (for example, variation of parameters).

The next example illustrates the method with its resulting algebraic
complexities.

EXAMPLE 5.6.2 **Using the Method of Undetermined Coefficients**

Let's consider the system $\dfrac{dx}{dt} = x + \sin t, \dfrac{dy}{dt} = t - y$ that we discussed at the beginning of this section. We have $\dot{X} = AX + B(t)$, where $A = \begin{bmatrix} 1 & 0 \\ 0 & -1 \end{bmatrix}$ and $B(t) = \begin{bmatrix} \sin t \\ t \end{bmatrix} = \sin t \begin{bmatrix} 1 \\ 0 \end{bmatrix} + t \begin{bmatrix} 0 \\ 1 \end{bmatrix}$.

The eigenvalues of A are 1 and -1, with corresponding eigenvectors $\begin{bmatrix} 1 \\ 0 \end{bmatrix}$ and $\begin{bmatrix} 0 \\ 1 \end{bmatrix}$. Thus the general solution of the homogeneous system can be written as

$$X_{\text{GH}} = c_1 e^t \begin{bmatrix} 1 \\ 0 \end{bmatrix} + c_2 e^{-t} \begin{bmatrix} 0 \\ 1 \end{bmatrix}.$$

(Verify the statements in this paragraph for yourself.)

Now we look for a particular solution of the original nonhomogeneous equation. First we compare the terms of $B(t)$ with the terms of X_{GH} to see whether there is any duplication. In this case, we see that the terms $\sin t$ and t are not terms that can be obtained just from X_{GH}. Because our system is equivalent to a single second-order differential equation, we realize that we must find a function that can combine with its own first and second derivatives to yield $B(t)$. We take a guess that X_{PNH} must look like $C \sin t + D \cos t + Et + F$, where C, D, E, and F are *vectors of constants*. Our trial solution for X_{PNH} consists of a linear combination of the functions $\sin t$ and t and their derivatives—a linear combination with *undetermined coefficients*.

Now let's substitute our guess into the nonhomogeneous system:

$$\overbrace{C \cos t - D \sin t + E}^{\dot{X}_{\text{PNH}}} = A\overbrace{\left(C \sin t + D \cos t + Et + F \right)}^{X_{\text{PNH}}} + \overbrace{\sin t \begin{bmatrix} 1 \\ 0 \end{bmatrix} + t \begin{bmatrix} 0 \\ 1 \end{bmatrix}}^{B(t)}$$

$$= AC \sin t + AD \cos t + AEt + AF + \sin t \begin{bmatrix} 1 \\ 0 \end{bmatrix} + t \begin{bmatrix} 0 \\ 1 \end{bmatrix}.$$

When we collect like terms, matching the coefficients of functions on each side, we get the following system:

(1) $C = AD$ [The coefficients of $\cos t$ must be equal.]

(2) $-D = AC + \begin{bmatrix} 1 \\ 0 \end{bmatrix}$ [The coefficients of $\sin t$ must be equal.]

(3) $0 = AE + \begin{bmatrix} 0 \\ 1 \end{bmatrix}$ [The coefficients of t must be equal.]

(4) $E = AF$ [The constant terms must be equal.].

Remembering that $A = \begin{bmatrix} 1 & 0 \\ 0 & -1 \end{bmatrix}$, we can solve equation (3) for E:

$$AE = -\begin{bmatrix} 0 \\ 1 \end{bmatrix} = \begin{bmatrix} 0 \\ -1 \end{bmatrix}, \text{ or } \begin{bmatrix} 1 & 0 \\ 0 & -1 \end{bmatrix}\begin{bmatrix} e_1 \\ e_2 \end{bmatrix} = \begin{bmatrix} 0 \\ -1 \end{bmatrix}, \text{ so } e_1 = 0 \text{ and } e_2 = 1. \text{ (Check}$$

this.) Now that we know E, we can use equation (4) to find F: $\begin{bmatrix} 0 \\ 1 \end{bmatrix} = \begin{bmatrix} 1 & 0 \\ 0 & -1 \end{bmatrix}\begin{bmatrix} f_1 \\ f_2 \end{bmatrix}$, so $f_1 = 0$ and $f_2 = -1$.

If we multiply both sides of (1) by A, we get $AC = A^2D = D$ (because $A^2 = I$, the 2×2 identity matrix), which we can substitute into equation (2): $-D = D + \begin{bmatrix} 1 \\ 0 \end{bmatrix}$, or $\begin{bmatrix} -1 \\ 0 \end{bmatrix} = 2D = \begin{bmatrix} 2d_1 \\ 2d_2 \end{bmatrix}$, so $d_1 = -\frac{1}{2}$ and $d_2 = 0$. *Make sure you follow all this.* Finally, we solve (1) for C: $\begin{bmatrix} c_1 \\ c_2 \end{bmatrix} = \begin{bmatrix} 1 & 0 \\ 0 & -1 \end{bmatrix}\begin{bmatrix} -1/2 \\ 0 \end{bmatrix}$, so $c_1 = -\frac{1}{2}$ and $c_2 = 0$.

We have determined all the coefficients. Putting the pieces together, we have

$$X_{\text{PNH}} = \begin{bmatrix} -1/2 \\ 0 \end{bmatrix}\sin t + \begin{bmatrix} -1/2 \\ 0 \end{bmatrix}\cos t + \begin{bmatrix} 0 \\ 1 \end{bmatrix}t + \begin{bmatrix} 0 \\ -1 \end{bmatrix} = \begin{bmatrix} -\dfrac{1}{2}(\sin t + \cos t) \\ t - 1 \end{bmatrix},$$

and we finally obtain

$$X_{\text{GNH}} = X_{\text{GH}} + X_{\text{PNH}}$$

$$= c_1 e^t\begin{bmatrix} 1 \\ 0 \end{bmatrix} + c_2 e^{-t}\begin{bmatrix} 0 \\ 1 \end{bmatrix} + \begin{bmatrix} -\dfrac{1}{2}(\sin t + \cos t) \\ t - 1 \end{bmatrix} = \begin{bmatrix} c_1 e^t - \dfrac{1}{2}(\sin t + \cos t) \\ t - 1 + c_2 e^{-t} \end{bmatrix}$$

as the general solution of the original nonhomogeneous equation.

Note that the system in this example is *uncoupled*—that is, each equation contains only one unknown function. Exercise 13 at the end of the section asks you to solve each equation separately to obtain the same answer as the one above. ◆

Practice in the technique of undetermined coefficients leads to a more systematic way of guessing a possible solution of the nonhomogeneous system. The second column of Table 5.2 indicates the component of X_{PNH} that corresponds to the matching component $b_i(t)$ of $B(t)$. If $b_i(t)$ is a sum of different functions, then it is a consequence of the Superposition Principle that the matching component of X_{PNH} is a sum of trial solutions.

There is an exception to the neatness of the table. If $b_i(t)$ contains terms that duplicate any corresponding parts of X_{GH}, then each corresponding trial term must be multiplied by t^m, where m is the smallest positive integer that eliminates the duplication.

In Example 5.6.2, we had $b_1(t) = \sin t$ and $b_2(t) = t$—a trigonometric function ($a \cos rt + b \sin rt$, with $a = 0$, $r = 1$, and $b = 1$) and a first-degree polynomial ($P_n(t) = a_n t^n + a_{n-1}t^{n-1} + \cdots + a_1 t + a_0$, where $n = 1$, $a_1 = 1$, and $a_0 = 0$). There was no duplication between X_{GH} and $B(t)$ because the terms making up X_{GH} are exponential functions. Consequently, our educated guess for X_{PNH} consisted of a linear combination of sine and cosine plus a first-degree polynomial.

Let's use the instant wisdom conferred by Table 5.2 in solving the next problem.

TABLE 5.2   Trial Particular Solutions for Nonhomogeneous Systems

$b_i(t)$	Form of Trial Solution
$c \neq 0$, a constant	K, a constant
$P_n(t) = a_n t^n + a_{n-1} t^{n-1} + \cdots + a_1 t + a_0$	$Q_n(t) = c_n t^n + c_{n-1} t^{n-1} + \cdots + c_1 t + c_0$
ce^{at}	Ke^{at}
$a \cos rt + b \sin rt$	$\alpha \cos rt + \beta \sin rt$
$e^{Rt}(a \cos rt + b \sin rt)$	$e^{Rt}(\alpha \cos rt + \beta \sin rt)$
$P_n(t)e^{at}$	$Q_n(t)e^{at}$

EXAMPLE 5.6.3   Undetermined Coefficients

Suppose we try to solve the system $\dfrac{dx}{dt} = y$, $\dfrac{dy}{dt} = 3y - 2x + 2t^2 + 3e^{2t}$. We can

write this system as $\dot{X} = AX + B(t)$, where $A = \begin{bmatrix} 0 & 1 \\ -2 & 3 \end{bmatrix}$ and $B(t) =$

$\begin{bmatrix} 0 \\ 2t^2 + 3e^{2t} \end{bmatrix} = (2t^2 + 3e^{2t})\begin{bmatrix} 0 \\ 1 \end{bmatrix}$. The eigenvalues of A are 1 and 2, with corre-

sponding eigenvectors $\begin{bmatrix} 1 \\ 1 \end{bmatrix}$ and $\begin{bmatrix} 1 \\ 2 \end{bmatrix}$. (*Verify this.*)

Now we know that the general solution of the homogeneous system is given by

$$X_{\text{GH}} = c_1 e^t \begin{bmatrix} 1 \\ 1 \end{bmatrix} + c_2 e^{2t} \begin{bmatrix} 1 \\ 2 \end{bmatrix}.$$

To find a particular solution of the nonhomogeneous system, we compare the terms of $B(t)$ with the terms of X_{GH} to see whether there is any duplication. In this example, ignoring constants, we see that e^{2t} appears in both X_{GH} and $B(t)$. We also recognize that the term t^2 in $B(t)$ is *not* found in X_{GH}. Using Table 5.2 and the description of how to handle duplicate terms, we guess that X_{PNH} must look like

$$Ct^2 + Dt + E + Fe^{2t} + Gte^{2t},$$

where C, D, E, F, and G are vectors of constants. Note that because there is a second-degree term, our trial particular solution contains a full quadratic polynomial and that multiplying e^{2t} by t eliminates the duplication.

If we substitute this guess into the nonhomogeneous system, we get

$$2Ct + D + 2Fe^{2t} + Ge^{2t} + 2Gte^{2t}$$

$$= A(Ct^2 + Dt + E + Fe^{2t} + Gte^{2t}) + (2t^2 + 3e^{2t})\begin{bmatrix} 0 \\ 1 \end{bmatrix}$$

$$= ACt^2 + ADt + AE + AFe^{2t} + AGte^{2t} + (2t^2 + 3e^{2t})\begin{bmatrix} 0 \\ 1 \end{bmatrix}$$

$$= \left(AC + \begin{bmatrix} 0 \\ 2 \end{bmatrix}\right)t^2 + ADt + AE + \left(AF + \begin{bmatrix} 0 \\ 3 \end{bmatrix}\right)e^{2t} + AGte^{2t}.$$

Matching the coefficients of like terms on each side, we get the system

(1) $0 = AC + \begin{bmatrix} 0 \\ 2 \end{bmatrix}$ [The coefficients of t^2 must be equal.]

(2) $2C = AD$ [The coefficients of t must be equal.]

(3) $D = AE$ [The constant terms must be equal.]

(4) $2F + G = AF + \begin{bmatrix} 0 \\ 3 \end{bmatrix}$ [The coefficients of e^{2t} must be equal.]

(5) $2G = AG$ [The coefficients of te^{2t} must be equal.].

Working through these equations (see Exercise 15), we find that $C = \begin{bmatrix} 1 \\ 0 \end{bmatrix}$, $D = \begin{bmatrix} 3 \\ 2 \end{bmatrix}$, $E = \begin{bmatrix} \frac{7}{2} \\ 3 \end{bmatrix}$, $F = \begin{bmatrix} 0 \\ 3 \end{bmatrix}$, and $G = \begin{bmatrix} 3 \\ 6 \end{bmatrix}$.

Now that we've determined the coefficients C, D, E, F, and G, we can construct the particular solution of the nonhomogeneous equation.

$$X_{\text{PNH}} = \begin{bmatrix} 1 \\ 0 \end{bmatrix} t^2 + \begin{bmatrix} 3 \\ 2 \end{bmatrix} t + \begin{bmatrix} \frac{7}{2} \\ 3 \end{bmatrix} + \begin{bmatrix} 0 \\ 3 \end{bmatrix} e^{2t} + \begin{bmatrix} 3 \\ 6 \end{bmatrix} te^{2t}.$$

Finally, we get the general solution of the nonhomogeneous equation.

$$\begin{aligned} X_{\text{GNH}} &= X_{\text{GH}} + X_{\text{PNH}} \\ &= c_1 e^t \begin{bmatrix} 1 \\ 1 \end{bmatrix} + c_2 e^{2t} \begin{bmatrix} 1 \\ 2 \end{bmatrix} + \begin{bmatrix} 1 \\ 0 \end{bmatrix} t^2 + \begin{bmatrix} 3 \\ 2 \end{bmatrix} t + \begin{bmatrix} \frac{7}{2} \\ 3 \end{bmatrix} + \begin{bmatrix} 0 \\ 3 \end{bmatrix} e^{2t} + \begin{bmatrix} 3 \\ 6 \end{bmatrix} te^{2t} \\ &= \begin{bmatrix} c_1 e^t + c_2 e^{2t} + t^2 + 3t + \frac{7}{2} + 3te^{2t} \\ c_1 e^t + 2c_2 e^{2t} + 2t + 3 + 3e^{2t} + 6te^{2t} \end{bmatrix} \\ &= \begin{bmatrix} c_1 e^t + (c_2 + 3t)e^{2t} + t^2 + 3t + \frac{7}{2} \\ c_1 e^t + (2c_2 + 3 + 6t)e^{2t} + 2t + 3 \end{bmatrix}. \end{aligned}$$

Of course, this means that $x(t) = c_1 e^t + (c_2 + 3t)e^{2t} + t^2 + 3t + 7/2$ and $y(t) = c_1 e^t + (2c_2 + 3 + 6t)e^{2t} + 2t + 3$ are the solutions of our system. You should check to see that these functions satisfy our original system.   ◆

When the nonhomogeneous system is also *autonomous*—that is, it has the form $\dot{X} = AX + B(t)$, where the entries of $B(t)$ are *constants*—we can analyze the stability of the system's solutions by finding the equilibrium point(s) (no longer the origin) and considering the eigenvalues and eigenvectors of the matrix A.

EXAMPLE 5.6.4   **Stability of an Autonomous Nonhomogeneous System**
We return to the system of Example 4.5.3:

$$\begin{aligned} \dot{x} &= 7y - 4x - 13 \\ \dot{y} &= 2x - 5y + 11. \end{aligned}$$

To find the equilibrium point(s), we solve the algebraic system

$$\begin{aligned} -4x + 7y &= 13 \\ 2x - 5y &= -11 \end{aligned}$$

to find that $(2, 3)$ is the only equilibrium point. (The details in this example are left as parts of Exercise 17.)

We can write our system of differential equations in the form

$$\dot{X} = AX + B(t) = \begin{bmatrix} -4 & 7 \\ 2 & -5 \end{bmatrix} \begin{bmatrix} x \\ y \end{bmatrix} + \begin{bmatrix} -13 \\ 11 \end{bmatrix}.$$ Because the eigenvalues of A are

$\lambda_1 = (\sqrt{57} - 9)/2$ and $\lambda_2 = -(\sqrt{57} + 9)/2$, both of which are negative real numbers, Table 5.1 at the end of Section 5.5 tells us that the equilibrium point $(2, 3)$ is a *sink*. (Go back to take another look at Figure 4.8.) ◆

Despite its limitations, the method of undetermined coefficients is very useful. In Chapter 6, we'll see another way of solving systems of nonhomogeneous linear equations, by means of the *Laplace transform*. This transform method is particularly useful in solving initial-value problems.

EXERCISES 5.6

1. Find the particular solution of the system in Example 5.6.2 that satisfies $x(0) = 0, y(0) = 1$.

2. Find the particular solution of the system in Example 5.6.3 that satisfies $x(0) = -1, y(0) = 2$.

Without using technology, find the general solution of each of the systems in Exercises 3–12. You may check your answers using a CAS.

3. $\dot{x} = y + 2e^t, \dot{y} = x + t^2$

4. $\dot{x} = y - 5\cos t, \dot{y} = 2x + y$

5. $\dot{x} = 3x + 2y + 4e^{5t}, \dot{y} = x + 2y$

6. $\dot{x} = 3x - 4y + e^{-2t}, \dot{y} = x - 2y - 3e^{-2t}$

7. $\dot{x} = 4x + y - e^{2t}, \dot{y} = y - 2x$ [*Hint:* Multiples of both e^{2t} and te^{2t} should appear in your guess for X_{PNH}.]

8. $\dot{x} = 2y - x + 1, \dot{y} = 3y - 2x$ [*Hint:* Multiples of both e^t and te^t should appear in your guess for X_{PNH}.]

9. $\dot{x} = 5x - 3y + 2e^{3t}, \dot{y} = x + y + 5e^{-t}$

10. $\dot{x} = x + y + 1 + e^t, \dot{y} = 3x - y$

11. $\dot{x} = 2x - y, \dot{y} = 2y - x - 5e^t \sin t$

12. $\dot{x} = x + 2y, \dot{y} = x - 5\sin t$

13. Consider each equation in Example 5.6.2 as a first-order linear equation and solve each equation separately, confirming that you get the same answer as in the worked-out example. (You may have to review Section 2.2 and the technique of integration by parts.)

14. **a.** Use technology to draw the phase portrait for the system in Example 5.6.2.

    b. Draw a graph of $x(t)$ against t.

    c. Draw a graph of $y(t)$ against t.

15. Assume that you have equations (1)–(5) in Example 5.6.3. Let $C = \begin{bmatrix} c_1 \\ c_2 \end{bmatrix}$, $D = \begin{bmatrix} d_1 \\ d_2 \end{bmatrix}$, and so on. Solve for the vectors C, D, E, F, and G in the order indicated below.

   a. Use equation (1) to show that $C = \begin{bmatrix} 1 \\ 0 \end{bmatrix}$.

   b. Use equation (2) to show that $D = \begin{bmatrix} 3 \\ 2 \end{bmatrix}$.

   c. Use equation (3) to show that $E = \begin{bmatrix} \frac{7}{2} \\ 3 \end{bmatrix}$.

   d. Assuming that G is not the zero vector, use equation (5) to derive a general form for G. (There is an arbitrary constant involved.)

   e. Substitute the general form for G found in part (d) into equation (4) to determine the concrete form of G. Then use this information to see that a convenient form for F is $\begin{bmatrix} 0 \\ 3 \end{bmatrix}$.

16. a. Use technology to draw the phase portrait for the system in Example 5.6.3.

   b. Draw the graph of $x(t)$ against t, assuming that $x(0) = 50$.

   c. Draw the graph of $y(t)$ against t, assuming that $y(0) = 100$.

17. Look at the system in Example 5.6.4.

   a. Show that the only equilibrium point is $(2, 3)$.

   b. Show that the eigenvalues of the matrix of coefficients A are

$$\lambda_1 = (\sqrt{57} - 9)/2 \quad \text{and} \quad \lambda_2 = -(\sqrt{57} + 9)/2.$$

   c. Find eigenvectors corresponding to λ_1 and λ_2.

   d. Express the general solution of the homogeneous system in terms of the eigenvalues and eigenvectors found in parts (b) and (c).

   e. Find a particular solution of the nonhomogeneous system.

   f. Put the answers to parts (d) and (e) together to get the general solution of the nonhomogeneous system. Then determine what happens as $t \to \infty$.

18. Newton's laws of motion give the following system as a model for the motion of an object falling under the influence of gravity:

$$\frac{dy}{dt} = v(t)$$

$$\frac{dv}{dt} = g - cv(t); \; y(0) = 0, \, v(0) = 0$$

for $0 \le t \le T$, where $y(T) = H$. Here $y(t)$ denotes the downward distance from the spot where the object was dropped to the place where the falling object is at time t; $v(t)$ is the velocity; g is the gravitational constant; and c is the *drag coefficient*, representing air resistance.

   a. Without using technology, solve this nonhomogeneous system for $y(t)$ and $v(t)$.

   b. Find $\lim\limits_{t \to \infty} v(t)$ and interpret your answer in physical terms.

19. A cold medication moving through the body can be modeled[3] by the IVP

$$\dot{x} = -k_1 x + I$$
$$\dot{y} = k_1 x - k_2 y; \ x(0) = 0, \ y(0) = 0,$$

where $x(t)$ and $y(t)$ are the amounts of medication in the gastrointestinal tract and the bloodstream, respectively, at time t measured in hours elapsed since the initial dosage. Here $I > 0$ is the constant dosage rate, and k_1, k_2 are positive transfer rates (out of the GI tract and bloodstream, respectively).

   a. Without using technology, solve the nonhomogeneous system for $x(t)$ and $y(t)$.

   b. Find $\lim\limits_{t \to \infty} x(t)$ and $\lim\limits_{t \to \infty} y(t)$

   c. Assume that the decongestant portion of a continuous-acting capsule (such as Contac) has $k_1 = 1.386/$hr and $k_2 = 0.1386/$hr and that the antihistamine portion has $k_1 = 0.6931/$hr and $k_2 = 0.0231/$hr. Also assume that $I = \frac{1}{6}$ (that is, 1 unit per 6 hours). Use technology to graph $x(t)$ against t and $y(t)$ against t for the decongestant on the same set of axes.

   d. Assuming the data given in part (c), use technology to graph $x(t)$ and $y(t)$ for the antihistamine on the same set of axes.

20. During World War I, the English scientist F. W. Lanchester (1868–1946) devised several mathematical models for the new art of aerial combat. These models have since been extended and applied to various modern conflicts. One model, describing the interaction of two conventional armies (as opposed to guerrilla forces or a mixture of conventional and guerrilla forces), is given by

$$\frac{dx}{dt} = -ay + f(t) - c$$
$$\frac{dy}{dt} = -bx + g(t) - d; \quad x(0) = \alpha, \ y(0) = \beta,$$

where $x(t)$ and $y(t)$ represent the strengths of the opposing forces at time t, a and b denote nonnegative loss rates, c and d are constant noncombat losses per day, and $f(t)$ and $g(t)$ denote reinforcement rates in number of combatants per day.

   a. Assuming that $f(t) = k$ and $g(t) = l$ (k and l constants) during a battle, determine the strength of each army at time t during the battle.

---

3. This model is based on the work of Edward Spitznagel of Washington University and was first communicated to me by Courtney Coleman, Harvey Mudd College.

b. If $\alpha > \dfrac{l-d}{b} > 0$ and $\beta > \dfrac{k-c}{a} > 0$ determine the conditions under which the y-force will be wiped out.

c. Assume that $a = 0.006$, $b = 0.008$, $c = d = 1000$, $k = 6000$, $l = 4000$, $\alpha = 90{,}000$, and $\beta = 200{,}000$, where c, d, k, and l are measured in combatants per day. Use technology to graph $x(t)$ and $y(t)$ for $0 \le t \le 50$. Then use the graphs to determine the time t^* when $x(t^*) = y(t^*)$. Which side is winning after 50 days?

5.7   GENERALIZATIONS: THE $n \times n$ CASE ($n \ge 3$)

MATRIX REPRESENTATION

We are going to extend our previous analysis of systems, first to 3×3 systems and then to nth-order linear systems. We can use matrix notation to represent a third-order system with constant coefficients

$$\dot{x}_1 = a_{11}x_1 + a_{12}x_2 + a_{13}x_3$$
$$\dot{x}_2 = a_{21}x_1 + a_{22}x_2 + a_{23}x_3$$
$$\dot{x}_3 = a_{31}x_1 + a_{32}x_2 + a_{33}x_3$$

symbolically, in the form $\dot{X} = AX$, where $X = \begin{bmatrix} x_1 \\ x_2 \\ x_3 \end{bmatrix}$, $\dot{X} = \begin{bmatrix} \dot{x}_1 \\ \dot{x}_2 \\ \dot{x}_3 \end{bmatrix}$, and

$$A = \begin{bmatrix} a_{11} & a_{12} & a_{13} \\ a_{21} & a_{22} & a_{23} \\ a_{31} & a_{32} & a_{33} \end{bmatrix}.$$

EXAMPLE 5.7.1   Matrix Representation of a 3 × 3 System

The system $\dot{x} = -2x + 4y - z$, $\dot{y} = 5x - y + 3z$, $\dot{z} = x + z$ can be written first in the usual vertical way

$$\begin{array}{ll} \dot{x} = -2x + 4y - z & \dot{x} = -2x + 4y - z \\ \dot{y} = 5x - y + 3z \quad \text{or} & \dot{y} = 5x - y + 3z \\ \dot{z} = x + z & \dot{z} = x + 0y + z \end{array}$$

and then more compactly as

$$\begin{bmatrix} \dot{x} \\ \dot{y} \\ \dot{z} \end{bmatrix} = \begin{bmatrix} -2 & 4 & -1 \\ 5 & -1 & 3 \\ 1 & 0 & 1 \end{bmatrix} \begin{bmatrix} x \\ y \\ z \end{bmatrix}.$$

◆

Eigenvalues and Eigenvectors

It is important to understand that the concepts of eigenvalue and eigenvector are valid for any system of n equations in n unknowns ($n \ge 2$). Specifically, given a system $\dot{X} = AX$, where X is a nonzero 3×1 column matrix (vector) and A is a

3×3 matrix, then an *eigenvalue* λ is a solution of the equation $AX = \lambda X$. Given an eigenvalue λ, an *eigenvector* associated with λ is a nonzero vector V that satisfies the equation $AV = \lambda V$.

The equation $AX = \lambda X$ can be expressed as $AX - \lambda X = \mathbf{0}$, where $\mathbf{0}$ denotes the 3×1 vector consisting entirely of zeros. This matrix equation is equivalent to the homogeneous algebraic system

$$\begin{aligned}
(a_{11} - \lambda)x_1 + a_{12}x_2 + a_{13}x_3 &= 0 \\
a_{21}x_1 + (a_{22} - \lambda)x_2 + a_{23}x_3 &= 0 \\
a_{31}x_1 + a_{32}x_2 + (a_{33} - \lambda)x_3 &= 0,
\end{aligned}$$

or

$$\begin{bmatrix} a_{11} - \lambda & a_{12} & a_{13} \\ a_{21} & a_{22} - \lambda & a_{23} \\ a_{31} & a_{32} & a_{33} - \lambda \end{bmatrix}\begin{bmatrix} x_1 \\ x_2 \\ x_3 \end{bmatrix} = \begin{bmatrix} 0 \\ 0 \\ 0 \end{bmatrix}. \tag{5.7.1}$$

Now the matrix of coefficients in (5.7.1) can be expressed as

$$\begin{bmatrix} a_{11} & a_{12} & a_{13} \\ a_{21} & a_{22} & a_{23} \\ a_{31} & a_{32} & a_{33} \end{bmatrix} - \begin{bmatrix} \lambda & 0 & 0 \\ 0 & \lambda & 0 \\ 0 & 0 & \lambda \end{bmatrix} = \begin{bmatrix} a_{11} & a_{12} & a_{13} \\ a_{21} & a_{22} & a_{23} \\ a_{31} & a_{32} & a_{33} \end{bmatrix} - \lambda\begin{bmatrix} 1 & 0 & 0 \\ 0 & 1 & 0 \\ 0 & 0 & 1 \end{bmatrix} = A - \lambda I,$$

so the equation $AX - \lambda X = \mathbf{0}$ can be written as $(A - \lambda I)X = \mathbf{0}$, where I is the 3×3 *identity matrix* consisting of ones down the main diagonal and zeros elsewhere. (The matrix I is such that $IX = X$ for any 3×1 vector X. See Section 5.1 for the 2×2 case.)

The equation $(A - \lambda I)X = \mathbf{0}$ represents a homogeneous algebraic system of three linear equations in three unknowns, and the theory of linear algebra indicates that there is a number Δ depending on the matrix of coefficients with the following important property:

The system (5.7.1) has only the zero solution $x_1 = x_2 = x_3 = 0$ if $\Delta \neq 0$. However, if $\Delta = 0$, then there is a solution x_1, x_2, x_3 with at least one of the x_i ($i = 1, 2, 3$) different from zero.

This number Δ is the *determinant* of the matrix of coefficients, denoted by $\det(A - \lambda I)$, and it is the extension to three dimensions of the determinant introduced in Section 5.2. (See Problem 12 of Exercises 5.1 and Problem 2 of Exercises 5.2 for the significance of the 2×2 determinant in the solution of a system of equations.) Therefore, $(A - \lambda I)X = \mathbf{0}$ has a nonzero solution X only if $\det(A - \lambda I) = 0$. An important fact is that $\det(A - \lambda I)$ is a third-degree polynomial in λ, called the **characteristic polynomial** of A, so the *eigenvalues of A are the roots of the characteristic equation* $\det(A - \lambda I) = 0$. There are algorithms for calculating determinants of 3×3 systems, but they are tedious and any graphing calculator or CAS can evaluate them. In particular, a CAS will provide characteristic polynomials, eigenvalues, and corresponding eigenvectors. Also, there are formulas for solving cubic equations, but these methods are more complicated than the quadratic formula, and it is advisable to use your calculator or computer to solve such equations.

Let's use technology in the next example to calculate determinants, eigenvalues, and eigenvectors for a three-dimensional system.

EXAMPLE 5.7.2   Eigenvalues and Eigenvectors via a CAS
Let's look at the matrix of coefficients in Example 5.7.1:

$$A = \begin{bmatrix} -2 & 4 & -1 \\ 5 & -1 & 3 \\ 1 & 0 & 1 \end{bmatrix}.$$

A computer algebra system provides the information that $\det(A) = -7$, the characteristic equation is $\lambda^3 + 2\lambda^2 - 20\lambda + 7 = 0$, and the eigenvalues (rounded to four decimal places) are $\lambda_1 = 3.3485$, $\lambda_2 = -5.7143$, and $\lambda_3 = 0.3658$. The corresponding representative eigenvectors are

$$V_1 = \begin{bmatrix} 2.3485 \\ 3.3903 \\ 1 \end{bmatrix}, \quad V_2 = \begin{bmatrix} -6.7143 \\ 6.4848 \\ 1 \end{bmatrix}, \quad \text{and} \quad V_3 = \begin{bmatrix} -0.6342 \\ -0.1251 \\ 1 \end{bmatrix}.$$

Don't be concerned if your CAS or calculator gives you eigenvectors that are different from these. You should check to see that each eigenvector you find is a constant multiple of one of the vectors V_1, V_2, and V_3 given above.   ◆

Linear Independence and Linear Dependence

At this point you should be asking yourself, "What do these eigenvalues and eigenvectors tell me about the system?" Just as in the 2×2 case, we can write the general solution of a 3×3 system in terms of the eigenvalues and eigenvectors of the matrix of coefficients. To see what's going on, we'll need a few formal notions that we have already seen in the 2×2 case. For example, given a number of vectors $v_1, v_2, \ldots, v_k$, a **linear combination** of these vectors is a vector that has the form $a_1 v_1 + a_2 v_2 + \cdots + a_k v_k$ for some choice of scalars $a_1, a_2, \ldots, a_k$. The collection of vectors is called **linearly independent** if the only way you can have $a_1 v_1 + a_2 v_2 + \cdots + a_k v_k = \mathbf{0}$ (the zero vector) is to have $a_1 = a_2 = \cdots = a_k = 0$. If you *could* find scalars a_i, not all zero, so that a linear combination of the vectors v_i was equal to the zero vector, then we say that the collection of vectors is **linearly dependent.** To see what linear dependence means, suppose that $a_1 v_1 + a_2 v_2 + \cdots + a_k v_k = \mathbf{0}$ and one of the scalars, say a_j, is not zero. Then we can write

$$a_1 v_1 + a_2 v_2 + \cdots + a_j v_j + \cdots + a_k v_k = \mathbf{0},$$

or

$$a_j v_j = -a_1 v_1 - a_2 v_2 - \cdots - a_{j-1} v_{j-1} - a_{j+1} v_{j+1} - \cdots - a_k v_k,$$

or

$$v_j = \left(-\frac{a_1}{a_j}\right)v_1 + \left(-\frac{a_2}{a_j}\right)v_2 + \cdots + \left(-\frac{a_{j-1}}{a_j}\right)v_{j-1}$$
$$+ \left(-\frac{a_{j+1}}{a_j}\right)v_{j+1} + \cdots + \left(-\frac{a_k}{a_j}\right)v_k.$$

What this last line tells us is that *if a collection of vectors is linearly dependent, then at least one of the vectors is a linear combination of the others.*

Let's see some examples of these concepts.

EXAMPLE 5.7.3   **Linearly Independent and Linearly Dependent Vectors**

The three vectors $\begin{bmatrix} 1 \\ 0 \\ 2 \end{bmatrix}$, $\begin{bmatrix} 0 \\ 1 \\ 2 \end{bmatrix}$, and $\begin{bmatrix} 1 \\ 2 \\ 0 \end{bmatrix}$ are linearly *independent* because the

equation

$$a_1 \begin{bmatrix} 1 \\ 0 \\ 2 \end{bmatrix} + a_2 \begin{bmatrix} 0 \\ 1 \\ 2 \end{bmatrix} + a_3 \begin{bmatrix} 1 \\ 2 \\ 0 \end{bmatrix} = \begin{bmatrix} 0 \\ 0 \\ 0 \end{bmatrix}$$

is equivalent to the algebraic system

$$\begin{aligned} a_1 \qquad\quad + \ a_3 &= 0 \\ a_2 + 2a_3 &= 0 \\ 2a_1 + 2a_2 \qquad\quad &= 0, \end{aligned}$$

which you can solve to find that $a_1 = a_2 = a_3 = 0$. (*Do the work!*)

On the other hand, the collection of vectors $\begin{bmatrix} 3 \\ 4 \\ -4 \end{bmatrix}$, $\begin{bmatrix} 0 \\ 1 \\ 2 \end{bmatrix}$, and $\begin{bmatrix} 1 \\ 2 \\ 0 \end{bmatrix}$ is lin-

early *dependent* because the vector equation

$$a_1 \begin{bmatrix} 3 \\ 4 \\ -4 \end{bmatrix} + a_2 \begin{bmatrix} 0 \\ 1 \\ 2 \end{bmatrix} + a_3 \begin{bmatrix} 1 \\ 2 \\ 0 \end{bmatrix} = \begin{bmatrix} 0 \\ 0 \\ 0 \end{bmatrix}$$

is equivalent to the algebraic system

$$\begin{aligned} 3a_1 \qquad\quad + \ a_3 &= 0 \\ 4a_1 + \ a_2 + 2a_3 &= 0 \\ -4a_1 + 2a_2 \qquad\quad &= 0, \end{aligned}$$

which has infinitely many solutions of the form $a_1 = K$, $a_2 = 2K$, and $a_3 = -3K$. In particular, we can let $K = 1$, so we have the nonzero solution $a_1 = 1$, $a_2 = 2$, and $a_3 = -3$. Note, for example, that we can write

$$\begin{bmatrix} 3 \\ 4 \\ -4 \end{bmatrix} = -2 \begin{bmatrix} 0 \\ 1 \\ 2 \end{bmatrix} + 3 \begin{bmatrix} 1 \\ 2 \\ 0 \end{bmatrix}. \qquad\qquad \blacklozenge$$

Now suppose that we have the system $\dot{X} = AX$, where X is a 3×1 vector and A is a 3×3 matrix of constants. *If A has three distinct real eigenvalues $\lambda_1, \lambda_2, \lambda_3$, then the theory of linear algebra tells us that the corresponding eigenvectors V_1, V_2, V_3 are linearly independent. Furthermore, the vectors $e^{\lambda_1 t}V_1, e^{\lambda_2 t}V_2, e^{\lambda_3 t}V_3$ are linearly independent, and the general solution of $\dot{X} = AX$ is*

$$X(t) = c_1 e^{\lambda_1 t}V_1 + c_2 e^{\lambda_2 t}V_2 + c_3 e^{\lambda_3 t}V_3, \tag{5.7.2}$$

where c_1, c_2, and c_3 are arbitrary constants. Compare this with (5.3.1).

EXAMPLE 5.7.4   Solving a 3 × 3 System via Eigenvalues and Eigenvectors
Consider the system

$$\begin{aligned} \dot{x} &= 4x && + z \\ \dot{y} &= && -2y \\ \dot{z} &= && -z. \end{aligned}$$

The matrix of coefficients is $A = \begin{bmatrix} 4 & 0 & 1 \\ 0 & -2 & 0 \\ 0 & 0 & -1 \end{bmatrix}$, and a CAS calculates the

eigenvalues to be $\lambda_1 = 4$, $\lambda_2 = -1$, and $\lambda_3 = -2$, with corresponding eigenvectors

$$V_1 = \begin{bmatrix} 1 \\ 0 \\ 0 \end{bmatrix}, \quad V_2 = \begin{bmatrix} 1 \\ 0 \\ -5 \end{bmatrix}, \quad \text{and} \quad V_3 = \begin{bmatrix} 0 \\ 1 \\ 0 \end{bmatrix}.$$

Note that these vectors must be linearly independent because the eigenvalues are distinct real numbers. Thus, by (5.7.2), the general solution of our system is given by

$$\begin{aligned} X(t) &= c_1 e^{4t} \begin{bmatrix} 1 \\ 0 \\ 0 \end{bmatrix} + c_2 e^{-t} \begin{bmatrix} 1 \\ 0 \\ -5 \end{bmatrix} + c_3 e^{-2t} \begin{bmatrix} 0 \\ 1 \\ 0 \end{bmatrix} \\ &= \begin{bmatrix} c_1 e^{4t} + c_2 e^{-t} \\ c_3 e^{-2t} \\ -5c_2 e^{-t} \end{bmatrix}. \end{aligned}$$

In this example, we could have noticed that the second and third differential equations making up our original system were separable. After solving each of these, we could have substituted for z in the first equation, which would then be a simple linear equation in x. (*Do this and compare your answer with the one given above.*)

A trajectory in x-y-z space (corresponding to the initial conditions $x(0) = 2$, $y(0) = 5$, and $z(0) = -10$) is shown in Figure 5.17, and the same trajectories in the t-z plane and the y-z plane are shown in Figures 5.18a and 5.18b, respectively.

Note that the graph of a solution of this system is really four-dimensional, a set of points of the form $(t, x(t), y(t), z(t))$. Therefore, what Figure 5.17 is showing is a *projection* of a four-dimensional curve onto three-dimensional x-y-z space.

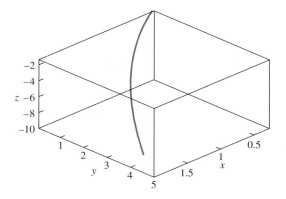

Figure 5.17

Solution of $\dot{x} = 4x + z,\ \dot{y} = -2y,\ \dot{z} = -z;\ x(0) = 2,\ y(0) = 5,\ z(0) = -10;\ 0 \le t \le 2$

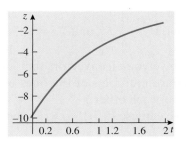

Figure 5.18a

Graph of $z(t),\ 0 \le t \le 2$

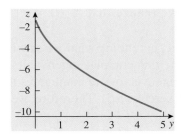

Figure 5.18b

Graph of $z(t)$ vs. $y(t),\ 0 \le t \le 2$ ◆

Accepting the fact that a 3×3 matrix has a cubic characteristic equation, we realize that we can have (1) three distinct real eigenvalues, (2) one distinct real eigenvalue and a different repeated real eigenvalue, (3) one repeated real eigenvalue, or (4) one real eigenvalue and a complex conjugate pair of eigenvalues. Possibilities 1 and 4 are handled easily by formula (5.7.2). However, when we have repeated eigenvalues, we must find linearly independent eigenvectors, sometimes by calculating one or more generalized eigenvectors. (Go back to

Example 5.4.2 and the discussion preceding it. Also see Exercise 14 at the end of this section.)

It should be clear how important the theory of linear algebra is to a full understanding of higher-order differential equations and their equivalent systems, but we will not investigate that theory further in this book.

The next example shows how techniques that we developed for two-dimensional systems in Section 5.5 can be extended to three-dimensional systems.

EXAMPLE 5.7.5   Solving a 3 × 3 System—Complex Eigenvalues

Look at the system $\dot{x} = x$, $\dot{y} = 2x + y - 2z$, $\dot{z} = 3x + 2y + z$. The matrix of coefficients is $A = \begin{bmatrix} 1 & 0 & 0 \\ 2 & 1 & -2 \\ 3 & 2 & 1 \end{bmatrix}$. A CAS provides the characteristic equation $\lambda^3 - 3\lambda^2 + 7\lambda - 5 = (\lambda - 1)(\lambda^2 - 2\lambda + 5) = 0$, which has roots $\lambda_1 = 1$, $\lambda_2 = 1 + 2i$, and $\lambda_3 = 1 - 2i$. A CAS also gives the corresponding eigenvectors $V_1 = \begin{bmatrix} 2 \\ -3 \\ 2 \end{bmatrix}$, $V_2 = \begin{bmatrix} 0 \\ i \\ 1 \end{bmatrix}$, and $V_3 = \begin{bmatrix} 0 \\ -i \\ 1 \end{bmatrix}$. (Remember that your calculator or CAS may give you eigenvectors that look different from these. Just check to see that yours are multiples of the ones used above. Also, note that V_2 and V_3 are conjugates of each other.)

Now we can use (5.7.2) to write the general solution in the form

$$X(t) = c_1 e^t \begin{bmatrix} 2 \\ -3 \\ 2 \end{bmatrix} + c_2 e^{(1+2i)t} \begin{bmatrix} 0 \\ i \\ 1 \end{bmatrix} + c_3 e^{(1-2i)t} \begin{bmatrix} 0 \\ -i \\ 1 \end{bmatrix}.$$

However, we realize that $X_1(t) = e^t \begin{bmatrix} 2 \\ -3 \\ 2 \end{bmatrix}$ is a solution of the system by itself.

Furthermore, extending what we saw in Section 5.5, we know that we need work only with the *first* complex eigenvalue-eigenvector pair, because the other eigenvalue and eigenvector are conjugates that produce the same solutions (see Problem 10 of Exercises 5.5). Therefore, we consider only

$$\tilde{X}(t) = e^{(1+2i)t} \begin{bmatrix} 0 \\ i \\ 1 \end{bmatrix} = e^t(\cos(2t) + i\sin(2t)) \begin{bmatrix} 0 \\ i \\ 1 \end{bmatrix}$$

$$= e^t \begin{bmatrix} 0 \\ -\sin(2t) + i\cos(2t) \\ \cos(2t) + i\sin(2t) \end{bmatrix} = e^t \begin{bmatrix} 0 \\ -\sin(2t) \\ \cos(2t) \end{bmatrix} + ie^t \begin{bmatrix} 0 \\ \cos(2t) \\ \sin(2t) \end{bmatrix}.$$

From this last expression we derive two linearly independent real-valued solutions of our system:

$$X_2(t) = \text{Re}\{\tilde{X}(t)\} = e^t \begin{bmatrix} 0 \\ -\sin(2t) \\ \cos(2t) \end{bmatrix} \text{ and } X_3(t) = \text{Im}\{\tilde{X}(t)\} = e^t \begin{bmatrix} 0 \\ \cos(2t) \\ \sin(2t) \end{bmatrix}.$$

Finally, the Superposition Principle tells us that

$$X(t) = c_1 X_1 + c_2 X_2 + c_3 X_3$$

$$= c_1 e^t \begin{bmatrix} 2 \\ -3 \\ 2 \end{bmatrix} + c_2 e^t \begin{bmatrix} 0 \\ -\sin(2t) \\ \cos(2t) \end{bmatrix} + c_3 e^t \begin{bmatrix} 0 \\ \cos(2t) \\ \sin(2t) \end{bmatrix}$$

$$= \begin{bmatrix} 2c_1 e^t \\ -3c_1 e^t - c_2 e^t \sin(2t) + c_3 e^t \cos(2t) \\ 2c_1 e^t + c_2 e^t \cos(2t) + c_3 e^t \sin(2t) \end{bmatrix}$$

$$= e^t \begin{bmatrix} 2c_1 \\ -3c_1 - c_2 \sin(2t) + c_3 \cos(2t) \\ 2c_1 + c_2 \cos(2t) + c_3 \sin(2t) \end{bmatrix}$$

is the real-valued general solution of the original system. If you use technology to solve this problem, be aware that your CAS may express the solution functions in a different but equivalent way.   ◆

Nonhomogeneous Systems

It is important to realize that we can also handle larger *nonhomogeneous* systems in this way, using the relationship explored in Section 5.6: $X_{\text{GNH}} = X_{\text{GH}} + X_{\text{PNH}}$. The method of undetermined coefficients becomes algebraically messier as the size of the system increases, and in Chapter 6 we'll examine a better way of handling such systems.

Generalization to n × n Systems

Everything we've done with 2 × 2 and 3 × 3 systems of equations can be generalized to $n \times n$ systems. We can express a homogeneous nth-order linear system with constant coefficients

$$\begin{aligned}
\dot{x}_1 &= a_{11}x_1 + a_{12}x_2 + a_{13}x_3 + \cdots + a_{1n}x_n \\
\dot{x}_2 &= a_{21}x_1 + a_{22}x_2 + a_{23}x_3 + \cdots + a_{2n}x_n \\
\dot{x}_3 &= a_{31}x_1 + a_{32}x_2 + a_{33}x_3 + \cdots + a_{3n}x_n \\
&\ \ \vdots \qquad\qquad \vdots \qquad\qquad \vdots \\
\dot{x}_n &= a_{n1}x_1 + a_{n2}x_2 + a_{n3}x_3 + \cdots + a_{nn}x_n
\end{aligned}$$

as $\dot{X} = AX$, where

$$X = \begin{bmatrix} x_1 \\ x_2 \\ x_3 \\ \vdots \\ x_n \end{bmatrix}, \quad \dot{X} = \begin{bmatrix} \dot{x}_1 \\ \dot{x}_2 \\ \dot{x}_3 \\ \vdots \\ \dot{x}_n \end{bmatrix}, \quad \text{and} \quad A = \begin{bmatrix} a_{11} & a_{12} & a_{13} & \cdots & a_{1n} \\ a_{21} & a_{22} & a_{23} & \cdots & a_{2n} \\ a_{31} & a_{32} & a_{33} & \cdots & a_{3n} \\ \vdots & \vdots & \vdots & \vdots & \vdots \\ a_{n1} & a_{n2} & a_{n3} & \cdots & a_{nn} \end{bmatrix}.$$

EXAMPLE 5.7.6   Matrix Form of a Four-Dimensional System

A compartmental analysis (see Section 2.2) of the processes involved in protein synthesis in animals and humans uses radioactive isotopes as tracers. A particular four-compartment model of this situation could lead to a system such as

$$\dot{x}_1 = -2x_1 + x_2$$
$$\dot{x}_2 = x_1 - 2x_2$$
$$\dot{x}_3 = x_1 + x_2 - x_3$$
$$\dot{x}_4 = x_3,$$

where $x_i(t)$ denotes the fraction of the total administered radioactivity attached to the material (albumen) in compartment i ($i = 1, 2, 3, 4$). The coefficients indicate flow rates of the radioactive material from compartment to compartment.

In matrix terms, we can express this system as

$$\begin{bmatrix} \dot{x}_1 \\ \dot{x}_2 \\ \dot{x}_3 \\ \dot{x}_4 \end{bmatrix} = \begin{bmatrix} -2 & 1 & 0 & 0 \\ 1 & -2 & 0 & 0 \\ 1 & 1 & -1 & 0 \\ 0 & 0 & 1 & 0 \end{bmatrix} \begin{bmatrix} x_1 \\ x_2 \\ x_3 \\ x_4 \end{bmatrix}.$$

◆

Extending the theory underlying the solution of algebraic systems of three linear equations in three unknowns to systems of n equations in n unknowns, we state that any $n \times n$ matrix has a determinant and that eigenvalues and eigenvectors can be defined for such square matrices. Specifically, given a system $\dot{X} = AX$, where X is an $n \times 1$ column matrix (vector) and A is an $n \times n$ matrix, an eigenvalue λ is a solution of the equation $\det(A - \lambda I) = 0$, where I is the $n \times n$ identity matrix consisting of ones down the main diagonal and zeros elsewhere. Given an eigenvalue λ, an eigenvector associated with λ is a nonzero vector V satisfying the equation $AV = \lambda V$.

The characteristic equation of an $n \times n$ matrix is an nth-degree polynomial. However, once a polynomial has degree greater than or equal to 5, there is no longer a general formula that gives the roots. In general, the only way to tackle such equations is to use *approximation* methods. A CAS—or even a graphing calculator—has various algorithms to do this.

Now suppose that we have the system $\dot{X} = AX$, where X is an $n \times 1$ vector and A is an $n \times n$ matrix of constants. *If A has n distinct real eigenvalues $\lambda_1, \lambda_2, \ldots, \lambda_n$, then the theory of linear algebra guarantees that the corresponding eigenvectors $V_1, V_2, \ldots, V_n$ are linearly independent. Furthermore, the vectors $e^{\lambda_1 t}V_1, e^{\lambda_2 t}V_2, \ldots, e^{\lambda_n t}V_n$ are linearly independent, and the general solution of $\dot{X} = AX$ is*

$$X(t) = c_1 e^{\lambda_1 t}V_1 + c_2 e^{\lambda_2 t}V_2 + \cdots + c_n e^{\lambda_n t}V_n, \qquad (5.7.3)$$

where $c_1, c_2, \ldots, c_n$ are arbitrary constants. You should expect the usual complications when there are repeated real roots, complex conjugate pairs of roots, and so forth.

If we investigate a mechanical system (Figure 5.19) consisting of two springs attached to two movable masses, the physics of the situation gives us a pair of second-order linear differential equations. In turn, this system of two equations can be expressed as a system of four first-order linear equations.

The following example assumes that we start from the equilibrium position by giving one mass an initial velocity. Most of the computational work will be done by a CAS.

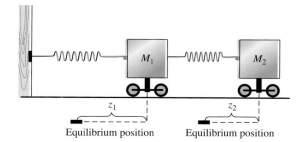

Figure 5.19
The spring-mass system for Example 5.7.7

EXAMPLE 5.7.7   A Four-Dimensional System from Mechanics

Let's consider the system

$$\frac{d^2 z_1}{dt^2} = -11z_1 + 6z_2$$

$$\frac{d^2 z_2}{dt^2} = -6z_2 + 6z_1,$$

where z_1 is the distance of mass 1 from its equilibrium position and z_2 is the distance of mass 2 from equilibrium. We'll assume the initial conditions $z_1(0) = 0$, $z_1'(0) = 0$, $z_2(0) = 0$, and $z_2'(0) = 1$.

Representation as a First-Order System

Introducing the new variables $x_1 = z_1$, $x_2 = \dfrac{dz_1}{dt}$, $x_3 = z_2$, and $x_4 = \dfrac{dz_2}{dt}$, we convert our pair of second-order equations into the four-dimensional system of first-order equations

$$\frac{dx_1}{dt} = x_2$$

$$\frac{dx_2}{dt} = -11x_1 + 6x_3$$

$$\frac{dx_3}{dt} = x_4$$

$$\frac{dx_4}{dt} = -6x_3 + 6x_1,$$

with $x_1(0) = x_2(0) = x_3(0) = 0$ and $x_3'(0) = x_4(0) = 1$.

Matrix Representation, Eigenvalues, Eigenvectors

We can express the last system as $\dfrac{d}{dt}X = \begin{bmatrix} 0 & 1 & 0 & 0 \\ -11 & 0 & 6 & 0 \\ 0 & 0 & 0 & 1 \\ 6 & 0 & -6 & 0 \end{bmatrix}\begin{bmatrix} x_1 \\ x_2 \\ x_3 \\ x_4 \end{bmatrix} = AX.$ A

CAS gives the characteristic equation of matrix A as $\lambda^4 + 17\lambda^2 + 30 = 0$, which we can factor as $(\lambda^2 + 15)(\lambda^2 + 2) = 0$, so the eigenvalues are $\lambda_1 = \sqrt{15}i$,

$\lambda_2 = -\sqrt{15}i$, $\lambda_3 = \sqrt{2}i$, and $\lambda_4 = -\sqrt{2}i$. The corresponding eigenvectors are

$$V_1 = \begin{bmatrix} 3 \\ 3\sqrt{15}i \\ -2 \\ -2\sqrt{15}i \end{bmatrix}, V_2 = \begin{bmatrix} 3 \\ -3\sqrt{15}i \\ -2 \\ 2\sqrt{15}i \end{bmatrix}, V_3 = \begin{bmatrix} 2 \\ 2\sqrt{2}i \\ 3 \\ 3\sqrt{2}i \end{bmatrix}, \text{ and } V_4 = \begin{bmatrix} 2 \\ -2\sqrt{2}i \\ 3 \\ -3\sqrt{2}i \end{bmatrix}$$

If you check this with a CAS, remember that you may get a different (but equivalent) form for the eigenvectors.

The General Solution

On the basis of our previous experience with complex conjugate pairs of eigenvalues and eigenvectors, we can just work with the pairs λ_1, V_1 and λ_3, V_3. First, we know that

$$\hat{X}(t) = e^{\lambda_1 t}V_1 = e^{\sqrt{15}it}\begin{bmatrix} 3 \\ 3\sqrt{15}i \\ -2 \\ -2\sqrt{15}i \end{bmatrix} = (\cos(\sqrt{15}t) + i\sin(\sqrt{15}t))\begin{bmatrix} 3 \\ 3\sqrt{15}i \\ -2 \\ -2\sqrt{15}i \end{bmatrix}$$

$$= \begin{bmatrix} 3\cos(\sqrt{15}t) \\ -3\sqrt{15}\sin(\sqrt{15}t) \\ -2\cos(\sqrt{15}t) \\ 2\sqrt{15}\sin(\sqrt{15}t) \end{bmatrix} + i\begin{bmatrix} 3\sin(\sqrt{15}t) \\ 3\sqrt{15}\cos(\sqrt{15}t) \\ -2\sin(\sqrt{15}t) \\ -2\sqrt{15}\cos(\sqrt{15}t) \end{bmatrix} = X_1(t) + iX_2(t),$$

where both $X_1(t)$ and $X_2(t)$ are linearly independent real-valued solutions of the system. Then we have

$$\tilde{X}(t) = e^{\lambda_3 t}V_3 = e^{\sqrt{2}it}\begin{bmatrix} 2 \\ 2\sqrt{2}i \\ 3 \\ 3\sqrt{2}i \end{bmatrix} = (\cos(\sqrt{2}t) + i\sin(\sqrt{2}t))\begin{bmatrix} 2 \\ 2\sqrt{2}i \\ 3 \\ 3\sqrt{2}i \end{bmatrix}$$

$$= \begin{bmatrix} 2\cos(\sqrt{2}t) \\ -2\sqrt{2}\sin(\sqrt{2}t) \\ 3\cos(\sqrt{2}t) \\ -3\sqrt{2}\sin(\sqrt{2}t) \end{bmatrix} + i\begin{bmatrix} 2\sin(\sqrt{2}t) \\ 2\sqrt{2}\cos(\sqrt{2}t) \\ 3\sin(\sqrt{2}t) \\ 3\sqrt{2}\cos(\sqrt{2}t) \end{bmatrix} = X_3(t) + iX_4(t),$$

where $X_3(t)$ and $X_4(t)$ are linearly independent real-valued solutions of the system. The general solution is

$$X(t) = c_1X_1 + c_2X_2 + c_3X_3 + c_4X_4$$

$$= c_1\begin{bmatrix} 3\cos(\sqrt{15}t) \\ -3\sqrt{15}\sin(\sqrt{15}t) \\ -2\cos(\sqrt{15}t) \\ 2\sqrt{15}\sin(\sqrt{15}t) \end{bmatrix} + c_2\begin{bmatrix} 3\sin(\sqrt{15}t) \\ 3\sqrt{15}\cos(\sqrt{15}t) \\ -2\sin(\sqrt{15}t) \\ -2\sqrt{15}\cos(\sqrt{15}t) \end{bmatrix} +$$

$$c_3 \begin{bmatrix} 2\cos(\sqrt{2}t) \\ -2\sqrt{2}\sin(\sqrt{2}t) \\ 3\cos(\sqrt{2}t) \\ -3\sqrt{2}\sin(\sqrt{2}t) \end{bmatrix} + c_4 \begin{bmatrix} 2\sin(\sqrt{2}t) \\ 2\sqrt{2}\cos(\sqrt{2}t) \\ 3\sin(\sqrt{2}t) \\ 3\sqrt{2}\cos(\sqrt{2}t) \end{bmatrix}.$$

Particular Solutions

Now the initial conditions $x_1(0) = x_2(0) = x_3(0) = 0$ and $x_3'(0) = x_4(0) = 1$ imply (Exercise 13) that $c_1 = c_3 = 0$ and $c_2 = -2\sqrt{15}/195$, $c_4 = 3\sqrt{2}/26$. Therefore,

$$z_1(t) = x_1(t) = \frac{3\sqrt{2}}{13}\sin(\sqrt{2}t) - \frac{2\sqrt{15}}{65}\sin(\sqrt{15}t)$$

and

$$z_2(t) = x_3(t) = \frac{9\sqrt{2}}{26}\sin(\sqrt{2}t) + \frac{4\sqrt{15}}{195}\sin(\sqrt{15}t). \qquad \blacklozenge$$

If we are interested in the *stability* of an $n \times n$ system rather than its exact solution, we can give a simplified version of the results we have seen for 2×2 systems (see Table 5.1 in Section 5.5): If A is an $n \times n$ matrix of constants, then the equilibrium solution $X = \mathbf{0}$ for the system $\dot{X} = AX$ is asymptotically stable (that is, it is a *sink*) if every eigenvalue of A has a negative real part and is unstable if A has at least one eigenvalue with a positive real part. Furthermore, if *all* eigenvalues have positive real parts, the n-dimensional origin $X = \mathbf{0}$ is a *source*; and if some eigenvalues have positive real parts and others have negative real parts, the equilibrium point is called a *saddle point*.

In the next chapter we'll learn another way to handle initial-value problems involving systems of linear equations. The method of *Laplace transforms*, especially when implemented by a CAS, is a powerful tool for solving various applied problems.

EXERCISES 5.7

For each of the systems in Exercises 1–4, (a) write the system in the form $\dot{X} = AX$; (b) use technology to find eigenvalues and representative eigenvectors; and (c) express the general solution as a single real-valued vector of functions.

1. $\dfrac{dx}{dt} = x - y + z$

$\dfrac{dy}{dt} = x + y - z$

$\dfrac{dz}{dt} = 2x - y$

2. $\dfrac{dx}{dt} = x - 2y - z$

$\dfrac{dy}{dt} = -x + y + z$

$\dfrac{dz}{dt} = x - z$

3. $\dfrac{dx}{dt} = 3x - y + z$ **4.** $\dfrac{dx}{dt} = 2x + y$

     $\dfrac{dy}{dt} = x + y + z$ $\dfrac{dy}{dt} = x + 3y - z$

     $\dfrac{dz}{dt} = 4x - y + 4z$ $\dfrac{dz}{dt} = 2y + 3z - x$

5. For the system in Exercise 1, use your CAS to plot the *x-y-z* space trajectory passing through the point $(0, 1, 0)$ when $t = 0$.

6. For the system in Exercise 4, use your CAS to plot the *x-y-z* space trajectory passing through the point $(1, 1, -1)$ when $t = 0$.

7. a. Solve the initial-value problem

$$\frac{dx}{dt} = z + y - x$$

$$\frac{dy}{dt} = z + x - y$$

$$\frac{dz}{dt} = x + y + z,$$

     $x(0) = 1,\ y(0) = -\frac{1}{3},\ z(0) = 0$.

b. Use the explicit solution found in part (a) to calculate $x(0.5)$, $y(0.5)$, and $z(0.5)$.

c. Use two or more numerical methods found in your CAS to approximate $x(0.5)$, $y(0.5)$, and $z(0.5)$. Compare the answers to each other and to the exact answers in part (a).

8. Solve the initial-value problem

$$\frac{dx}{dt} = y + z$$

$$\frac{dy}{dt} = z + x$$

$$\frac{dz}{dt} = x + y,$$

     $x(0) = -1,\ y(0) = 1,\ z(0) = 0$.

9. There are three tanks (see the accompanying figure) that pump fluid back and forth in the following way: Tank A pumps fluid into tank B at a rate of 2% of its volume per hour and also back into itself at a rate of 1% of the volume per hour. Tank B pumps into itself, tank A, and tank C, all at a rate of 2% of its volume per hour. Tank C pumps into tank B at a rate of 2% of its volume per hour and back into itself at the rate of 3% of its volume per hour. Assuming that the initial volumes in tanks A, B, and C are 23000, 1000, and 1000 liters, respectively, describe the changes in volume of fluid in each tank as functions of time. (Use technology only to find the eigenvalues and the corresponding eigenvectors.)

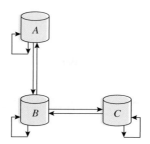

10. Suppose that you have a *double pendulum*—that is, one pendulum suspended from another, as shown in the figure. The laws of physics, after a simplifying change of variables, give us the following system as a model for small oscillations about the equilibrium position:

$$\dot{x} = y$$
$$\dot{y} = -x + \alpha u$$
$$\dot{u} = v$$
$$\dot{v} = x - u.$$

Here $\alpha = (m_2/m_1)(1 + m_2/m_1)^{-1}$, $x = \theta_1$, $u = \theta_2$, and y and v are the angular velocities $\dot{\theta}_1$ and $\dot{\theta}_2$, respectively. For this problem, let $\alpha = 0.3$.

a. Express the system in matrix form.

b. Use technology to find the eigenvalues of the system.

c. Use technology to find eigenvectors corresponding to the eigenvalues found in part (b).

d. Find the general real-valued solution of the system.

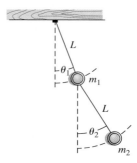

11. Consider the pair of 50-gallon tanks shown on the following page. Initially, tank I is full of compound B and tank II is full of compound C. Start to introduce compound A into each tank at the rates shown in the figure.

a. Let $x_1(t)$ and $x_2(t)$ denote the amount of compound A in tanks I and II, respectively. Similarly, define $y_1(t)$, $y_2(t)$, $z_1(t)$, and $z_2(t)$ for the amounts of compounds B and C in tanks I and II. Now write a system of six nonhomogeneous differential equations describing the flow of the various substances into and out of tanks I and II, expressing any fractions in decimal form. Be sure to write initial conditions.

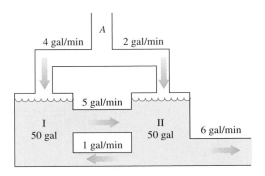

b. Use technology to solve the IVP expressed in part (a).

c. Use technology to graph $x_1(t)$, $y_1(t)$, and $z_1(t)$ against t, all on the same set of axes.

d. Use technology to graph $x_2(t)$, $y_2(t)$, and $z_2(t)$ against t, all on the same set of axes.

12. Consider the system

$$\ddot{x} = 2x + \dot{x} + y$$
$$\dot{y} = 4x + 2y.$$

a. Convert this system to a system of three first-order equations, $\dot{Y} = AY$.

b. Use technology to find the eigenvalues of the matrix A in part (a).

c. Use technology to find two linearly independent eigenvectors corresponding to the eigenvalues found in part (b).

d. Take $W = \begin{bmatrix} t \\ 1 \\ -1 - 2t \end{bmatrix}$ as a third eigenvector that is independent of the two found in part (c) and give the general solution of $\dot{Y} = AY$.

e. Find the general solution $x(t)$, $y(t)$ of the original system.

13. In Example 5.7.7, use the initial conditions to show that $c_1 = c_3 = 0$ and $c_2 = -2\sqrt{15}/195$, $c_4 = 3\sqrt{2}/26$.

14. Suppose you have a system $\dot{X} = AX$, where A is a 3×3 matrix that has an eigenvalue λ of multiplicity 3 and corresponding eigenvector V. Then it can be shown that the general solution of the system can be written as $c_1 X_1 + c_2 X_2 + c_3 X_3$, where $X_1 = e^{\lambda t}V$, $X_2 = e^{\lambda t}(W + tV)$, $X_3 = e^{\lambda t}\left(U + tW + \dfrac{t^2}{2}V\right)$, W satisfies $(A - \lambda I)W = V$, and U satisfies $(A - \lambda I)U = W$.

a. Find the repeated eigenvalue and representative eigenvector for the system

$$x' = x + y + z$$
$$y' = 2x + y - z$$
$$z' = -3x + 2y + 4z.$$

b. Use the method described above and technology to write the general solution of this system.

15. Find the general solution of the nonautonomous, nonhomogeneous system

$$\frac{dx}{dt} = 2t$$

$$\frac{dy}{dt} = 3x + 2t$$

$$\frac{dz}{dt} = x + 4y + t$$

    a. by using ideas from Exercise 14, followed by the technique of *undetermined coefficients*. (See Section 5.6.)

    b. by solving the first equation and then substituting this solution in the second equation, and so forth.

5.8 SUMMARY

By using matrices and their properties, any autonomous system of linear equations with constant coefficients

$$\dot{x}_1 = a_{11}x_1 + a_{12}x_2 + a_{13}x_3 + \cdots + a_{1n}x_n$$
$$\dot{x}_2 = a_{21}x_1 + a_{22}x_2 + a_{23}x_3 + \cdots + a_{2n}x_n$$
$$\dot{x}_3 = a_{31}x_1 + a_{32}x_2 + a_{33}x_3 + \cdots + a_{3n}x_n$$
$$\vdots \qquad \vdots \qquad\qquad \vdots$$
$$\dot{x}_n = a_{n1}x_1 + a_{n2}x_2 + a_{n3}x_3 + \cdots + a_{nn}x_n$$

can be written in the compact form $\dot{X} = AX$, where

$$X = \begin{bmatrix} x_1 \\ x_2 \\ x_3 \\ \vdots \\ x_n \end{bmatrix}, \quad \dot{X} = \begin{bmatrix} \dot{x}_1 \\ \dot{x}_2 \\ \dot{x}_3 \\ \vdots \\ \dot{x}_n \end{bmatrix}, \quad \text{and} \quad A = \begin{bmatrix} a_{11} & a_{12} & a_{13} & \cdots & a_{1n} \\ a_{21} & a_{22} & a_{23} & \cdots & a_{2n} \\ a_{31} & a_{32} & a_{33} & \cdots & a_{3n} \\ \vdots & \vdots & \vdots & \vdots & \vdots \\ a_{n1} & a_{n2} & a_{n3} & \cdots & a_{nn} \end{bmatrix}.$$

For two-dimensional systems in which the matrix of coefficients A has a nonzero determinant, the origin is the only equilibrium point. The qualitative behavior (stability) of such a linear system is completely determined by the eigenvalues and eigenvectors of A. If the system has two real eigenvalues λ_1 and λ_2, with $\lambda_1 \neq \lambda_2$, and V_1 and V_2 are the corresponding (linearly independent) eigenvectors, then the general solution of the system is given by $X(t) = c_1 e^{\lambda_1 t}V_1 + c_2 e^{\lambda_2 t}V_2$. If the system has two real and equal eigenvalues, we can try to find two linearly independent eigenvectors corresponding to the single eigenvalue. If we can't find two such eigenvectors, we can start with one eigenvector V and calculate a *generalized eigenvector* W so that the general solution of the system can be written in the form $X(t) = c_1 e^{\lambda t}V + c_2[te^{\lambda t}V + e^{\lambda t}W]$. Finally, if the system has a pair of complex conjugate eigenvalues, we can still write the solution as $X(t) = c_1 e^{\lambda_1 t}V_1 + c_2 e^{\lambda_2 t}V_2$, but we have to use *Euler's formula*,

$e^{p+qi} = e^p(\cos(q) + i\sin(q))$, to simplify this expression and wind up with *real-valued* solutions. Specifically, first we get a solution of the form $X(t) = X_1(t) + iX_2(t)$, where $X_1(t)$ and $X_2(t)$ are real-valued matrix (vector) functions called the *real part* and the *imaginary part*, respectively, of $X(t)$. Then the real-valued general solution is $X(t) = C_1X_1(t) + C_2X_2(t)$, where C_1 and C_2 are real numbers. Using these forms for the general solution of our system, we can analyze the stability of the system qualitatively in terms of eigenvalues and eigenvectors. These results were summarized in Table 5.1 at the end of Section 5.5.

We may have a *nonhomogeneous* system

$$
\begin{aligned}
\dot{x}_1 &= a_{11}x_1 + a_{12}x_2 + a_{13}x_3 + \cdots + a_{1n}x_n + b_1(t) \\
\dot{x}_2 &= a_{21}x_1 + a_{22}x_2 + a_{23}x_3 + \cdots + a_{2n}x_n + b_2(t) \\
\dot{x}_3 &= a_{31}x_1 + a_{32}x_2 + a_{33}x_3 + \cdots + a_{3n}x_n + b_3(t) \\
&\vdots \qquad\qquad \vdots \qquad\qquad\qquad \vdots \\
\dot{x}_n &= a_{n1}x_1 + a_{n2}x_2 + a_{n3}x_3 + \cdots + a_{nn}x_n + b_n(t),
\end{aligned}
$$

which can be written as $\dot{X} = AX + B(t)$, with

$$
X = \begin{bmatrix} x_1 \\ x_2 \\ x_3 \\ \vdots \\ x_n \end{bmatrix}, \quad
\dot{X} = \begin{bmatrix} \dot{x}_1 \\ \dot{x}_2 \\ \dot{x}_3 \\ \vdots \\ \dot{x}_n \end{bmatrix},
$$

$$
A = \begin{bmatrix}
a_{11} & a_{12} & a_{13} & \cdots & a_{1n} \\
a_{21} & a_{22} & a_{23} & \cdots & a_{2n} \\
a_{31} & a_{32} & a_{33} & \cdots & a_{3n} \\
\vdots & \vdots & \vdots & \vdots & \vdots \\
a_{n1} & a_{n2} & a_{n3} & \cdots & a_{nn}
\end{bmatrix}, \quad \text{and} \quad
B(t) = \begin{bmatrix} b_1(t) \\ b_2(t) \\ b_3(t) \\ \vdots \\ b_n(t) \end{bmatrix}.
$$

In this situation, we know that *the general solution, X_{GNH}, of a linear nonhomogeneous system is obtained by finding a particular solution, X_{PNH}, of the nonhomogeneous system and adding it to the general solution, X_{GH}, of the homogeneous system: $X_{\text{GNH}} = X_{\text{GH}} + X_{\text{PNH}}$.* The method of *undetermined coefficients* can be used to make an intelligent guess about the particular solution if the entries of vector $B(t)$ contain terms that are constants, exponential functions, sines, cosines, polynomials, or any sum or product of such terms. For other kinds of functions making up $B(t)$, X_{PNH} must be found using some other technique (for example, *variation of parameters*).

Although we started with a thorough analysis of the equilibrium points and the stability of the system near these points for *two*-dimensional systems of equations with constant coefficients, we saw eventually that the concepts of eigenvalue and eigenvector were meaningful for systems of n equations. Specifically, given a system $\dot{X} = AX$, where X is an $n \times 1$ column matrix (vector) and A is an $n \times n$ matrix, an eigenvalue λ is a solution of the equation $\det(A - \lambda I) = 0$, where I is the $n \times n$ identity matrix consisting of ones down the main diagonal and zeros elsewhere. We know that $\det(A - \lambda I)$ is an nth-degree polynomial in λ. Given an eigenvalue λ, an eigenvector associated with λ is a nonzero vector V satisfying the equation $AV = \lambda V$.

For values of n greater than 3, we lose the ordinary intuitive geometric interpretation of our results. Also, when n is greater than or equal to 5, there is no general procedure we can follow to solve the characteristic equations. We must use approximation methods, and technology becomes crucial here. The question of the multiplicity of eigenvalues leads to complicated linear-algebra considerations, and the general vector form of the solution of a system is difficult to describe without delving more deeply into linear algebra.

PROJECT 5-1

A Vicious Circle

There are three species of omnivores on an island: xaccoons, yadgers, and zoyotes. Xaccoons eat zoyotes, zoyotes prey on yadgers, and yadgers find xaccoons delicious. Given the species' individual birth rates and predation rates, we can set up the following system:

$$\dot{x} = 21x - 9y$$
$$\dot{y} = 2y - 4z$$
$$\dot{z} = 6z - 7x.$$

Here $x(t)$, $y(t)$, and $z(t)$ denote the populations of the three species (in an obvious way) at time t, where t is in centuries.

Long before human beings arrived on the island, the three species lived in a state of equilibrium. At time $t = 0$, shortly after humans discovered the island and disrupted the equilibrium by hunting, chopping down trees, and so on, there were 300 xaccoons, 598 yadgers, and 323 zoyotes. (This is no longer the equilibrium state.)

a. Write the system in matrix terms, and use technology to find the eigenvalues and eigenvectors of the system.

b. Solve the system IVP manually, using the results of part (a). (You may use technology to solve an algebraic system of three equations in three unknowns.)

c. What were the equilibrium populations before humans arrived on the scene? (Look at the solution in part (b) as $t \to -\infty$.)

d. Substitute the answers found in part (c) into the three differential equations making up the system. For each population, what does the result say about its birth rate compared to its loss by predation during the period of equilibrium?

e. Graph $x(t)$ against t, $0 \le t \le 0.1$. Graph $y(t)$ against t, for $0 \le t \le 0.1$ and then for $0 \le t \le 0.2$. Graph $z(t)$ against t for $0 \le t \le 0.16$. (Use technology to obtain these graphs.)

f. Which is the most acutely endangered species? After how many years will it become extinct? At the time of this species' extinction, what will be the populations of the surviving species?

PROJECT 5-2

Go with the Flow

The setup for a complicated mixing system is shown in the accompanying figure. We have three tanks interconnected as indicated by pipes whose capacity of flow is 5 gal/min. Initially, tank I contains 20 gallons of red paint, tank II contains 30 gallons of yellow paint, and tank III contains 40 gallons of blue paint. What will be the mixture of paints in each tank at the end of 5 minutes? *Use technology to solve this problem.*

  Comments: Let x_i = the amount of red paint in tank i at time t, y_i = the amount of yellow paint in tank i at time t, and z_i = the amount of blue paint in tank i at time t. Then the initial conditions say that at $t = 0$, $x_1 = 20$, $x_2 = 0$, $x_3 = 0$; $y_1 = 0$, $y_2 = 30$, $y_3 = 0$; and $z_1 = 0$, $z_2 = 0$, $z_3 = 40$. Furthermore, the flow pattern in the diagram, together with the observation that the total volume in each tank is constant—20, 30, and 40 gallons, respectively—leads to a system of nine differential equations. *However, the system of three equations containing the variables x_1, x_2, and x_3 is exactly the same as the system containing the y_i and the system containing the z_i.* The only difference lies in the initial conditions. Physically, the obvious symmetry of the tanks and the flow pattern explain this. Therefore, you need solve only one of the three-dimensional systems and use it to find all three sets of variables by varying the initial conditions.

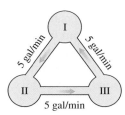

6 | The Laplace Transform

6.0 INTRODUCTION

The idea of a *transform,* or *transformation,* is a very important one in mathematics and problem solving in general. Faced with a difficult problem, it is often a good idea to change it in some way into an easier problem, solve that easier problem, and then take your solution and apply it to your original problem. One of the first examples of this process that you have seen involves the idea of a *logarithm.* When John Napier and others developed logarithms in the early 1600s, it was as an aid to calculation. Given a difficult multiplication problem, one could transform it into an addition problem, perform the addition, and then transform one's answer back into the answer to the original problem. For example, if one wanted to multiply 8743 by 2591, one could apply the natural logarithm to this product, getting the sum $\ln(8743) + \ln(2591) = 9.07600865918\ldots + 7.85979918056\ldots = 16.9358078397.\ldots$ Then one would determine (from a "log table" in those dark ages) the number whose natural logarithm is $16.9358078397.\ldots$ That number should be the original product. The process of going from the sum of logs back to the original product can be called an *inverse* transformation. Of course, we recognize that the inverse of the logarithmic transformation is the exponential transformation:

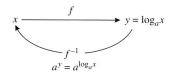

Another important familiar transform converts a nonnegative integrable function $f(x)$ into the *area function* $A(x)$ as follows: $f(x) \to A(x) = \displaystyle\int_a^x f(t)\,dt.$

271

What this is saying is that if f is a nonnegative function whose integral is defined for $x \geq a$, then $A(x)$ represents the area under the graph of f from $t = a$ to $t = x$.

Back in Section 2.2 you encountered a transformation when you introduced the *integrating factor* into a linear equation. In multiplying the equation by the appropriate exponential factor, you transformed the left-hand side into an exact derivative, which could then be integrated to yield the unknown function. You solved the equation by changing it into an equivalent form that was easier to deal with. You might think of these transform methods in terms of confronting a shark. You wouldn't want to meet up with a shark in the ocean, but if that shark were dragged (transformed) onto a beach, you would be able to deal with it (to avoid its snapping jaws) more easily. *Same shark, different environment.*

The famous mathematical tool known as the *Laplace transform*, which is named for the great French mathematician Pierre-Simon de Laplace (1749–1827) but was probably used earlier by Euler, is useful to us because it removes derivatives from differential equations and replaces them with *algebraic* expressions. In this way, differential equations are replaced by algebraic equations. This transformation turns out to be particularly useful when we are dealing with initial-value problems, nonhomogeneous equations with discontinuous forcing terms, and systems of differential equations. The downside is that the use of the Laplace transform is restricted to the solution of *linear* differential equations and *linear* systems of equations.

6.1   THE LAPLACE TRANSFORM OF SOME IMPORTANT FUNCTIONS

We start by assuming that $f(t)$ is a function that is defined for $t \geq 0$. The **Laplace transform** of this function, $\mathcal{L}[f(t)]$, is defined as

$$\mathcal{L}[f(t)] = \int_0^\infty f(t)e^{-st}dt, \tag{6.1.1}$$

when this improper integral exists. Note that after you've integrated with respect to t, the result will have the parameter s in it—that is, this integral is a *function* of the parameter s, so we can write $\mathcal{L}[f(t)] = F(s)$.

Before we give some examples, let's just examine the integral in (6.1.1). From basic calculus, we know that the improper integral is defined as $\lim_{b\to\infty} \int_0^b f(t)e^{-st}dt$ when this limit exists. There are two important requirements here. First, the ordinary Riemann integral $\int_0^b f(t)e^{-st}dt$ must exist for every $b > 0$; and then the limit must exist as $b \to \infty$. Both requirements are taken care of if we stick to *continuous* or *piecewise continuous* functions $f(t)$ for which there exist positive constants M and K such that $|f(t)| < e^{Mt}$ for all $t \geq K$. This says that the function f doesn't grow faster than an exponential function, so the integrand $f(t)e^{-st}$ in (6.1.1) behaves like the function $e^{Mt} \cdot e^{-st} = e^{-(s-M)t}$ for values of s greater than M and for t large enough. The improper integral of this kind of function converges. (See Appendix A.6 for basic definitions and examples.)

Now suppose that $f(t) \equiv 1$. Then $F(s) = \mathscr{L}[1] = \int_0^\infty 1 \cdot e^{-st}dt = \lim_{b \to \infty} \int_0^b e^{-st}dt =$

$\lim_{b \to \infty} \dfrac{e^{-st}}{-s}\Big|_0^b = \dfrac{1}{-s}\left(\lim_{b \to \infty}(e^{-sb} - 1)\right) = \dfrac{1}{s}$. From this you can see that the Laplace

transform of a constant function $f(t) \equiv C$ is $\dfrac{C}{s}$ for $s > 0$. (*Yes?*)

Next we can find $\mathscr{L}[t]$ by using integration by parts and the value of $\mathscr{L}[1]$. For $s > 0$,

$$\mathscr{L}[t] = \int_0^\infty te^{-st}dt = \lim_{b \to \infty} \int_0^b te^{-st}dt$$

$$= \lim_{b \to \infty}\left\{\frac{te^{-st}}{-s}\Big|_0^b - \int_0^b \frac{e^{-st}}{-s}dt\right\} = 0 + \frac{1}{s}\mathscr{L}[1] = \frac{1}{s^2}.$$

Similarly, we can show that $\mathscr{L}[t^2] = \dfrac{2}{s^3}$ and $\mathscr{L}[t^3] = \dfrac{6}{s^4}$ for $s > 0$. (See Exercises 1 and 2 at the end of this section.) In general, for all integers $n \geq 0$, it can be shown that

$$\mathscr{L}[t^n] = \frac{n!}{s^{n+1}}(s > 0), \tag{6.1.2}$$

where $0!$ is defined to be 1.

Now, from the properties of integrals, we can see that

$$\mathscr{L}[c \cdot f(t)] = c\mathscr{L}[f(t)],$$

where c is any real constant, and that

$$\mathscr{L}[f(t) + g(t)] = \mathscr{L}[f(t)] + \mathscr{L}[g(t)],$$

whenever the Laplace transforms of both f and g exist. Any transformation that satisfies the last two properties is called a **linear operator** or a **linear transformation**. (See Section 2.2.) If c_1 and c_2 are constants, then we can combine the two properties to write

$$\mathscr{L}[c_1 f(t) + c_2 g(t)] = c_1 \mathscr{L}[f(t)] + c_2 \mathscr{L}[g(t)]. \tag{6.1.3}$$

Extending (6.1.3), we can see how to calculate the Laplace transform of any *polynomial function*:

$$\mathscr{L}[c_0 + c_1 t + c_2 t^2 + \cdots + c_k t^k + \cdots + c_n t^n]$$
$$= \mathscr{L}[c_0] + \mathscr{L}[c_1 t] + \mathscr{L}[c_2 t^2] + \cdots + \mathscr{L}[c_k t^k] + \cdots + \mathscr{L}[c_n t^n]$$
$$= c_0 \mathscr{L}[1] + c_1 \mathscr{L}[t] + c_2 \mathscr{L}[t^2] + \cdots + c_k \mathscr{L}[t^k] + \cdots + c_n \mathscr{L}[t^n]$$
$$= \frac{c_0}{s} + \frac{c_1}{s^2} + \frac{2c_2}{s^3} + \frac{6c_3}{s^4} + \cdots + \frac{k!c_k}{s^{k+1}} + \cdots + \frac{n!c_n}{s^{n+1}} \quad (s > 0).$$

If a is a constant, let us find the Laplace transform of $f(t) = e^{at}$, an important function for us because of its frequent appearance in differential equations. By definition,

$$\mathcal{L}[e^{at}] = \int_0^\infty e^{at} e^{-st} dt = \lim_{b \to \infty} \int_0^b e^{(a-s)t} dt$$

$$= \lim_{b \to \infty} \frac{e^{(a-s)t}}{(a-s)} \bigg|_0^b = \frac{1}{a-s} \left(\lim_{b \to \infty} (e^{(a-s)b} - 1) \right) = -\frac{1}{a-s} = \frac{1}{s-a} \quad \text{for } s > a.$$

Why is this assumption about s needed? In what step is the assumption crucial?

To have tools with which to handle a variety of differential equations, we have to stock our warehouse with different Laplace transforms. Another basic function we should deal with is $\sin at$, where a is a constant. This transform requires two integrations by parts:

For $s > 0$, $\mathcal{L}[\sin at] =$

$$\int_0^\infty \sin at \, e^{-st} dt = \lim_{b \to \infty} \sin at \frac{e^{-st}}{-s} \bigg|_0^b - \int_0^\infty a \cos at \frac{e^{-st}}{-s} dt = \frac{a}{s} \int_0^\infty \cos at \, e^{-st} dt$$

$$= \frac{a}{s} \left(\lim_{b \to \infty} \cos at \frac{e^{-st}}{-s} \bigg|_0^b - \int_0^\infty -a \sin at \frac{e^{-st}}{-s} dt \right) = \frac{a}{s} \left(\frac{1}{s} - \frac{a}{s} \mathcal{L}[\sin at] \right),$$

so that

$$\left(1 + \frac{a^2}{s^2} \right) \mathcal{L}[\sin at] = \frac{a}{s^2} \quad \text{and} \quad \mathcal{L}[\sin at] = \frac{a}{s^2 + a^2}.$$

Using one of the steps from this result, it is easy to show that $\mathcal{L}[\cos at] = \dfrac{s}{s^2 + a^2}$.

(See Exercise 3.)

To help set the stage for a type of applied differential equation problem that can be handled neatly by using the Laplace transform, let's find $\mathcal{L}[f(t)]$ for the piecewise continuous function defined as follows:

$$f(t) = \begin{bmatrix} t & \text{for } 0 \le t \le 2 \\ 4 - t & \text{for } 2 \le t \le 4 \\ 0 & \text{for } t \ge 4. \end{bmatrix}$$

You should sketch the graph of this function. All we have to do is split the integral in definition (6.1.1) into three pieces, one corresponding to the interval $[0, 2]$, another corresponding to $[2, 4]$, and the last matching $[4, \infty)$:

$$\mathcal{L}[f(t)] = \int_0^2 te^{-st} dt + \int_2^4 (4 - t)e^{-st} dt + \int_4^\infty 0 \cdot e^{-st} dt$$

$$= \frac{1 - e^{-2s} - 2se^{-2s}}{s^2} + \frac{e^{-4s} - e^{-2s} + 2se^{-2s}}{s^2} = \frac{1 + e^{-4s} - 2e^{-2s}}{s^2}$$

for $s > 0$. (Carry out all the integrations yourself!)

Finally, before we can apply Laplace transforms to the solution of differential equations, we have to know the transforms of f', f'', and higher-order derivatives. So suppose that $F(s) = \mathscr{L}[f(t)]$ exists for $s > c$. Then we have, for $s > c$,

$$\mathscr{L}[f'(t)] = \int_0^\infty f'(t)e^{-st}dt$$

$$= \lim_{b \to \infty} f(t)e^{-st}\Big|_0^b + \int_0^\infty sf(t)e^{-st}dt = -f(0) + s\mathscr{L}[f(t)],$$

which we can write as

$$\mathscr{L}[f'(t)] = s\,\mathscr{L}[f(t)] - f(0). \tag{6.1.4}$$

Note that in this derivation, we are assuming that $f(b)e^{-sb} \to 0$ as $b \to \infty$.

Now if we assume that $f'(b)e^{-sb}$ also tends to 0 as $b \to \infty$, we can apply formula (6.1.4) twice, first with f replaced by f', to get

$$\mathscr{L}[f''(t)] = -f'(0) + s\mathscr{L}[f'(t)] = -f'(0) + s[s\mathscr{L}[f(t)] - f(0)]$$

$$= -f'(0) + s^2\mathscr{L}[f(t)] - sf(0),$$

so we can write

$$\mathscr{L}[f''(t)] = s^2\mathscr{L}[f(t)] - sf(0) - f'(0) \quad \text{(for } s > c\text{)}. \tag{6.1.5}$$

In general, for any positive integer n, if the *nth* derivative is continuous (or piecewise continuous), and all the lower-order derivatives are continuous and have the proper growth rate, then

$$\mathscr{L}[f^{(n)}(t)] = s^n\mathscr{L}[f(t)] - \sum_{i=1}^{n} s^{n-i}f^{(i-1)}(0)$$

$$= s^n\mathscr{L}[f(t)] - s^{n-1}f(0) - s^{n-2}f'(0) - \cdots - sf^{(n-2)}(0) - f^{(n-1)}(0). \tag{6.1.6}$$

It is important to note that *this last formula implies that a linear differential equation with constant coefficients will be transformed into a purely algebraic equation—that is, an equation without derivatives.* (Recognize that $f^{(k)}(0)$ is a *number* for $k \geq 0$.)

As a hint of what we'll be doing in the next few sections, we'll convert a differential equation into an algebraic expression by using the Laplace transform.

EXAMPLE 6.1.1   **The Laplace Transform of a Differential Equation**
Let's look at the initial-value problem

$$x'' + 3x' + 2x = 12e^{2t}; \, x(0) = 1, \, x'(0) = -1.$$

We're going to apply the Laplace transform to both sides of the equation and substitute the initial conditions where appropriate:

$$\mathscr{L}[x'' + 3x' + 2x] = \mathscr{L}[12e^{2t}],$$

or, using the linearity of the transform—(6.1.3),

$$\mathscr{L}[x''] + 3\mathscr{L}[x'] + 2\mathscr{L}[x] = 12\mathscr{L}[e^{2t}].$$

We have already calculated the Laplace transform of an exponential function. This, together with formulas (6.1.4) and (6.1.5), allows us to write

$$\{s^2 \mathcal{L}[x(t)] - sx(0) - x'(0)\} + 3\{s\mathcal{L}[x(t)] - x(0)\} + 2\mathcal{L}[x(t)] = \frac{12}{s-2}.$$

Now we substitute the given initial conditions to get

$$\{s^2\mathcal{L}[x(t)] - s + 1\} + 3\{s\mathcal{L}[x(t)] - 1\} + 2\mathcal{L}[x(t)] = \frac{12}{s-2}.$$

Finally, collecting like terms, we find that

$$(s^2 + 3s + 2)\mathcal{L}[x(t)] = \frac{12}{s-2} + s + 2 = \frac{s^2 + 8}{s-2},$$

so we can solve for $\mathcal{L}[x(t)]$:

$$\mathcal{L}[x(t)] = \frac{s^2 + 8}{s-2} \cdot \frac{1}{s^2 + 3s + 2} = \frac{s^2 + 8}{(s-2)(s^2 + 3s + 2)}$$

$$= \frac{s^2 + 8}{(s-2)(s+2)(s+1)}.$$

Now what? We have an unknown function, the solution of an IVP, whose Laplace transform is known. If we can *reverse* the process and figure out what function has this Laplace transform, we can solve our original initial-value problem. This is what we'll focus on in the next section. ◆

EXERCISES 6.1

Find the Laplace transform of the functions in Exercises 1–8.

1. $f(t) = t^2$

2. $g(t) = t^3$

3. $h(t) = \cos at$, where a is a constant

4. $F(t) = te^{at}$, where a is a constant

5. $G(t) = \dfrac{e^{at} - e^{bt}}{a - b}$, where a and b are constants, $a \neq b$

6. $H(t) = 2t^3 - 7t^2 + 5t - 17$

7. $A(t) = \begin{cases} 1 & \text{for } 0 \leq t \leq 1 \\ 2 - t & \text{for } 1 \leq t < 2 \\ 0 & \text{for } 2 \leq t \end{cases}$

8. $B(t) = \begin{cases} 1 & \text{for } 0 \leq t \leq 2\pi \\ \cos t & \text{for } 2\pi \leq t \leq 7\pi/2 \\ 0 & \text{for } 7\pi/2 \leq t < \infty \end{cases}$

*In Exercises 9–13, find the Laplace transform of the solution of each initial-value problem, assuming that the Laplace transform exists in each case. (*Do not try to solve the IVPs.*)*

9. $y' - y = 0$; $y(0) = 1$

10. $y' + y = e^{-x}$; $y(0) = 1$

11. $y'' + 4y' + 4y = 0$; $y(0) = 1, y'(0) = 1$

12. $y'' - y' - 2y = 5\sin x$; $y(0) = 1, y'(0) = -1$

13. $y''' - 2y'' + y' = 2e^x + 2x$; $y(0) = 0, y'(0) = 0, y''(0) = 0$

14. If $f(t) = e^{t^2}$ for $t \geq 0$, show that there are no constants M and K such that $|f(t)| < e^{Mt}$ for all $t \geq K$. Thus show that the Laplace transform of $f(t)$ doesn't exist. [*Hint:* $e^{t^2} < e^{Mt}$ implies that $t^2 < Mt$ for t large enough.]

15. Use definition (6.1.1) and the fact that $\displaystyle\int_0^\infty e^{-x^2}dx = \frac{\sqrt{\pi}}{2}$ to find the Laplace transform of $f(t) = \dfrac{1}{\sqrt{t}} = t^{-\frac{1}{2}}$. [*Hint:* Make the substitution $t = \dfrac{u^2}{s}$.]

16. Apply formula (6.1.4) to the function $f''(t)$ to show that
$$\mathcal{L}[f'''(t)] = s^3\mathcal{L}[f(t)] - s^2f(0) - sf'(0) - f''(0).$$

17. Suppose that a is any real number and that $F(s) = \mathcal{L}[f(t)]$. Show that $\mathcal{L}[e^{at}f(t)] = F(s - a)$ for $s > a$. (This is usually called the *First Shift Formula*. See Section 6.3 for the *Second Shift Formula*.)

6.2   THE INVERSE TRANSFORM AND THE CONVOLUTION

THE INVERSE LAPLACE TRANSFORM

Recall that in Example 6.1.1 we took an initial-value problem, applied the Laplace transform, and then wound up with the Laplace transform of the solution of the IVP. Now we would like to *reverse* the transformation process so that, given $\mathcal{L}[f(t)]$ as a function of the parameter s, we can find $f(t)$. This involves the idea of the **inverse Laplace transform, $\mathcal{L}^{-1}$**.

Now think back to the concept of inverse of a *function*. When you first encountered the inverse in calculus or precalculus, you may have worked with both the formal definition and the graphical interpretation in terms of a "horizontal line test." In any case, the important idea is that to have an inverse function f^{-1} we must guarantee that for any element in the *range* of the original function f, there is one and only one corresponding element in the *domain* of f. Another way of saying this is that *a function has an inverse if and only if it is a one-to-one function*.

For our purposes, the important fact is that *if the Laplace transforms of the continuous functions f and g exist and are equal for $s \geq c$ (c, a constant), then f(t) = g(t) for all $t \geq 0$.* This says that a continuous function can be uniquely recovered from its Laplace transform. Letting $\mathcal{L}[f(t)] = F(s)$, we can express the definition of the inverse Laplace transform as follows:

$$\mathcal{L}^{-1}[F] = f \quad \text{if and only if} \quad \mathcal{L}[f] = F. \tag{6.2.1}$$

It is easily verified (see Exercise 1) that the inverse Laplace transform is a linear transformation:

$$\mathcal{L}^{-1}[c_1F(t) + c_2G(t)] = c_1\mathcal{L}^{-1}[F(t)] + c_2\mathcal{L}^{-1}[G(t)]. \tag{6.2.2}$$

Now how do we find the inverse Laplace transform in practice? It turns out that the relationship between calculating a Laplace transform and determining its inverse is very similar to that between differentiation and antidifferentiation.

TABLE 6.1   Some Laplace Transforms

$f(t)$	$F(s) = \mathcal{L}[f(t)]$
1. $t^n \ (n = 0, 1, 2, \ldots)$	$\dfrac{n!}{s^{n+1}}, s > 0$
2. e^{at}	$\dfrac{1}{s - a}, s > a$
3. $\sin at$	$\dfrac{a}{s^2 + a^2}, s > 0$
4. $\cos at$	$\dfrac{s}{s^2 + a^2}, s > 0$
5. $e^{at} \sin bt$	$\dfrac{b}{(s - a)^2 + b^2}, s > a$
6. $e^{at} \cos bt$	$\dfrac{s - a}{(s - a)^2 + b^2}, s > a$
7. $t \sin at$	$\dfrac{2as}{(s^2 + a^2)^2}, s > 0$
8. $t \cos at$	$\dfrac{s^2 - a^2}{(s^2 + a^2)^2}, s > 0$
9. $f'(t)$	$sF(s) - f(0)$
10. $f''(t)$	$s^2F(s) - sf(0) - f'(0)$
11. $e^{at}f(t)$	$F(s - a), s > a$

What this means is that in calculus, the indefinite integral of a function f answers the question "What is a function whose derivative is f?" (Note that in calculus the answer to this question is not unique.) Just as a list of differentiation formulas helps us to construct a list of antidifferentiation formulas (integrals), so will a table of Laplace transforms aid us in finding inverses. In the examples that follow, we will use the information in Table 6.1. Some of these transforms were derived in Section 6.1; others were given as problems in Exercises 6.1.

Now let's return to Example 6.1.1 and solve the initial-value problem using Laplace transforms and the inverse Laplace transform.

EXAMPLE 6.2.1   Solving an IVP Using the Inverse Laplace Transform
The IVP was $x'' + 3x' + 2x = 12e^{2t}$, $x(0) = 1$, $x'(0) = -1$, and we found that

$$\mathcal{L}[x(t)] = \frac{s^2 + 8}{(s - 2)(s + 2)(s + 1)}.$$

If we tried to work with the given expression for the transform (the single rational expression in s), we would have a tough time figuring out what function $x(t)$ might have this as its Laplace transform. This expression doesn't seem to correspond to any of the forms in the last column of Table 6.1.

However, we can use the partial-fractions technique to express the transform as the sum of three simpler terms, each of which matches an entry in the table:

$$\frac{s^2 + 8}{(s - 2)(s + 2)(s + 1)} = \frac{1}{s - 2} + \frac{3}{s + 2} - \frac{3}{s + 1}.$$

You should be able to see, for example, that the term $\dfrac{3}{s+2}\left(=\dfrac{3}{s-(-2)}\right)$ is the Laplace transform of $3e^{-2t}$. Applying the inverse transform to each side of

$$\mathscr{L}[x(t)] = \frac{1}{s-2} + \frac{3}{s+2} - \frac{3}{s+1}$$

and using (6.2.1) and the linearity of $\mathscr{L}^{-1}$, we see that

$$x(t) = \mathscr{L}^{-1}[\mathscr{L}[x(t)]] = \mathscr{L}^{-1}\left[\frac{1}{s-2} + \frac{3}{s+2} - \frac{3}{s+1}\right]$$

$$= \mathscr{L}^{-1}\left[\frac{1}{s-2}\right] + 3\mathscr{L}^{-1}\left[\frac{1}{s+2}\right] - 3\mathscr{L}^{-1}\left[\frac{1}{s+1}\right]$$

$$= e^{2t} + 3e^{-2t} - 3e^{-t},$$

where we have used formula 2 from Table 6.1 three times to evaluate the right-hand side. ◆

The alternative to using the Laplace transform to solve the initial-value problem in our last example is to go back to the method first explained in Section 4.2: First find the general solution of the homogeneous equation $x'' + 3x' + 2x = 0$; then find a particular solution of the nonhomogeneous equation $x'' + 3x' + 2x = 12e^{2t}$; finally, add these two solutions together to get the general solution of the original *nonhomogeneous* equation. And even then we would not be finished, because we would have to use the initial conditions to determine the two arbitrary constants in the general solution. Note that the Laplace transform method enables us to handle the nonhomogeneous equation and initial conditions all at once.

Now let's see what the Laplace transform method does in an important applied problem that we first saw as Example 2.2.5. (Also see Problems 25–27 in Exercises 2.2.)

EXAMPLE 6.2.2 Solving a Circuit Problem via the Laplace Transform

The current I flowing in a particular electrical circuit can be described by the initial-value problem $L\dfrac{dI}{dt} + RI = v_0 \sin(\omega t)$, $I(0) = 0$. Here L, R, v_0, and ω are positive constants.

First we apply the Laplace transform to each side of the differential equation:

$$\mathscr{L}\left[L\frac{dI}{dt} + RI\right] = \mathscr{L}[v_0 \sin(\omega t)]$$

$$L\mathscr{L}\left[\frac{dI}{dt}\right] + R\mathscr{L}[I(t)] = v_0\mathscr{L}[\sin(\omega t)]$$

$$sL\mathscr{L}[I(t)] - LI(0) + R\mathscr{L}[I(t)] = v_0\left(\frac{\omega}{s^2 + \omega^2}\right)$$

$$(Ls + R)\mathscr{L}[I(t)] - LI(0) = v_0\left(\frac{\omega}{s^2 + \omega^2}\right)$$

$$L\left(s + \frac{R}{L}\right)\mathscr{L}[I(t)] = v_0\left(\frac{\omega}{s^2 + \omega^2}\right)$$

so that we have $\mathcal{L}[I(t)] = \left(\dfrac{v_0}{L}\right) \cdot \omega \cdot \dfrac{1}{\left(s + \dfrac{R}{L}\right)(s^2 + \omega^2)}$. To find the inverse

Laplace transform, we have to use the method of partial fractions on the right-hand side:

$$\frac{1}{\left(s + \dfrac{R}{L}\right)(s^2 + \omega^2)} = \frac{A}{\left(s + \dfrac{R}{L}\right)} + \frac{Bs + C}{s^2 + \omega^2}.$$

With a little sweat, we find that

$$\frac{1}{\left(s + \dfrac{R}{L}\right)(s^2 + \omega^2)} = \frac{\dfrac{1}{\left(\dfrac{R^2}{L^2} + \omega^2\right)}}{s + \dfrac{R}{L}} + \frac{-\dfrac{1}{\left(\dfrac{R^2}{L^2} + \omega^2\right)}s + \dfrac{\left(\dfrac{R}{L}\right)}{\left(\dfrac{R^2}{L^2} + \omega^2\right)}}{s^2 + \omega^2}$$

and

$$\mathcal{L}[I(t)] = \left(\frac{v_0}{L}\right) \cdot \omega \cdot \frac{1}{\left(s + \dfrac{R}{L}\right)(s^2 + \omega^2)}$$

$$= \left(\frac{v_0}{L}\right) \cdot \frac{\omega}{\left(\dfrac{R^2}{L^2} + \omega^2\right)} \left\{ \frac{1}{s + \dfrac{R}{L}} + \frac{-s}{s^2 + \omega^2} + \frac{\dfrac{R}{L}}{s^2 + \omega^2} \right\}.$$

Check out the last three equalities. The final step is to apply the inverse Laplace transform to both sides of this last equation and then use formulas 2, 3, and 4 from Table 6.1.

$$I(t) = \left(\frac{v_0}{L}\right) \frac{\omega}{\left(\dfrac{R^2}{L^2} + \omega^2\right)} \left\{ \mathcal{L}^{-1}\left[\frac{1}{s + \dfrac{R}{L}}\right] - \mathcal{L}^{-1}\left[\frac{s}{s^2 + \omega^2}\right] + \frac{R}{L}\mathcal{L}^{-1}\left[\frac{1}{s^2 + \omega^2}\right] \right\}$$

$$= \left(\frac{v_0}{L}\right) \frac{\omega}{\left(\dfrac{R^2}{L^2} + \omega^2\right)} \left\{ e^{-\frac{R}{L}t} - \cos(\omega t) + \frac{R}{L}\frac{1}{\omega}\sin(\omega t) \right\}$$

$$= \left(\frac{v_0}{L}\right) \frac{1}{\left(\dfrac{R^2}{L^2} + \omega^2\right)} \left\{ \omega e^{-\frac{R}{L}t} - \omega\cos(\omega t) + \frac{R}{L}\sin(\omega t) \right\}.$$

Compare this solution to the one obtained in Example 2.2.5.   ◆

EXAMPLE 6.2.3   **Solving an IVP Using the Inverse Laplace Transform**

Let's look at the initial-value problem $\ddot{x} - 2\dot{x} = e^t(t - 3)$, $x(0) = 2 = \dot{x}(0)$. As before, we take Laplace transforms of both sides and use the table. Letting $\mathcal{L}[x(t)] = X(s)$, we get

$$\{s^2 X(s) - sx(0) - \dot{x}(0)\} - 2\{sX(s) - x(0)\} = \mathcal{L}[e^t(t - 3)]. \qquad (\#)$$

To evaluate the right-hand side, first note that

$$\mathcal{L}[t - 3] = \frac{1}{s^2} - \frac{3}{s} = F(s),$$

so if we use entry 11 in Table 6.1 (with $a = 1$), we get

$$\mathcal{L}[e^t(t - 3)] = F(s - 1) = \frac{1}{(s - 1)^2} - \frac{3}{s - 1} = \frac{4 - 3s}{(s - 1)^2}.$$

If we return to (#) and put in our initial conditions, we get

$$s(s - 2)\, X(s) = 2s - 2 + \frac{4 - 3s}{(s - 1)^2}$$

$$= \frac{2(s - 1)^3 + (4 - 3s)}{(s - 1)^2} = \frac{(s - 2)(2s^2 - 2s - 1)}{(s - 1)^2}.$$

Therefore, we conclude that

$$X(s) = \frac{2s^2 - 2s - 1}{s(s - 1)^2} = \frac{3}{s - 1} - \frac{1}{(s - 1)^2} - \frac{1}{s}.$$

We go to the table to find the function $x(t) = \mathcal{L}^{-1}[X(t)]$, where we have to use entries 1 and 11 for the second term. The solution of the IVP is $x(t) = 3e^t - te^t - 1$. (*Be sure to check that this is the solution.*)   ◆

THE CONVOLUTION

In each of the last three examples, applying the Laplace transform yielded an expression for $\mathcal{L}[f(t)]$ that seemed to involve the product of two or more transforms. Because we didn't know any way to find the inverse transform of such products, we had to resort to the messiness of a partial-fraction decomposition. This, at least, enabled us to use the linearity of the inverse transform.

There is, however, a way to deal with this problem—a method that involves a special product of functions. The **convolution** of two functions f and g is the integral

$$(f * g)(t) = \int_0^t f(r)g(t - r)\,dr,$$

provided that the integral exists for $t > 0$. For example, the convolution of $\cos t$ and t is

$$(\cos t) * t = \int_0^t (\cos r)(t - r) dr = \int_0^t t \cos r \, dr - \int_0^t r \cos r \, dr$$

$$= t \int_0^t \cos r \, dr - \int_0^t r \cos r \, dr = 1 - \cos t$$

after integration by parts for the second integral. For this example, you should verify that $(\cos t) * (t) = (t) * (\cos t)$.

Convolution has important algebraic properties (see Exercise 7 at the end of this section), but the most significant property for us right now is that *the Laplace transform of a convolution of two functions is equal to the product of the Laplace transforms of these two functions.* More precisely, suppose that f and g are two functions whose Laplace transforms exist. Let $F(s) = \mathscr{L}[f(t)]$ and $G(s) = \mathscr{L}[g(t)]$. Then the **Convolution Theorem** says that

$$\mathscr{L}[(f * g)(t)] = \mathscr{L}\left[\int_0^t f(r) g(t - r) dr\right] = \mathscr{L}[f(t)] \cdot \mathscr{L}[g(t)] = F(s) \cdot G(s).$$

Now let's revisit part of Example 6.2.2 to see how the convolution property helps us.

EXAMPLE 6.2.4   **Example 6.2.2 Revisited**

How can we find $\mathscr{L}^{-1}\left[\dfrac{1}{\left(s + \dfrac{R}{L}\right)(s^2 + \omega^2)}\right]$?

The expression inside the brackets is the product of two transforms F and G:

$$F(s) \, G(s) = \left[\frac{1}{\left(s + \dfrac{R}{L}\right)(s^2 + \omega^2)}\right],$$

where $F(s) = \dfrac{1}{\left(s + \dfrac{R}{L}\right)}$ and $G(s) = \dfrac{1}{(s^2 + \omega^2)}$. Entries 2 and 3 of Table 6.1 tell us

that $f(t) = \mathscr{L}^{-1}[F(s)] = e^{-\frac{R}{L}t}$ and $g(t) = \mathscr{L}^{-1}[G(s)] = \dfrac{1}{\omega} \sin(\omega t)$. Then the Convolution Theorem enables us to conclude that

$$\mathscr{L}^{-1}[F(s) \, G(s)] = \int_0^t f(r) g(t - r) dr,$$

or

$$\mathscr{L}^{-1}\left[\frac{1}{\left(s + \dfrac{R}{L}\right)(s^2 + \omega^2)}\right] = \int_0^t e^{-\frac{R}{L}r} \frac{1}{\omega} \sin \omega(t - r) dr$$

$$= \frac{1}{\omega} \int_0^t e^{-\frac{R}{L}r} \sin \omega(t - r) dr.$$

A computer algebra system evaluates this integral as

$$\frac{L\left(\omega L e^{\left(-\frac{LR}{t}\right)} - \omega L \cos(\omega t) + R \sin(\omega t)\right)}{(R^2 + \omega^2 L^2)\omega}.$$

A bit of algebra should show you that this corresponds to the inverse transform found in Example 6.2.2 via partial fractions. (*Do the work.*)   ◆

Integral Equations and Integro-Differential Equations

The Convolution Theorem is certainly useful in solving differential equations, but it can also help us solve **integral equations**—equations involving an integral of the unknown function—and **integro-differential equations**—those involving both a derivative and an integral of the unknown function.

EXAMPLE 6.2.5   The Convolution Theorem and an Integral Equation

A store manager finds that the proportion of merchandise that remains unsold at time t after she has bought the merchandise is given by $f(t) = e^{-1.5t}$. She wants to find the rate at which she should purchase the merchandise so that the stock in the store remains constant.

Suppose that the store starts off by buying an amount A of the merchandise at time $t = 0$ and buys at a rate $r(t)$ subsequently. Over a short time interval $u \leq t \leq u + \Delta u$, an amount $r(t) \cdot \Delta u$ is bought by the store, and at time t the portion of this remaining unsold is $e^{-1.5(t-u)}r(u)\,\Delta u$. Then the amount of previously purchased merchandise remaining unsold at time t is given by

$$Ae^{-1.5t} + \int_0^t e^{-1.5(t-u)}r(u)\,du.$$

Because this is the total stock of the store and the store manager wants it to remain constant at its initial value, we must have

$$A = Ae^{-1.5t} + \int_0^t e^{-1.5(t-u)}r(u)\,du,$$

and the required restocking rate $r(t)$ is the solution of this integral equation.

If we look carefully at the integral on the right-hand side of this last equation, we should recognize something familiar about its form: It looks like a convolution—in fact, it is $e^{-1.5t} * r(t)$. Now we can rewrite the integral equation in the form

$$A = Ae^{-1.5t} + (e^{-1.5t} * r(t)).$$

Taking the Laplace transform of each side and letting $R(s) = \mathcal{L}[r(t)]$, we get

$$\mathcal{L}[A] = A\,\mathcal{L}[e^{-1.5t}] + \mathcal{L}[e^{-1.5t} * r(t)] = \frac{A}{s + 1.5} + \frac{1}{s + 1.5} \cdot R(s),$$

$$\frac{A}{s} = \frac{A}{s + 1.5} + \frac{1}{s + 1.5} \cdot R(s),$$

$$(s + 1.5)\left(\frac{A}{s} - \frac{A}{s + 1.5}\right) = R(s),$$

$$\frac{1.5A}{s} = R(s).$$

Applying the inverse Laplace transform to each side, we find that $r(t) = 1.5\,A$. That is, the restocking rate should be a constant one-and-a-half times the original amount bought. (*Check that this is a solution of our original integral equation.*)   ◆

EXAMPLE 6.2.6   An Integro-Differential Equation

The following integro-differential equation can also be solved using the properties of the Laplace transform.

$$\frac{dx}{dt} + x(t) - \int_0^t x(r)\sin(t - r)dr = -\sin t, \quad x(0) = 1.$$

As in the previous example, we recognize that the integral in our equation represents a convolution, this time $(x * \sin)(t)$. Therefore, taking the Laplace transform of each side of the equation, we get

$$\mathscr{L}[dx/dt] + \mathscr{L}[x(t)] - \mathscr{L}[(x * \sin)(t)] = \mathscr{L}[-\sin t],$$

or, using formula 10 in Table 6.1 and the Convolution Theorem,

$$[s\,\mathscr{L}[x(t)] - x(0)] + \mathscr{L}[x(t)] - \mathscr{L}[x(t)] \cdot \mathscr{L}[\sin t] = -\frac{1}{s^2 + 1},$$

which becomes

$$[s\,\mathscr{L}[x(t)] - 1] + \mathscr{L}[x(t)] - \mathscr{L}[x(t)] \cdot \frac{1}{s^2 + 1} = -\frac{1}{s^2 + 1}.$$

This simplifies to

$$\left(\frac{s^3 + s^2 + s}{s^2 + 1}\right)\mathscr{L}[x(t)] = \frac{s^2}{s^2 + 1},$$

so we wind up with $\mathscr{L}[x(t)] = \dfrac{s^2}{s^3 + s^2 + s} = \dfrac{s}{s^2 + s + 1}.$

A bit of clever algebra shows us that

$$\frac{s}{s^2 + s + 1} = \frac{s + \frac{1}{2}}{(s + \frac{1}{2})^2 + (\frac{\sqrt{3}}{2})^2} - \frac{\frac{1}{2}}{(s + \frac{1}{2})^2 + (\frac{\sqrt{3}}{2})^2}$$

$$= \frac{s - (-\frac{1}{2})}{(s - (-\frac{1}{2}))^2 + (\frac{\sqrt{3}}{2})^2} - \frac{1}{\sqrt{3}}\frac{\frac{\sqrt{3}}{2}}{(s - (-\frac{1}{2}))^2 + (\frac{\sqrt{3}}{2})^2}.$$

Using formulas 5 and 6 to invert this transform, we find that

$$x(t) = e^{-t/2}\cos\left(\frac{\sqrt{3}t}{2}\right) - \frac{1}{\sqrt{3}}e^{-t/2}\sin\left(\frac{\sqrt{3}t}{2}\right).$$

(Checking that this is the solution involves a bit of work, but try it!)   ◆

THE LAPLACE TRANSFORM AND TECHNOLOGY

Most computer algebra systems have built-in Laplace transform and inverse transform capabilities. In particular, some systems (for example, *Maple*) have sophisticated differential equation solvers with a "laplace" option for initial-value problems. If you have such an option at your command, learn to use it. However, realize that all the machinery is under the covers, so you have to develop an understanding of what the system is really doing.

Be aware that some computer algebra systems (such as *Mathematica* and MATLAB) can find Laplace transforms and their inverses but have no direct way of solving a linear IVP with these tools. In this case, you have to apply the Laplace transform to the differential equation, solve for the transform $\mathcal{L}[x(t)]$ of the solution algebraically (via a *solve* command or by hand), use technology to find the inverse transform $\mathcal{L}^{-1}[\mathcal{L}[x(t)]]$, and finally substitute the initial conditions.

Determine what your options are in using technology to solve IVPs via the Laplace transform. Some of the exercises that follow will help you do this.

EXERCISES 6.2

1. If $F(s)$ and $G(s)$ are the Laplace transforms of $f(t)$ and $g(t)$, respectively, and c_1 and c_2 are constants, show that

$$\mathcal{L}^{-1}[c_1 F(s) + c_2 G(s)] = c_1 \mathcal{L}^{-1}[F(s)] + c_2 \, \mathcal{L}^{-1}[G(s)].$$

2. Find the inverse Laplace transform of $\dfrac{a}{s^2(s^2 + a^2)}$.

3. Find the inverse Laplace transform of $\dfrac{1}{s(s^2 + 2s + 2)}$.

4. Find the inverse Laplace transform of $\dfrac{2s - 10}{s^2 - 4s + 20}$.

5. **a.** Show that the Laplace transform of $t^n f(t)$ is $(-1)^n F^{(n)}(s)$, where $F(s) = \mathcal{L}[f(t)]$.

   b. Use the result of part (a) and the derivative of the function $F(s) = \ln\left(2 + \dfrac{3}{s}\right)$, $s > 0$, to find its inverse Laplace transform.

6. Find the convolution $f * g$ of each of the following pairs of functions.
   a. $f(t) = t^2, g(t) = 1$
   b. $f(t) = t, g(t) = e^{-t}$ for $t \geq 0$
   c. $f(t) = t^2, g(t) = (t^2 + 1)$ for $t \geq 0$
   d. $f(t) = \cos t, g(t) = \cos t$ [*Hint:* For part (d) you need some trigonometric identities.]

7. Prove the following properties of the convolution of functions.
   a. $f * g = g * f$ Commutativity
   b. $(f * g) * h = f * (g * h)$ Associativity

c. $f * (g + h) = f * g + f * h$ Distributivity

d. $f * 0 = 0$, but $f * 1 \neq f$ and $f * f \neq f^2$ in general. (In particular, $1 * 1 \neq 1$.)

8. a. Using property (b) of Exercise 7, find $1 * 1 * 1$.

b. Find $1 * t * t^2$.

9. Use the Convolution Theorem to find the Laplace transform of

$$f(t) = \int_0^t \cos(t - r) \sin r \, dr.$$

10. Find the Laplace transform of $h(t) = \int_0^t e^{t-v} \sin v \, dv$.

11. Find the solution of the IVP $y'' + 3y' + 2y = 4t^2$; $y(0) = 0, \, y'(0) = 0$.

12. Solve the IVP $y'' + 4y' + 4y = e^{-2x}$; $y(0) = 0, \, y'(0) = 1$.

13. Solve the IVP $y''' - 2y'' + y' = 2e^x + 2x$; $y(0) = 0, \, y'(0) = 0, \, y''(0) = 0$.

14. Solve the IVP $y'' + 6y' + 9y = H(x)$; $y(0) = 0, \, y'(0) = 0$, where $H(x)$ is a known function of x. [*Hint:* Use the Convolution Theorem.]

15. An electrical circuit that is initially unforced but is plugged into a particular alternating voltage source at time $t = \pi$ can be modeled by the IVP

$$Q'' + 2Q' + 2Q = \begin{cases} 0 & \text{for } 0 \leq t < \pi \\ -\sin t & \text{for } t \geq \pi, \end{cases}$$

with $Q(0) = 0$ and $Q'(0) = 1$. Solve the IVP for $Q(t)$, the charge on the capacitor at time t.

16. The equation $v \dfrac{dv}{ds} = v \cos(2s) - v^2$ describes the velocity v of a piston moving into an oil-filled cylinder under a variable force. Here s is the distance moved in time t.

a. Rewrite the given equation as a linear equation with constant coefficients.

b. Assuming that $v(0) = 0$, use the Laplace transform to solve for v as a function of s. Is there a singular solution?

c. Use technology to graph the nontrivial solution found in part (b) for $0 \leq s \leq 20$

17. Solve the integral equation for f: $f(t) = 4t + \int_0^t f(t - r) \sin r \, dr$

18. Solve for g: $g(t) - t = -\int_0^t (t - r)g(r)dr$

19. Solve for x: $\dot{x}(t) = 1 - \int_0^t x(t - r)e^{-2r}dr, \, x(0) = 1$

20. Solve the integro-differential equation

$$\dot{y} + y + \int_0^t y(u)du = 1, \text{ with } y(0) = 0.$$

21. Solve the equation $\dot{x} - 4x + 4 \int_0^t x(u)\,du = t^3 e^{2t}$, with $x(0) = 0$.

22. Solve the equation $f''(x) + \int_0^x e^{2(x-y)} f'(y)\,dy = 1$; $f(0) = 0, f'(0) = 0$.

6.3 TRANSFORMS OF DISCONTINUOUS FUNCTIONS

Differential equations are often used to model complex systems. In some situations, models have to deal with abrupt changes in these systems. For instance, in the circuit problem described in Example 2.2.5 (or Example 6.2.2), we have the equation $L\dfrac{dI}{dt} + RI = v_0 \sin(\omega t)$, where the right-hand side (the forcing term) represents a continuous alternating-current source. Now suppose that the voltage $E(t)$ were applied for only a short period of time and then discontinued. Mathematically, this means that the forcing term would have the form

$$f(t) = \begin{cases} E(t) & \text{for } 0 \le t \le a \\ 0 & \text{for } t > a. \end{cases}$$

Perhaps we have a switch that we can open and close so that the voltage is applied, removed, and then applied again:

$$g(t) = \begin{cases} E(t) & \text{for } 0 \le t \le a \\ 0 & \text{for } a < t < b \\ E(t) & \text{for } t \ge b. \end{cases}$$

Problem 29 in Exercises 2.2, in which advertising expenditure is terminated after a certain period of time, is another illustration of this kind of behavior. The common element here is abrupt change. In mathematical terms, we are dealing with *piecewise continuous functions*.

THE HEAVISIDE (UNIT STEP) FUNCTION

In Section 6.1 we saw a simple example of the Laplace transform applied to a piecewise continuous function. We computed the transform directly from the definition, breaking the integral up into two parts. This can be tedious if there are several intervals involved in the definition of the function. Now we are going to see how these kinds of functions can be expressed in such a way that the Laplace transform method doesn't have to consider separate intervals.

We start with the **unit step function** U defined by

$$U(t) = \begin{cases} 0 & \text{if } t < 0 \\ 1 & \text{if } t \ge 0. \end{cases}$$

(This is sometimes called the **Heaviside (unit step) function** for the English electrical engineer and applied mathematician Oliver Heaviside (1850–1925), who developed many of these ideas.) We can say that the function U is "off" ($= 0$) for negative values of t and "on" ($= 1$) for values of t greater than or equal to 0. This "switching" aspect makes U an important building block in modeling abrupt changes.

It follows that

$$U(t - a) = \begin{cases} 0 & \text{if } t < a \\ 1 & \text{if } t \geq a. \end{cases}$$

The function has a jump discontinuity at $t = a$. Figure 6.1 shows $U(t - 3)$ for $t \geq 0$.

The nice thing is that these step functions can be used to express a piecewise continuous function in terms of a single formula. For example, if

$$f(t) = \begin{cases} A(t) & \text{for } t < a \\ B(t) & \text{for } t \geq a, \end{cases}$$

then we can see that $f(t) = A(t) + U(t - a)[B(t) - A(t)]$: If $t < a$, then $U(t - a) = 0$, so $f(t) = A(t)$; whereas if $t \geq a$, then we have $U(t - a) = 1$, so $f(t) = A(t) + [B(t) - A(t)] = B(t)$. (*Okay?*)

You can see that this technique can be extended to functions such as

$$g(t) = \begin{cases} A(t) & \text{for } a \leq t < b \\ B(t) & \text{for } b \leq t < c \\ C(t) & \text{for } c \leq t < d. \end{cases}$$

We can write $g(t) = U(t - a)A(t) + U(t - b)[B(t) - A(t)] + U(t - c)[C(t) - B(t)]$. You should be sure that you see how this works.

When we are solving differential equations that model abrupt changes, the following result comes in handy: *If $\mathcal{L}[f(t)]$ exists for $s > c$ and if $a > 0$, then*

$$\mathcal{L}[f(t - a) \, U(t - a)] = e^{-as} \, \mathcal{L}[f(t)] \text{ for } s > c. \tag{6.3.1a}$$

This result is usually called the *Second Shift Formula*—see Problem 17 in Exercises 6.1 for the *First Shift Formula*.

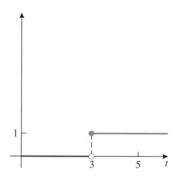

Figure 6.1
Graph of $U(t - 3), t \geq 0$

Alternatively, we can write (6.3.1a) as

$$f(t - a) U(t - a) = \mathcal{L}^{-1}[e^{-as} \mathcal{L}[f(t)]]. \tag{6.3.1b}$$

Formula (6.3.1a) follows from a straightforward calculation:

$$\mathcal{L}[f(t - a)U(t - a)] = \int_0^{\infty} f(t - a)U(t - a)e^{-st}dt = \int_a^{\infty} f(t - a)e^{-st}dt$$

$$= \int_0^{\infty} f(u)e^{-s(u+a)}du = e^{-sa} \mathcal{L}[f(t)],$$

where we have made the substitution $t - a = u$ in the second integral.

The next example shows how to use the Laplace transform of a unit step function to solve an initial-value problem.

EXAMPLE 6.3.1   An IVP with a Discontinuous Forcing Term

Let's look at the initial-value problem

$$x'(t) + x = \begin{cases} t & \text{for } 0 \le t < 4 \\ 1 & \text{for } 4 \le t \end{cases} \quad x(0) = 1.$$

Using the unit step function, we can write the differential equation as

$$x'(t) + x = t + (1 - t)U(t - 4) = t - (t - 4)U(t - 4) - 3U(t - 4).$$

(Note that in order to use formula (6.3.1b) later, we have to use algebra to convert the term $(1 - t)U(t - 4)$ into the form $f(t - a)U(t - a)$.) Now we apply the Laplace transform to both sides of the equation to get

$$\mathcal{L}[x'(t)] + \mathcal{L}[x(t)] = \mathcal{L}[t] - \mathcal{L}[(t - 4)U(t - 4)] - 3 \mathcal{L}[U(t - 4)],$$

or, using (6.1.4), entry 1 in Table 6.1, and then formula (6.3.1a) twice,

$$s \mathcal{L}[x(t)] - 1 + \mathcal{L}[x(t)] = \frac{1}{s^2} - e^{-4s} \mathcal{L}[t] - 3e^{-4s} \mathcal{L}[1],$$

so

$$(s + 1) \mathcal{L}[x(t)] = 1 + \frac{1}{s^2} - e^{-4s}\left(\frac{1}{s^2} + \frac{3}{s}\right).$$

Therefore,

$$\mathcal{L}[x(t)] = \frac{1}{s + 1} + \frac{1}{s^2(s + 1)} - e^{-4s}\left(\frac{3s + 1}{s^2(s + 1)}\right)$$

$$= [\text{by partial fractions}] \frac{2}{s + 1} - \frac{1}{s} + \frac{1}{s^2} - e^{-4s}\left(\frac{2}{s} + \frac{1}{s^2} - \frac{2}{s + 1}\right).$$

Finally, applying the inverse transform to both sides and using (6.3.1b), we get

$$x(t) = 2e^{-t} - 1 + t - \mathcal{L}^{-1}\left[e^{-4s}\left(\frac{2}{s} + \frac{1}{s^2} - \frac{2}{s + 1}\right)\right]$$

$$= 2e^{-t} - 1 + t - \left[U(t) \mathcal{L}^{-1}\left(\frac{2}{s} + \frac{1}{s^2} - \frac{2}{s + 1}\right)\right] \tag{*}$$

(where t must be replaced by $t - 4$ within the brackets before we're finished)

$$= 2e^{-t} - 1 + t - [U(t) (2 + t - 2e^{-t})] \qquad (*)$$
$$= 2e^{-t} - 1 + t - U(t - 4) (t - 2 - 2e^{-t+4})$$
$$= \begin{cases} 2e^{-t} + t - 1 & \text{for } 0 \le t < 4 \\ 2e^{-t} + 2e^{-t+4} + 1 & \text{for } 4 \le t. \end{cases}$$ ◆

The next example shows the application of the Laplace transform and the unit step function to an important type of applied problem.

EXAMPLE 6.3.2   A Cantilever Beam Problem

A wooden beam the ends of which are considered to be at $x = 0$ and $x = L$ on a horizontal axis will "give" (that is, bend) when a vertical load, given by $W(x)$ per unit length, acts on the beam (Figure 6.2). (See Problem 25 in Exercises 1.2.)

Then $y(x)$, the amount of bending, or deflection, in the direction of the load force at the point x, satisfies the differential equation $\dfrac{d^4 y}{dx^4} = \dfrac{W(x)}{EI}$ for $0 < x < L$.

Here E and I are constants that describe characteristics of the beam. The graph of $y(x)$ is called the *deflection curve* or *elastic curve*.

Now suppose that we have a *cantilever beam*—one that is clamped at the end $x = 0$ and free at the end $x = L$—and that this beam carries a load per unit length given by

$$W(x) = \begin{cases} W_0 & \text{for } 0 < x < \dfrac{L}{2} \\ 0 & \text{for } \dfrac{L}{2} < x < L. \end{cases}$$

Then engineering mechanics shows that finding the deflection amounts to solving the boundary-value problem

$$\dfrac{d^4 y}{dx^4} = \dfrac{W(x)}{EI} \text{ (for } 0 < x < L); \quad y(0) = 0, \, y'(0) = 0, \, y''(L) = 0, \, y'''(L) = 0.$$

(In physics terms, the quantities $y''(L)$ and $y'''(L)$ are called the *bending moment* and the *shear force*, respectively.)

First of all, note that up to now we have applied the technique of Laplace transforms only to initial-value problems, not to boundary-value problems

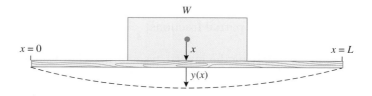

Figure 6.2

A Loaded Beam

(BVPs). Second, to use the Laplace transform, we must assume that $y(x)$ and $W(x)$ are defined on the interval $(0, \infty)$ rather than just on $(0, L)$. This means that we should extend the definition of $W(x)$ as follows:

$$W(x) = \begin{cases} W_0 & \text{for } 0 < x < \dfrac{L}{2} \\ 0 & \text{for } x > \dfrac{L}{2}. \end{cases}$$

We can write this function in terms of the unit step function as

$$W(x) = W_0\left\{U(x) - U\left(x - \frac{L}{2}\right)\right\}.$$

Now take the Laplace transform of each side of our fourth-order equation, letting $Y = Y(s) = \mathcal{L}[y(x)]$ for convenience. Using (6.1.6), we find that

$$s^4Y - s^3y(0) - s^2y'(0) - sy''(0) - y'''(0) = \frac{W_0}{EI}\left\{\frac{1 - e^{-sL/2}}{s}\right\}.$$

Note that in formula (6.1.6) the second and third derivatives of y are evaluated at 0. However, our BVP gives us the values of these derivatives at L. Letting $y''(0) = C_1$ and $y'''(0) = C_2$, we can use all the boundary conditions as given and solve the last equation for Y:

$$Y = \frac{C_1}{s^3} + \frac{C_2}{s^4} + \frac{W_0}{EIs^5}\{1 - e^{-sL/2}\}.$$

Using the inverse transform, we find that

$$y(x) = \frac{C_1x^2}{2!} + \frac{C_2x^3}{3!} + \frac{W_0}{EI}\frac{x^4}{4!} - \frac{W_0}{EI}\frac{\left(x - \dfrac{L}{2}\right)^4}{4!}U\left(x - \frac{L}{2}\right),$$

which is equivalent to

$$y(x) = \begin{cases} \dfrac{C_1x^2}{2} + \dfrac{C_2x^3}{6} + \dfrac{W_0}{24EI}x^4 & \text{for } 0 \le x < \dfrac{L}{2} \\ \dfrac{C_1x^2}{2} + \dfrac{C_2x^3}{6} + \dfrac{W_0}{24EI}x^4 - \dfrac{W_0}{24EI}\left(x - \dfrac{L}{2}\right)^4 & \text{for } x \ge \dfrac{L}{2}. \end{cases}$$

Now we use the conditions $y''(L) = 0$ and $y'''(L) = 0$ to find that $C_1 = \dfrac{W_0L^2}{8EI}$ and $C_2 = -\dfrac{W_0L}{2EI}$. (*Be sure to go through the calculations for yourself.*)

Finally, we can write our deflection function as

$$y(x) = \begin{cases} \dfrac{W_0L^2}{16EI}x^2 - \dfrac{W_0L}{12EI}x^3 + \dfrac{W_0}{24EI}x^4 & \text{for } 0 \le x < \dfrac{L}{2} \\ \dfrac{W_0L^2}{16EI}x^2 - \dfrac{W_0L}{12EI}x^3 + \dfrac{W_0}{24EI}x^4 - \dfrac{W_0}{24EI}\left(x - \dfrac{L}{2}\right)^4 & \text{for } \dfrac{L}{2} < x < L. \end{cases} \quad \blacklozenge$$

EXERCISES 6.3

In Exercises 1–4, (a) sketch the graph of each function $f(t)$ and (b) write each function as a sum of multiples of the unit step function $U(t)$.

1. $f(t) = \begin{cases} 1 & \text{for } 1 \leq t < 2 \\ 0 & \text{elsewhere} \end{cases}$

2. $f(t) = \begin{cases} 1 & \text{for } 1 \leq t < 2 \\ -2 & \text{for } 2 \leq t < 3 \\ 0 & \text{elsewhere} \end{cases}$

3. $f(t) = \begin{cases} t & \text{for } 0 \leq t < 2 \\ 4 - t & \text{for } 2 \leq t < 4 \\ 0 & \text{elsewhere} \end{cases}$

4. $f(t) = \begin{cases} t & \text{for } 0 \leq t < 2 \\ t - 2 & \text{for } 2 \leq t < 4 \\ 0 & \text{elsewhere} \end{cases}$

5. Show that $\mathcal{L}[tU(t - a)] = (1 + as)s^{-2}e^{-as}$ for $a > 0$.

6. Calculate $\mathcal{L}[t^2U(t - 1)]$.

7. Use formula (6.3.1a) to compute the Laplace transform of the function in Exercise 1.

8. Use formula (6.3.1a) to compute the Laplace transform of the function in Exercise 2.

9. Use formula (6.3.1a) to compute the Laplace transform of the function in Exercise 3.

10. Use formula (6.3.1a) to compute the Laplace transform of the function in Exercise 4.

11. Suppose that the fish population in a large lake is growing too rapidly, and the local authorities decide to give out fishing licenses that allow a total of h fish to be caught per day over a 30-day period. A model for such a situation could be

$$P'(t) = kP(t) - \begin{cases} h & \text{for } 0 \leq t \leq 30 \\ 0 & \text{for } t > 30 \end{cases},$$

where $P(t)$ denotes the number of fish in the lake at time t (in days) and k is a positive constant describing the natural growth rate of the fish population.

    a. Use technology and the Laplace transform to find an expression for $P(t)$ if $P(0) = A$.

    b. Find a relation among A, h, and k that guarantees that exactly 330 days after the end of the 30-day fishing season, the fish population will once more be at the level A.

Solve the IVPs in Exercises 12–17 by writing each discontinuous forcing function as a linear combination of unit step functions and then using the Laplace transform.

12. $4y' - 5y = \begin{cases} 0 & \text{for } t < 0 \\ -30t & \text{for } 0 \leq t < 1; \\ 0 & \text{for } t \geq 1 \end{cases}$ $y(0) = 2$

13. $4y' + 5y = \begin{cases} 0 & \text{for } t < 0 \\ \sin 8t & \text{for } 0 \leq t \leq 2; \\ 0 & \text{for } t > 2 \end{cases}$ $y(0) = 1$

14. $y'' + 5y' + 2y = \begin{cases} 0 & \text{for } t < 0 \\ 8 & \text{for } 0 \le t \le 1; \\ 0 & \text{for } t > 1 \end{cases}$ $y(0) = 0, y'(0) = 0$

15. $3y'' + 3y' + 2y = \begin{cases} 0 & \text{for } t < 0 \\ 5 & \text{for } 0 \le t \le 5; \\ 0 & \text{for } t > 5 \end{cases}$ $y(0) = 0, y'(0) = 0$

16. $y' - 3y = f(t); y(0) = 0$, where the graph of $f(t)$ is

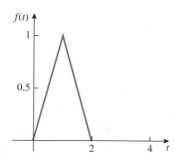

17. $y' + y = g(t); y(0) = 0$, where the graph of $g(t)$ is

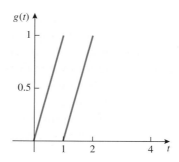

18. Problem 28 of Exercises 2.2 concerns the population of Botswana from 1975 to 1990 under certain basic assumptions. Now consider the situation that occurs if we start with a population of 0.755 million people in 1975 ($t = 0$) and assume that births and deaths, as well as immigration and emigration, balance each other until 1977 ($t = 2$). In 1977 an emigration pattern begins in such a way that the population $P(t)$ can be described by the equation

$$P' - kP = \begin{cases} 0 & \text{for } 0 < t < 2 \\ -a(t - 2) & \text{for } t \ge 2 \end{cases}$$

with $P(0) = 0.755$, $k = 0.0355$, and $a = 1.60625 \times 10^{-3}$.

a. Express the function on the right-hand side of the equation in terms of the unit step function.

b. Use technology and the Laplace transform to solve for $P(t)$, expressing the answer as a step function.

   c. Graph the solution on the interval $0 \leq t \leq 35$ and explain what the graph means in terms of the population of Botswana.

19. The IVP $y'' + 3y' + 2y = W(t)$; $y(0) = 0$, $y'(0) = 0$ represents a damped spring-mass system subjected to a *square wave* forcing term given by $W(t) = U(t - 1) - U(t - 2)$.

   a. Graph $W(t)$.

   b. Without using technology, solve the IVP when $W(t)$ is not present in the system. (That is, make the right-hand side of the differential equation zero.)

   c. Without using technology, solve the given IVP (that is, with $W(t)$ as the forcing term).

   d. Use technology to graph the solutions to parts (b) and (c) on the same set of axes. What difference does the forcing term make?

6.4   TRANSFORMS OF IMPULSE FUNCTIONS— THE DIRAC DELTA FUNCTION

In the last section we dealt with situations in which some abrupt change occurred. To describe it in general terms, we were dealing with systems acted on by some external force that was applied suddenly. But although the change was sudden, force was assumed to have been applied for some measurable period of time. Now we want to examine problems in which there is an external force of large magnitude applied suddenly for a very short period of time. For example, think about a baseball being hit by a major-league player. The time of contact of ball with bat is very brief, but enough force can be applied to send that horsehide soaring into the stands. More dramatic instances of this phenomenon include an electrical surge caused by a power line that is suddenly struck by lightning and a population that is growing at a certain rate until some sudden disaster strikes the community.

Mathematically, we can start to approach this idea by considering a piecewise continuous function that looks like

$$\delta_b(t) = \begin{cases} \dfrac{1}{b} & \text{for } 0 \leq t \leq b \\ 0 & \text{for } t > b. \end{cases}$$

Here we must assume that $\delta_b(t)$, which is pronounced "delta sub b of t," does not exist if $b = 0$. This function can represent a force of magnitude $1/b$ applied for a time period of length b (see Figure 6.3).

First of all, note that $\displaystyle\int_0^\infty \delta_b(t)\,dt = \int_0^b \frac{1}{b}\,dt = 1$ for all values of $b > 0$. Now look at what happens as we allow the value of b to get smaller and smaller. This situation describes a force whose magnitude $1/b$ is getting larger and larger over a

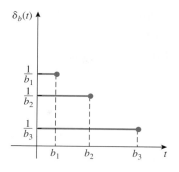

Figure 6.3
The graph of $\delta_b(t)$

shorter and shorter interval of time $(0, b)$. *Can you see what's going on?* More precisely, the unusual nature of this discontinuous function led various physicists, mathematicians, and engineers to consider the limiting behavior of $\delta_b(t)$ as $b \to 0$. In particular, they defined $\delta(t)$ as follows:

$$\delta(t) = \lim_{b \to 0} \delta_b(t) = \begin{cases} \infty & \text{for } t = 0 \\ 0 & \text{for } t \neq 0. \end{cases}$$

This "function" δ is called the **unit impulse function** or the **Dirac delta function** [named for the English-Belgian theoretical physicist Paul A. M. Dirac (1902–1984), who won the Nobel prize in 1933 with E. Schrödinger for his work on quantum theory]. More generally, we can define

$$\delta(t - a) = \lim_{b \to 0} \delta_b(t - a) = \begin{cases} \infty & \text{for } t = a \\ 0 & \text{for } t \neq a. \end{cases}$$

In the mathematically precise sense of the word, this limit does not exist and so does not define a function. However, such *generalized functions*, or *distributions*, can be put on a firm mathematical foundation and are very useful in modern physics and engineering theory. Before we look at examples of the delta function's use in solving differential equations, we should try to calculate its Laplace transform. The only reasonable way to do this is to make the formal assumption that

$$\mathcal{L}[\delta(t - a)] = \lim_{b \to 0} \mathcal{L}[\delta_b(t - a)].$$

(Mathematically, this raises an important theoretical question of whether

$$\lim_{b \to 0} \mathcal{L}[\delta_b(t - a)] = \mathcal{L}[\lim_{b \to 0} \delta_b(t - a)].$$

This question is beyond the scope of this course and will be ignored.)

Now let's write $\delta_b(t - a)$ in terms of the unit step function, as we did for functions in Section 6.3:

$$\delta_b(t - a) = \frac{1}{b}[U(t - a) - U(t - (a + b))].$$

If we use the linearity of the Laplace transform together with formula (6.3.1a)—taking $f(t - a) \equiv 1$—we get

$$\mathcal{L}[\delta(t - a)] = \lim_{b \to 0} \mathcal{L}[\delta_b(t - a)]$$

$$= \lim_{b \to 0} \frac{1}{b} \left\{ \frac{e^{-sa}}{s} - \frac{e^{-s(a+b)}}{s} \right\} = \lim_{b \to 0} e^{-sa} \left\{ \frac{1 - e^{-sb}}{bs} \right\}$$

$$= e^{-sa} \lim_{b \to 0} \left\{ \frac{1 - e^{-sb}}{bs} \right\} = e^{-sa},$$

where we have used L'Hôpital's rule to evaluate the indeterminate form in this last limit.

Because we have shown that

$$\mathcal{L}[\delta(t - a)] = e^{-sa}, \tag{6.4.1a}$$

it seems reasonable to take $a = 0$ and conclude that

$$\mathcal{L}[\delta(t)] = 1. \tag{6.4.1b}$$

Now let's see how to solve a differential equation involving an impulse function. You may want to review the spring-mass problems in Examples 4.5.5–4.5.8.

EXAMPLE 6.4.1   Solving an ODE That Involves the Dirac Delta Function
A mass attached to a spring is released from rest 1 meter below the equilibrium position for the spring-mass system and begins to move up and down. After 3 seconds, the mass is struck by a hammer in a downward direction. The undamped system is governed by the IVP

$$\frac{d^2x}{dt^2} + 9x = 3\delta(t - 3); \quad x(0) = 1, \frac{dx}{dt}(0) = 0,$$

where $x(t)$ denotes the displacement from equilibrium at time t, and we want to determine a formula for $x(t)$. (Note that the impulse force applied at $t = 3$ has magnitude 3.)

Let $X = X(s) = \mathcal{L}[x(t)]$. Then, taking the Laplace transform of both sides of our ODE and using (6.1.5) and (6.4.1a) with $a = 3$, we get

$$s^2X - s + 9X = 3e^{-3s},$$

so we can solve for X:

$$X = \frac{s}{s^2 + 9} + e^{-3s} \frac{3}{s^2 + 9}.$$

Applying the inverse transform yields

$$x(t) = \cos 3t + \sin 3(t - 3)U(t - 3)$$
$$= \begin{cases} \cos 3t & \text{for } t < 3 \\ \cos 3t + \sin 3(t - 3) & \text{for } 3 \leq t. \end{cases}$$

Figure 6.4 is the graph of $x(t)$, where the solid curve shows the displacement of the mass if the hammer had not hit it.

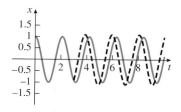

Figure 6.4

Graph of $x(t) = \begin{cases} \cos 3t & \text{for } t < 3 \\ \cos 3t + \sin 3(t - 3) & \text{for } 3 \le t \end{cases}$

EXERCISES 6.4

Solve the IVPs in Exercises 1–8.

1. $y'' = \delta(t - a);\quad y(0) = 0, y'(0) = 0$
2. $y' + 8y = \delta(t - 1) + \delta(t - 2);\quad y(0) = 0$
3. $2y'' + y' + 2y = \delta(t - 5);\quad y(0) = 0, y'(0) = 0$
4. $y'' + 2y' + y = 2\delta(t - 1);\quad y(0) = 1, y'(0) = 1$
5. $y'' + 6y' + 109y = \delta(t - 1) - \delta(t - 7);\quad y(0) = 0, y'(0) = 0$
6. $y'' + y = 1 + \delta(t - 2\pi);\quad y(0) = 1, y'(0) = 0$
7. $y'' + y = 4\delta(t - \frac{3}{2}\pi);\quad y(0) = 0, y'(0) = 1$
8. $y^{(iv)} - y = \delta(t - 1);\quad y(0) = 0, y'(0) = 0, y''(0) = 0, y'''(0) = 0$

9. A uniform beam of length L carries a load W concentrated at $x = \dfrac{L}{2}$. The beam is embedded at its left end and is free at its right end. The deflection $y(x)$ is governed by the equation $EI\dfrac{d^4y}{dx^4} = W\delta\left(x - \dfrac{L}{2}\right)$, where $y(0) = 0$, $y'(0) = 0$, $y''(L) = 0$, and $y'''(L) = 0$. Use the Laplace transform to determine the deflection $y(x)$.

10. If, at time $t = a$, the upper end of an undamped spring-mass system is jerked upward suddenly and returned to its original position, the equation that models the situation is $mx'' + kx = kH\delta(t - a); x(0) = x_0, x'(0) = x_1$, where m is the mass, k is the spring constant, and H is a constant.
    a. Solve the IVP manually, with $x(0) = 0 = x'(0)$.
    b. Use the solution found in part (a) to explain the significance of the constant H.
    c. Choose a value for H such that the mass achieves a prescribed displacement from equilibrium A for $t \ge a$.

11. Suppose we have the equation $y'' + ay' + by = f(t)$, where a and b are constants and f is a piecewise continuous function whose Laplace transform exists. Show that the effect of replacing $f(t)$ by $f(t) + c\delta(t)$, where c is a constant, is the same as increasing the initial value of $y'(0)$ by the constant c.

12. If the function $g(t)$ is continuous at a, show that $\int_0^\infty \delta(t - a)g(t)dt = g(a)$.

13. a. Show that $\mathcal{L}[\delta(t - a)f(t)] = e^{-as}f(a)$.

   b. Use the result in part (a) to solve the IVP $y'' + 2y' + y = \delta(t - 1)t$; $y(0) = 0$, $y'(0) = 0$.

14. Show that if a, b, and c are constants, the solution $x(t)$ of the linear IVP

$$x''(t) + ax'(t) + bx(t) = \delta(t - c); \quad x(0) = 0, x'(0) = 0$$

is $x(t) = k(t - c)\, U(t - c)$, where $k(t) = \mathcal{L}^{-1}\left[\dfrac{1}{s^2 + as + b}\right]$.

6.5   TRANSFORMS AND SYSTEMS OF LINEAR DIFFERENTIAL EQUATIONS

We have seen what the Laplace transform does to a single linear equation with constant coefficients. It should be easy to see that when initial conditions are given, the Laplace transform converts a system of linear differential equations with constant coefficients into a system of simultaneous algebraic equations. Then we can solve the algebraic equations for the *transformed* solution functions. Finally, applying the inverse transform to these functions gives us the solutions of the original system of linear ODEs.

Conceptually, this process is easy. The algebraic details, however, may be something else. Problems of this kind make us appreciate the availability of technology.

EXAMPLE 6.5.1   Solving a Linear System via the Laplace Transform

Let's start with the system

$$\frac{dx}{dt} = -3x + y$$
$$\frac{dy}{dt} = x - 3y,$$

where we want the solutions $x(t)$ and $y(t)$ that satisfy $x(0) = 2$ and $y(0) = 3$. (This system was discussed briefly in Example 1.2.9.)

Applying the Laplace transform to each side of each equation gives us the system

$$s\mathcal{L}[x(t)] - x(0) = -3\mathcal{L}[x(t)] + \mathcal{L}[y(t)]$$
$$s\mathcal{L}[y(t)] - y(0) = \mathcal{L}[x(t)] - 3\mathcal{L}[y(t)].$$

Inserting the initial conditions and simplifying the resulting equations, we get the system

$$(s + 3)\mathcal{L}[x(t)] - \mathcal{L}[y(t)] = 2$$
$$(s + 3)\mathcal{L}[y(t)] - \mathcal{L}[x(t)] = 3.$$

Now we solve this last system for $\mathscr{L}[x(t)]$ and $\mathscr{L}[y(t)]$ just as we would solve any algebraic system of two equations in two unknowns. (To simplify things, you could let $\mathscr{L}[x(t)] = X$ and $\mathscr{L}[y(t)] = Y$.) For instance, we can eliminate the variable $\mathscr{L}[y(t)]$ by multiplying the first equation by $(s + 3)$ and then adding the result to the second equation. When the dust settles, we get

$$\{(s + 3)^2 - 1\}\mathscr{L}[x(t)] = 2(s + 3) + 3,$$

so we find that

$$\mathscr{L}[x(t)] = \frac{2s + 9}{(s + 3)^2 - 1} = \frac{2s + 9}{[(s + 3) + 1][(s + 3) - 1]}$$

$$= \frac{2s + 9}{(s + 4)(s + 2)} = \frac{-\dfrac{1}{2}}{s + 4} + \frac{\dfrac{5}{2}}{s + 2} = \frac{-\dfrac{1}{2}}{s - (-4)} + \frac{\dfrac{5}{2}}{s - (-2)}$$

and

$$x(t) = -\frac{1}{2}\mathscr{L}^{-1}\left[\frac{1}{s - (-4)}\right] + \frac{5}{2}\mathscr{L}^{-1}\left[\frac{1}{s - (-2)}\right]$$

$$= -\frac{1}{2}e^{-4t} + \frac{5}{2}e^{-2t}.$$

Now we could go through this process again to eliminate $\mathscr{L}[x(t)]$ and solve for $y(t)$ this time (see Exercise 1)—or we could just substitute our solution for $x(t)$ in the first equation of our original system and solve for y:

$$y(t) = \frac{dx}{dt} + 3x = \frac{d}{dx}\left(-\frac{1}{2}e^{-4t} + \frac{5}{2}e^{-2t}\right) + 3\left(-\frac{1}{2}e^{-4t} + \frac{5}{2}e^{-2t}\right)$$

$$= 2e^{-4t} - 5e^{-2t} - \frac{3}{2}e^{-4t} + \frac{15}{2}e^{-2t} = \frac{1}{2}e^{-4t} + \frac{5}{2}e^{-2t}. \qquad \blacklozenge$$

The next example shows how we can handle a system of two second-order linear equations. In particular, note that we don't have to write this as a system of four first-order equations. The Laplace transform technique works directly on higher-order derivatives via formula (6.1.6) or, in this case, by (6.1.5).

EXAMPLE 6.5.2   A System of Second-Order Equations

The system IVP we want to solve is

$$\frac{d^2x}{dt^2} - 4x + \frac{dy}{dt} = 0$$

$$-4\frac{dx}{dt} + \frac{d^2y}{dt^2} + 2y = 0,$$

with $x(0) = 0$, $x'(0) = 1$, $y(0) = -1$, and $y'(0) = 2$.

Applying the Laplace transform to each side of each equation, we get

$$\mathscr{L}[x''(t)] - 4\mathscr{L}[x(t)] + \mathscr{L}[y'(t)] = 0$$
$$-4\mathscr{L}[x'(t)] + \mathscr{L}[y''(t)] + 2\mathscr{L}[y(t)] = 0.$$

Using (6.1.4) and (6.1.5), we can write this last system as

$$s^2\mathcal{L}[x(t)] - x'(0) - sx(0) - 4\mathcal{L}[x(t)] + s\mathcal{L}[y(t)] - y(0) = 0$$
$$-4s\mathcal{L}[x(t)] + 4x(0) + s^2\mathcal{L}[y(t)] - y'(0) - sy(0) + 2\mathcal{L}[y(t)] = 0.$$

Now we insert the initial conditions and simplify the resulting equations to get

$$(s^2 - 4)\mathcal{L}[x(t)] + s\mathcal{L}[y(t)] = 0$$
$$-4s\mathcal{L}[x(t)] + (s^2 + 2)\mathcal{L}[y(t)] = 2 - s. \qquad (**)$$

As in the previous example, we can solve these equations by realizing that they constitute a system of ordinary algebraic equations in the unknowns $\mathcal{L}[x(t)]$ and $\mathcal{L}[y(t)]$. If we multiply the first equation of (* *) by $4s$, multiply the second by $s^2 - 4$, and then add the resulting equations, we obtain

$$(s^4 + 2s^2 - 8)\mathcal{L}[y(t)] = -s^3 + 2s^2 + 4s - 8,$$

so

$$\mathcal{L}[y(t)] = \frac{-s^3 + 2s^2 + 4s - 8}{s^4 + 2s^2 - 8} = \frac{-s^3 + 2s^2 + 4s - 8}{(s^2 + 4)(s^2 - 2)} \qquad (***)$$

$$= \frac{-s^3 + 2s^2 + 4s - 8}{(s^2 + 4)(s + \sqrt{2})(s - \sqrt{2})}$$

$$= \frac{1}{6}\left[\frac{1 + \sqrt{2}}{s + \sqrt{2}} + \frac{1 - \sqrt{2}}{s - \sqrt{2}} - \frac{8(s - 2)}{s^2 + 4}\right]$$

$$= \frac{1}{6}\left[\frac{1 + \sqrt{2}}{s + \sqrt{2}} + \frac{1 - \sqrt{2}}{s - \sqrt{2}} - 8\frac{s}{s^2 + 2^2} + 8\frac{2}{s^2 + 2^2}\right].$$

Using entries 2, 3, and 4 of the table of transforms (Table 6.1), we see that

$$y(t) = \frac{1}{6}[(1 + \sqrt{2})e^{-\sqrt{2}t} + (1 - \sqrt{2})e^{\sqrt{2}t} - 8\cos 2t + 8\sin 2t].$$

To find $\mathcal{L}[x(t)]$, we can go back to system (* *) and eliminate $\mathcal{L}[y(t)]$, or we can substitute expression (* * *) for $\mathcal{L}[y(t)]$ in either equation of (* *) and solve for $\mathcal{L}[x(t)]$. Let's try the latter method.

Using (* * *) and the first equation in (* *), we find that

$$(s^2 - 4)\mathcal{L}[x(t)] + s\left(\frac{-s^3 + 2s^2 + 4s - 8}{(s^2 + 4)(s^2 - 2)}\right) = 0.$$

Solving for $\mathcal{L}[x(t)]$, we get

$$\mathcal{L}[x(t)] = -s\left(\frac{-s^3 + 2s^2 + 4s - 8}{(s^2 - 4)(s^2 + 4)(s^2 - 2)}\right) = \frac{s(s - 2)^2(s + 2)}{(s - 2)(s + 2)(s^2 + 4)(s^2 - 2)}$$

$$= \frac{s(s - 2)}{(s^2 + 4)(s + \sqrt{2})(s - \sqrt{2})}$$

$$= -\frac{1}{12}\left[\frac{2 + \sqrt{2}}{s + \sqrt{2}} + \frac{2 - \sqrt{2}}{s - \sqrt{2}} - 4\left(\frac{s + 2}{s^2 + 4}\right)\right]$$

$$= -\frac{1}{12}\left[\frac{2 + \sqrt{2}}{s + \sqrt{2}} + \frac{2 - \sqrt{2}}{s - \sqrt{2}} - 4\left(\frac{s}{s^2 + 2^2} + \frac{2}{s^2 + 2^2}\right)\right].$$

Formulas 2, 3, and 4 from Table 6.1 tell us that

$$x(t) = -\frac{1}{12}[(2 + \sqrt{2})e^{-\sqrt{2}t} + (2 - \sqrt{2})e^{\sqrt{2}t} - 4\cos 2t - 4\sin 2t].$$

You should confirm that these are the solutions to the original IVP.   ◆

EXERCISES 6.5

1. Eliminate $\mathcal{L}[x(t)]$ from the algebraic system
$$(s + 3)\mathcal{L}[x(t)] - \mathcal{L}[y(t)] = 2$$
$$(s + 3)\mathcal{L}[y(t)] - \mathcal{L}[x(t)] = 3$$
and then solve for $y(t)$. (See Example 6.5.1.)

Solve the IVPs in Exercises 2–11 by using the Laplace transform.

2. $\{x' = 2x - 3y, y' = y - 2x\}$; $x(0) = 8, y(0) = 3$

3. $\{x' = 12x + 5y, y' = -6x + y\}$; $x(0) = 0, y(0) = 1$

4. $\{x' = -2x + y, y' = -9x + 4y\}$; $x(0) = 5, y(0) = -3$

5. $\{x' = -6x + 2y, y' = -7x + 3y\}$; $x(0) = 1, y(0) = 0$

6. $\{x' = x + y, y' = -4x + y\}$; $x(0) = 1, y(0) = 1$

7. $\{x' + y' = -3x - 2y + e^{-2t}, 2x' + y' = -2x - y + 1\}$; $x(0) = 0, y(0) = 0$

8. $\{x' = x - y - e^{-t}, y' = 2x + 3y + e^{-t}\}$; $x(0) = 1, y(0) = 0$

9. $\{x' + y' = x, y' + z' = x, z' + x' = x\}$; $x(0) = 1, y(0) = 1, z(0) = 1$

10. $\{x'' + y' = 4x, 4x' - y'' = 9y\}$; $x(0) = 0, x'(0) = 1, y(0) = -1, y'(0) = 2$

11. $\{x'' - y' = -t + 1, x' - x + 2y' = 4e^t\}$; $x(0) = 0, x'(0) = 1, y(0) = 0$

12. In determining the concentration of a chemical in a system that consists of two compartments separated by a membrane, we get the system of equations

$$\dot{x} = ay - bx$$
$$\dot{y} = bx - ay - \beta y,$$

subject to the conditions $x(0) = x^*$ and $y(0) = y^*$, where x^* and y^* are constants. (Here x and y represent the masses of the chemical in compartments 1 and 2, respectively, at any time t, and the constants a, b, and β are positive constants of proportionality related to the rate of flow of the chemical from one compartment to another.)

a. Solve this system of equations using Laplace transforms.

b. Letting $p = \frac{1}{2}(b + a + \beta)$ and $q = \frac{1}{2}\sqrt{(b + a + \beta)^2 - 4\beta b}$, show that q is a (positive) *real* number and that $p > q$.

c. Using the solution found in part (a) and the results of part (b), show that the chemical masses x and y approach zero steadily.

13. The system

$$mx'' = -k_1(x - a\theta) - k_2(x + a\theta)$$
$$mr^2\theta'' = k_1a(x - a\theta) - k_2a(x + a\theta)$$

models the motion of a slab of mass m mounted on two springs, as shown in the accompanying figure. Here x is the vertical displacement of the center of mass, and θ is the angle shown. The constant r represents the radius of gyration of the slab about the appropriate axis through the center of mass. Use the Laplace transform and technology to solve the system for x and θ if $m = 1$, $k_1 = 1$, $k_2 = 2$, $a = 1$, $r = 1$, $x(0) = 1$, $x'(0) = 0$, $\theta(0) = 0.1$, and $\theta'(0) = 0$.

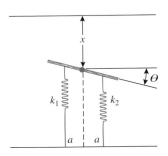

14. The system consisting of two pendulums connected by a spring (see the accompanying figure) has its motion approximated by the system of equations

$$mx'' + m\omega_0^2 x = -k(x - y)$$
$$my'' + m\omega_0^2 y = -k(y - x),$$

where L is the length of each pendulum, g is the gravitational constant, and $\omega_0^2 = g/L$. Use the Laplace transform and technology to solve this system with $m = 1$, $L = 5$, $g = 32$, $k = 2$, and the initial conditions $x(0) = 0$, $x'(0) = 2$, $y(0) = 0$, and $y'(0) = 2$.

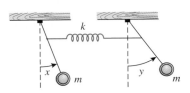

15. The circuit that follows is described by the system

$$L_1 \dot{I}_1 + R_1(I_1 - I_2) = v(t)$$
$$L_2 \dot{I}_2 + R_2 I_2 + R_1(I_2 - I_1) = 0.$$

Determine I_1 and I_2 when the switch is closed if $L_1 = L_2 = 2$ henrys, $R_1 = 3$ ohms, $R_2 = 8$ ohms, and $v(t) = 6$ volts. Assume that $I_1(0) = I_2(0) = 0$.

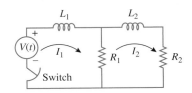

6.6 QUALITATIVE ANALYSIS VIA THE LAPLACE TRANSFORM

Back in Chapter 5 (specifically, in Sections 5.2–5.5), we analyzed autonomous two-dimensional systems of linear equations and their equivalent single second-order homogeneous equations by means of eigenvalues and eigenvectors. This qualitative analysis, which was very neat and very satisfying, depended on the roots of polynomial equations (the *characteristic equations*). However, the eigenvalue method didn't stretch quite far enough to handle general *nonhomogeneous* systems (Section 5.6).

HOMOGENEOUS EQUATIONS

Despite the overwhelming emphasis on the Laplace transform as a tool for obtaining exact, closed-form solutions, it turns out that the transform can provide insight into the *qualitative* nature of a solution as well. In fact, because the Laplace transform treats nonhomogeneous equations in essentially the same way as it treats homogeneous equations (with slightly messier algebra), the Laplace transform in effect gives us an extension of eigenvalue analysis to nonhomogeneous equations. Let's examine this by analyzing nonhomogeneous second-order equations of the form

$$c_2 x'' + c_1 x' + c_0 x = f(t), \tag{6.6.1}$$

where c_2, c_1, and c_0 are constants and $c_2 \neq 0$. For the sake of simplicity and clarity, we'll start with $f(t) \equiv 0$, the *homogeneous* case.

Taking the Laplace transform of both sides of (6.6.1)—with $f(t) \equiv 0$—we get

$$c_2 \mathcal{L}[x''(t)] + c_1 \mathcal{L}[x'(t)] + c_0 \mathcal{L}[x(t)] = 0,$$

which becomes

$$c_2 \{s^2 \mathcal{L}[x(t)] - sx(0) - x'(0)\} + c_1 \{s\mathcal{L}[x(t)] - x(0)\} + c_0 \mathcal{L}[x(t)] = 0,$$

or, after simplifying,

$$(c_2 s^2 + c_1 s + c_0) \mathcal{L}[x(t)] - (c_2 s + c_1) x(0) - c_2 x'(0) = 0.$$

Solving, we find that

$$\mathcal{L}[x(t)] = \frac{(c_2 s + c_1) x(0)}{c_2 s^2 + c_1 s + c_0} + \frac{c_2 x'(0)}{c_2 s^2 + c_1 s + c_0}. \tag{6.6.2}$$

We should note something significant about the denominator, $c_2 s^2 + c_1 s + c_0$, of the Laplace transform of the solution. It is the *characteristic polynomial* corresponding to the second-order differential equation $c_2 x'' + c_1 x' + c_0 x = 0$ or to the equivalent system

$$x_1' = x_2$$
$$x_2' = \left(-\frac{c_0}{c_2}\right) x_1 - \left(\frac{c_1}{c_2}\right) x_2.$$

Interesting! If the right-hand side of equation (6.6.2) is expressed as a single fraction (rational function) with no common factors in the numerator and denominator, then the zeros of the characteristic polynomial—the values of s that make the denominator zero—are called the **poles,** or **singularities,** of the transform $\mathcal{L}[x(t)]$.

Now suppose that λ_1 and λ_2 are the zeros of the characteristic polynomial—the *eigenvalues* of the system. Let's see what happens if both zeros are *real* numbers. For convenience, also assume that $\lambda_1 \neq \lambda_2$.

First we can divide through by c_2 and write $c_2 s^2 + c_1 s + c_0 = 0$ in the equivalent form $s^2 + \left(\dfrac{c_1}{c_2}\right)s + \left(\dfrac{c_0}{c_2}\right) = 0$; then we can write $s^2 + \left(\dfrac{c_1}{c_2}\right)s + \left(\dfrac{c_0}{c_2}\right) = (s - \lambda_1)(s - \lambda_2)$. Now, returning to (6.6.2), we can apply the inverse transform to each side to get

$$
\begin{aligned}
x(t) &= \mathcal{L}^{-1}\left[\frac{(c_2 s + c_1)x(0)}{c_2 s^2 + c_1 s + c_0}\right] + \mathcal{L}^{-1}\left[\frac{c_2 x'(0)}{c_2 s^2 + c_1 s + c_0}\right] \\[2mm]
&= \mathcal{L}^{-1}\left[\frac{(c_2 s + c_1)x(0)}{c_2\left(s^2 + \left(\frac{c_1}{c_2}\right)s + \left(\frac{c_0}{c_2}\right)\right)}\right] + \mathcal{L}^{-1}\left[\frac{c_2 x'(0)}{c_2\left(s^2 + \left(\frac{c_1}{c_2}\right)s + \left(\frac{c_0}{c_2}\right)\right)}\right] \\[2mm]
&= \mathcal{L}^{-1}\left[\frac{\frac{1}{c_2}(c_2 s + c_1)x(0)}{\left(s^2 + \left(\frac{c_1}{c_2}\right)s + \left(\frac{c_0}{c_2}\right)\right)}\right] + \mathcal{L}^{-1}\left[\frac{\frac{1}{c_2}c_2 x'(0)}{\left(s^2 + \left(\frac{c_1}{c_2}\right)s + \left(\frac{c_0}{c_2}\right)\right)}\right] \\[2mm]
&= \mathcal{L}^{-1}\left[\frac{\left(s + \frac{c_1}{c_2}\right)x(0)}{(s - \lambda_1)(s - \lambda_2)}\right] + \mathcal{L}^{-1}\left[\frac{x'(0)}{(s - \lambda_1)(s - \lambda_2)}\right] \\[2mm]
&= x(0)\mathcal{L}^{-1}\left[\frac{\left(s + \frac{c_1}{c_2}\right)}{(s - \lambda_1)(s - \lambda_2)}\right] + x'(0)\mathcal{L}^{-1}\left[\frac{1}{(s - \lambda_1)(s - \lambda_2)}\right] \\[2mm]
&= x(0)\mathcal{L}^{-1}\left[\frac{\frac{c_1 + c_2\lambda_1}{c_2(\lambda_1 - \lambda_2)}}{s - \lambda_1} - \frac{\frac{c_1 + c_2\lambda_2}{c_2(\lambda_1 - \lambda_2)}}{s - \lambda_2}\right] + x'(0)\mathcal{L}^{-1}\left[\frac{1}{\lambda_1 - \lambda_2}\left(\frac{1}{s - \lambda_1} - \frac{1}{s - \lambda_2}\right)\right] \\[2mm]
&= \frac{x(0)}{\lambda_1 - \lambda_2}\mathcal{L}^{-1}\left[\frac{\lambda_1 + \frac{c_1}{c_2}}{s - \lambda_1} - \frac{\lambda_2 + \frac{c_1}{c_2}}{s - \lambda_2}\right] + \frac{x'(0)}{\lambda_1 - \lambda_2}\mathcal{L}^{-1}\left[\frac{1}{s - \lambda_1} - \frac{1}{s - \lambda_2}\right] \\[2mm]
&= A\mathcal{L}^{-1}\left[\frac{1}{s - \lambda_1}\right] - B\mathcal{L}^{-1}\left[\frac{1}{s - \lambda_2}\right] + C\mathcal{L}^{-1}\left[\frac{1}{s - \lambda_1}\right] - D\mathcal{L}^{-1}\left[\frac{1}{s - \lambda_2}\right] \\[2mm]
&= K_1\mathcal{L}^{-1}\left[\frac{1}{s - \lambda_1}\right] + K_2\mathcal{L}^{-1}\left[\frac{1}{s - \lambda_2}\right] = K_1 e^{\lambda_1 t} + K_2 e^{\lambda_2 t},
\end{aligned}
$$

where $A, B, C, D, K_1,$ and K_2 are constants. (*Check the last few lines carefully.*)

Stability

Imitating the qualitative analysis we did in Chapter 5 for real eigenvalues (see Table 5.1 at the end of Section 5.5 for a summary), we can see that if λ_1 and λ_2 are unequal and *positive*, then the origin is a *source*. If λ_1 and λ_2 are unequal and *neg-*

ative, then the origin is a *sink*. If λ_1 and λ_2 have different signs, then the origin is a *saddle point*.

Now suppose that the zeros of our characteristic polynomial are complex numbers: $\lambda_1 = p + qi$, $\lambda_2 = p - qi$. (Remember that complex roots of a quadratic equation occur in complex conjugate pairs.) Then we can write

$$s^2 + \left(\frac{c_1}{c_2}\right)s + \left(\frac{c_0}{c_2}\right) = (s - \lambda_1)(s - \lambda_2) = [s - (p + qi)][s - (p - qi)]$$
$$= [(s - p) - qi][(s - p) + qi] = (s - p)^2 + q^2.$$

Now when we express our solution $x(t)$ in terms of the inverse Laplace transform of functions, we will have $(s - p)^2 + q^2$ in all the denominators and either constants or constant multiples of s in the numerators of these functions. Looking at entries 5 and 6 in the transform table (Table 6.1, in Section 6.2), we realize that the inverse Laplace transforms we will get are constant multiples of either $e^{pt}\sin qt$ or $e^{pt}\cos qt$.

The next example will help us understand these ideas better.

EXAMPLE 6.6.1   A Qualitative Analysis via the Laplace Transform
Let's look at the equation $x'' + 3x' + 5x = 0$. The Laplace transform of this equation is

$$\mathscr{L}[x''] + 3\mathscr{L}[x'] + 5\mathscr{L}[x] = 0$$
$$\{s^2\mathscr{L}[x] - sx(0) - x'(0)\} + 3\{s\mathscr{L}[x] - x(0)\} + 5\mathscr{L}[x] = 0$$
$$(s^2 + 3s + 5)\mathscr{L}[x] - (s + 3)x(0) - x'(0) = 0,$$

so

$$\mathscr{L}[x] = \frac{(s + 3)x(0) + x'(0)}{s^2 + 3s + 5} = \frac{(s + 3)x(0)}{s^2 + 3s + 5} + \frac{x'(0)}{s^2 + 3s + 5}.$$

The characteristic polynomial $s^2 + 3s + 5$ has complex conjugate zeros $-\frac{3}{2} + \frac{\sqrt{11}}{2}i$ and $-\frac{3}{2} - \frac{\sqrt{11}}{2}i$. Because the real part is negative, we expect the solution to oscillate with decreasing amplitude. Figure 6.5 shows x against t, with $x(0) = 3$ and $x'(0) = 20$.

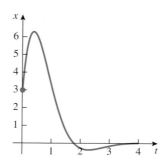

Figure 6.5

Graph of $x(t) = e^{-3t/2}\left(3\cos\left(\frac{\sqrt{11}}{2}t\right) + \frac{49\sqrt{11}}{11}\sin\left(\frac{\sqrt{11}}{2}t\right)\right)$, $0 \le t \le 4$

NONHOMOGENEOUS EQUATIONS

When we look at the nonhomogeneous version of (6.6.1), we find that

$$X(s) = \mathcal{L}[x(t)] = \frac{\mathcal{L}[f(t)]}{P(s)} + \frac{Q(s)}{P(s)} = \frac{F(s)}{P(s)} + \frac{Q(s)}{P(s)}, \tag{6.6.3}$$

where $P(s)$ is the characteristic polynomial $c_2 s^2 + c_1 s + c_0$ and $Q(s)$ is the linear polynomial $\{c_2 x(0)\}s + \{c_1 x(0) + c_2 x'(0)\}$. (*Verify this.*)

This observation can be expanded to the case of the general nth-order linear equation with constant coefficients. In this situation, $P(s)$ is the nth-degree characteristic polynomial and $Q(s)$ is a polynomial in s of degree $n - 1$. The coefficients of $Q(s)$ consist of combinations of products of the coefficients in the equation and the n initial conditions.

If we let $W(s) = \dfrac{1}{P(s)}$, we can write (6.6.3) as

$$X(s) = W(s)F(s) + W(s)Q(s). \tag{6.6.4}$$

Applying the inverse Laplace transform to each side of (6.6.4), we see that

$$x(t) = \mathcal{L}^{-1}[W(s)F(s)] + \mathcal{L}^{-1}[W(s)Q(s)],$$

which expresses the output $x(t)$ of the system as a superposition of two outputs—the first due to the input $f(t)$ and the second due to the initial conditions.

Let's look at a problem we've seen before, as Examples 6.1.1 and 6.2.1.

EXAMPLE 6.6.2   Qualitative Analysis of a Nonhomogeneous Equation
Consider the nonhomogeneous initial-value problem

$$x'' + 3x' + 2x = 12e^{2t}; \quad x(0) = x_0, x'(0) = x_1.$$

The Laplace transform of this equation is

$$\mathcal{L}[x''] + 3\mathcal{L}[x'] + 2\mathcal{L}[x] = 12\mathcal{L}[e^{2t}]$$

$$\{s^2\mathcal{L}[x] - sx(0) - x'(0)\} + 3\{s\mathcal{L}[x] - x(0)\} + 2\mathcal{L}[x] = \frac{12}{s - 2}$$

$$(s^2 + 3s + 2)\mathcal{L}[x] - (s + 3)x_0 - x_1 = \frac{12}{s - 2},$$

so

$$\mathcal{L}[x] = \frac{\dfrac{12}{s - 2}}{s^2 + 3s + 2} + \frac{(s + 3)x_0 + x_1}{s^2 + 3s + 2},$$

the form indicated in (6.6.3).

The characteristic polynomial $s^2 + 3s + 2$ has negative real zeros -1 and -2, so the second term in the last equation contributes a part of the solution that decays (tends to 0) as $t \to \infty$. (*Why?*) The first term has an additional pole, at 2. If we imagine the partial-fractions version of the Laplace transform of the solution, written with denominators $s - 2$ and $s^2 + 3s + 2$, we realize that the only terms that

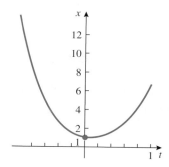

Figure 6.6
Graph of $x(t) = e^{2t} + 3e^{-2t} - 3e^{-t}, -1 \le t \le 1$

appear in the solution are e^{2t}, e^{-t}, and e^{-2t}. Therefore, for increasing values of t, the solution is dominated by the term containing e^{2t}. Figure 6.6 shows x against t, with $x(0) = x_0 = 1$ and $x'(0) = x_1 = -1$, as in Examples 6.1.1 and 6.2.1. ◆

Transfer Functions and Impulse Response Functions

If we consider the nonhomogeneous second-order equation (6.6.1) with initial conditions $x(0) = 0$, $x'(0) = 0$, then equation (6.6.3) becomes

$$\mathcal{L}[x(t)] = \frac{1}{c_2 s^2 + c_1 s + c_0} \cdot \mathcal{L}[f(t)],$$

or

$$\frac{X(s)}{F(s)} = \frac{1}{c_2 s^2 + c_1 s + c_0}, \tag{6.6.5}$$

where $F(s) = \mathcal{L}[f(t)]$ and $X(s) = \mathcal{L}[x(t)]$. A system with these initial conditions is sometimes described as "relaxed" or at rest until $t = 0$.

In certain areas of engineering (for example, those dealing with feedback and control systems), this ratio (6.6.5) of the Laplace transform of the *output* to the Laplace transform of the *input* is called the **transfer function** of the system modeled by equation (6.6.1) with all initial values zero. The inverse Laplace transform of the transfer function is called the **impulse response function** for the system, because in physical terms, it describes (for example) the solution when a spring-mass system is struck by a hammer. (See Example 6.4.1, for instance, and Exercise 12 on the following page.) The analysis of this transfer function provides a picture of what can be called the **response** of the system. The values of s that make the denominator of (6.6.5) zero are called **poles** or **singularities** of the transfer function. On the basis of our analysis of eigenvalues in Chapter 5 and our discussion in this section, you should see that the nature of the poles (real, complex, positive, and so on) determines the behavior of the system. For example, in this second-order situation, the system could be undamped, overdamped, or under-damped, or the response of the system could grow without bound.

EXERCISES 6.6

Suppose that $X(s) = \mathcal{L}[x(t)]$ is the Laplace transform of the solution of a linear differential equation. For each transform in Exercises 1–7, determine the qualitative behavior of $x(t)$ for large values of t without finding the inverse of the transform. (That is, determine whether $x(t)$ oscillates, goes to 0, or becomes unbounded as t becomes large.)

1. $X(s) = \dfrac{2}{3s + 5}$

2. $X(s) = \dfrac{4}{s^2 - 1}$

3. $X(s) = \dfrac{1}{s^2 + 2s + 10}$

4. $X(s) = \dfrac{s + 2}{s^2 + 4}$

5. $X(s) = \dfrac{2s + 6}{s^2 + 6s + 18}$

6. $X(s) = \dfrac{2s + 5}{s^2 + 3s + 2}$

7. $X(s) = \dfrac{s}{s^4 + 5s^2 + 4}$

In Exercises 8–11, (a) compute the Laplace transform of each solution, (b) find the poles of the Laplace transform of the solution, and (c) discuss the behavior of the solution (oscillatory, unbounded, and so on) without solving the equation.

8. $x'' - x = 0$; $\quad x(0) = 0$, $x'(0) = 1$

9. $\ddot{x} + 2\dot{x} + 2x = e^{-t/10}$; $\quad x(0) = 4$, $\dot{x}(0) = 1$

10. $x'' + 2x' + 2x = e^{-2t} \sin 4t$; $\quad x(0) = 2$, $x'(0) = -2$

11. $2\ddot{x} + 7\dot{x} + 3x = 2\cos t$; $\quad x(0) = 1$, $x'(0) = 0$

12. Consider the initial-value problem

$$c_2 x'' + c_1 x' + c_0 x = \delta(t); \quad x(0) = x'(0) = 0,$$

where $\delta(t)$ denotes the unit impulse function (Section 6.4). Show that the transfer function of the system is $X(s) = \dfrac{1}{c_2 s^2 + c_1 s + c_0}$.

13. Show that if I is an interval containing the origin and f is continuous on I, then the unique solution to the IVP

$$c_2 x'' + c_1 x' + c_0 x = f(t); \quad x(0) = x_0, x'(0) = x_1$$

is given by $(r * f)(t) + x_H(t)$, where $R = \dfrac{X(s)}{F(s)}$, $r = \mathcal{L}^{-1}\{R\}(t)$ is the response function, and $x_H(t)$ is the unique solution to the homogeneous equation $c_2 x'' + c_1 x' + c_0 x = 0$; $x(0) = x_0, x'(0) = x_1$. (Of course, $*$ denotes convolution.)

14. Suppose that a linear system is described by the equation

$$\ddot{x} + 2\dot{x} + 5x = f(t); \quad x(0) = 2, \dot{x}(0) = -2.$$

a. Find the transfer function for the system.

b. Find the impulse response function.

c. Give a formula for the solution of the IVP. (Use the result of Exercise 13. Your answer should contain an integral.)

15. If a linear system is governed by the initial-value problem

$$y'' - y' - 6y = g(t); \quad y(0) = 1, \, y'(0) = 8,$$

    a. Find the transfer function for the system.

    b. Find the impulse response function.

    c. Find a formula for the solution of the IVP. (Use the result of Exercise 13. Your answer should contain an integral.)

16. Consider the initial-value problem

$$y'' + 2y' + 2y = \sin(\alpha t); \quad y(0) = 0, \, y'(0) = 0.$$

    a. Find the transfer function for the system.

    b. Find the impulse response function.

    c. Find a formula for the solution of the IVP. (Use the result of Exercise 13. Your answer should contain an integral.)

17. Consider the first-order system $a_1 x' + a_0 x = f(t)$, where a_1 and a_0 are constants, $a_1 \neq 0$.

    a. Find the transfer function of this system.

    b. Show that the transfer function of a constant-coefficient first-order system can be written as $W(s) = \dfrac{c}{1 + Ts}$, where c is a constant and T is a constant related to the exponential function component of the solution.

6.7   SUMMARY

Transformation methods are important examples of how we can change difficult problems into problems that can be handled more easily. If $f(t)$ is a function that is integrable for $t \geq 0$, then the **Laplace transform** of f is defined by

$$\mathcal{L}[f(t)] = \int_0^\infty f(t)e^{-st}dt,$$

when this improper integral exists. The integral will exist if we stick to *continuous* or *piecewise continuous* functions $f(t)$ for which there exist positive constants M and K such that $|f(t)| < e^{Mt}$ for all $t \geq K$. Note that this integral is a function of the parameter s, so we can write $\mathcal{L}[f(t)] = F(s)$.

Using basic properties of integrals, we can see that

$$\mathcal{L}[c \cdot f(t)] = c \cdot \mathcal{L}[f(t)],$$

where c is any real constant, and that

$$\mathcal{L}[f(t) + g(t)] = \mathcal{L}[f(t)] + \mathcal{L}[g(t)],$$

whenever the Laplace transforms of both f and g exist. Any transformation that satisfies the last two properties is called a *linear transformation*. If c_1 and c_2 are constants, we can combine the two properties to write

$$\mathcal{L}[c_1 f(t) + c_2 g(t)] = c_1 \mathcal{L}[f(t)] + c_2 \mathcal{L}[g(t)].$$

Table 6.1 in Section 6.2 gives the Laplace transform of some important classes of functions, including power functions, exponentials, trigonometric functions, and multiples of these. There are also important formulas for the Laplace transforms of f', f'', and higher derivatives. The Laplace transform method enables us to handle a linear nonhomogeneous equation with initial conditions all at once.

Once we have calculated the Laplace transform of a function—in particular, once we have transformed a differential equation into an algebraic equation—we have to be able to reverse the process to gain information about the original problem. An important fact is that *if the Laplace transforms of the continuous functions f and g exist and are equal for $s \geq c$ (c a constant), then $f(t) = g(t)$ for all $t \geq 0$.* This says that a continuous function can be uniquely recovered from its Laplace transform. Letting $\mathcal{L}[f(t)] = F(s)$, we can express the definition of the **inverse Laplace transform** as follows:

$$\mathcal{L}^{-1}[F] = f \quad \text{if and only if} \quad \mathcal{L}[f] = F.$$

It can be shown that the inverse Laplace transform is a linear transformation:

$$\mathcal{L}^{-1}[c_1 F(t) + c_2 G(t)] = c_1 \mathcal{L}^{-1}[F(t)] + c_2 \mathcal{L}^{-1}[G(t)].$$

In trying to find the inverse transform of an expression that is the product of two or more transforms, we encounter the idea of the *convolution* of two functions. The **convolution** of two functions f and g is the integral

$$(f * g)(t) = \int_0^t f(r) g(t - r) dr,$$

provided that the integral exists for $t > 0$. This product has important algebraic properties, and one of the most useful is that *the Laplace transform of a convolution of two functions is equal to the product of the Laplace transforms of these two functions.* More precisely, suppose that f and g are two functions whose Laplace transforms exist. Let $F(s) = \mathcal{L}[f(t)]$ and $G(s) = \mathcal{L}[g(t)]$. Then the **Convolution Theorem** says that

$$\mathcal{L}[(f * g)(t)] = \mathcal{L}\left[\int_0^t f(r) g(t - r) dr\right] = \mathcal{L}[f(t)] \cdot \mathcal{L}[g(t)] = F(s) \cdot G(s).$$

By using the **unit step function** (or **Heaviside function**) U, defined by

$$U(t) = \begin{cases} 0 & \text{for } t < 0 \\ 1 & \text{for } t \geq 0, \end{cases}$$

we can model systems in which there are abrupt changes. Mathematically, this means that we can express *piecewise continuous functions* in a simple way, using $U(t)$ as a basic building block.

When we are solving differential equations that model abrupt changes, the following result comes in handy. If $\mathcal{L}[f(t)]$ exists for $s > c$ and if $a > 0$, then

$$\mathcal{L}[f(t - a) U(t - a)] = e^{-as} \mathcal{L}[f(t)] \quad \text{for } s > c.$$

Alternatively, we can write this last formula as

$$f(t - a) U(t - a) = \mathcal{L}^{-1}[e^{-as} \mathcal{L}[f(t)]].$$

If we want to consider problems in which there is an external force of large magnitude applied suddenly for a very short period of time, we need the idea of the **unit impulse function,** or **Dirac delta function,** defined as

$$\delta(t) = \lim_{b \to 0} \delta_b(t) = \begin{cases} \infty & \text{for } t = 0 \\ 0 & \text{for } t \neq 0, \end{cases}$$

where

$$\delta_b(t) = \begin{cases} \dfrac{1}{b} & \text{for } 0 \leq t \leq b \\ 0 & \text{for } t > b. \end{cases}$$

We can show that $\mathcal{L}[\delta(t - a)] = e^{-sa}$. In particular, $\mathcal{L}[\delta(t)] = 1$.

When initial conditions are given, the Laplace transform converts a system of linear differential equations with constant coefficients to a system of simultaneous algebraic equations. Then we can solve the algebraic equations for the transformed solution functions. Finally, applying the inverse transform to these functions gives us the solutions of the original system of linear ODEs. However neat this sounds conceptually, the algebraic details are often quite messy, and technology comes in handy.

Despite the overwhelming emphasis on the Laplace transform as a tool for obtaining exact, closed-form solutions, it turns out that the transform can provide insight into the *qualitative* nature of a solution as well. In certain applied areas, when we are considering the important second-order equation $c_2 x'' + c_1 x' + c_0 x = f(t)$, the ratio $\dfrac{X(s)}{F(s)} = \dfrac{1}{c_2 s^2 + c_1 s + c_0}$, where $F(s) = \mathcal{L}[f(t)]$ and $X(s) = \mathcal{L}[x(t)]$, is called the **transfer function** of the system modeled by the equation with all initial values zero. The inverse Laplace transform of the transfer function is called the **impulse response function** for the system and describes the solution (and the system) in a way that is similar to the qualitative techniques used in Chapter 5. The analysis of this transfer function provides a picture of what can be called the **response** of the system. The values of s that make the denominator of the transfer function zero are called **poles** or **singularities** of the transfer function. On the basis of the analysis of eigenvalues in Chapter 5 and the discussion in this section, you should see that the algebraic nature of the poles determines the behavior of the system.

PROJECT 6-1

Current Affairs

The current, i, flowing through the capacitor in the accompanying diagram is modeled by the integro-differential equation

$$L\frac{di}{dt} + Ri + \frac{1}{C}\int_0^t i(r)\,dr = E(t)$$

with $i(0) = 0$ and $\dfrac{di}{dt}(0) = 0$. The voltage is supplied by a battery with an electro-motive force (emf) of E_0.

Suppose that the switch is open initially ($t = 0$). At time $t = 1$, the switch is closed and remains closed until time $t = 2$, when the switch is opened again. Assume that $E_0 = 100$ volts, $L = 1$ henry, $R = 30$ ohms, and $C = 0.005$ farad. Determine a formula for the current $i(t)$ by going through steps (a)–(e).

a. Express $E(t)$ in terms of the unit step function $U(t)$.

b. Show that $\mathcal{L}\left[\int_0^t f(r)\,dr\right] = \dfrac{1}{s}F(s)$, where $F(s) = \mathcal{L}[f(t)]$. (If you are not familiar with double integrals, skip this part and just use the result in subsequent parts.)

c. Apply the Laplace transform to the integro-differential equation and solve for $I(s) = \mathcal{L}[i(t)]$.

d. Write the expression for $I(s)$ found in part (c) in terms of partial fractions.

e. Determine $i(t) = \mathcal{L}^{-1}[I(s)]$. [*Hint:* Use formula (6.3.1a).]

f. Graph the function $i(t)$ found in part (d).

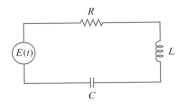

7 | Systems of Nonlinear Differential Equations

7.0 INTRODUCTION

We have discussed various nonlinear equations throughout previous chapters, especially in Chapters 2 and 3, treating them numerically, graphically, and analytically. In general, we can't expect to find the explicit (closed-form) solution of a nonlinear equation, so we are forced to rely on qualitative and computational methods rather than on purely analytical techniques. This complexity is magnified when we address *systems* of nonlinear equations.

In Chapter 5 we analyzed the *stability* of systems of linear differential equations—that is, the behavior of such systems near equilibrium points—and saw that this stability could be described completely in terms of the eigenvalues and eigenvectors of the system. This kind of analysis can be done for nonlinear systems, but it is not quite so satisfactory and complete. One way of carrying out this study is to examine how closely we can approximate (in some sense) a nonlinear system by a linear system and then apply the linear theory.

The modern qualitative theory of stability discussed in Chapter 5 and in this chapter originated in the late 1800s with the work of the French mathematician Henri Poincaré (1854–1912), who was studying nothing less than whether the solar system was a stable system. The equations involved in Poincaré's study of celestial mechanics could not be solved explicitly, so he and others developed implicit (qualitative) methods to deal with the complicated problems of planetary motion.

7.1 EQUILIBRIA OF NONLINEAR SYSTEMS

Recall that an *equilibrium point* of a differential equation or a system of differential equations is a constant solution. If we look at the two (somewhat similar) equations (1) $y' = -y$ and (2) $y' = -y(1 - y)$, we will see some important differences between linear and nonlinear equations.

Equation (1) is linear but, more fundamentally, *separable*, so it is easy to find the general solution: $y = Ce^{-t}$, where C is an arbitrary constant. (We recognize that $C = y(0)$, the initial state of the system being modeled by the equation.)

Now equation (2) is nonlinear and separable, and its general solution is

$$y = \frac{Ce^{-t}}{1 - C + Ce^{-t}},$$ where $C = y(0)$. (*Verify the solutions to both equations.*)

Let's examine some typical solution curves for equation (1). Figure 7.1 shows that there is only one equilibrium solution, $y \equiv 0$, and this is a *sink*. (Review Section 2.5 if necessary.) If an object described by the equation starts off at zero (that is, if $C = 0$), it remains at zero for all time. If the object's initial state is not zero, then the object will approach the solution $y \equiv 0$ as its asymptotically stable solution (or sink).

On the other hand, Figure 7.2 shows the same kind of information for equation (2). For such a nonlinear equation there can be more than one equilibrium solution, in this case $y \equiv 0$ and $y \equiv 1$. Also note that some solutions of a nonlinear equation may "blow up in finite time"—that is, become unbounded as t approaches some finite value. For our equation, if $y(0) = C$, where $C > 1$, then there is a vertical asymptote at $t = \ln\left(\dfrac{C}{C - 1}\right)$. [Where does the denominator of the general solution to equation (2) vanish?] In contrast, all solutions of a linear

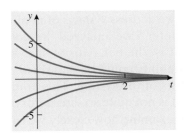

Figure 7.1

Solutions of $y' = -y$; $y(0) = 5, 3, 1, -2, -4$

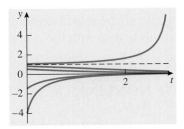

Figure 7.2

Solutions of $y' = -y(1 - y)$

equation or a system of linear equations are defined for all values of the independent variable. Finally, looking closely at the behavior of solutions with different initial values, we see that the solutions starting off above 1 behave differently from those solutions with initial values less than 1. The equilibrium solution $y \equiv 0$ is a *sink* if $y(0) < 1$ and $y \equiv 1$ is a *source* if $y(0) > 1$. Furthermore, for solutions with initial values C greater than 1, the line $t = \ln\left(\dfrac{C}{C-1}\right)$ is a vertical asymptote.

The last three types of behavior cannot occur when we are dealing with a linear equation. You should expect that the situation with nonlinear *systems* is appropriately complicated.

Let's look at an example of a nonlinear system and its behavior near its equilibrium points.

EXAMPLE 7.1.1   Stability of a Nonlinear System

The nonlinear system

$$x' = x - x^2 - xy$$
$$y' = -y - y^2 + 2xy$$

represents two populations interacting in a predator-prey relationship. This is essentially a Lotka-Volterra system (See Section 4.5, especially Example 4.5.4) with "crowding" terms (the squared terms) added for both species.

To calculate the equilibrium points of this system, we solve the system $\{x' = 0, y' = 0\}$, which is the same as the nonlinear algebraic system

(A) $x(1 - x - y) = 0$
(B) $y(-1 - y + 2x) = 0.$

Clearly the origin, $x = y = 0$, is an equilibrium point. Logically, there are only three other cases to examine: (1) $x = 0$, $y \neq 0$; (2) $x \neq 0$, $y = 0$; and (3) $x \neq 0$, $y \neq 0$. Assuming case 1, we can eliminate equation (A) and examine (B), which becomes $y(-1 - y) = 0$. Because $y \neq 0$, we conclude that $-1 - y = 0$, or $y = -1$. Thus our second equilibrium point is $(0, -1)$. Moving to case 2, we can ignore equation (B) and focus on (A), which now looks like $x(1 - x) = 0$. Because we are assuming in case 2 that $x \neq 0$, we can see that $x = 1$, which gives us the third equilibrium point $(1, 0)$. Finally, if $x \neq 0$ and $y \neq 0$, our system of algebraic equations becomes

(A2) $x + y = 1$
(B2) $y - 2x = -1.$

(We have divided out x and y in (A) and (B) and then rearranged the terms of each equation.) Subtracting (B2) from (A2) gives us $3x = 2$, or $x = \frac{2}{3}$. Substituting this value of x in (A2) yields $y = \frac{1}{3}$. Therefore, the last equilibrium point is $(\frac{2}{3}, \frac{1}{3})$.

In terms of a population problem, the only interesting equilibrium point is the last one we found. (*Why is this so?*) If we look at a slope field for the original system of nonlinear differential equations near the point $(\frac{2}{3}, \frac{1}{3})$, we see some interesting behavior (Figure 7.3a).

The apparent spiraling of solutions into the equilibrium point can be seen more clearly if we show some (numerically generated) solution curves (Figure

7.3b). Figure 7.3b represents a predator-prey population that is stabilizing. If the units are thousands of creatures, then the X population is heading for a steady population of about 667, whereas the Y population has 333 as its stable value.

Mathematically, however, we should look at the entire phase portrait to understand the complex behavior of nonlinear systems. We'll return for a detailed analysis in Example 7.2.3.

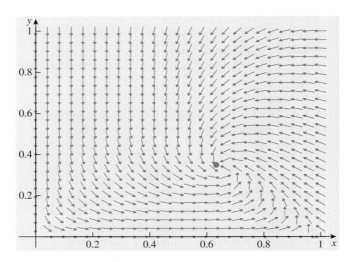

Figure 7.3a

Slope field for $x' = x - x^2 - xy$, $y' = -y - y^2 + 2xy$ near $\left(\dfrac{2}{3}, \dfrac{1}{3}\right)$

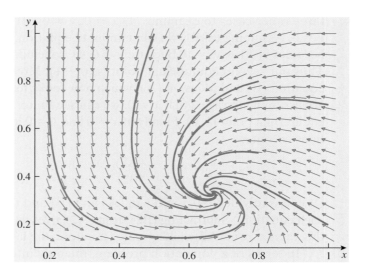

Figure 7.3b

Phase portrait for $x' = x - x^2 - xy$, $y' = -y - y^2 + 2xy$ near $\left(\dfrac{2}{3}, \dfrac{1}{3}\right)$

$(x(0), y(0)) = (0.2, 1), (0.8, 0.8), (0.8, 0.5), (1, 0.7), (1, 0.2), (0.5, 1)$ ◆

EXERCISES 7.1

Find all equilibrium points for each of the systems in Exercises 1–10, using technology if necessary.

1. $x' = -x + xy$, $y' = -y + 2xy$ **2.** $x' = x - xy$, $y' = y - xy$

3. $x' = x^2 - y^2$, $y' = x - xy$ **4.** $x' = 1 - y^2$, $y' = 1 - x^2$

5. $x' = x + y + 2xy$, $y' = -2x + y + y^3$

6. $x' = y(1 - x^2)$, $y' = -x(1 - y^2)$

7. $x' = x - x^2 - xy$, $y' = 3y - xy - 2y^2$

8. $x' = 1 - y$, $y' = x^2 - y^2$

9. $x' = (1 + x) \sin y$, $y' = 1 - x - \cos y$ [*Hint:* Graph the two equations on the same set of axes.]

10. $x' = 3y - e^x$, $y' = 2x - y$ [*Hint:* There are two equilibrium points. Use your CAS to approximate these points.]

11. A two-mode laser produces two different kinds of photons, whose numbers are n_1 and n_2. The equations modeling the rates of photon production are

$$\dot{n}_1 = G_1 N n_1 - k_1 n_1$$
$$\dot{n}_2 = G_2 N n_2 - k_2 n_2,$$

where $N(t) = N_0 - a_1 n_1 - a_2 n_2$ is the number of excited atoms. The parameters G_1, G_2, k_1, k_2, a_1, a_2, and N_0 are all positive. Use a CAS "solve" command to find all equilibrium points of the system.

12. A *chemostat* is a device for growing and studying bacteria by supplying nutrients and maintaining convenient levels of the bacteria in a culture. (See Project 2-2 at the end of Chapter 2.) One model of a chemostat is the nonlinear system

$$\frac{dN}{dt} = a_1 \left(\frac{C}{1 + C} \right) N - N$$
$$\frac{dC}{dt} = -\left(\frac{C}{1 + C} \right) N - C + a_2,$$

where $N(t)$ denotes the bacterial density at time t, $C(t)$ denotes the concentration of nutrient, and a_1, a_2 are positive parameters. Use technology to find all equilibrium solutions (N^*, C^*) of the system.

13. In the absence of damping and any external force, the motion of a pendulum is described by the equation $\dfrac{d^2\theta}{dt^2} + \dfrac{g}{L} \sin\theta = 0$, where θ is the angle between the pendulum and the downward vertical, g is the acceleration due to gravity, and L is the length of the pendulum.

a. Write this equation as a system of two first-order equations.

b. Describe all equilibrium points of the system.

7.2   LINEAR APPROXIMATION AT EQUILIBRIUM POINTS

One important aspect of *linear* systems is that the behavior of solutions near an equilibrium point ("local" behavior) tells you the behavior of solutions in the entire phase plane. However, even though most of the "nice" properties of linear systems are not present when we analyze nonlinear systems, we may be able to understand the local behavior of nonlinear systems by a process of *linearization* or *linear approximation*. This means that we try to replace the original nonlinear system by a linear system that is "close" or near an equilibrium point. Remember that in Section 3.1 we first discussed *Euler's method*, which involved approximating solution curves by tangent lines.

To see how this might work, let's go back to the nonlinear equation $y' = -y(1 - y) = -y + y^2$ discussed in Section 7.1. We know that $y = 0$ is an equilibrium point. Now note that for values of y close to zero, y^2 is smaller than y. For example, if $y = 0.00001$, then $y^2 = 0.0000000001$. Then, dropping the squared (nonlinear) terms, we can guess that the linear equation $y' = -y$ is a good approximation for the original equation and that the behavior of this last equation near $y = 0$ should tell us how $y' = -y(1 - y)$ behaves near $y = 0$. A comparison of the solution curves near $y = 0$ in Figure 7.1 and Figure 7.2 shows us that this is true. However, it should also be clear that we would be wrong to base our analysis of $y' = -y(1 - y)$ on $y' = -y$ for *all* initial values.

If we want to analyze the behavior of $y' = -y(1 - y)$ near its other equilibrium point, $y = 1$, we can use a simple change of variable: Let $y = 1 + z$, so that studying the behavior of $y' = -y(1 - y)$ near $y = 1$ is the same as analyzing the behavior of the equation near $z = 0$. (*Make sure you see this.*) With this change of variable, we get the new equation $z' = -y(1 - y) = (-1 - z)(-z) = z + z^2$. Using the same reasoning as before, we can take $z' = z$ as a good linear approximation near $z = 0$. This last equation has the general solution $z = Ce^t$, so solutions of $z' = z + z^2$ move *away from* $z = 0$ as t increases. But because $y = 1 + z$, solutions of $y' = -y(1 - y)$ near $y = 1$ curve away from $y = 1$, behavior we can verify by looking at Figure 7.2.

As another example, take the second-order nonlinear equation $\dfrac{d^2x}{dt^2} + \dfrac{g}{L}\sin x = 0$, which describes the swinging of a pendulum (where x is the angle the pendulum makes with the vertical, g is the acceleration due to gravity, and L is the pendulum's length). This equation is not easy to deal with analytically, so what is often done is to remove the nonlinearity by a substitution. For small values of x (that is, for an oscillation of small amplitude), $\sin x \approx x$, so we can replace our original nonlinear equation by the *linear* equation that approximates it: $\dfrac{d^2x}{dt^2} + \dfrac{g}{L}x = 0$. This approximate pendulum model has the same mathematical behavior as the undamped spring-mass system; see equation (4.5.4) in Section 4.5. Despite our success in approximating a nonlinear equation by one that is linear, this is a limited victory. For example, the analysis of the linear approximations implies that all solutions are defined for all values of t, but this is clearly not the case for the nonlinear equation. The next example illustrates the failure of linearization more dramatically.

EXAMPLE 7.2.1   Linearization Can Mislead

Let's look at the system

$$\dot{x} = y + ax(x^2 + y^2)$$
$$\dot{y} = -x + ay(x^2 + y^2),$$

where a is a given real number.

Clearly the origin $(x, y) = (0, 0)$ is an equilibrium point regardless of the value of the parameter a. In our example the obvious linearized system is

$$\dot{x} = y$$
$$\dot{y} = -x$$

(Look back at the spring-mass system analyzed in Example 4.5.5.) This can be written in the form $\dot{X} = AX$, where $X = \begin{bmatrix} x \\ y \end{bmatrix}$ and $A = \begin{bmatrix} 0 & 1 \\ -1 & 0 \end{bmatrix}$. The characteristic polynomial is $\lambda^2 + 1 = 0$, so that the eigenvalues of A are purely imaginary: i and $-i$. From Table 5.1 in Section 5.5, we conclude that the origin is a *stable center* of the linearized system. *But this is the wrong conclusion with respect to the original nonlinear system*, as the phase portrait of the original nonlinear system near $(0, 0)$ shows. This portrait (Figure 7.4) corresponds to $a = -1$.

It seems that the trajectories spiral in toward the equilibrium point, indicating that the origin is actually a *spiral sink* for the nonlinear system. However, appearances can be deceiving, and Exercise 1 at the end of this section suggests a way of proving this claim about the origin.

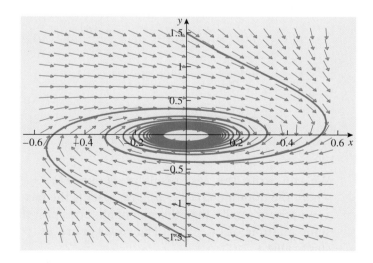

Figure 7.4
Trajectories of $\dot{x} = y - x(x^2 + y^2)$, $\dot{y} = -x - y(x^2 + y^2)$, $0 \le t \le 60$
$(x(0), y(0)) = (0, 1.5), (0, -1.5)$

You should suspect that the stability of the original system depends on the value of the parameter *a*. If $a = 0$, for example, then the nonlinear portion of the system disappears, leaving us with a purely linear system—in fact, the same system analyzed in Example 4.5.5 (with $\beta = 1$). As we've said, the origin is a stable center for this linear system, every trajectory closing perfectly after one cycle. Exercise 1 asks you to explore these ideas further.   ◆

To summarize, we can look at this last example as a linear system "perturbed" (disturbed or knocked off kilter) by a nonlinear component. We can write this system as $\dot{X} = \begin{bmatrix} 0 & 1 \\ -1 & 0 \end{bmatrix} X + \begin{bmatrix} f(x, y) \\ g(x, y) \end{bmatrix}$, where *f* and *g* are nonlinear functions of *x* and *y*. As we'll see later, if the nonlinear perturbation is "nice" enough, the behavior of the whole system can be predicted from the behavior of its linear portion near equilibrium points.

ALMOST LINEAR SYSTEMS

If we want to make all this talk about linear approximation mathematically sound, we have to remind ourselves of some basic calculus facts. (See Appendix A.1 for additional information.) Back in Section 3.1 we discussed *local linearity*, the idea that if we "zoom" in on a point on a curve $y = f(x)$, the curve looks like a straight line—in fact, like a piece of the tangent line drawn to the curve at that point. More precisely, for values of the independent variable *x* close to $x = a$, we can write $f(x) \approx f(a) + f'(a)(x - a)$. You should recognize that this expression consists of the first two terms of an *n*th-degree $(n \geq 1)$ *Taylor polynomial approximation* of *f* near $x = a$—or, equivalently, the first two terms of the *Taylor series* expansion of *f* in a neighborhood of $x = a$:

$$f(x) = f(a) + f'(a)(x - a) + \frac{f''(a)}{2!}(x - a)^2 + \frac{f'''(a)}{3!}(x - a)^3 + \cdots$$
$$+ \frac{f^{(n)}}{n!}(x - a)^n + \cdots$$
$$= f(a) + f'(a)(x - a) + (x - a)^2 \left\{ \frac{f''(a)}{2!} + \frac{f'''(a)}{3!}(x - a) + \cdots \right.$$
$$\left. + \frac{f^{(n)}}{n!}(x - a)^{n-2} + \cdots \right\}.$$

We can write this last result as $f(x) \approx f(a) + f'(a)(x - a) + O((x - a)^2)$, where the notation $O((x - a)^2)$ represents the fact that if *x* is close to *a* (so that $x - a$ is very small), then the sum of all terms past the second will be bounded by some multiple of $(x - a)^2$. (The series in braces, $\{\cdots\}$, converges to some constant value.)

Now assume that we have a general nonlinear autonomous system of the form

$$\dot{x} = F(x, y)$$
$$\dot{y} = G(x, y) \qquad\qquad (7.2.1)$$

which has the origin as an equilibrium point—that is, $F(0, 0) = 0$ and $G(0, 0) = 0$. This last assumption is just for convenience as we develop some methodology. If we can write F as $ax + by + f(x, y)$ and G as $cx + dy + g(x, y)$, where f and g are nonlinear functions, then we can express the system in the form

$$\dot{X} = \begin{bmatrix} a & b \\ c & d \end{bmatrix} X + \begin{bmatrix} f(x, y) \\ g(x, y) \end{bmatrix}.$$

If the nonlinear functions f and g are "small enough" (in a sense to be explained later) that their effect is negligible, then we can call our system "almost linear" and see that near the origin, our nonlinear system behaves essentially like the linear system $\dot{X} = \begin{bmatrix} a & b \\ c & d \end{bmatrix} X$—that is, like the system

$$\dot{x} = ax + by$$
$$\dot{y} = cx + dy.$$

Earlier we recalled that the tangent line $y = f(a) + f'(a)(x - a)$ gives the best linear approximation of a single-variable function f near $x = a$. For $F(x, y)$, a function of *two* variables, the best approximation near a point (a, b) is provided by the *tangent plane* given by the approximation formula

$$F(x, y) \approx F(a, b) + \frac{\partial F}{\partial x}(a, b)(x - a) + \frac{\partial F}{\partial y}(a, b)(y - b), \qquad (7.2.2)$$

where $\frac{\partial F}{\partial x}(a, b)$ and $\frac{\partial F}{\partial y}(a, b)$ denote the partial derivatives of F evaluated at the point (a, b). For example, if we want to approximate $F(x, y) = x^3 + y^3$ near the point $(1, 1)$, we calculate

$$F(1, 1) = 1^3 + 1^3 = 2$$
$$\frac{\partial F}{\partial x} = 3x^2, \quad \frac{\partial F}{\partial x}(1, 1) = 3(1)^2 = 3$$
$$\frac{\partial F}{\partial y} = 3y^2, \quad \frac{\partial F}{\partial y}(1, 1) = 3(1)^2 = 3,$$

so the equation of the tangent plane is $z = 2 + 3(x - 1) + 3(y - 1)$.

You should think of the right-hand side of equation (7.2.2) as the first-degree Taylor polynomial approximation of F, the linear terms in x and y of the two-variable Taylor series expansion of F. This approximation ignores the rest of the series consisting of the terms in x and y of the second degree and higher, which we can denote by $f(x, y)$. Thus, in our last example, we can write $x^3 + y^3 \approx 2 + 3(x - 1) + 3(y - 1)$ near $(1, 1)$ or $x^3 + y^3 = 2 + 3(x - 1) + 3(y - 1) + f(x, y)$ near $(1, 1)$. (See Appendix A.8 for more information on this.)

If we choose the point (a, b) to be the origin, then we can rewrite (7.2.1) as

$$\dot{x} = F(0, 0) + \frac{\partial F}{\partial x}(0, 0)x + \frac{\partial F}{\partial y}(0, 0)y + f(x, y)$$
$$\dot{y} = G(0, 0) + \frac{\partial G}{\partial x}(0, 0)x + \frac{\partial G}{\partial y}(0, 0)y + g(x, y)$$

or (remembering that we have assumed $F(0, 0) = G(0, 0) = 0$), as

$$\dot{x} = ax + by + f(x, y)$$
$$\dot{y} = cx + dy + g(x, y), \qquad (7.2.3)$$

where $a = \dfrac{\partial F}{\partial x}(0, 0)$, $b = \dfrac{\partial F}{\partial y}(0, 0)$, $c = \dfrac{\partial G}{\partial x}(0, 0)$, and $d = \dfrac{\partial G}{\partial y}(0, 0)$.

The technical definition of the "smallness" of f and g near the origin is that

$$\lim_{(x, y) \to (0, 0)} \left(\frac{f(x, y)}{\sqrt{x^2 + y^2}} \right) = \lim_{(x, y) \to (0, 0)} \left(\frac{g(x, y)}{\sqrt{x^2 + y^2}} \right) = 0. \qquad (7.2.4)$$

The limits in (7.2.4) just say that near the origin, f and g are small in comparison to $r = \sqrt{x^2 + y^2}$, which is the radial distance of the point (x, y) from the origin.

Now we can define an **almost linear system** as a nonlinear system (7.2.3) that satisfies (7.2.4). In this situation, the linear part

$$\dot{x} = ax + by$$
$$\dot{y} = cx + dy \qquad (7.2.5)$$

is called the **associated linear system** or **linear approximation** about the equilibrium point $(0, 0)$.

EXAMPLE 7.2.2   A Linear Approximation

Let's examine the behavior of the following system near the origin:

$$\dot{x} = x + 2y + x \cos y$$
$$\dot{y} = -y - \sin y.$$

First of all, we can see that $(0, 0)$ is an equilibrium point for the system. Now we must find the associated linear system, which is not obvious because $x \cos y$ and $-\sin y$ actually contain linear terms that must be combined with the linear terms already visible in the original system.

Substituting the Taylor (or Maclaurin) expansions for $\cos y$ and $\sin y$ in the given equations and collecting terms, we have

$$\dot{x} = x + 2y + x\left(1 - \frac{y^2}{2!} + \frac{y^4}{4!} - \cdots\right) = 2x + 2y + x\left(-\frac{y^2}{2!} + \frac{y^4}{4!} - \cdots\right)$$
$$\dot{y} = -y - \left(y - \frac{y^3}{3!} + \frac{y^5}{5!} - \cdots\right) = -2y - \left(-\frac{y^3}{3!} + \frac{y^5}{5!} - \cdots\right).$$

Thus the associated linear system is

$$\dot{x} = 2x + 2y$$
$$\dot{y} = -2y,$$

or $\dot{X} = \begin{bmatrix} 2 & 2 \\ 0 & -2 \end{bmatrix} X = AX$. The characteristic equation of this linear system is given by $\lambda^2 - 4 = 0$, so the eigenvalues are $\lambda = -2$ and $\lambda = 2$.

Table 5.1 in Section 5.5 tells us that two real eigenvalues opposite in sign indicate that we have a *saddle point*. Figure 7.5a shows the slope field for the system; Figure 7.5b shows some trajectories for the nonlinear system around the origin. Figure 7.5c shows trajectories around the origin for the associated linear system.

We can see from these phase portraits that the linear approximation captures the behavior of the nonlinear system near the origin.

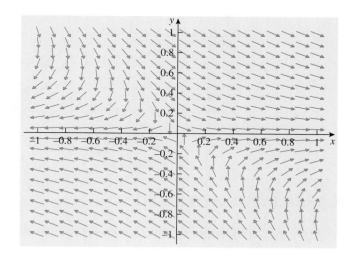

Figure 7.5a
Slope field for $\dot{x} = x + 2y + x\cos y$, $\dot{y} = -y - \sin y$

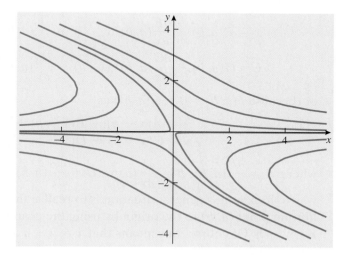

Figure 7.5b
Trajectories for $\dot{x} = x + 2y + x\cos y$, $\dot{y} = -y - \sin y$

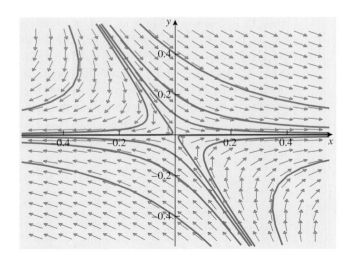

Figure 7.5c
Trajectories for $\dot{x} = 2x + 2y$, $\dot{y} = -2y$ ◆

THE POINCARÉ-LIAPUNOV THEOREM

More generally, suppose that (a, b) is an equilibrium point for the system

$$\dot{x} = F(x, y)$$
$$\dot{y} = G(x, y),$$

which means that $F(a, b) = 0 = G(a, b)$. Using the tangent plane approximation formula (7.2.2), we can rewrite this system as

$$\dot{x} = F(a, b) + \frac{\partial F}{\partial x}(a, b)(x - a) + \frac{\partial F}{\partial y}(a, b)(y - b) + f(x, y)$$

$$\dot{y} = G(a, b) + \frac{\partial G}{\partial x}(a, b)(x - a) + \frac{\partial G}{\partial y}(a, b)(y - b) + g(x, y)$$

or (because $F(a, b) = 0 = G(a, b)$) as

$$\dot{x} = A(x - a) + B(y - b) + f(x, y)$$
$$\dot{y} = C(x - a) + D(y - b) + g(x, y), \qquad (7.2.6)$$

where $A = \dfrac{\partial F}{\partial x}(a, b)$, $B = \dfrac{\partial F}{\partial y}(a, b)$, $C = \dfrac{\partial G}{\partial x}(a, b)$, and $D = \dfrac{\partial G}{\partial y}(a, b)$. Another way to look at this general situation is to realize that we are translating the equilibrium point (a, b) to the origin by using the change of variables $u = x - a$ and $v = y - b$. Of course, this means that $x = u + a$ and $y = v + b$, so that we can rewrite (7.2.6) as

$$\dot{u} = Au + Bv + f(u, v)$$
$$\dot{v} = Cu + Dv + g(u, v),$$

which has $(0, 0)$ as an equilibrium point. Note that this says that any equilibrium point $(a^*, b^*) \neq (0, 0)$ can be transformed to the origin for the purpose of analyzing the stability of the system. Therefore, we can state an important stability result for nonlinear systems in terms of an equilibrium point at the origin:

Suppose we have the nonlinear autonomous system

$$\begin{aligned} \dot{x} &= ax + by + f(x, y) \\ \dot{y} &= cx + dy + g(x, y), \end{aligned} \qquad (7.2.7)$$

where $ad - bc \neq 0$, $\displaystyle\lim_{(x, y) \to (0, 0)} \left(\frac{f(x, y)}{\sqrt{x^2 + y^2}} \right) = \lim_{(x, y) \to (0, 0)} \left(\frac{g(x, y)}{\sqrt{x^2 + y^2}} \right) = 0$, and the origin is an equilibrium point. If λ_1 and λ_2 are the eigenvalues of the associated linear system

$$\begin{aligned} \dot{x} &= ax + by \\ \dot{y} &= cx + dy, \end{aligned} \qquad (7.2.8)$$

then the equilibrium points of the two systems, (7.2.7) and (7.2.8), are related as follows:

(A) *If the eigenvalues λ_1 and λ_2 are **not** equal real numbers or are **not** pure imaginary numbers, then the trajectories of the almost linear system (7.2.7) near the equilibrium point $(0, 0)$ behave the same way as the trajectories of the associated linear system (7.2.8) near the origin. That is, we can use the appropriate table entries given in Section 5.5 to determine whether the origin is a node, a saddle point, or a spiral point of both systems.*

(B) *If λ_1 and λ_2 are real and equal, then the origin is either a node or a spiral point of both systems. Furthermore, if $\lambda_1 = \lambda_2 < 0$, then the origin is asymptotically stable; and if $\lambda_1 = \lambda_2 > 0$, then the origin is an unstable equilibrium point.*

(C) *If λ_1 and λ_2 are pure imaginary numbers, then the equilibrium point $(0, 0)$ is either a center or a spiral point of the nonlinear system. Also, this spiral point may be asymptotically stable, stable, or unstable.*

This important result was discovered by Poincaré and the Russian mathematician A. M. Liapunov (1857–1918). The next example shows how to use the Poincaré-Liapunov theorem.

EXAMPLE 7.2.3   An Application of the Poincaré-Liapunov Theorem

Let's return to the system in Example 7.1.1:

$$\begin{aligned} x' &= x - x^2 - xy \\ y' &= -y - y^2 + 2xy. \end{aligned}$$

We saw that there were four equilibrium points: $(0, 0)$, $(0, -1)$, $(1, 0)$, and $(\frac{2}{3}, \frac{1}{3})$.

Near the origin, because the terms x^2, y^2, and xy are smaller than the terms x and y, we can replace the nonlinear system by its associated linear system

$$x' = x$$
$$y' = -y.$$

The eigenvalues of this linear system are -1 and 1. According to part (A) of the Poincaré-Liapunov result, the trajectories of the nonlinear system should behave the same way as the trajectories of this associated linear system. Table 5.1 in Section 5.5 tells us that the origin is a *saddle point* for both systems.

If we want to examine what happens near the equilibrium point $(0, -1)$, we make the change of variables $u = x - 0 = x$ and $v = y - (-1) = y + 1$ so that we can rewrite the original system as

$$u' = x' = u - u^2 - u(v - 1) = 2u - u^2 - uv$$
$$v' = y' = -(v - 1) - (v - 1)^2 + 2u(v - 1) = -2u + v - v^2 + 2uv.$$

Then the associated linear system is

$$u' = 2u$$
$$v' = -2u + v,$$

with eigenvalues 1 and 2. (*Check this.*) Now result (A) and the table in Section 5.5 tell us that the equilibrium point $(0, -1)$ is a *source* for the nonlinear system.

The equilibrium point $(1, 0)$ leads us to make the change of variables $u = x - 1$ and $v = y - 0 = y$, so that the nonlinear system is transformed into

$$u' = -u - v - u^2 - uv$$
$$v' = v - v^2 + 2uv,$$

with associated linear system

$$u' = -u - v$$
$$v' = v.$$

Be sure to check the details for yourself. The eigenvalues for this last system are -1 and 1, so that $(1, 0)$ is a *saddle point* for both the nonlinear system and its associated linear system.

Finally, we look at the equilibrium point $(\frac{2}{3}, \frac{1}{3})$. The transformation $u = x - \frac{2}{3}$, $v = y - \frac{1}{3}$ leads to the system

$$u' = \frac{-2u}{3} - \frac{2v}{3} - u^2 - uv$$

$$v' = \frac{2u}{3} - \frac{v}{3} - v^2 + 2uv.$$

The linear approximation is given by

$$u' = \frac{-2u}{3} - \frac{2v}{3}$$

$$v' = \frac{2u}{3} - \frac{v}{3},$$

which has eigenvalues $-\dfrac{1}{2} + \dfrac{\sqrt{15}}{6}i$ and $-\dfrac{1}{2} - \dfrac{\sqrt{15}}{6}i$. Therefore, from result (A) and Table 5.1, we know that $(\frac{2}{3}, \frac{1}{3})$ is a *spiral sink*. Look back at Figures 7.3a and 7.3b to see this clearly.

Figure 7.6a shows some trajectories near the origin. Figure 7.6b illustrates the behavior of the system near the equilibrium point $(0, -1)$, a source.

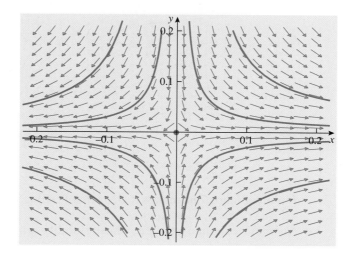

Figure 7.6a
Trajectories of $x' = x - x^2 - xy$, $y' = -y - y^2 + 2xy$ near the origin

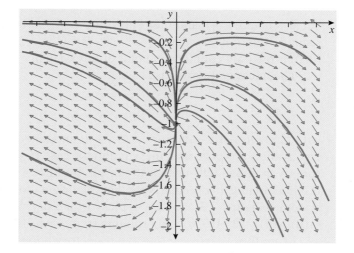

Figure 7.6b
Trajectories of $x' = x - x^2 - xy$, $y' = -y - y^2 + 2xy$ near $(0, -1)$

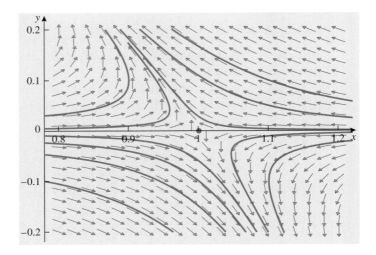

Figure 7.6c
Trajectories of $x' = x - x^2 - xy$, $y' = -y - y^2 + 2xy$ near $(1, 0)$

Finally, Figure 7.6c makes it clear that $(1, 0)$ is indeed a saddle point. ◆

Now let's examine a system whose stability is not so clear.

EXAMPLE 7.2.4   Another Application of Poincaré-Liapunov
The system we'll investigate is

$$\dot{x} = -x^3 - y$$
$$\dot{y} = x - y^3.$$

Set $\dot{x} = 0$ and $\dot{y} = 0$ and then substitute $y = -x^3$ from the first equation into the second equation. We get $x + x^9 = 0$, or $x(1 + x^8) = 0$, so $x = 0$. It follows that $(0, 0)$ is the only equilibrium point of this system.

The linearized system is

$$\dot{x} = -y$$
$$\dot{y} = x,$$

with characteristic equation $\lambda^2 + 1 = 0$ and eigenvalues $-i$ and i. Because the eigenvalues are pure imaginary numbers, case (C) of the Poincaré-Liapunov result given above tells us that the origin is either a center or a spiral point of the original nonlinear system. (Note that the origin is a *center* of the associated linear system.) Figure 7.7 shows a typical trajectory, in this case with initial state $(x(0), y(0)) = (-0.5, 0)$ and t running from -9 to 100.

From this we can see that the trajectory appears to spiral in toward the origin—that is, the equilibrium point is *asymptotically stable*. We could have seen this analytically by defining the function

$$d(t) = \sqrt{x^2(t) + y^2(t)},$$

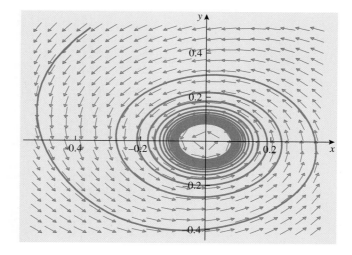

Figure 7.7
Trajectories of $\dot{x} = -x^3 - y$, $\dot{y} = x - y^3$ near the origin

which gives the distance from any point $(x(t), y(t))$ on a trajectory to $(0, 0)$. Differentiating this function and then substituting from our original equations, we get

$$\dot{d}(t) = \frac{1}{2}[x^2(t) + y^2(t)]^{-1/2}(2x(t)\dot{x}(t) + 2y(t)\dot{y}(t))$$

$$= \frac{x(t)(-x^3(t) - y(t)) + y(t)(x(t) - y^3(t))}{\sqrt{x^2(t) + y^2(t)}} = -\frac{x^4(t) + y^4(t)}{\sqrt{x^2(t) + y^2(t)}} < 0.$$

This says that the distance between points on the trajectory and the origin is *decreasing* with time—that is, the trajectory is always moving closer and closer to the origin. ◆

The next example shows another type of behavior.

EXAMPLE 7.2.5 **Yet Another Application of Poincaré-Liapunov**
The system

$$\dot{x} = 2x - 6x^2y$$
$$\dot{y} = 2y + x$$

has the origin as its only equilibrium point. *Check this for yourself.* The linearization of this system is

$$\dot{x} = 2x$$
$$\dot{y} = 2y + x,$$

which has the characteristic equation $\lambda^2 - 4\lambda + 4 = (\lambda - 2)^2 = 0$. Because the eigenvalues are positive and equal, we use result (B) to conclude that the origin is an *unstable equilibrium point, a source.* Figure 7.8 shows this.

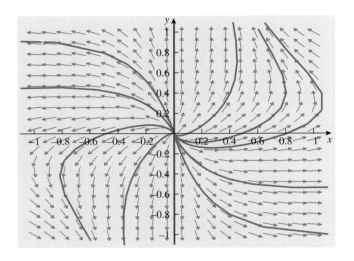

Figure 7.8
Trajectories of $\dot{x} = 2x - 6x^2y$, $\dot{y} = 2y + x$ near the origin   ◆

EXERCISES 7.2

1. Let's return to the system in Example 7.2.1:

$$\dot{x} = y + ax(x^2 + y^2)$$
$$\dot{y} = -x + ay(x^2 + y^2).$$

   a. Introduce polar coordinates defined by $x = r(t)\cos\theta(t)$, $y = r(t)\sin\theta(t)$. Note that $x^2 + y^2 = r^2$ and use the Chain Rule to show that $x\dot{x} + y\dot{y} = r\dot{r}$.

   b. In the expression for $r\dot{r}$ found in part (a), substitute for $\dot{x}$ and $\dot{y}$ using the equations in the system, and show that $\dot{r} = ar^3$ for $r > 0$.

   c. Show that $\theta = \arctan(y/x)$ and that $\dot{\theta} = \dfrac{x\dot{y} - y\dot{x}}{r^2}$. Substitute for $\dot{x}$ and $\dot{y}$ in this last formula to see that $\dot{\theta} = -1$.

   d. The results of parts (b) and (c) show that our original system is equivalent to the system $\{\dot{r} = ar^3, \dot{\theta} = -1\}$. The second equation says that all trajectories rotate around the origin with constant angular velocity 1. Recognizing that the first equation describes the radial distance from the origin to a point on the trajectory (see the function $d(t)$ introduced in Example 7.2.4), examine what happens to $r(t)$ as $t \to \infty$ in the three cases $a < 0$, $a = 0$, and $a > 0$. What does this say about the nature of the equilibrium point at the origin? Sketch a trajectory (in the *x-y* plane) for each of the three cases.

   In Exercises 2–11, (a) verify that $(0, 0)$ *is an equilibrium point, (b) show that the system is almost linear, and (c) discuss the type and stability of the origin by examining the associated linear system.*

2. $x' = 3x + y + xy$, $y' = 2x + 2y - 2xy^2$
3. $x' = x - y + x^2$, $y' = x + y$
4. $x' = x - xy - 8x^2$, $y' = -y + xy$
5. $x' = -4x + y - xy^3$, $y' = x - 2y + 3x^2$
6. $x' = 3 \sin x + y$, $y' = 4x + \cos y - 1$
7. $x' = x - y$, $y' = 1 - e^x$
8. $x' = -3x - y - xy$, $y' = 5x + y + xy^3$
9. $x' = y(1 - x^2)$, $y' = -x(1 - y^2)$
10. $x' = -x + x^3$, $y' = -2y$
11. $x' = -2x + 3y + xy$, $y' = -x + y - 2xy^2$

12. The *Brusselator* is a simple model of a hypothetical chemical oscillator that first appeared in a 1968 paper by Belgian scientists I. Prigogine (a Nobel laureate) and R. Lefever and was named after their home country's capital. One version of the model is

$$\dot{x} = 1 - (a + 1)x + bx^2y$$
$$\dot{y} = ax - bx^2y,$$

where x and y are concentrations of chemicals and a and b are positive parameters.

   a. Use technology, if necessary, to find the only equilibrium solution of this system.
   b. Linearize the system about the equilibrium point found in part (a).
   c. Find the eigenvalues of the associated linear system. (Technology could be useful here.)
   d. Using your answers from part (c) and the Poincaré-Liapunov result, discuss the nature of the equilibrium solutions for the following cases: (1) $a = 3, b = 1$; (2) $a = 2, b = 7$; (3) $a = 1, b = 4$.

13. A woman rows a boat across a river a units wide occupying the strip $0 \le x \le a$ in the x-y plane, always rowing toward a fixed point on one bank, say $(0, 0)$. She rows at a constant speed u relative to the water, and the river flows at a constant speed v. The situation can be modeled by the equations

$$\dot{x} = -\frac{ux}{\sqrt{x^2 + y^2}}, \quad \dot{y} = v - \frac{uy}{\sqrt{x^2 + y^2}},$$

where (x, y) describes the coordinates of the boat.

   a. Use technology to sketch the phase portrait of the system for $u > v$. (Pick some reasonable values of u and v.) What kind of equilibrium point is the origin? What happens to the boat?
   b. Use technology to sketch the phase portrait of the system for $u < v$. [Just reverse the values of u and v used in part (a).] What kind of equilibrium point is the origin? What happens to the boat?

7.3 TWO IMPORTANT EXAMPLES OF NONLINEAR EQUATIONS AND SYSTEMS

Now that we know something about the behavior of nonlinear systems, we can apply this knowledge to the analysis of some important nonlinear equations and systems of nonlinear equations.

THE LOTKA-VOLTERRA EQUATIONS

As we saw in the discussion preceding Example 4.5.4, the nonlinear *Lotka-Volterra equations* describe a wide class of problems in mathematical ecology and cannot in general be solved in closed form. Now we will look at this system from the Poincaré-Liapunov point of view.

EXAMPLE 7.3.1   The Lotka-Volterra Equations Revisited
The Lotka-Volterra equations are

$$\dot{x} = ax - bxy$$
$$\dot{y} = -cy + dxy,$$

where $a, b, c,$ and d are positive constants.

The equilibrium points for this system are solutions of the algebraic system

$$ax - bxy = x(a - by) = 0$$
$$-cy + dxy = y(-c + dx) = 0.$$

Clearly $x = y = 0$ is a solution—that is, the origin $(0, 0)$ is an equilibrium point. It should also be clear from these last equations that if either x or y is zero, then the other variable must also be zero. Therefore, if there are any other equilibrium points, we must have $x \neq 0$ and $y \neq 0$. In the first algebraic equation, if $x \neq 0$, then we must have $a - by = 0$, so $y = a/b$. From the second equation, we see that if $y \neq 0$, then $-c + dx = 0$, so $x = c/d$. Thus the only equilibrium points for the Lotka-Volterra system are $(0, 0)$ and $(c/d, a/b)$.

Near the origin, we can replace our original system by the associated linear system

$$\dot{x} = ax$$
$$\dot{y} = -cy,$$

which can be written in matrix form as $\dot{X} = AX$, where $A = \begin{bmatrix} a & 0 \\ 0 & -c \end{bmatrix}$ and $X = \begin{bmatrix} x \\ y \end{bmatrix}$. Now the characteristic equation of A is $\lambda^2 + (c - a)\lambda - ac = 0$, so the eigenvalues are a and $-c$. Because the eigenvalues are real and of opposite signs, Table 5.1 in Section 5.5 indicates that the origin is a *saddle point* for the linearized system. The Poincaré-Liapunov result tells us that $(0, 0)$ is also a saddle point for our original nonlinear system.

To study the behavior of the system near the equilibrium point $(c/d, a/b)$, we transform the system by defining $u = x - c/d$ and $v = y - a/b$. Then our original system becomes

$$\dot{u} = a\left(u + \frac{c}{d}\right) - b\left(u + \frac{c}{d}\right)\left(v + \frac{a}{b}\right)$$
$$\dot{v} = -c\left(v + \frac{a}{b}\right) + d\left(u + \frac{c}{d}\right)\left(v + \frac{a}{b}\right),$$

which simplifies to

$$\dot{u} = \left(-\frac{bc}{d}\right)v - buv$$
$$\dot{v} = \left(\frac{ad}{b}\right)u + duv.$$

The associated linear system is given by $\dot{X} = AX$, where $A = \begin{bmatrix} 0 & -\dfrac{bc}{d} \\ \dfrac{ad}{b} & 0 \end{bmatrix}$. The

characteristic polynomial here is $\lambda^2 + ac = 0$, so the eigenvalues are $\lambda_1 = \sqrt{ac}\,i$ and $\lambda_2 = -\sqrt{ac}\,i$. Because we have pure imaginary eigenvalues, part (C) of the Poincaré-Liapunov result tells us that $(c/d, a/b)$ is either a *center* or a *spiral point* for the nonlinear system. (The table in Section 5.5 indicates that $(c/d, a/b)$ is a stable center for the associated linear system, but this doesn't have to be true for our nonlinear system.) Let $a = b = c = d = 1$. Then Figure 7.9a shows the slope field for the nonlinear system near the equilibrium point $(c/d, a/b) = (1, 1)$, and Figure 7.9b depicts some trajectories near $(1, 1)$.

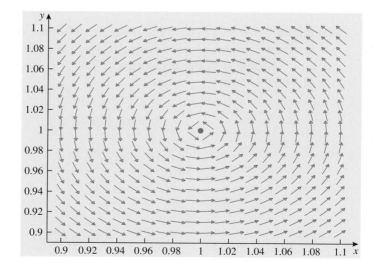

Figure 7.9a

Slope field of $\dot{x} = x - xy$, $\dot{y} = -y + xy$ near $(1, 1)$

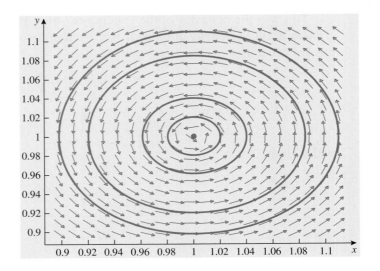

Figure 7.9b
Trajectories of $\dot{x} = x - xy$, $\dot{y} = -y + xy$ near $(1, 1)$

These figures suggest (but do not *prove*) that the equilibrium point $(1, 1)$ is a *stable center* for the nonlinear system. (Exercise 5 proposes some investigations in this direction.)   ◆

THE UNDAMPED PENDULUM

After our ecological field trip, let's return to the world of physics and look at the motion of a simple pendulum. In Section 7.2, we saw that the second-order nonlinear equation $\dfrac{d^2\theta}{dt^2} + \dfrac{g}{L}\sin\theta = 0$ describes the motion of an **undamped pendulum**—that is, a pendulum under the influence of gravity with no friction or air resistance impeding its movement. Here θ is the angle the pendulum makes with the vertical, g is the acceleration due to gravity, and L is the pendulum's length (Figure 7.10).

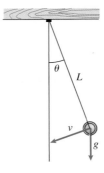

Figure 7.10
The undamped pendulum

EXAMPLE 7.3.2 **The Undamped Pendulum: A Poincaré-Liapunov Analysis**

Letting $x = \theta$ and $y = \dot{\theta} = \dot{x}$, we can express the single equation $\dfrac{d^2\theta}{dt^2} + \dfrac{g}{L}\sin\theta = 0$ as the nonlinear system

$$\dot{x} = y$$

$$\dot{y} = -\frac{g}{L}\sin x.$$

The first thing we have to do is find the equilibrium points of this system. (This was Problem 16 in Exercises 4.5.) Clearly, any equilibrium point (x, y) must have $y = 0$. The equation $-\dfrac{g}{L}\sin x = 0$ has solutions $x = n\pi, n = 0, \pm 1,$ $\pm 2, \ldots$. Thus all points of the form $(n\pi, 0)$ for $n = 0, \pm 1, \pm 2, \ldots$ are equilibrium points for the system describing the pendulum's swing. Because the sine function has period 2π—that is, $\sin(x + 2k\pi) = \sin x$ for any integer k—the second equation in the system remains the same for angles differing by integer multiples of 2π. Thus there is no physical difference in the system for such angles. (*Think about this in physical terms.*) Now all the equilibrium point first coordinates that are *even* multiples of π differ from 0 by multiples of 2π, so we can just study what happens near $(0, 0)$. (For example, the point $(-8\pi, 0)$ is the same as $(0 + (-4) \cdot 2\pi, 0)$.) Similarly, all the equilibrium point first coordinates that are *odd* multiples of π differ from π by multiples of 2π, so we can just see what happens to the system near $(\pi, 0)$. (For example, $(17\pi, 0)$ is the same as $(\pi + (8) \cdot 2\pi, 0)$.) Therefore, by analyzing the behavior of the system near the points $(0, 0)$ and $(\pi, 0)$, we can understand the behavior near *any* of the infinite number of equilibrium points.

Near the origin we can replace $\sin x$ by its Taylor series expansion, so our system can be written as

$$\dot{x} = y$$

$$\dot{y} = -\frac{g}{L}\sin x = -\frac{g}{L}\left(x - \frac{x^3}{3!} + \frac{x^5}{5!} - \frac{x^7}{7!} + \cdots\right)$$

and we can see that the linearization of our system is given by

$$\dot{x} = y$$

$$\dot{y} = -\frac{g}{L}x.$$

In matrix form, this becomes $\dot{X} = \begin{bmatrix} 0 & 1 \\ -\dfrac{g}{L} & 0 \end{bmatrix} X$, with characteristic equation

$\lambda^2 + \dfrac{g}{L} = 0$ and pure imaginary eigenvalues $\lambda = \pm\sqrt{g/L}\,i$. Once again, part (C)

of the Poincaré-Liapunov result points to either a *center* or a *spiral point*. Intuitively, we should realize that this is like the situation with the undamped spring-mass system: In the absence of any kind of resistance, the object will continue to move periodically about its equilibrium state. In our case, we would expect the pendulum to swing back and forth indefinitely. (Compare the pendulum's associated linear system with system (4.5.5) in Example 4.5.5.) Figure 7.11 shows the phase portrait of the nonlinear system with $g = L$ near the origin. Note what the figure tells us. If the pendulum starts with $x_0 = \theta_0$ anywhere between 0 and π and we release the weight at the end (the *bob*), then the pendulum will swing in a clockwise (*negative*) direction toward the vertical position and go past the vertical ($x = \theta = 0$) until it makes the same initial angle on the other side. At this point in time ($x = x_0 = -\theta_0$), the pendulum starts its journey back to the vertical position and then goes past it until $x = \theta_0$ once more. The variable y represents the *angular velocity*, which is zero as we release the pendulum, becomes negative as the velocity increases in a negative (clockwise) direction, attains its maximum as the pendulum swings through the vertical position, and then decreases as the pendulum approaches $x = x_0 = -\theta_0$. At this point, the pendulum begins its swing back toward the center and ultimately back to its initial position, its velocity increasing and decreasing appropriately. (We'll deal with the curves at the very top and bottom of Figure 7.11 shortly.)

Now let's examine the pendulum's behavior near the equilibrium point $(\pi, 0)$. The transformation $u = x - \pi, v = y - 0$ results in the nonlinear system

$$\dot{u} = v$$

$$\dot{v} = -\frac{g}{L}\sin(u + \pi) = -\frac{g}{L}(-\sin u) = \frac{g}{L}\sin u,$$

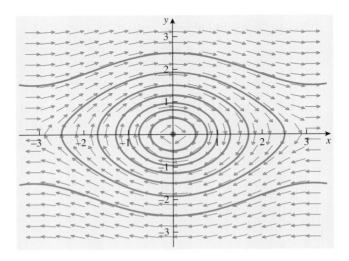

Figure 7.11

Trajectories of $\dot{x} = y, \dot{y} = -\sin x$ near the origin

with the associated linear system

$$\dot{u} = v$$

$$\dot{v} = \frac{g}{L}u.$$

Make sure you understand how we arrived here. This linear system has the characteristic equation $\lambda^2 - \frac{g}{L} = 0$ and eigenvalues $\pm\sqrt{\frac{g}{L}}$. We look to part (A) of our stability statement (and Table 5.1 in Section 5.5) to see that the equilibrium point $(\pi, 0)$ is a *saddle point*. Figure 7.12 (again with $g = L$) focuses on the system's behavior near this point.

As neat as this seems to be, we've brushed past something we haven't explained yet: the strange curves at the very top and bottom of Figure 7.11. If we take a wider look at the entire phase portrait (Figure 7.13), this strangeness becomes more evident.

Clearly, if the initial velocity imparted to the undamped pendulum is low enough, the pendulum swings indefinitely back and forth about its equilibrium point $(0, 0)$. Physically, this equilibrium position corresponds to the pendulum at rest $(y = \dot{\theta} = 0)$ and hanging straight down $(x = \theta = 0)$. If we give the pendulum a high enough initial velocity, it will whirl up and over the top—over and over again in the absence of any friction or air resistance. Its velocity will vary periodically, attaining its minimum at *odd* multiples of π (when its position is straight up) and its maximum at *even* multiples of π (when it is moving through the straight-down position).

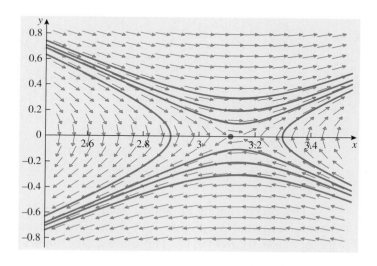

Figure 7.12
Trajectories of $\dot{x} = y$, $\dot{y} = -\sin x$ near $(\pi, 0)$

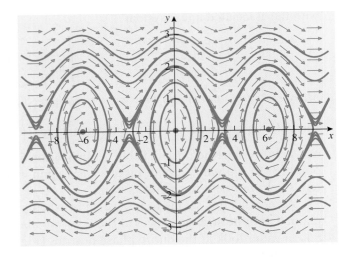

Figure 7.13
Phase portrait of $\dot{x} = y$, $\dot{y} = -\sin x$

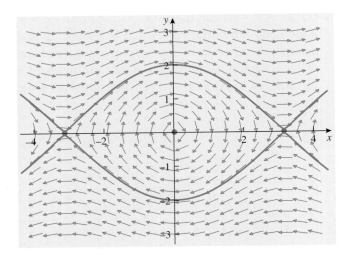

Figure 7.14
Separatrices connecting $(-\pi, 0)$ and $(\pi, 0)$

The curves joining the saddle points (odd multiples of π on the x-axis) need careful explanation. Figure 7.14 focuses on the curves connecting the saddle points $(-\pi, 0)$ *and* $(\pi, 0)$. These are called **separatrices** (the plural of **separatrix**); they separate the regions of "normal" behavior from each other. (More technically, they are called **heteroclinic trajectories** or **saddle connections.**) As we've indicated before, the saddle points represent a pendulum pointed straight up and at rest. Physically, then, these heteroclinic trajectories describe the fact that the pendulum slows down just as it approaches the upside-down position.

Exercises 12 and 13 suggest a more analytic way of understanding the undamped pendulum's behavior.   ◆

EXERCISES 7.3

Find the nontrivial equilibrium point for each of the Lotka-Volterra systems in Exercises 1–4. You may need technology.

1. $\dot{x} = 3x - 2xy$, $\dot{y} = -y + 4xy$

2. $\dot{x} = 0.1x - 0.2xy$, $\dot{y} = -0.5y + 0.3xy$

3. $\dot{x} = 0.005x - 0.02xy$, $\dot{y} = -0.3y + 0.4xy$

4. $\dot{x} = x - 2xy$, $\dot{y} = -3y + 4xy$

5. Consider the Lotka-Volterra equations for $a = b = c = d = 1$. Figure 7.9b shows some trajectories corresponding to this situation. Without relying on the graph, we want to show that the trajectories are closed curves—that is, that the equilibrium point $(1, 1)$ is a *stable center*.

   a. Show that the slope field for $\dfrac{dy}{dx}$ is symmetric about the line $y = x$. [*Hint:* Look at what happens when you interchange x and y in the slope equation.]

   b. Argue that if you start at some point $P = (x, y)$ on the line $y = x$ and travel along the trajectory once around the point $(1, 1)$, you wind up back at the same point P, so that the curve is closed.

6. Consider the Lotka-Volterra equations (Example 7.3.1) for $a = b = c = d = 1$. To develop some confidence in the power of numerical methods, use whatever Runge-Kutta algorithm and step size your instructor suggests to approximate the solution to the initial-value problem with $x(0) = 1$ and $y(0) = 2$ over the interval $[0, 1]$.

7. Consider the system first examined in Example 4.5.4:

$$\dot{x} = 0.2x - 0.002xy$$
$$\dot{y} = -0.1y + 0.001xy$$

   a. Find the equilibrium points for the system.

   b. Plot the trajectory corresponding to the initial conditions $x(0) = 100$ and $y(0) = 300$. Interpret these initial values and the shape of the trajectory in terms of the predator and prey populations. (Choose the interval $[0, 55]$ for your independent variable t.)

   c. Use the graph of the trajectory found in part (b) to estimate the maximum and minimum values of the populations x and y.

   d. Find the slope equation $\dfrac{dy}{dx}$ and solve it (implicitly) using the initial conditions given in part (b).

   e. Use technology to plot the solution found in part (d), using ranges for x and y consistent with your answers to part (c).

8. Recall that the Lotka-Volterra system (Example 7.3.1) has the nontrivial equilibrium point $(c/d, a/b)$. To understand the direction of any trajectory for the Lotka-Volterra equations without relying on a graph provided by

technology, divide the first quadrant of the x-y plane into four subquadrants via the lines $x = c/d$ and $y = a/b$. (Sketch this situation.)

a. Show that for $x > c/d$ and $y > a/b$, you have $\dot{x} < 0$ and $\dot{y} > 0$.

b. Show that for $x < c/d$ and $y > a/b$, you have $\dot{x} < 0$ and $\dot{y} < 0$.

c. Show that for $x < c/d$ and $y < a/b$, you have $\dot{x} > 0$ and $\dot{y} < 0$.

d. Show that for $x > c/d$ and $y < a/b$, you have $\dot{x} > 0$ and $\dot{y} > 0$.

e. From the results of parts (a)–(d), conclude that any point $(x(t), y(t))$ on a trajectory for the Lotka-Volterra equations moves in a *counterclockwise* direction.

9. Recall that in Example 7.3.1, the Lotka-Volterra equations $\dot{x} = ax - bxy$, $\dot{y} = -cy + dxy$ were linearized to $\dot{u} = (-bc/d)v$, $\dot{v} = (ad/b)u$ near the equilibrium point $(c/d, a/b)$.

a. Find the slope equation $\dfrac{du}{dv}$ and conclude that the linear system has a solution that satisfies $ad^2u^2 + b^2cv^2 = K$, where K is a positive constant.

b. Rewrite the solution in part (a) in terms of the original variables x and y, and show that you get the equation of an ellipse with center at $(c/d, a/b)$ and with axes parallel to the axes of the x-y plane.

c. Compute the derivative of each equation of the linearized system to get the equations $\ddot{u} = -acu$, $\ddot{v} = -acv$—uncoupled second-order linear equations of the form $\ddot{w} = -Rw$.

d. Show that the solution of the linearized system is a pair of functions $(u(t), v(t))$ with the same period $2\pi/\sqrt{ac}$.

10. Focus on the equation for the *predator* population, $\dfrac{dy}{dt} = -cy + dxy$.

a. Divide the equation by y and integrate between the initial time t_0 and some arbitrary time t.

b. Assuming that the predator population is periodic (see Figure 4.9 or 4.10b, for example) with period T, let $t = t_1$ in part (a) so that $t_1 - t_0 = T$ and $y(t_1) = y(t_0)$. Show that the *average value* of the *prey* population is c/d, the same as the equilibrium population of the prey. (Recall that the average value of a function f on the interval $[a, b]$ is defined as

$$\frac{1}{b-a}\int_a^b f(r)\,dr.)$$ [*Hint:* Note that $\dot{y}/y = -c + dx$ and integrate from 0 to T, using the periodicity of $\ln|y(t)|$.]

11. Assuming the result of Problem 10(b), and that the average value of the predator population $y(t)$ is a / b, the equilibrium population of the predator, and also assuming that both predator and prey populations have the same period T, show that the average value of $x(t) \cdot y(t)$ equals the average value of $x(t)$ times the average value of $y(t)$. [*Hint:* $(\dot{y} + cy)/d = xy$.]

12. Consider the simplified pendulum equation used in Figure 7.11, Example 7.3.2 : $\dfrac{d^2\theta}{dt^2} + \sin\theta = 0$. You're going to show (analytically) that this equation has periodic solutions—that is, that there are closed trajectories in the phase plane corresponding to the system version of the equation.

    a. Show that this equation is equivalent to the system

    $$\frac{dx}{dt} = y$$
    $$\frac{dy}{dt} = -\sin x.$$

    b. Show that any trajectory in the phase plane is a solution of $\dfrac{dy}{dx} = -\dfrac{\sin x}{y}$.

    c. Solve the equation in part (b).

    d. Show that there are closed trajectories in the x-y plane and hence that the undamped pendulum problem has periodic solutions. [*Hint:* Find suitable values of the constant of integration you get in part (c).]

13. When you find the general solution of the equation $\dfrac{dy}{dx} = -\dfrac{\sin x}{y}$ [as in part (b) of the previous exercise], you have an arbitrary constant C.

    a. What values of C give the wavy trajectories at the top and bottom of Figure 7.13?

    b. What values of C give the separatrices, as in Figure 7.14?

14. For small values of θ, $\sin\theta \approx \theta$, so the linearized equation of the undamped pendulum is $\ddot{\theta} + \dfrac{g}{L}\theta = 0$. Work with this equation and the initial conditions $\theta(0) = 0$, $\dot{\theta}(0) = 2$.

    a. Find $\theta(t)$ if the length of the pendulum is 8 feet. (Take $g = 32$ ft/sec^2.)

    b. What is the period of the function found in part (a)?

    c. If the pendulum is part of a clock that ticks once for each time the pendulum makes a complete swing, how many ticks does the clock make in one minute?

    d. How is the motion of the pendulum affected when the length is changed to $L = 4$?

15. The equation $\ddot{\theta} + k\dot{\theta} + \sin\theta = 0$ describes a particular *damped* pendulum—that is, a pendulum with friction or air resistance. Here k is a positive constant, the coefficient of friction.

    a. Convert this second-order equation to a system of first-order equations.

    b. Use technology to produce the phase portrait when $k = 0.1$.

    c. Use technology to produce the phase portrait when $k = 0.5$.

    d. Compare the phase portraits in parts (b) and (c) and give a physical interpretation of what you see.

7.4   VAN DER POL'S EQUATION AND LIMIT CYCLES

VAN DER POL'S EQUATION

The next example deals with a famous equation that arose when radios were first developed. The original context was the study of certain electrical circuits containing a vacuum tube ("triode generator"), but the work also has had significant biological applications. The pioneering experiments and the first theoretical analysis were conducted by Dutch electrical engineer Balthasar van der Pol (1889–1959) and others in the 1920s.

EXAMPLE 7.4.1   The van der Pol Equation

The **van der Pol equation** (or **van der Pol oscillator**)

$$x'' + \varepsilon(x^2 - 1)x' + x = 0, \tag{7.4.1}$$

where ε is a positive parameter, can also be interpreted in terms of a spring-mass system with nonlinear resistance. (See Exercise 1.) Equation (7.4.1) can be written as the system

$$\begin{aligned} x_1' &= x_2 \\ x_2' &= -x_1 + \varepsilon x_2(1 - x_1^2). \end{aligned} \tag{7.4.2}$$

The first thing we have to do is find the equilibrium points of (7.4.2). To get a sense of how this system behaves, let's assume that $\varepsilon = 1$. (See Exercises 2 and 3, which ask you to consider other values of ε.) The linearized version of the nonlinear system (7.4.2) is then

$$\begin{aligned} x_1' &= x_2 \\ x_2' &= -x_1 + x_2 \end{aligned}$$

with characteristic equation $\lambda^2 - \lambda + 1 = 0$ and eigenvalues $(1 \pm \sqrt{3i})/2$. This implies (*Why?*) that both the nonlinear system (7.4.1) and its linear approximation (7.4.2) have a *spiral source* at the origin. However, this particular system exhibits some new, characteristically nonlinear behavior. Figure 7.15a shows the phase portrait of the nonlinear system (7.4.1) near $(0, 0)$.

What is happening here is that several paths starting near the origin spiral *outward* from the origin (as expected) toward a particular closed curve, whereas other trajectories starting farther away from $(0, 0)$ also seem to be approaching this closed curve asymptotically (that is, as $t \to \infty$). Reasonably enough, such a closed trajectory is called a **stable limit cycle.** A stable limit cycle can also be described as a periodic trajectory that attracts other nearby trajectories, whereas an **unstable limit cycle** *repels* nearby trajectories. It is important to note that *linear* systems never have limit cycles. (See the discussion of limit cycles following this example.) Note that the phase portrait of the linearized system (Figure 7.15b) shows no limit cycle, only the spiraling away from the origin.

Figure 7.15c, on page 344, shows a plot of x against t with the initial conditions $x(0) = 0.5$, $x'(0) = -0.5$. This graph reflects the eventual periodicity of the

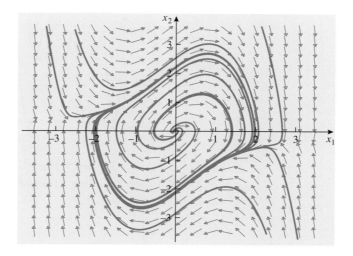

Figure 7.15a
Phase portrait of $x_1' = x_2$, $x_2' = -x_1 + x_2(1 - x_1^2)$ near the origin

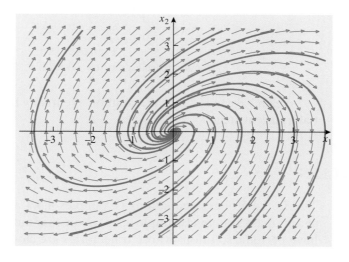

Figure 7.15b
Phase portrait of $x_1' = x_2$, $x_2' = -x_1 + x_2$ near the origin

solution and the fact that the spirals work their way *outward* (through increasing values of *t*) to the limit cycle. The solution shows *transient* behavior (temporary or short-lived behavior) at the beginning, before settling into its periodic pattern.

On the other hand, if we choose an initial point $(x(0) = -3, x'(0) = -5)$ in a region that appears to be *outside* the limit cycle shown in Figure 7.15a, we see the solution behavior shown by Figure 7.15d. This illustrates how a spiral finds its way *inward* to the limit cycle.

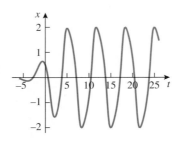

Figure 7.15c

Plot of $x(t)$ against t, $x(0) = x'(0) = 0.5$, $-6 \leq t \leq 26$

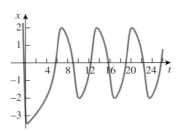

Figure 7.15d

Plot of $x(t)$ against t, $x(0) = -3$, $x'(0) = -5$, $-0.6 \leq t \leq 26$

Again we can see that the solution eventually becomes periodic, after an initial transient stage. ◆

LIMIT CYCLES

The Lotka-Volterra equations (Examples 4.5.4 and 7.3.1) and the undamped spring-mass system (Example 4.5.5) show that autonomous systems sometimes have periodic solutions whose trajectories are closed curves in the phase plane. As we have seen, the van der Pol oscillator, which can be described as a negatively damped nonlinear oscillator (look at the form of the equation), has solutions whose *limiting* behavior (as $t \to \infty$) is that of a finite periodic solution. As we have indicated, such a nontrivial isolated closed trajectory is called a *limit cycle*. Here "nontrivial" means that the solution curve is not a single point, and "isolated" refers to the fact that no trajectory sufficiently near the limit cycle is also closed.

In general, a *linear* system $\dot{X} = AX$ may have closed trajectories, but they won't be isolated: If $X(t)$ is a periodic solution, then so is $cX(t)$ for any nonzero constant c. (*Can you show this?*) Therefore, for instance, by choosing $c = (1 - 1/k)$ $(k = 1, 2, 3, \ldots)$ we see that $X(t)$ is being crowded by a one-parameter family of closed trajectories. (In this way the closed trajectories shown in Figure 7.9b of

Section 7.3 are not isolated and so could not possibly be limit cycles. You can get trajectories as close to each other as you wish.)

Every trajectory that begins sufficiently near a limit cycle approaches it either for $t \to \infty$ or for $t \to -\infty$. Graphically, this means that such a trajectory either winds itself around the limit cycle or unwinds *from* it. A limit cycle is called **semistable** if trajectories approach one side of it while pulling away from the other side.

As one researcher has put it,

> The stable limit cycle is the basic model for all self-sustained oscillators—those which return, or recover, to some fundamental periodic orbit when perturbed from it. The stable oscillations, "beating" of the human heart (which returns to some normal rate after we raise it by sprinting), cycles of predator-prey systems, and various electrical circuits are three among myriad examples. Business cycles and certain periodic outbreaks of social unrest . . . are, quite possibly, others.[1]

Let's look at other examples of this important phenomenon, the limit cycle. Van der Pol's equation exhibited a *stable* limit cycle, but the next example shows a different type of behavior.

EXAMPLE 7.4.2   An Unstable Limit Cycle

Let's examine the autonomous system

$$\dot{x} = -y + x(x^2 + y^2 - 1)$$
$$\dot{y} = x + y(x^2 + y^2 - 1).$$

The presence of the algebraic form $x^2 + y^2$, with its suggestion of circularity (rotation), tips us off that we may be able to see things more clearly if we switch to *polar coordinates* (see Appendix B.1). Making the substitutions $x = r \cos\theta = r(t) \cos\theta(t)$, $y = r \sin\theta = r(t) \sin\theta(t)$, and $\theta = \arctan(y/x)$, we have $x^2 + y^2 = r^2$. (You may have seen this sort of substitution in evaluating certain integrals in calculus class—or in Problem 1 of Exercises 7.2.) A few algebraic manipulations (see Exercise 4) give us the polar coordinate form of the system we started with:

$$\dot{r} = (r^2 - 1)r \quad (r \geq 0)$$
$$\dot{\theta} = 1.$$

This system describes the motion of an object in terms of its radial distance $r = r(t)$ from the origin and its (constant) angular velocity $\dot{\theta}$ in a counterclockwise direction. Figure 7.16 illustrates this in general.

Because the equations are independent (or *uncoupled*), each involving only one dependent variable, we can analyze them separately. We can look at the first equation as a first-order nonlinear equation and consider its phase portrait in the manner of Section 2.4 (Figure 7.17). Recalling that r is nonnegative, we see that the only equilibrium solutions are $r \equiv 0$ and $r \equiv 1$. Note that the first equation tells us that if $r < 1$, then $\dot{r} < 0$, so the trajectory's distance from the origin is decreasing—that is, the trajectory is approaching the origin and moving away from

1. J. M. Epstein, *Nonlinear Dynamics, Mathematical Biology, and Social Science* (Addison-Wesley, 1997): 121.

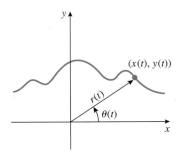

Figure 7.16

Motion described in terms of radial distance
and angular velocity

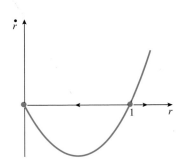

Figure 7.17

Phase portrait of $\dot{r} = (r^2 - 1)r, r \geq 0$

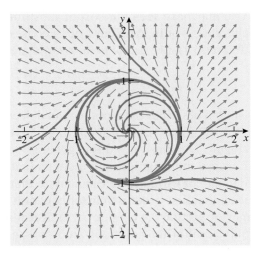

Figure 7.18

Phase portrait of $\dot{r} = (r^2 - 1)r, r \geq 0$, in the x-y phase plane

the unit circle ($r \equiv 1, 0 \leq \theta \leq 2\pi$); whereas if $r > 1$, we have $\dot{r} > 0$, so trajectories are also repelled by the unit circle.

From this phase portrait we can see that $r \equiv 0$ is a *sink* and $r \equiv 1$ is a *source*. We could have used the *Derivative Test* of Section 2.5 to see this. (Also see Exercise 5 at the end of this section.) In particular, the origin is a *sink* for the system in its original rectangular coordinate form. Figure 7.18 shows the phase portrait in *x-y* space.

From this we see that the unit circle is an *unstable limit cycle*. ◆

Now we're ready for something a bit more complicated, but rewarding.

EXAMPLE 7.4.3   A System with Two Limit Cycles
Let's look at the system

$$\dot{r} = r(r - 1)(r - 2)$$
$$\dot{\theta} = 1.$$

As in the previous example, the system describes the motion of an object in terms of its radial distance $r = r(t)$ from the origin and its (constant) angular velocity $\dot{\theta}$ in a counterclockwise direction. Let's look at the phase portrait of the first equation (Figure 7.19), whose equilibrium solutions are $r \equiv 0$, $r \equiv 1$, and $r \equiv 2$. As we can see, $r \equiv 0$ is a source, $r \equiv 1$ is a sink, and $r \equiv 2$ is a source. What this tells us is that the system has two circular limit cycles: one stable ($r \equiv 1$) and one unstable ($r \equiv 2$). Trajectories starting out inside the unit circle approach the unit circle as $t \to \infty$, as do trajectories with initial point inside the ring formed by the two circles $r \equiv 1$ and $r \equiv 2$. Any trajectory starting outside the circle of radius 2 moves farther away as $t \to \infty$.

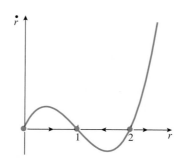

Figure 7.19

Phase portrait of $\dot{r} = r(r - 1)(r - 2), r \geq 0$ ◆

The next example illustrates a third kind of limit cycle.

EXAMPLE 7.4.4   A Semistable Limit Cycle

What kind of behavior is shown by the following system?

$$\dot{r} = r(r - 1)^2$$
$$\dot{\theta} = 1.$$

The phase portrait for the first equation (Figure 7.20) tells the story.

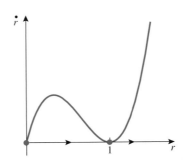

Figure 7.20

Phase portrait of $\dot{r} = r(r - 1)^2, r \geq 0$

The equilibrium point $r \equiv 0$ is a source, whereas $r \equiv 1$ is a node because $\dot{r} > 0$ for $0 < r < 1$ and for $r > 1$ as well. The graphical interpretation of this fact is that the unit circle described by $r \equiv 1$ is a *semistable limit cycle*. Trajectories approach the unit circle from inside it, whereas trajectories that start outside escape the unit circle. ◆

Of course, a nonlinear equation or system may have *no* limit cycles (isolated nonconstant periodic solutions). Because nonlinear equations and systems are usually too difficult to solve, other methods—qualitative methods—have been developed to determine the existence or nonexistence of limit cycles. These methods involve advanced mathematical ideas that we won't discuss in detail in this book. Exercises 11–16 illustrate a *negative* criterion due to the Swedish mathematician Ivar Bendixson (1861–1935).

EXERCISES 7.4

1. In the discussion of the van der Pol equation (7.4.1), we noted that it can be interpreted as a spring-mass system with nonlinear resistance. Specifically, the term $\varepsilon(x^2 - 1)$ represents a *variable* damping coefficient.

   a. Explain why $\varepsilon(x^2 - 1) < 0$ when $-1 < x < 1$, so that damping is *negative* for the small oscillations corresponding to $-1 < x < 1$. (This means that small-amplitude oscillations are *amplified* if they become too small.)

   b. Explain why $\varepsilon(x^2 - 1) > 0$ when $|x| > 1$, so that damping is *positive* for the large oscillations corresponding to $|x| > 1$. (This means that large-amplitude oscillations are made to *decay* if they become too large.)

2. Use technology to draw phase portraits of the van der Pol equation for $\varepsilon = \frac{1}{4}, \frac{3}{2}$, and 3.

3. Consider the van der Pol equation in the system form (7.4.2), where $x_1(0) = 1$ and $x_2(0) = 0$.

   a. For $\varepsilon = \frac{1}{4}$, graph the trajectory in the x_1-x_2 plane. Then graph $x_1(t)$ against t and $x_2(t)$ against t on different sets of axes. Use technology.

   b. For $\varepsilon = 4$, graph the trajectory in the x_1-x_2 plane. Then graph $x_1(t)$ against t and $x_2(t)$ against t on different sets of axes. Use technology.

   c. Comment on the differences between the graphs in part (a) and the graphs in part (b).

4. Go back to Example 7.4.2 and look at the trigonometric substitutions suggested there. You're going to verify the polar coordinate form of the system equations.

   a. Use the Chain Rule to show that $r\dot{r} = x\dot{x} + y\dot{y}$.

   b. Show that $\dot{\theta} = -\dfrac{1}{x^2 + y^2}(y\dot{x} - x\dot{y})$, or $-r^2\dot{\theta} = (y\dot{x} - x\dot{y})$.

   c. Show that $x\dot{x} + y\dot{y} = (x^2 + y^2)(x^2 + y^2 - 1) = r^2(r^2 - 1)$. [*Hint:* Multiply the first equation in the system by x and the second equation by y, and then add the results.]

   d. Use parts (a) and (c) to conclude that $\dot{r} = r(r^2 - 1)$.

   e. Use part (b) and the general method in part (c) to show that $\dot{\theta} = 1$.

5. Consider again the uncoupled system (polar coordinate form) in Example 7.4.2.

   a. Solve for $r(t)$.

   b. Solve for $\theta(t)$.

   c. Use your answers to parts (a) and (b) to construct $x(t)$ and $y(t)$.

   d. Construct trajectories in the x-y plane by graphing $x(t)$ and $y(t)$ on the same set of axes, for different constants of integration (the same for each $x(t)$, $y(t)$ pair) and different values of t (positive and negative).

6. Follow the directions given in Exercise 5 for the system in Example 7.4.3.

7. Consider the system $\{\dot{r} = r(1 - r^2), \dot{\theta} = 1\}$.

   a. Show that this is equivalent to the system

   $$\dot{x} = x - y - x(x^2 + y^2)$$
   $$\dot{y} = x + y - y(x^2 + y^2),$$

   where $x = r(t) \cos\theta(t)$ and $y = r(t) \sin\theta(t)$.

   b. Use either form of the system to determine its unique limit cycle.

8. Consider the system $\{\dot{r} = r(4 - r^2), \dot{\theta} = 1\}$, where $x(t) = r(t) \cos\theta(t)$ and $y = r(t) \sin\theta(t)$. Given the initial conditions $x(0) = 0.1$, $y(0) = 0$, sketch the graph of $x(t)$ *without finding an explicit expression for $x(t)$.* (*Hint:* Study Example 7.4.2 carefully.)

9. The system

$$\dot{x} = -y - y^2$$
$$\dot{y} = \tfrac{1}{2}x - \tfrac{1}{5}y + xy - \tfrac{6}{5}y^2$$

was discovered by the Chinese mathematician Tung Chin Chu in the late 1950s in his investigation of a famous unsolved problem on limit cycles.

a. Find the equilibrium point(s) of this system.

b. Use technology to draw a phase portrait for each equilibrium point, focusing on the region around that point. (It's a bit tricky to get a good phase portrait for this problem. Be patient.)

c. Using the phase portrait(s), identify and describe any limit cycle(s) you see with the term *stable* or *unstable*.

10. Find all limit cycles of the system

$$\dot{r} = r(r - 1)(r - 2)^2(r - 3)$$
$$\dot{\theta} = 1$$

and identify them as stable, unstable, or semistable.

Suppose we have an autonomous system $\{\dot{x} = f(x, y), \dot{y} = g(x, y)\}$, *where f and g have continuous first partial derivatives in some region R of the phase plane that doesn't have any "holes." Then* Bendixson's theorem *(or* negative criterion*) states that if* $\dfrac{\partial f}{\partial x} + \dfrac{\partial g}{\partial y}$ *is always positive or always negative at points of R, then the system has no periodic solutions in R. For example, the system* $\{\dot{x} = xy^2,$ $\dot{y} = x^2 + 8y\}$ *has no limit cycles anywhere, because* $\dfrac{\partial(xy^2)}{\partial x} + \dfrac{\partial(x^2 + 8y)}{\partial y} = y^2 + 8 > 0$ *for all values of x and y in the plane.*

   Use Bendixson's criterion to show that the systems in Exercises 11–14 have no limit cycles in the phase plane.

11. $\{\dot{x} = x + 2xy + x^3, \dot{y} = -y^2 + x^2y\}$

12. $\{\dot{x} = x^3 + x + 7y, \dot{y} = x^2y\}$

13. $\{\dot{x} = -2x - x\sin y, \dot{y} = -x^2y^3\}$

14. $\{\dot{x} = -x + 2y^3 - 2y^4, \dot{y} = -x - y + xy\}$

15. Show that the system

$$\dot{x} = 12x + 10y + x^2y + y\sin y - x^3$$
$$\dot{y} = x + 14y - xy^2 - y^3$$

has no periodic solution in the disk $x^2 + y^2 \le 8$.

16. A mechanical system with variable damping can be modeled by the equation

$$\ddot{x} + a(x)\dot{x} + b(x) = 0,$$

where $a(x)$ is a positive function.

a. Write this equation in system form.

b. Use Bendixson's criterion to show that this mechanical system has no nonconstant periodic solution.

7.5 SUMMARY

Nonlinear differential equations and systems of nonlinear equations are rarely handled satisfactorily by finding closed-form solutions. In particular, we can't analyze the *stability* of systems of nonlinear equations as easily as we analyzed the stability of linear systems in Chapter 5. The modern study of nonlinear phenomena relies heavily on the qualitative methods pioneered by H. Poincaré and A. Liapunov at the end of the nineteenth century and in the beginning of the twentieth century. Current technology implements the power of these qualitative techniques.

One of the differences between linear and nonlinear equations is that a nonlinear equation may have more than one equilibrium solution. Another difference is that a solution of a nonlinear equation may "blow up in finite time"—that is, become unbounded as t approaches some finite value. A third difference is that a nonlinear equation or system may be extremely sensitive to initial conditions. A slight change in an initial value may lead to drastic changes in the behavior of the solution or solutions.

A point (a^*, b^*) is an *equilibrium point* of the general nonlinear autonomous system

$$\dot{x} = F(x, y)$$
$$\dot{y} = G(x, y)$$

if $F(a^*, b^*) = 0 = G(a^*, b^*)$. If the origin is an equilibrium point, and the functions F and G are "nice" enough, we may be able to write our system in the form

$$\dot{x} = ax + by + f(x, y)$$
$$\dot{y} = cx + dy + g(x, y),$$

where f and g are nonlinear functions and $a = \dfrac{\partial F}{\partial x}(0, 0)$, $b = \dfrac{\partial F}{\partial y}(0, 0)$, $c = \dfrac{\partial G}{\partial x}(0, 0)$, and $d = \dfrac{\partial G}{\partial y}(0, 0)$. More generally, if $(a, b) \neq (0, 0)$ is an equilibrium point for the system, we can rewrite the system as

$$\dot{x} = A(x - a) + B(y - b) + f(x, y)$$
$$\dot{y} = C(x - a) + D(y - b) + g(x, y),$$

where f and g are nonlinear and $A = \dfrac{\partial F}{\partial x}(a, b)$, $B = \dfrac{\partial F}{\partial y}(a, b)$, $C = \dfrac{\partial G}{\partial x}(a, b)$, and $D = \dfrac{\partial G}{\partial y}(a, b)$.

Another way to look at this general situation is to realize that we are translating the equilibrium point (a, b) to the origin by using the change of variables $u = x - a$ and $v = y - b$. Of course, this means that $x = u + a$ and $y = v + b$, so we can rewrite the last system as

$$\dot{u} = Au + Bv + f(u, v)$$
$$\dot{v} = Cu + Dv + g(u, v)$$

which has $(0, 0)$ as an equilibrium point. Note that this says that any equilibrium point $(a, b) \neq (0, 0)$ can be transformed to the origin for the purpose of analyzing the stability of the system.

A nonlinear autonomous system

$$\dot{x} = ax + by + f(x, y)$$
$$\dot{y} = cx + dy + g(x, y),$$

where $ad - bc \neq 0$, $\displaystyle\lim_{(x, y) \to (0, 0)} \left(\frac{f(x, y)}{\sqrt{x^2 + y^2}} \right) = \lim_{(x, y) \to (0, 0)} \left(\frac{g(x, y)}{\sqrt{x^2 + y^2}} \right) = 0$, and the origin is an equilibrium point, is called an **almost linear system,** and the reduced system

$$\dot{x} = ax + by$$
$$\dot{y} = cx + dy$$

is called the **associated linear system** (or **linear approximation**) about the origin.

An important qualitative result discovered by Poincaré and Liapunov states that if λ_1 and λ_2 are the eigenvalues of the associated linear system, then the equilibrium points of the two systems are related as follows:

(A) If the eigenvalues λ_1 and λ_2 are *not* equal real numbers or are *not* pure imaginary numbers, then the trajectories of the almost linear system near the equilibrium point $(0, 0)$ behave the same way as the trajectories of the associated linear system near the origin. That is, we can use the appropriate entries given in Table 5.1 (Section 5.5) to determine whether the origin is a node, a saddle point, or a spiral point of both systems.

(B) If λ_1 and λ_2 are real and equal, then the origin is either a node or a spiral point of both systems. Furthermore, if $\lambda_1 = \lambda_2 < 0$, then the origin is asymptotically stable; and if $\lambda_1 = \lambda_2 > 0$, then the origin is an unstable equilibrium point.

(C) If λ_1 and λ_2 are pure imaginary numbers, then the equilibrium point $(0, 0)$ is either a center or a spiral point of the nonlinear system. Also, this spiral point may be asymptotically stable, stable, or unstable.

In situations (B) and (C), further analysis is necessary to determine the nature of the equilibrium points.

The **Lotka-Volterra equations,** the **undamped pendulum,** and the **van der Pol equation** provide important examples of nonlinear systems and their analyses. In particular, the van der Pol oscillator exhibits uniquely nonlinear behavior in having a **stable limit cycle,** an isolated closed trajectory that (in this case) serves as an asymptotic limit for all other trajectories as $t \to \infty$. Some limit cycles, called **unstable limit cycles,** repel nearby trajectories. Finally, if trajectories near a limit cycle approach it from one side while being repelled from the other side, the cycle is called **semistable.**

PROJECT 7-1

Butterflies in Space

In 1963, E. N. Lorenz, an MIT meteorology professor, published a report[2] on the nonlinear system

$$\dot{x} = -\sigma x + \sigma y$$
$$\dot{y} = rx - y - xz$$
$$\dot{z} = xy - bz,$$

where σ, r, and b are positive parameters.

The equations arose from a model of a layer in the earth's atmosphere, heated from below by the ground that has absorbed sunlight, and cooled from above as it loses heat into space.

a. Show that if $0 < r \leq 1$, then the only equilibrium point is $(0, 0, 0)$.

b. Show that if $r > 1$, then there are three equilibrium points: $(0, 0, 0)$, $(\sqrt{b(r-1)}, \sqrt{b(r-1)}, r-1)$, and $(-\sqrt{b(r-1)}, -\sqrt{b(r-1)}, r-1)$.

c. Let $b = \frac{8}{3}$, $r = 28$, and $\sigma = 10$ (values used by Lorenz in his initial experiments). For these values of b, r, and σ, linearize the system about the equilibrium points $(\pm\sqrt{b(r-1)}, \pm\sqrt{b(r-1)}, r-1)$.

d. Use technology to show that the characteristic equation of the linearized system found in part (c) is $\lambda^3 + (b + \sigma + 1)\lambda^2 + b(r + \sigma)\lambda + 2\sigma b(r-1) = 0$. Show that the characteristic polynomial has a negative real root $\lambda_1 \approx -13.85$ and complex conjugate roots with positive real parts ≈ 0.09.

e. Use part (d) and the table at the end of Section 5.5 to conclude that the two nonzero equilibrium points given in part (c) are *saddle points* of the system.

f. With $b = \frac{8}{3}$, $r = 28$, and $\sigma = 10$, use technology to plot $x(t)$ against t, $y(t)$ against t, and $z(t)$ against t for $0 \leq t \leq 10$.

g. With $b = \frac{8}{3}$, $r = 28$, and $\sigma = 10$, use technology to plot y against x, z against y, and z against x.

2. E. N. Lorenz, "Deterministic Nonperiodic Flow," *J. Atmos. Sci.* 20 (1963): 130–141.

APPENDIX A

Some Calculus Concepts and Results

Appendix A is intended to offer either a brief review of, or an introduction to, selected key ideas of calculus.

1. LOCAL LINEARITY: THE TANGENT LINE APPROXIMATION

The concept of **local linearity** says that if the function f is differentiable (that is, if it has a derivative) at $x = a$ and we "zoom in" on the point $(a, f(a))$ on the graph of $y = f(x)$, then the portion of the curve that surrounds the point looks very much like a straight line—at least to the naked eye. Another way of saying this is to say that the tangent line at the point $(a, f(a))$ is a good approximation to the curve for values of x close to a. Figure A1 illustrates this.

Using the point-slope formula from algebra, we can write the equation of this tangent line as $y = f(a) + f'(a)(x - a)$, so we can express this **tangent line approximation** as

$$f(x) \approx f(a) + f'(a)(x - a)$$

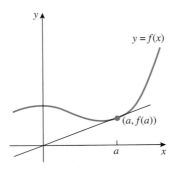

Figure A1

for x close to a. As x takes on values farther away from a, we expect the **absolute error** $|E(x)| = |f(x) - f(a) - f'(a)(x - a)|$ to become larger.

For example, the equation of the tangent line drawn to the sine curve at the origin is $y = \sin(0) + \cos(0)(x - 0) = x$. This says that near the origin, $\sin x \approx x$. One consequence of this is that $\dfrac{\sin x}{x} \approx 1$ for values of x near (but not equal to) zero, so we get the famous result $\lim\limits_{x \to 0} \dfrac{\sin x}{x} = 1$.

2. THE CHAIN RULE

You should know the rules for finding the derivatives of power functions, polynomials, exponential functions, logarithms, and trigonometric and inverse trigonometric functions. You may also have learned about differentiating certain combinations of exponential functions called *hyperbolic functions*. You should know the *Product Rule* and the *Quotient Rule* for differentiation, as well as how to deal with *implicit functions*.

The Chain Rule applies to *composite functions*. Suppose, for example, that a quantity z depends on a quantity y and that the quantity y depends on the value of quantity x. Using function notation, we can write this as follows: $z = f(y)$, $y = g(x)$, so $z = f(g(x))$. This says that ultimately, z depends on (is a function of) x. The **Chain Rule** tells us how a change in the value of x affects the value of z. In Leibniz notation,

$$\frac{dz}{dx} = \frac{dz}{dy} \cdot \frac{dy}{dx}.$$

This form is useful in many applied problems and in Chapter 4, where the *phase plane* is introduced.

EXAMPLE
If $z = y^{57}$ and $y = \sin x$, then

$$\frac{dz}{dx} = \frac{dz}{dy} \cdot \frac{dy}{dx} = (57y^{56}) \cdot \cos x = 57 \sin^{56}x \cos x. \qquad \blacklozenge$$

You may have learned another way to see the Chain Rule: If $z = f(g(x))$, then $z' = f'(g(x)) \cdot g'(x)$. This alternative point of view uses the idea of an "inside" function and an "outside" function. Try this on the last example, where the 57th-power function is outside and the sine function is inside.

3. THE TAYLOR POLYNOMIAL/TAYLOR SERIES

To extend the idea of the tangent line approximation, we look for a polynomial P_n of degree n that approximates a function f as closely as possible on an interval about a point $x = a$. What this means mathematically is that we want the polynomial to satisfy the following closeness conditions: $P_n(a) = f(a)$, $P_n'(a) = f'(a)$, $P_n''(a) = f''(a)$, $P_n'''(a) = f'''(a), \ldots$, and $P_n^{(n)}(a) = f^{(n)}(a)$. For a given function f, a point $x = a$, and degree n, the polynomial that satisfies all these conditions is given by the formula

$$P_n(x) = f(a) + f'(a)(x - a) + \frac{f''(a)}{2!}(x - a)^2$$

$$+ \frac{f'''(a)}{3!}(x - a)^3 + \cdots + \frac{f^{(n)}(a)}{n!}(x - a)^n.$$

This is called the **Taylor polynomial of degree n** about $x = a$, and we can write $f(x) \approx P_n(x)$ for x close to a. The closeness of the approximation depends on both the value of x and the value of n. In general, the closer the value of x is to the value a and the higher the degree n, the better the approximation.

If we consider what happens to a Taylor polynomial as we let n get larger and larger, we arrive at the idea of the (infinite) **Taylor series:**

$$P(x) = \lim_{n \to \infty} P_n(x) = \lim_{n \to \infty} \sum_{k=0}^{n} \frac{f^{(k)}(a)}{k!}(x - a)^k$$

$$= f(a) + f'(a)(x - a) + \frac{f''(a)}{2!}(x - a)^2$$

$$+ \frac{f'''(a)}{3!}(x - a)^3 + \cdots + \frac{f^{(n)}(a)}{n!}(x - a)^n + \cdots.$$

More precisely, suppose that f is a function with derivatives of all order in some interval $(a - r, a + r)$. Then the Taylor series given above represents the function f on the interval $(a - r, a + r)$ if and only if $\lim_{n \to \infty} R_n(x) = 0$, where $R_n(x)$ is the remainder in Taylor's formula:

$$R_n(x) = f(x) - \left(f(a) + f'(a)(x - a) + \frac{f''(a)}{2!}(x - a)^2 \right.$$

$$\left. + \frac{f'''(a)}{3!}(x - a)^3 + \cdots + \frac{f^{(n)}(a)}{n!}(x - a)^n \right)$$

$$= \frac{f^{(n+1)}(c)}{(n + 1)!}(x - a)^{n+1} \quad \text{for some point } c \text{ in } (a - r, a + r).$$

Here are some Taylor series that occur often in applications:

$$e^x = 1 + x + \frac{x^2}{2!} + \frac{x^3}{3!} + \frac{x^4}{4!} + \cdots + \frac{x^n}{n!} + \cdots$$

$$\sin x = x - \frac{x^3}{3!} + \frac{x^5}{5!} - \frac{x^7}{7!} + \cdots + (-1)^k \frac{x^{2k+1}}{(2k + 1)!} + \cdots$$

$$\cos x = 1 - \frac{x^2}{2!} + \frac{x^4}{4!} - \frac{x^6}{6!} + \cdots + (-1)^k \frac{x^{2k}}{(2k)!} + \cdots$$

$$\ln(1 + x) = x - \frac{x^2}{2} + \frac{x^3}{3} - \frac{x^4}{4} + \cdots + (-1)^{k+1} \frac{x^k}{k} + \cdots$$

$$\frac{1}{1 - x} = 1 + x + x^2 + x^3 + \cdots + x^n + \cdots.$$

Although the first three series are valid ("converge") for any value of x, the logarithmic series is valid only on the interval $(-1, 1]$. The last series, a *geometric series*, converges for $|x| < 1$. In Appendix C.4, we'll see how Euler used the exponential series to arrive at a formula for the complex exponential function.

A Taylor series is a special type of *power series*. We can differentiate or integrate a power series term by term for values of x within its interval of convergence. If the power series $\sum_{n=0}^{\infty} a_n x^n$ converges to $S(x)$ for x in some interval I, then

$$S'(x) = \sum_{n=0}^{\infty} n a_n x^{n-1} = a_1 + 2a_2 x + 3a_3 x^2 + \cdots + n a_n x^{n-1} + \cdots$$

and

$$\int_0^x S(t)\,dt = \int_0^x \left(\sum_{n=0}^{\infty} a_n t^n \right) dt = \sum_{n=0}^{\infty} \int_0^x a_n t^n dt = \sum_{n=0}^{\infty} \frac{a_n}{n+1} x^{n+1}$$

$$= a_0 x + \frac{a_1}{2} x^2 + \frac{a_2}{3} x^3 + \cdots + \frac{a_n}{n+1} x^{n+1} + \cdots .$$

In Appendix D, we will show how to solve certain differential equations using power series methods.

4. THE FUNDAMENTAL THEOREM OF CALCULUS (FTC)

A very important connection between derivatives and integrals is expressed by the **Fundamental Theorem of Calculus (FTC).** This result comes in two flavors:

(A) If $f(x)$ is continuous on the closed interval $[a, b]$ and if $F(x)$ is any antiderivative of $f(x)$ on this interval, then

$$\int_a^b f(x)\,dx = F(b) - F(a).$$

(B) Let $f(x)$ be defined and continuous on a closed interval $[a, b]$ and define the function $G(x)$ on this interval:

$$G(x) = \int_a^x f(t)\,dt.$$

Then $G(x)$ is differentiable there with derivative $f(x)$: $G'(x) = f(x)$.

Version A simplifies the whole business of finding the value of a definite integral: Just find an antiderivative of the integrand. Of course, as you know, this isn't always as simple as it sounds. At least half of your calculus course was probably devoted to techniques of substitution and integration by parts, trying to recognize integrands as derivatives resulting from the Product Rule, the Chain Rule, and so forth.

A slight twist on version A tells us that if we integrate f', the *rate* function over $[a, b]$, we get the *total change* in f, the *amount* function over the same interval. For example, if $s(t)$, $v(t)$, and $a(t)$ denote the position, velocity, and acceleration, respectively, of a moving object at time t, then we have

$$\int v(t)\, dt = s(t) + C \quad \text{and} \quad \int a(t)\, dt = v(t) + K,$$

where C and K denote arbitrary constants. Consequently, we can write

$$\int_a^b v(t)\, dt = \int_a^b s'(t)\, dt = s(b) - s(a),$$

which says that if we integrate the *velocity* function, we get the *total change in position* of a moving object as t changes from a to b. If we integrate the *speed* function—the absolute value of the velocity—we get the *total distance* traveled by the object. (See Example 1.2.8 in the text.)

As useful as version A is in solving differential equations, version B extends the notion of differentiation (and therefore of integration) to functions defined by integrals. (See Problems 10 and 11 in Exercises 1.2.)

EXAMPLE

Suppose that $Q(x) = \displaystyle\int_{-2}^{x} \cos(u^2)\, du$. Then $Q'(x) = \cos(x^2)$, $Q''(x) = -2x \sin(x^2)$, and so on.

◆

5. PARTIAL FRACTIONS

An important and useful result from algebra says that every rational function (quotient of polynomial functions), no matter how complicated, comes from adding simpler fractions. For example, the function

$$\frac{8x + 1}{x^2 - x - 6}$$

comes from the following addition of simpler pieces:

$$\frac{3}{x + 2} + \frac{5}{x - 3} = \frac{3(x - 3) + 5(x + 2)}{(x + 2)(x - 3)} = \frac{8x + 1}{x^2 - x - 6}.$$

In calculus, when we have an integrand that is a rational function, we can reverse this addition process to find the simpler fractions, fractions that we can integrate easily. Thus, for example,

$$\int \frac{8x + 1}{x^2 - x - 6}\, dx = \int \frac{3}{x + 2}\, dx + \int \frac{5}{x - 3}\, dx = 3\ln|x + 2| + 5\ln|x - 3| + C.$$

In this example, the algebraic challenge is to find constants A and B such that

$$\frac{8x + 1}{x^2 - x - 6} = \frac{A}{x + 2} + \frac{B}{x - 3}. \tag{*}$$

The fractions $A/(x + 2)$ and $B/(x - 3)$ are called **partial fractions** because each contributes a piece of the whole. In particular, the denominators $x + 2$ and $x - 3$ are parts (factors) of the original denominator $x^2 - x - 6$. The numbers A and B are called **undetermined coefficients** (see Section 5.6 and Appendix D). To find A and B, we clear equation (*) of fractions by multiplying both sides by $x^2 - x - 6$. The result is

$$8x + 1 = A(x - 3) + B(x + 2).$$

This is supposed to be an identity in x. If we let $x = 3$, we find that $8(3) + 1 = 0 + 5B$, or $B = 5$. Similarly, letting $x = -2$, we get $8(-2) + 1 = -5A + 0$, so $A = 3$.

This technique works for a rational function in lowest terms whose denominator can be factored into distinct linear factors. More complicated denominators can also be handled by this kind of algebraic method, but you can find a more detailed discussion in your calculus text. Most computer algebra systems can produce such "partial-fraction decompositions" and can evaluate integrals with integrands that are rational functions. Partial fractions are particularly useful in Section 2.1 and in Chapter 6.

6. IMPROPER INTEGRALS

In dealing with the definite (Riemann) integral $\int_a^b f(x)\,dx$, we make two basic assumptions: (1) The interval $[a, b]$ is finite, and (2) the integrand f is bounded (that is, does not become infinite) on the closed interval $[a, b]$. If we violate one or both of these assumptions, we encounter a type of **improper integral.**

First let us assume that we want to consider the interval $[a, \infty)$ or $(-\infty, b]$, where a and b are real numbers. We can define

$$\int_a^\infty f(x)\,dx = \lim_{B \to \infty} \int_a^B f(x)\,dx \quad \text{or} \quad \int_{-\infty}^b f(x)\,dx = \lim_{A \to \infty} \int_{-A}^b f(x)\,dx$$

provided that each limit exists. If the limit exists, we say that the improper integral **converges.** Otherwise, we say that the improper integral **diverges.** Finally,

$$\int_{-\infty}^\infty f(x)\,dx = \lim_{A \to \infty} \int_{-A}^c f(x)\,dx + \lim_{B \to \infty} \int_c^B f(x)\,dx$$

provided that each limit on the right-hand side exists individually. Here c is an arbitrary real number. It is *not* correct to define $\int_{-\infty}^\infty f(x)\,dx$ as $\lim_{C \to \infty} \int_{-C}^C f(x)\,dx$.

EXAMPLE

$$\int_1^\infty \frac{dx}{1 + x^2} = \lim_{B \to \infty} \arctan x \Big|_1^B = \lim_{B \to \infty} (\arctan B - \arctan 1)$$

$$= \lim_{B \to \infty} \left(\arctan B - \frac{\pi}{4} \right) = \lim_{B \to \infty} \arctan B - \frac{\pi}{4}$$

$$= \frac{\pi}{2} - \frac{\pi}{4} = \frac{\pi}{4}.$$

EXAMPLE

Consider $\int_0^\infty \sin x \, dx$. The limit

$$\lim_{B \to \infty} \int_0^B \sin x \, dx = \lim_{B \to \infty} (-\cos(B) + \cos(0)) = -\lim_{B \to \infty} \cos(B) + 1$$

doesn't exist because $\cos(B)$ oscillates from -1 to 1 as B tends to infinity. ◆

When we are dealing with this first type of improper integral, for which the interval is not finite, sometimes a form of *L'Hôpital's Rule* comes in handy: Suppose that as $x \to a$, where a is $\pm\infty$, $f(x) \to \pm\infty$ and $g(x) \to \pm\infty$. If $\lim_{x \to a} \dfrac{f'(x)}{g'(x)} = L$, where L is either a real number or $\pm\infty$, then $\lim_{x \to a} \dfrac{f(x)}{g(x)} = L$.

EXAMPLE

Consider Euler's *gamma function*, defined by $\Gamma(x) = \int_0^\infty t^{x-1} e^{-t} dt$. Integration by parts tells us that

$$\Gamma(x) = -t^{x-1} e^{-t}\big]_0^\infty - \int_0^\infty (x-1) t^{x-2} (-e^{-t}) dt$$

$$= \lim_{c \to \infty} \frac{-t^{c-1}}{e^c} + (x-1) \int_0^\infty t^{x-2} e^{-t} dt$$

$$= (x-1) \cdot \Gamma(x-1),$$

where we have used L'Hôpital's Rule several times in evaluating the limit. (Successive differentiations of the numerator and denominator of $\dfrac{-t^{c-1}}{e^c}$ eventually give us $-(c-1)!$ in the numerator, whereas the denominator remains e^c, so the limit as c tends to infinity is 0.) Note that because $\Gamma(1) = \Gamma(2) = 1$, we can conclude that $\Gamma(x+1) = x \cdot (x-1) \cdot (x-2) \cdot (x-3) \cdots 3 \cdot 2 \cdot 1 = x!$ when x is an integer, so the gamma function provides a generalization of $n!$ to the case in which n is not an integer. (See Appendix D.3, especially footnote 4.) ◆

Now let's suppose that f is defined and finite on the interval $[a, b]$ except at the endpoint b. Then the integral $\int_a^b f(x) dx$ is improper, and we define it as

$$\int_a^b f(x) \, dx = \lim_{B \to b^-} \int_a^B f(x) \, dx$$

provided that this *left-hand limit* (or *limit from the left*) exists. Similarly, if f is unbounded at the endpoint a, then we define

$$\int_a^b f(x) \, dx = \lim_{A \to a^+} \int_A^b f(x) \, dx$$

provided that this *right-hand limit* (or *limit from the right*) exists.

EXAMPLE

The function $1/\sqrt{1-x^2}$ is unbounded at $x = 1$ (and at $x = -1$). The improper integral of this function on the interval $[0, 1]$ converges:

$$\int_0^1 \frac{dx}{\sqrt{1-x^2}} = \lim_{B \to 1^-} \int_0^B \frac{dx}{\sqrt{1-x^2}} = \lim_{B \to 1^-} \arcsin(B) - \arcsin(0) = \frac{\pi}{2}. \qquad \blacklozenge$$

Another possibility is that the function f is defined and finite on $[a, b]$ except at a point ξ *inside* the interval. The improper integral is then defined as

$$\int_a^b f(x)\,dx = \lim_{c \to \xi^-} \int_a^c f(x)\,dx + \lim_{d \to \xi^+} \int_d^b f(x)\,dx$$

provided that both one-sided limits exist.

EXAMPLE

$$\begin{aligned}
\int_0^2 \frac{dx}{(x-1)^{2/3}} &= \lim_{c \to 1^-} \int_0^c \frac{dx}{(x-1)^{2/3}} + \lim_{d \to 1^+} \int_d^2 \frac{dx}{(x-1)^{2/3}} \\
&= \lim_{c \to 1^-} 3(x-1)^{1/3}\Big|_0^c + \lim_{d \to 1^+} 3(x-1)^{1/3}\Big|_d^2 \\
&= \lim_{c \to 1^-} [3(c-1)^{1/3} - 3(-1)^{1/3}] + \lim_{d \to 1^+} [3(1)^{1/3} - 3(d-1)^{1/3}] \\
&= 3 + 3 = 6. \qquad \blacklozenge
\end{aligned}$$

7. FUNCTIONS OF SEVERAL VARIABLES / PARTIAL DERIVATIVES

Sometimes we encounter functions that depend on more than one independent variable. For example, the area of a rectangle depends on both its length and its width. We can express this relationship as $A = f(l, w) = l \cdot w$. In general, if there are two independent variables (x and y) and one dependent variable (z), we can express this situation as $z = f(x, y)$. In words, the variable z depends on (is a function of) the variables x and y. This means that changes in the value of either x or y (or both) will lead to changes in z. The *instantaneous rate of change of z with respect to x* is given by the **partial derivative of z with respect to x,** which is defined by the formula

$$\frac{\partial z}{\partial x} = \lim_{h \to 0} \frac{f(x+h, y) - f(x, y)}{h}.$$

Similarly, the *instantaneous rate of change of z with respect to y* is given by the **partial derivative of z with respect to y,** which is defined by the formula

$$\frac{\partial z}{\partial y} = \lim_{h \to 0} \frac{f(x, y+h) - f(x, y)}{h}.$$

What this means in terms of practical calculation is that to find $\dfrac{\partial z}{\partial x}$, you just treat y as a constant and differentiate with respect to x as usual. For $\dfrac{\partial z}{\partial y}$, you treat x as a constant and regard y as the "live" variable.

EXAMPLE

If $z = f(x, y) = x^2 y^2 - 3xy^3 + 5x^4 y^2$, then

$$\frac{\partial z}{\partial x} = 2xy^2 - 3y^3 + 20x^3 y^2 \quad \text{and} \quad \frac{\partial z}{\partial y} = 2x^2 y - 9xy^2 + 10x^4 y. \qquad \blacklozenge$$

EXAMPLE

Suppose $w = e^{2x+3y} \sin(xy)$. Then, using the Product Rule and the Chain Rule, we find that

$$\frac{\partial w}{\partial x} = e^{2x+3y} \cos(xy)\, y + 2e^{2x+3y} \sin(xy)$$

and

$$\frac{\partial w}{\partial y} = e^{2x+3y} \cos(xy)\, x + 3e^{2x+3y} \sin(xy). \qquad \blacklozenge$$

In general, if you have a function of n variables, $z = f(x_1, x_2, x_3, \ldots, x_n)$, then you can define the partial derivative of z with respect to x_k and calculate it by treating x_k as the only true variable, the other x_i ($i \neq k$) being treated as constants. You can define higher derivatives and *mixed* derivatives in the obvious way: $\dfrac{\partial^2 z}{\partial x_i \partial x_k}, \dfrac{\partial^n z}{\partial x_k^n}$, and so on.

EXAMPLE

Using $z = f(x, y) = x^2 y^2 - 3xy^3 + 5x^4 y^2$ and the results of the first example, we have

$$\frac{\partial^2 z}{\partial x^2} = \frac{\partial}{\partial x}\left(\frac{\partial z}{\partial x}\right) = \frac{\partial}{\partial x}(2xy^2 - 3y^3 + 20x^3 y^2) = 2y^2 + 60x^2 y^2$$

$$\frac{\partial^2 z}{\partial y^2} = \frac{\partial}{\partial y}\left(\frac{\partial z}{\partial y}\right) = \frac{\partial}{\partial y}(2x^2 y - 9xy^2 + 10x^4 y) = 2x^2 - 18xy + 10x^4$$

$$\frac{\partial^2 z}{\partial x \partial y} = \frac{\partial}{\partial x}\left(\frac{\partial z}{\partial y}\right) = \frac{\partial}{\partial x}(2x^2 y - 9xy^2 + 10x^4 y) = 4xy - 9y^2 + 40x^3 y,$$

and so forth. $\qquad \blacklozenge$

8. THE TANGENT PLANE; THE TAYLOR EXPANSION OF $f(x, y)$

In Section 1 of this appendix, we saw that the tangent line $y = f(a) + f'(a)(x - a)$ gives the best linear approximation of a single-variable function f near $x = a$. For $F(x, y)$, a function of *two* variables, the best approximation near a point (a, b) is provided by the **tangent plane** given by the approximation formula

$$F(x, y) \approx F(a, b) + \frac{\partial F}{\partial x}(a, b)(x - a) + \frac{\partial F}{\partial y}(a, b)(y - b),$$

where $\dfrac{\partial F}{\partial x}(a, b)$ and $\dfrac{\partial F}{\partial y}(a, b)$ denote the partial derivatives evaluated at the point (a, b).

EXAMPLE

Let's calculate the tangent plane approximation of the function $F(x, y) = x^3 - x^2y^2 + y^3$ near the point $(a, b) = (1, 2)$. We have $\dfrac{\partial F}{\partial x} = 3x^2 - 2xy^2$ and $\dfrac{\partial F}{\partial y} = -2x^2y + 3y^2$, so $F(1, 2) = 1^3 - 1^2 2^2 + 2^3 = 5, \dfrac{\partial F}{\partial x}(1, 2) = 3(1)^2 - 2(1)(2)^2 = -5$, and $\dfrac{\partial F}{\partial y}(1, 2) = -2(1)^2(2) + 3(2)^2 = 8$. Putting these results into the tangent plane formula, we get

$$\begin{aligned}
F(x, y) &\approx F(1, 2) + \frac{\partial F}{\partial x}(1, 2)(x - 1) + \frac{\partial F}{\partial y}(1, 2)(y - 2) \\
&= 5 - 5(x - 1) + 8(y - 2).
\end{aligned}$$

Figure A2 shows the three-dimensional picture of the surface and its tangent plane.

For points (x, y) close to $(1, 2)$, the values of z on the tangent plane are close to the values of z on the surface defined by $z = F(x, y)$. ◆

We can define a full Taylor series expansion of a function of several variables, but for this text we need only the idea of the tangent plane (linear) approximation.

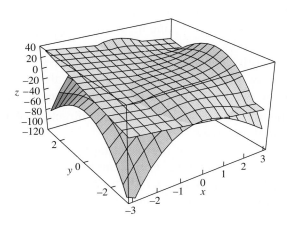

Figure A2
Tangent plane to the surface $z = x^3 - x^2y^2 + y^3$ at $(1, 2)$

APPENDIX B

Vectors and Matrices

Appendix B is intended to offer either a brief review of, or an introduction to, the basic ideas of vector and matrix algebra needed in this text.

1. VECTORS AND VECTOR ALGEBRA; POLAR COORDINATES

A **vector** is a quantity that has both magnitude (size) and direction. In mathematics and physics, there are two usual ways to represent a vector: (1) As an ordered pair of real numbers, written (x, y) or $\begin{bmatrix} x \\ y \end{bmatrix}$; and (2) as an *arrow* from the origin (usually) of the x-y plane to a point (x, y) or $\begin{bmatrix} x \\ y \end{bmatrix}$. The numbers x and y are called the **components** or **coordinates** of the vector. As shown in Chapter 5, we can also consider vectors with complex-number coordinates and vectors whose components are functions. For the sake of simplicity in this appendix, we'll work with vectors whose components are real numbers. Also, we will usually write vectors horizontally.

Both ways of looking at a vector are shown in Figure B1. In the second (the "arrow") view, the vector $v = (x, y)$ is always the hypotenuse of a right triangle, so by the Pythagorean Theorem its **length**—denoted $|v|$—is given by the expression $\sqrt{x^2 + y^2}$. Of course, the direction of a vector is indicated by the direction of the arrow.

Vectors—often representing forces of various kinds—can be combined to indicate interactions. For example, you can add two vectors as follows: If $v_1 = (x_1, y_1)$ and $v_2 = (x_2, y_2)$, then $v_1 + v_2 = (x_1, y_1) + (x_2, y_2) = (x_1 + x_2, y_1 + y_2)$. Subtraction is similar. You can also multiply a vector by a real number (or a complex number or a function), which is called a **scalar** in this situation. To do this, just

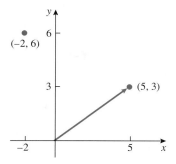

Figure B1

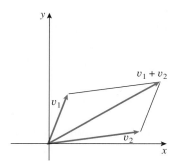

Figure B2
The Parallelogram Law

multiply each component of the vector by the scalar: If $v = (x, y)$ and r is any real number, then $rv = (rx, ry)$. Because the components of vectors are real numbers, we should expect the usual rules of algebra to apply. If v_1, v_2, and v_3 are vectors and r is a scalar, then $v_1 + v_2 = v_2 + v_1$ [commutative property], $v_1 + (v_2 + v_3) = (v_1 + v_2) + v_3$ [associative property], and $r(v_1 + v_2) = rv_1 + rv_2$ [distributive property]. There is a **zero vector,** denoted by $\mathbf{0} = (0, 0)$, such that $v_1 + \mathbf{0} = v_1 = \mathbf{0} + v_1$ [additive identity].

Geometrically, the addition or subtraction of vectors is captured by the **Parallelogram Law** (see Figure B2).

Another way of representing a vector in two-dimensional space is by using **polar coordinates** (Figure B3). If we have a vector corresponding to the point (x, y), then we can describe it using its *length r* (its radial distance from the origin) and the angle θ that the arrow makes with the positive *x*-axis, measured in a counterclockwise direction. As we saw above, the length is given by the formula $r = \sqrt{x^2 + y^2}$.

As we look at Figure B3, simple trigonometry tells us that $x = r \cos\theta$, $y = r \sin\theta$, and $\theta = \arctan\left(\dfrac{y}{x}\right)$, $x \neq 0$. As indicated in some of the examples in

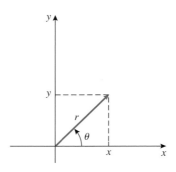

Figure B3
Polar representation of a vector

Chapter 7, the polar representation of vectors may be more natural in problems involving expressions that look like x^2, y^2, $x^2 + y^2$, and so on.

There is no reason to restrict our definition of vectors to two dimensions. In three-dimensional space, a vector is an ordered triplet, (x, y, z) or $\begin{bmatrix} x \\ y \\ z \end{bmatrix}$, of real numbers or an arrow joining the origin $(0, 0, 0)$ to the point (x, y, z). In general, an ***n*-dimensional vector** is an ordered n-tuple, $(x_1, x_2, x_3, \ldots, x_n)$ or $\begin{bmatrix} x_1 \\ x_2 \\ x_3 \\ \vdots \\ x_n \end{bmatrix}$, of real numbers. The coordinate-by-coordinate arithmetic/algebra of vectors generalizes to any dimension in the obvious way.

Given a set of vectors $\{V_1, V_2, \ldots, V_m\}$, any vector of the form

$$c_1 V_1 + c_2 V_2 + \cdots + c_m V_m,$$

where $c_1, c_2, \ldots, c_m$ are scalars, is called a **linear combination** of the set of vectors. The set of vectors $\{V_1, V_2, \ldots, V_m\}$ is called **linearly independent** if the only way we can have

$$c_1 V_1 + c_2 V_2 + \cdots + c_m V_m = \mathbf{0} \quad \text{(the zero vector)}$$

is if $c_1 = c_2 = \cdots = c_m = 0$. Otherwise, the set of vectors is **linearly dependent.** Linear dependence implies that at least one vector in the set can be expressed as a linear combination of the others.

EXAMPLE

We will determine whether the following vectors are linearly independent:

$$V_1 = \begin{bmatrix} 1 \\ 1 \\ 0 \\ 0 \end{bmatrix}, V_2 = \begin{bmatrix} 1 \\ 0 \\ 1 \\ 0 \end{bmatrix}, V_3 = \begin{bmatrix} 0 \\ 0 \\ 1 \\ 1 \end{bmatrix}, V_4 = \begin{bmatrix} 0 \\ 1 \\ 0 \\ 1 \end{bmatrix}.$$

The statement $c_1 V_1 + c_2 V_2 + c_3 V_3 + c_4 V_4 = \mathbf{0}$ is equivalent to the system of algebraic equations

$$
\begin{aligned}
c_1 + c_2 &= 0 \\
c_1 + c_4 &= 0 \\
 + c_2 + c_3 &= 0 \\
 + c_3 + c_4 &= 0.
\end{aligned}
$$

This system is not difficult to solve by hand using substitution or elimination, but we can also use the capability of a graphing calculator or CAS to solve such systems of equations. In any case, we find that $c_1 = 1$, $c_2 = -1$, $c_3 = 1$, and $c_4 = -1$ is a solution. Because the scalars are not all zero, we conclude that the four vectors are *linearly dependent*. Note, for example, that we can write the first vector as a linear combination of the remaining vectors: $V_1 = V_2 - V_3 + V_4$. ◆

2. MATRICES AND BASIC MATRIX ALGEBRA

A **matrix** (the plural is **matrices**) is simply a rectangular arrangement (array) of numbers or other mathematical objects (such as functions) and is usually denoted by a capital letter. It can be considered a generalization of a vector. For example, we can have the matrix

$$
A = \begin{bmatrix} 0 & -4 & 1/2 & 9 \\ \pi & 14/5 & -0.15 & 2 \\ 7 & \sqrt{3} & 0 & -3 \end{bmatrix}.
$$

The numbers or objects making up a matrix are called its **elements** or **entries.** Most of the time we'll use real numbers, although complex numbers and even functions can appear as entries of matrices (as they can for components of vectors).

One way to describe a matrix is by indicating how many rows and columns it has. Matrix A above has 3 rows and 4 columns and is called a *3 by 4 matrix*, or a *3 × 4 matrix*. A matrix with m rows and n columns is called an ***m by n matrix* (*m ×
n matrix*).** Note that each row or column of a matrix can be considered a vector. An $n \times 1$ matrix is called a **column vector,** whereas a $1 \times n$ matrix is called a **row vector.** Two matrices are called **equal** if they have the same number of rows and columns and their corresponding elements are equal. For example, we can write

$$
\begin{bmatrix} 1 & 0 & -5/3 \\ 1/\sqrt{2} & 3 & 0.25 \end{bmatrix} = \begin{bmatrix} 7/7 & 0 & -15/9 \\ \sqrt{2}/2 & 15/5 & 1/4 \end{bmatrix}.
$$

You can add and subtract matrices of the same shape by adding or subtracting their corresponding elements, but (for example) you can't add a 3 by 4 matrix and a 4 by 3 matrix or subtract one of these from the other. If we take the matrix A that we have already defined and introduce the matrix B, so that we have

$$
A = \begin{bmatrix} 0 & -4 & 1/2 & 9 \\ \pi & 14/5 & -0.15 & 2 \\ 7 & \sqrt{3} & 0 & -3 \end{bmatrix} \quad \text{and} \quad B = \begin{bmatrix} -3 & 5 & -1/2 & 4 \\ 3/4 & 6/5 & 0.65 & 8 \\ -9 & \sqrt{2} & 8 & 3 \end{bmatrix},
$$

then

$$A + B = \begin{bmatrix} 0 + (-3) & -4 + 5 & 1/2 + (-1/2) & 9 + 4 \\ \pi + 3/4 & 14/5 + 6/5 & -0.15 + 0.65 & 2 + 8 \\ 7 + (-9) & \sqrt{3} + \sqrt{2} & 0 + 8 & -3 + 3 \end{bmatrix}$$

$$= \begin{bmatrix} -3 & 1 & 0 & 13 \\ \pi + 3/4 & 4 & 0.5 & 10 \\ -2 & \sqrt{3} + \sqrt{2} & 8 & 0 \end{bmatrix}$$

and

$$A - B = \begin{bmatrix} 0 - (-3) & -4 - 5 & 1/2 - (-1/2) & 9 - 4 \\ \pi - 3/4 & 14/5 - 6/5 & -0.15 - 0.65 & 2 - 8 \\ 7 - (-9) & \sqrt{3} - \sqrt{2} & 0 - 8 & -3 - 3 \end{bmatrix}$$

$$= \begin{bmatrix} 3 & -9 & 1 & 5 \\ \pi - 3/4 & 8/5 & -0.8 & -6 \\ 16 & \sqrt{3} - \sqrt{2} & -8 & -6 \end{bmatrix}.$$

The role of zero in matrix algebra is played by the **zero matrix** of the appropriate size—the matrix all of whose entries are zero.

We can also multiply a matrix by a number (or even a function) called a **scalar,** as in the case for vectors. We just multiply every element of the matrix by that scalar:

$$-5 \cdot \begin{bmatrix} 3 & -2 & 0 \\ -7 & 4 & 1/3 \\ 5 & -6 & \sqrt{2} \end{bmatrix} = \begin{bmatrix} -5(3) & -5(-2) & -5(0) \\ -5(-7) & -5(4) & -5(1/3) \\ -5(5) & -5(-6) & -5(\sqrt{2}) \end{bmatrix}$$

$$= \begin{bmatrix} -15 & 10 & 0 \\ 35 & -20 & -5/3 \\ -25 & 30 & -5\sqrt{2} \end{bmatrix}.$$

We've just multiplied a 3 by 3 matrix—one type of **square matrix**—by the scalar -5.

3. LINEAR TRANSFORMATIONS AND MATRIX MULTIPLICATION

The really interesting thing about matrix arithmetic and algebra is how we *multiply* matrices. The natural thing to do—take two matrices with the same shape and multiply their corresponding elements—is not meaningful in the theory of linear algebra. Instead, there is a *row-by-column* process that looks strange at first but becomes more natural when you see its applications.

To motivate the multiplication of matrices, let's return to elementary algebra for a moment and look at a system of two equations in two unknowns:

$$-2x + 3y = 5$$
$$x - 4y = -2.$$

In connection with this system, we can think of a point (x, y) in the plane transformed into another point as follows:

$$T(x, y) = (-2x + 3y, x - 4y).$$

For example,

$$T(1, 0) = (-2(1) + 3(0), 1 - 4(0)) = (-2, 1)$$
$$T(-4, 5) = (-2(-4) + 3(5), -4 - 4(5)) = (23, -24)$$

and

$$T(-2.8, -0.2) = (-2(-2.8) + 3(-0.2), -2.8 - 4(-0.2)) = (5, -2).$$

Note that this last calculation says that the ordered pair $(x, y) = (-2.8, -0.2)$ is a solution of our system of linear equations.

Geometrically, the point $(1, 0)$ has been moved to the location $(-2, 1)$, the point $(-4, 5)$ has been changed to $(23, -24)$, and the point $(-2.8, -0.2)$ has been transformed into $(5, -2)$. If we think of a point (x, y) as defining a vector, then the transformation stretches (or shrinks) the vector and rotates it through some angle θ until it becomes another vector. Figure B4 shows this interpretation of the effect of T on the vector $(1, 0)$.

More abstractly, we should be able to see that T is a **linear transformation** of points (x, y) in the plane to other points $(\hat{x}, \hat{y})$ in the plane: If $u = (x_1, y_1)$ and $v = (x_2, y_2)$, then $T(c_1 u + c_2 v) = T(c_1 u) + T(c_2 v) = c_1 T(u) + c_2 T(v)$ for any constants c_1 and c_2.

Matrix notation was invented by the English mathematician Arthur Cayley precisely to describe linear transformations. If $T(x, y) = (\hat{x}, \hat{y})$, where

$$ax + by = \hat{x}$$
$$cx + dy = \hat{y},$$

then we can pick out the coefficients $a, b, c,$ and d and write them in a square array $A = \begin{bmatrix} a & b \\ c & d \end{bmatrix}$ called a matrix. If we know what variables $x, y, \hat{x},$ and $\hat{y}$ we're using,

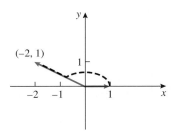

Figure B4
The effect of $T(x, y) = (-2x + 3y, x - 4y)$
on the vector $(1, 0)$

then knowing this **matrix of coefficients** enables us to understand what T is doing to points in the plane. We can focus on these variables by introducing the vectors $X = \begin{bmatrix} x \\ y \end{bmatrix}$ and $\hat{X} = \begin{bmatrix} \hat{x} \\ \hat{y} \end{bmatrix}$. Now we can write our system of equations compactly as

$$\begin{bmatrix} a & b \\ c & d \end{bmatrix} \begin{bmatrix} x \\ y \end{bmatrix} = \begin{bmatrix} \hat{x} \\ \hat{y} \end{bmatrix},$$

or $AX = \hat{X}$. To make sense, the "product" of A and X must be the column matrix $\begin{bmatrix} ax + by \\ cx + dy \end{bmatrix}$, which leads to a *row-by-column multiplication*:

$$\begin{bmatrix} a & b \end{bmatrix} \begin{bmatrix} x \\ y \end{bmatrix} = ax + by \quad \text{and} \quad \begin{bmatrix} c & d \end{bmatrix} \begin{bmatrix} x \\ y \end{bmatrix} = cx + dy.$$

Furthermore, the multiplication of two matrices of the appropriate sizes can be interpreted as a *composition of transformations*—one transformation followed by another.

EXAMPLE

Suppose that we have two linear transformations defined by

$$M(x, y) = (x + 2y, 3x + 4y) \quad \text{and} \quad P(x, y) = (-2x, x + 3y).$$

Then

$$\begin{aligned} (M \circ P)(x, y) = M(P(x, y)) &= M(-2x, x + 3y) \\ &= (-2x + 2(x + 3y), 3(-2x) + 4(x + 3y)) \\ &= (6y, -2x + 12y). \end{aligned}$$

In particular, $(M \circ P)(1, 1) = (6, 10)$.

The matrices of coefficients for the transformations M and P look like $M = \begin{bmatrix} 1 & 2 \\ 3 & 4 \end{bmatrix}$ and $P = \begin{bmatrix} -2 & 0 \\ 1 & 3 \end{bmatrix}$, so the composition $M \circ P$ takes the form of a *product of 2×2 matrices*:

$$\begin{aligned} \begin{bmatrix} 1 & 2 \\ 3 & 4 \end{bmatrix} \begin{bmatrix} -2 & 0 \\ 1 & 3 \end{bmatrix} \begin{bmatrix} x \\ y \end{bmatrix} &= \begin{bmatrix} 1 & 2 \\ 3 & 4 \end{bmatrix} \begin{bmatrix} -2x \\ x + 3y \end{bmatrix} \\ &= \begin{bmatrix} -2x + 2(x + 3y) \\ 3(-2x) + 4(x + 3y) \end{bmatrix} = \begin{bmatrix} 6y \\ -2x + 12y \end{bmatrix} \end{aligned}$$

and when $\begin{bmatrix} x \\ y \end{bmatrix} = \begin{bmatrix} 1 \\ 1 \end{bmatrix}$,

$$\begin{bmatrix} 1 & 2 \\ 3 & 4 \end{bmatrix} \begin{bmatrix} -2 & 0 \\ 1 & 3 \end{bmatrix} \begin{bmatrix} 1 \\ 1 \end{bmatrix} = \begin{bmatrix} 6 \\ 10 \end{bmatrix}.$$

You should check to see that $(M \circ P)(x, y) \neq (P \circ M)(x, y)$ or, equivalently, that

$$\begin{bmatrix} 1 & 2 \\ 3 & 4 \end{bmatrix} \begin{bmatrix} -2 & 0 \\ 1 & 3 \end{bmatrix} \neq \begin{bmatrix} -2 & 0 \\ 1 & 3 \end{bmatrix} \begin{bmatrix} 1 & 2 \\ 3 & 4 \end{bmatrix}.$$

◆

Looking at the last example, we see that transforming the vector (x, y) by P and then by M is equivalent to transforming the vector by the single transformation $T(x, y) = (6y, -2x + 12y)$. In matrix terms, we can express the effect of the composition $M \circ P$ as

$$\begin{bmatrix} 0 & 6 \\ -2 & 12 \end{bmatrix} \begin{bmatrix} x \\ y \end{bmatrix} = \begin{bmatrix} 6y \\ -2x + 12y \end{bmatrix}.$$

Note what we get when we add the results of multiplying each element of the first *row* of the matrix associated with M, $[1 \quad 2]$, by the corresponding element of the first *column* of the matrix associated with P, $\begin{bmatrix} -2 \\ 1 \end{bmatrix}$: $(1)(-2) + (2)(1) = 0$, which happens to be the first row, first column element of the matrix corresponding to $M \circ P$. Similarly, for example, combining the second row of the matrix associated with M, $(3 \quad 4)$, and the first column of the matrix associated with P, $\begin{bmatrix} -2 \\ 1 \end{bmatrix}$, we get the element in the second row, first column of $M \circ P$: $(3)(-2) + (4)(1) = -2$. In this way we can describe the matrix for $M \circ P$ as the **product** of the matrix representing M and the matrix representing P.

In general, if A and B are both 2×2 matrices, the element in row i and column j of the product matrix $C = AB$ is found by adding the products of each element of row i of matrix A and the corresponding element in column j of matrix B. For example, here's what the matrix product corresponding to $M \circ P$ in the last example looks like in full:

$$\begin{bmatrix} 1 & 2 \\ 3 & 4 \end{bmatrix} \begin{bmatrix} -2 & 0 \\ 1 & 3 \end{bmatrix} = \begin{bmatrix} (1)(-2) + (2)(1) & (1)(0) + (2)(3) \\ (3)(-2) + (4)(1) & (3)(0) + (4)(3) \end{bmatrix} = \begin{bmatrix} 0 & 6 \\ -2 & 12 \end{bmatrix}.$$

You should be able to calculate the matrix product corresponding to $P \circ M$. You'll notice that the *order* of composition/multiplication counts: The matrix corresponding to $M \circ P$ is not necessarily the matrix corresponding to $P \circ M$. In general, **matrix multiplication is not commutative:** *If A and B are two matrices that can be multiplied (see the next paragraph), then $AB \neq BA$ in general.*

This situation of one function or transformation followed by another is the motivation for matrix multiplication. The general multiplication of matrices remains the row-by-column procedure described for 2×2 matrices. In order for us to calculate the matrix product $C = AB$, the number of columns of A must be the same as the number of rows of B. Let $C = AB$, where A is $m \times r$ and B is $r \times n$. Then the product is a matrix with m rows and n columns:

$$A \quad \cdot \quad B \quad = \quad C.$$
$$m \times r \quad r \times n \quad m \times n$$

Thus if A is a 3 by 5 matrix and B is a 5 by 7 matrix, you can find the product AB, which will be a 3 by 7 matrix. However, the product BA does *not* make sense because the number of columns of B (7) does not equal the number of rows of A (3).

If the sizes of A and B are compatible as described above, then c_{ij}, the element in row i and column j of the product matrix C, is just the sum of the products of each element of row i of matrix A and the corresponding element in column j of matrix B. Letting a_{ik} denote the entry in row i and column k of matrix A, and letting b_{kj} denote the element in row k and column j of matrix B, we can write the last sentence more concisely as

$$c_{ij} = \sum_{k=1}^{r} a_{ik}b_{kj} = a_{i1}b_{1j} + a_{i2}b_{2j} + \cdots + a_{ir}b_{rj}. \tag{B.1}$$

Schematically, we can represent this matrix multiplication as follows:

$$
\begin{bmatrix}
c_{11} & c_{12} & \cdots & c_{1j} & \cdots & c_{1n} \\
c_{21} & c_{22} & \cdots & c_{2j} & \cdots & c_{2n} \\
\vdots & \vdots & \vdots & \vdots & \vdots & \vdots \\
c_{i1} & c_{i2} & \cdots & \boxed{c_{ij}} & \cdots & c_{in} \\
\vdots & \vdots & \vdots & \vdots & \vdots & \vdots \\
c_{m1} & c_{m2} & \cdots & c_{mj} & \cdots & c_{mn}
\end{bmatrix} =
$$

$$
\begin{bmatrix}
a_{11} & a_{12} & \cdots & a_{1r} \\
a_{21} & a_{22} & \cdots & a_{2r} \\
\vdots & \vdots & \vdots & \vdots \\
\boxed{a_{i1} \quad a_{i2} \quad \cdots \quad a_{ir}} \\
\vdots & \vdots & \vdots & \vdots \\
a_{m1} & a_{m2} & \cdots & a_{mr}
\end{bmatrix}
\begin{bmatrix}
b_{11} & b_{12} & \cdots & b_{1j} & \cdots & b_{1n} \\
b_{21} & b_{22} & \cdots & b_{2j} & \cdots & b_{2n} \\
\vdots & \vdots & \vdots & \vdots & \vdots & \vdots \\
b_{r1} & b_{r2} & \cdots & b_{rj} & \cdots & b_{rn}
\end{bmatrix}
$$

Here are some more examples of matrix multiplication.

EXAMPLE

$$
\begin{bmatrix} 2 & -3 & 0 \\ 4 & 0 & 1 \end{bmatrix} \cdot \begin{bmatrix} 1 & 2 \\ 3 & 4 \\ 5 & 6 \end{bmatrix} = \begin{bmatrix} 2(1) - 3(3) + 0(5) & 2(2) - 3(4) + 0(6) \\ 4(1) + 0(3) + 1(5) & 4(2) + 0(4) + 1(6) \end{bmatrix}
$$

$$
= \begin{bmatrix} -7 & -8 \\ 9 & 14 \end{bmatrix},
$$

$$
\begin{bmatrix} \pi & -2 & 6 \\ 0 & 4 & 1 \\ -3 & 5 & 7 \end{bmatrix} \cdot \begin{bmatrix} 2 & -3 & 0 \\ 9 & 2 & -6 \\ 2 & 1 & 4 \end{bmatrix}
$$

$$
= \begin{bmatrix}
\pi(2) - 2(9) + 6(2) & \pi(-3) - 2(2) + 6(1) & \pi(0) - 2(-6) + 6(4) \\
0(2) + 4(9) + 1(2) & 0(-3) + 4(2) + 1(1) & 0(0) + 4(-6) + 1(4) \\
-3(2) + 5(9) + 7(2) & -3(-3) + 5(2) + 7(1) & -3(0) + 5(-6) + 7(4)
\end{bmatrix}
$$

$$
= \begin{bmatrix}
2\pi - 6 & -3\pi + 2 & 36 \\
38 & 9 & -20 \\
53 & 26 & -2
\end{bmatrix}.
$$

◆

A particularly important and useful 2×2 matrix is the **identity matrix** $I = \begin{bmatrix} 1 & 0 \\ 0 & 1 \end{bmatrix}$. You should check to see that this matrix plays the same role in matrix algebra that the number 1 plays in arithmetic—that is, $I \cdot A = A \cdot I$ for any 2×2 matrix A. If matrix A is $2 \times n$, then $I \cdot A = A$, but $A \cdot I$ is not defined unless $n = 2$. Similarly, if A is an $n \times 2$ matrix, then $A \cdot I = A$, but $I \cdot A$ is not defined unless $n = 2$. In general, for any positive integer n, the $n \times n$ matrix with ones on the main diagonal (upper left corner to lower right corner) and zeros elsewhere serves as the identity matrix I for $n \times n$ matrix multiplication.

Given an $n \times n$ matrix A, the $n \times n$ matrix B is called the (*multiplicative*) *inverse* of A if $AB = I = BA$. If an inverse of A exists, then it is unique and is denoted by A^{-1}.

With the definitions we have seen, matrix addition and multiplication satisfy all the familiar basic rules of algebra—except for commutativity. For example, we have the *associative law for multiplication*: If A is an $m \times r$ matrix, B is an $r \times s$ matrix, and C is an $s \times n$ matrix, then $A(BC) = (AB)C$, an $m \times n$ matrix. We also have the *distributive law*: If A is an $m \times r$ matrix and B and C are $r \times n$ matrices, then $A(B + C) = AB + AC$ (which is an $m \times n$ matrix).

Let's prove the distributive law in the situation where A is an $m \times r$ matrix and B and C are $r \times 1$ matrices (vectors). We expect the product $A(B + C)$ to be a vector having m rows. Now suppose that a_{ik} denotes the element in row i, column k of A, that b_k and c_k are the elements in row k ($k = 1, 2, \ldots, r$), and that p_i is the element in row i of the product $A(B + C)$. Then by equation (B.1), we have (for $i = 1, 2, \ldots, m$)

$$p_i = \sum_{k=1}^{r} a_{ik}(b_k + c_k) = \sum_{k=1}^{r} (a_{ik}b_k + a_{ik}c_k) = \sum_{k=1}^{r} a_{ik}b_k + \sum_{k=1}^{r} a_{ik}c_k$$
$$= \text{(the entry in row } i \text{ of } AB) + \text{(the entry in row } i \text{ of } AC),$$

so we have shown that $A(B + C) = AB + AC$.

4. EIGENVALUES AND EIGENVECTORS

As we saw in the previous section, if A is an $n \times n$ matrix and X is a nonzero $n \times 1$ vector $\begin{bmatrix} x_1 \\ x_2 \\ \vdots \\ x_n \end{bmatrix}$, we can consider the multiplication of X by matrix A in the form AX as somehow transforming or changing the vector X. If there is a scalar λ such that $AX = \lambda X$, then λ is called an **eigenvalue** of A, and the vector X is called an **eigenvector** corresponding to λ. Geometrically, we're saying that *an eigenvector is a nonzero vector that gets changed into a constant multiple of itself.*

For example, identifying a vector $X = \begin{bmatrix} x \\ y \end{bmatrix}$ with the point (x, y) in the familiar Cartesian coordinate system, we can see that an eigenvector is a point (not the origin) such that it and its transformed self lie on the same straight line through the origin. The direction of an eigenvector is either unchanged (if $\lambda > 0$) or reversed (if $\lambda < 0$) when the vector is multiplied by A. The matrix equation $AX = \lambda X$ is like the functional equation $f(x) = \lambda x$, which represents a straight line through the origin with slope λ.

If we start with the assumption that $AX = \lambda X$, then $AX - \lambda X = \mathbf{0}$ (the zero vector), and the distributive property of matrix multiplication allows us to write $(A - \lambda I)X = \mathbf{0}$. (We must write $A - \lambda I$ instead of $A - \lambda$ because it doesn't make sense to subtract a number from a matrix.) If we can find an inverse for $A - \lambda I$ —that is, an $n \times n$ matrix B such that $(A - \lambda I)B = I = B(A - \lambda I)$—then we can divide the factor $(A - \lambda I)$ out of the matrix equation $(A - \lambda I)X = \mathbf{0}$ to get $X = \mathbf{0}$, the $n \times 1$ vector all of whose elements are 0. Therefore, remembering that an eigenvector was defined as a *nonzero* vector, we see that the only interesting situation occurs when the matrix $A - \lambda I$ does *not* have an inverse. (*Do you follow the logic?*)

The equation $(A - \lambda I)X = \mathbf{0}$ represents a homogeneous system of n algebraic linear equations in n unknowns, and the theory of linear algebra indicates that there is a number Δ, depending on the matrix $A - \lambda I$, with the following important property: If $\Delta \neq 0$, then the system $(A - \lambda I)X = \mathbf{0}$ has only the zero solution $x_1 = x_2 = \cdots = x_n = 0$. However, if $\Delta = 0$, then there is a solution

$$X = \begin{bmatrix} x_1 \\ x_2 \\ \vdots \\ x_n \end{bmatrix}$$ with at least one of the x_i different from zero. This number Δ is the

determinant of the matrix $A - \lambda I$, denoted by $\det(A - \lambda I)$. Therefore $(A - \lambda I)X = \mathbf{0}$ has a nonzero solution X only if $\det(A - \lambda I) = 0$. For any $n \times n$ linear system (homogeneous or not), the nature of the solutions depends on (that is, is *determined* by) whether the determinant is zero. The determinant is often calculated by means of successive operations on the rows and/or columns of the matrix. Rather than spend time learning tedious algorithms for finding determinants, you should learn how to get these numbers from your CAS. Even graphing calculators will give you a determinant if the matrix is not too large. From a more abstract point of view, a determinant is just a special kind of function from a set of square matrices to the real numbers.

For now, let's see how a determinant arises in solving a simple system of algebraic equations.

EXAMPLE

Suppose we want to solve the following system of two equations in two unknowns:

$$2x - 3y = 2$$
$$x + 4y = -5.$$

We can use the method of *elimination* to solve this system. For example, we can subtract twice the second equation from the first equation to eliminate the variable x and get $-11y = 12$, or $y = -\dfrac{12}{11}$. Then we can substitute this value of y in the second equation and solve for x. We get $x = -\dfrac{7}{11}$. Note that when we solve this particular system by elimination, the denominator of each component of the solution is 11.

Now write the system in matrix form:

$$\begin{bmatrix} 2 & -3 \\ 1 & 4 \end{bmatrix} \begin{bmatrix} x \\ y \end{bmatrix} = \begin{bmatrix} 2 \\ -5 \end{bmatrix}.$$

What do we get if we take the matrix of coefficients, multiply the *main-diagonal* (upper left, lower right) elements 2 and 4, and then subtract the product of the other diagonal elements -3 and 1? We get $(2)(4) - (-3)(1) = 11$. *Surprise!* The number calculated this way is the determinant of the coefficient matrix. In solving any system of linear equations in two unknowns, you always wind up dividing by the determinant—if it's not zero. *Cramer's Rule*, which you may have seen in a college algebra course, is a general $n \times n$ linear system solution formula that uses determinants. ◆

For a larger system, a CAS or graphing calculator provides important information about a system easily. Let's use technology in the next example to calculate the determinant, eigenvalues, and eigenvectors for a three-dimensional system.

EXAMPLE

Suppose we have a system with the matrix of coefficients

$$A = \begin{bmatrix} 2 & 2 & -6 \\ 2 & -1 & -3 \\ -2 & -1 & 1 \end{bmatrix}.$$

A CAS (*Maple* in this case) tells us that $\det(A) = 24$ and that the eigenvalues are $\lambda_1 = 6$, $\lambda_2 = -2 = \lambda_3$. The corresponding (linearly independent) eigenvectors are

$$\begin{bmatrix} -2 \\ -1 \\ 1 \end{bmatrix}, \begin{bmatrix} 1 \\ -2 \\ 0 \end{bmatrix}, \text{ and } \begin{bmatrix} 0 \\ 3 \\ 1 \end{bmatrix}.$$

If you try this example using your own CAS, the eigenvectors may not look like those here, but each should be a constant multiple of one of those given in the last paragraph. ◆

APPENDIX C

Complex Numbers

1. COMPLEX NUMBERS: THE ALGEBRAIC VIEW

Historically, the need for complex numbers arose when people tried to solve equations such as $x^2 + 1 = 0$ and realized that there was no real number that satisfied this equation. The basic element in the expansion of the number system is the **imaginary unit,** $i = \sqrt{-1}$. There is an interesting pattern to the powers of i: $i^1 = i, i^2 = -1, i^3 = -i, i^4 = 1, i^5 = i, i^6 = -1, i^7 = -i, i^8 = 1, \ldots$. You can use this repetition in groups of four, for example, to calculate a high power of i: $i^{338} = (i^2)^{169} = (-1)^{169} = -1$. A **complex number** is any expression of the form $x + yi$, where x and y are real numbers. If you have a complex number $z = x + yi$, then x is called the **real part**—denoted $\operatorname{Re}(z)$—and y is called the **imaginary part**—denoted $\operatorname{Im}(z)$—of the complex number. (Note that despite its name, y is a real number.) In particular, any real number x is a member of the family of complex numbers because it can be written as $x + 0 \cdot i$. Any complex number of the form $yi \; (= 0 + yi)$ is called a **pure imaginary number.**

Complex numbers can be added and subtracted in a reasonable way by combining real parts and imaginary parts as follows:

$$(a + bi) + (c + di) = (a + c) + (b + d)i$$

and

$$(a + bi) - (c + di) = (a - c) + (b - d)i.$$

You can also multiply complex numbers as you would multiply any binomials in algebra, remembering to replace i^2 whenever it occurs by -1:

$$(a + bi) \cdot (c + di) = ac + adi + bci + bdi^2 = (ac - bd) + (ad + bc)i.$$

Division of complex numbers is a bit trickier. If $z = x + yi$ is a complex number, then its **complex conjugate,** $\bar{z}$, is defined as follows: $\bar{z} = x - yi$. (You just reverse the sign of the imaginary part.) The complex conjugate is important in division because $z \cdot \overline{z} = x^2 + y^2$, a real number. (*Check this out.*) In division of complex numbers, the conjugate plays much the same role as the conjugate you learned to use in algebra to simplify fractions. For example, in algebra, if you were asked to simplify the fraction $\dfrac{3}{\sqrt{5}}$, you would "rationalize the denominator" as follows:

$$\frac{3}{\sqrt{5}} = \frac{3}{\sqrt{5}} \cdot \frac{\sqrt{5}}{\sqrt{5}} = \frac{3\sqrt{5}}{5}.$$

Another example from algebra makes the similarity between conjugates more obvious:

$$\frac{2 + \sqrt{3}}{3 - \sqrt{2}} = \frac{2 + \sqrt{3}}{3 - \sqrt{2}} \cdot \frac{3 + \sqrt{2}}{3 + \sqrt{2}} = \frac{6 + 2\sqrt{2} + 3\sqrt{3} + \sqrt{6}}{9 - 2}$$
$$= \frac{6 + 2\sqrt{2} + 3\sqrt{3} + \sqrt{6}}{7}.$$

In this last example, $3 + \sqrt{2}$ is the conjugate of $3 - \sqrt{2}$; when you multiply these conjugates, the radical sign disappears, leaving you with the integer 7. Now if we have to divide two complex numbers, we use the complex conjugate to get the answer, the quotient, to look like a complex number. For example,

$$\frac{2 + 3i}{3 + 5i} = \frac{2 + 3i}{3 + 5i} \cdot \frac{3 - 5i}{3 - 5i} = \frac{21 - i}{9 + 25} = \frac{21}{34} - \frac{1}{34}i.$$

In general, if $z = a + bi$ and $w = c + di$, then

$$\frac{z}{w} = \frac{a + bi}{c + di} = \frac{a + bi}{c + di} \cdot \frac{c - di}{c - di} = \frac{ac + bd}{c^2 + d^2} + \frac{bc - ad}{c^2 + d^2}i.$$

If z and w are complex numbers, you should be able to see that $\overline{\overline{z}} = z$, $\overline{(z + w)} = \overline{z} + \overline{w}, \overline{z \cdot w} = \overline{z} \cdot \overline{w}$, and $\overline{\left(\dfrac{z}{w}\right)} = \dfrac{\overline{z}}{\overline{w}}$ for $w \neq 0$. Also, $\mathrm{Re}(z) = \dfrac{z + \overline{z}}{2}$ and $\mathrm{Im}(z) = \dfrac{z - \overline{z}}{2i}$.

The important algebraic rules of commutativity, associativity, and distributivity work for complex numbers. Furthermore, all the properties in this section extend to vectors and matrices with complex-number entries. For example, if

$$V = \begin{bmatrix} c_1 \\ c_2 \\ \vdots \\ c_n \end{bmatrix}$$ is a vector with complex components, then $\overline{V} = \overline{\begin{bmatrix} c_1 \\ c_2 \\ \vdots \\ c_n \end{bmatrix}} = \begin{bmatrix} \overline{c_1} \\ \overline{c_2} \\ \vdots \\ \overline{c_n} \end{bmatrix}$. If

$A = (a_{ij})$ represents a matrix with entry a_{ij} in row i and column j, then $\overline{A} = \overline{(a_{ij})} = (\overline{a}_{ij})$.

2.   COMPLEX NUMBERS: THE GEOMETRIC VIEW

The geometric interpretation of complex numbers occurred at roughly the same time to three people: the Norwegian surveyor and map maker Caspar Wessel (1745–1818), the French-Swiss mathematician Jean Robert Argand (1768–1822), and Karl Friedrich Gauss (1777–1855), the German mathematician-astronomer-physicist.

The idea here is to represent a complex number using the familiar Cartesian coordinate system, making the horizontal axis the **real axis** and the vertical axis the **imaginary axis.** Such a system is called the **complex plane.** For example, Figure C1 shows how the complex number $3 + 2i$ would be represented as a point in this way.

If we join this point to the origin with a straight line, we get a vector. (See Appendix B.1.) The sum of $z = a + bi$ and $w = c + di$ corresponds to the point (or vector) $(a + c, b + d)$. This implies that the addition/subtraction of complex numbers corresponds to the Parallelogram Law of vector algebra (Figure C2).

The **modulus,** or **absolute value,** of the complex number $z = x + yi$, denoted by $|z|$, is the nonnegative real number defined by the equation $|z| = \sqrt{x^2 + y^2}$. The number $|z|$ represents the distance between the origin and the point (x, y) in the complex plane, the length of the vector representing the complex number $z = x + yi$. Note that $|z|^2 = z \cdot \bar{z}$.

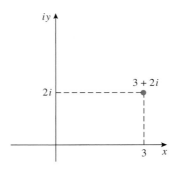

Figure C1

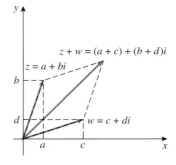

Figure C2

3.   THE QUADRATIC FORMULA

Given the quadratic equation $ax^2 + bx + c = 0$, where a, b, and c are real numbers with $a \neq 0$, the solutions are given by the **quadratic formula:**

$$x = \frac{-b \pm \sqrt{b^2 - 4ac}}{2a}.$$

The expression inside the radical sign, $b^2 - 4ac$, is called the **discriminant** and enables you to discriminate among the possibilities for solutions. If $b^2 - 4ac > 0$, the quadratic formula yields two real solutions. If $b^2 - 4ac = 0$, you get a single repeated solution—a solution of *multiplicity two.* Finally, if $b^2 - 4ac < 0$, the quadratic formula produces two complex numbers as solutions, a **complex conjugate pair.** To see this last situation, suppose that $b^2 - 4ac = -q$, where q is a positive real number. Then the solution formula looks like

$$x = \frac{-b \pm \sqrt{-q}}{2a} = \frac{-b \pm \sqrt{q(-1)}}{2a} = \frac{-b \pm \sqrt{q}i}{2a},$$

so the two solutions are $x_1 = -\dfrac{b}{2a} + \dfrac{\sqrt{q}}{2a}i$ and $x_2 = -\dfrac{b}{2a} - \dfrac{\sqrt{q}}{2a}i$, which are complex conjugates of each other.

4.   EULER'S FORMULA

Around 1740, while studying differential equations of the form $y'' + y = 0$, Euler discovered his famous formula for complex exponentials:

$$e^{iy} = \cos y + i \sin y.$$

If $z = x + iy$, then we have

$$e^z = e^{x+iy} = e^x e^{iy} = e^x(\cos y + i \sin y).$$

Without fully understanding the way infinite series work, Euler just substituted the complex number iy in the series for e^x (see Appendix A.3) and then separated real and imaginary parts:

$$\begin{aligned} e^{iy} &= 1 + iy + \frac{(iy)^2}{2!} + \frac{(iy)^3}{3!} + \frac{(iy)^4}{4!} + \frac{(iy)^5}{5!} + \cdots \\ &= 1 + iy - \frac{y^2}{2!} - i\frac{y^3}{3!} + \frac{y^4}{4!} + i\frac{y^5}{5!} - \cdots \\ &= \underbrace{\left(1 - \frac{y^2}{2!} + \frac{y^4}{4!} - \cdots\right)}_{\cos y} + i\underbrace{\left(y - \frac{y^3}{3!} + \frac{y^5}{5!} - \cdots\right)}_{\sin y} = \cos y + i \sin y. \end{aligned}$$

APPENDIX D

Series Solutions of Differential Equations

Appendix D supplements the treatment of linear equations in Chapters 5 and 6.

1. POWER SERIES SOLUTIONS OF FIRST-ORDER EQUATIONS

In Chapters 5 and 6 we discussed solutions for second- and higher-order linear equations with constant coefficients. The methods we discuss in this appendix can be applied to equations—not necessarily linear—with *variable* coefficients, equations that in general do not yield closed-form solutions. Among these are equations important in many areas of applied mathematics.

As an illustration of the key idea, we'll solve a simple first-order equation.

EXAMPLE

Consider the equation $y' = 1 - xy$. We make the fundamental assumption that a solution y can be expanded in a power series (Taylor series, Maclaurin series)

$$y(x) = a_0 + a_1 x + a_2 x^2 + a_3 x^3 + \cdots + a_n x^n + \cdots$$

that converges in some interval. (See Appendix A.3 for basics.) We have chosen an interval around the origin.

Then, because a convergent power series can be differentiated term by term within its interval of convergence (see Appendix A.3),

$$y'(x) = a_1 + 2a_2 x + 3a_3 x^2 + \cdots + na_n x^{n-1} + \cdots.$$

Substituting these last two series in the differential equation, we have

$$
\begin{aligned}
a_1 &+ 2a_2 x + 3a_3 x^2 + \cdots + na_n x^{n-1} + \cdots \\
&= 1 - x\{a_0 + a_1 x + a_2 x^2 + a_3 x^3 + \cdots + a_n x^n + \cdots\} \\
&= 1 - a_0 x - a_1 x^2 - a_2 x^3 - a_3 x^4 - \cdots - a_n x^{n+1} - \cdots.
\end{aligned}
$$

380

Because these power series are equal, coefficients of equal powers of x on both sides must be equal. (This is really the *method of undetermined coefficients* that we saw in Section 5.6 of the text.) Therefore, we have

$$a_1 = 1, 2a_2 = -a_0, 3a_3 = -a_1, 4a_4 = -a_2, 5a_5 = -a_3, \ldots, n\, a_n = -a_{n-2}, \ldots,$$

so

$$a_1 = 1, a_2 = -\frac{a_0}{2}, a_3 = -\frac{a_1}{3} = -\frac{1}{3}, a_4 = -\frac{a_2}{4} = \frac{-\dfrac{a_0}{2}}{4}$$

$$= \frac{a_0}{2 \cdot 4}, a_5 = -\frac{a_3}{5} = \frac{1}{3 \cdot 5}, \ldots, a_n = -\frac{a_{n-2}}{n}, \ldots.$$

These formulas, in which we define later coefficients by relating them to earlier coefficients, are called **recurrence** (or **recursion**) **relations.** If we look carefully, we see that for odd indices (subscripts), the pattern is

$$a_1 = 1, a_3 = -\frac{1}{3}, a_5 = \frac{1}{3 \cdot 5}, a_7 = -\frac{1}{3 \cdot 5 \cdot 7}, \ldots.$$

Similarly, for even indices we find the pattern

$$a_0 = \text{ arbitrary}, a_2 = -\frac{a_0}{2}, a_4 = \frac{a_0}{2 \cdot 4}, a_6 = -\frac{a_0}{2 \cdot 4 \cdot 6}, \ldots.$$

In general, the pattern is

$$a_{2k} = \frac{(-1)^k a_0}{2 \cdot 4 \cdot 6 \cdot \cdots \cdot (2k)} \quad \text{for } k = 1, 2, 3, \ldots;$$

$$a_{2k+1} = \frac{(-1)^k}{1 \cdot 3 \cdot 5 \cdot 7 \cdot \cdots \cdot (2k+1)} \quad \text{for } k = 0, 1, 2, \ldots.$$

Therefore, we can write the power series form of the solution as

$$y(x) = a_0 + x - \frac{a_0}{2}x^2 - \frac{1}{1 \cdot 3}x^3 + \left(\frac{a_0}{2 \cdot 4}\right)x^4 + \frac{1}{1 \cdot 3 \cdot 5}x^5 + \cdots$$

$$= \left(x - \frac{x^3}{1 \cdot 3} + \frac{x^5}{1 \cdot 3 \cdot 5} - \cdots\right) + a_0\left(1 - \frac{x^2}{2} + \frac{x^4}{2 \cdot 4} - \cdots\right),$$

where $a_0 = y(0)$ is the arbitrary constant that we expect in the general solution of a first-order equation.

To approximate $y(x)$ for a value of x close to zero, we just substitute the value in the series, taking as many terms of this series as are needed to guarantee the accuracy we wish. ◆

If you solved the linear equation in the last example using the technique of integrating factors (see Section 2.2), you would get the answer

$$y = e^{x^2/2} \int e^{-x^2/2} dx + Ce^{x^2/2},$$

which can't be expressed in a more elementary way. If you integrated the power series representation of $e^{-x^2/2}$ term by term, multiplied by the series form of $e^{x^2/2}$, and then added the series for $Ce^{x^2/2}$, you would get the same series solution we found (after collecting terms).

In using this power series method, sometimes you can recognize the series in your solution as a representation of an elementary function. Try using the method on the equation $y' = ay$, where a is a constant, for example. You should recognize the series solution as the Taylor series representation of Ce^{ax} about the origin.

All computer algebra systems have the ability to work with series expansions, usually truncating the series after a fixed number of terms that the user can control. However, not all systems can give you a power series solution of an ODE directly. For example, *Maple* has a very useful power series package, *powseries*, and the *dsolve* command (in the package *DEtools*) has a *series* option; but *Mathematica* and MATLAB require the user to do much more work in finding a series solution.

2. SERIES SOLUTIONS OF SECOND-ORDER LINEAR EQUATIONS: ORDINARY POINTS

In this section we'll examine second-order linear equations of the form

$$a(t)y'' + b(t)y' + c(t)y = 0, \qquad \text{(D.1)}$$

where $a(t)$, $b(t)$, and $c(t)$ are polynomial functions. We divide through by $a(t)$ and write equation (D.1) in the *standard form*

$$y'' + P(t)y' + Q(t)y = 0, \qquad \text{(D.2)}$$

where $P(t) = \dfrac{b(t)}{a(t)}$ and $Q(t) = \dfrac{c(t)}{a(t)}$.

A point t_0 is called an **ordinary point** of equation (D.2) if both P and Q can be expanded in power series centered at t_0 that converge for every t in an open interval containing t_0. Functions that have such power series representations are called **analytic** at the point t_0. If t_0 is not an ordinary point, it is called a **singular point** of the equation.

EXAMPLE
The point $t = 0$ is an ordinary point of the equation $(t + 2)y'' + t^2 y' + y = 0$ because each of the functions $P(t) = \dfrac{t^2}{t + 2}$ and $Q(t) = \dfrac{1}{t + 2}$ has its own power series expansion that converges near $t = 0$:

$$Q(t) = \frac{1}{2} - \frac{t}{4} + \frac{t^2}{8} - \frac{t^3}{16} + \cdots \quad \text{and} \quad P(t) = \frac{t^2}{2} - \frac{t^3}{4} + \frac{t^4}{8} - \frac{t^5}{16} + \cdots.$$

(See the *geometric series* in Appendix A.3.) However, $t = -2$ is a singular point because the denominators of $P(t)$ and $Q(t)$ are zero at $t = -2$.   ◆

Let's apply the undetermined coefficient method of the last section to a famous second-order linear equation near an ordinary point. The equation is named for the English mathematician Sir George Bidell Airy (1801–1892), who did pioneering work in elasticity and in partial differential equations.

EXAMPLE

Airy's equation, $y'' + xy = 0$, which occurs in the study of optics and quantum physics, cannot be solved in terms of elementary functions. You can think of the equation as describing a spring-mass system in which the stiffness of the spring is increasing with time. (Maybe the room containing the system is getting colder.)

Noting that $x = 0$ is an ordinary point of this equation, we assume that we can write a solution as

$$y(x) = a_0 + a_1 x + a_2 x^2 + a_3 x^3 + a_4 x^4 + \cdots + a_n x^n + \cdots.$$

Then

$$y'(x) = a_1 + 2a_2 x + 3a_3 x^2 + 4a_4 x^3 + \cdots + na_n x^{n-1} + \cdots$$

and

$$y''(x) = 2a_2 + 6a_3 x + 12a_4 x^2 + \cdots + n(n-1)a_n x^{n-2} + \cdots.$$

Substituting in the differential equation, we get

$$(2a_2 + 6a_3 x + 12a_4 x^2 + \cdots + n(n-1)a_n x^{n-2} + \cdots)$$
$$+ x(a_0 + a_1 x + a_2 x^2 + a_3 x^3 + a_4 x^4 + \cdots + a_n x^n + \cdots) = 0.$$

Collecting terms, we can write this last equation as

$$2a_2 + (6a_3 + a_0)x + (12a_4 + a_1)x^2 + \cdots$$
$$+ (n(n-1)a_n + a_{n-3})x^{n-2} + \cdots = 0.$$

Equating coefficients of equal powers of x, we see that the last equation implies that

$$2a_2 = 0, \text{ or } a_2 = 0; 6a_3 + a_0 = 0, \text{ or } a_3 = -\frac{a_0}{2 \cdot 3};$$

$$12a_4 + a_1 = 0, \text{ or } a_4 = -\frac{a_1}{3 \cdot 4}; 20a_5 + a_2 = 0, \text{ or } a_5 = -\frac{a_2}{4 \cdot 5},$$

and so forth, so we can see the recurrence relation as $a_n = -\dfrac{a_{n-3}}{(n-1) \cdot n}$ for $n = 3, 4, 5, \ldots$. Note that a_0 and a_1 are arbitrary and that the coefficients are connected by jumps of three in the subscripts. In particular we have $0 = a_2 = a_5 = a_8 = \cdots = a_{2+3k} = \cdots$. Also, we can see the pattern when the subscript is a multiple of 3:

$$a_3 = -\frac{a_0}{2 \cdot 3} \quad a_6 = -\frac{a_3}{5 \cdot 6} = \frac{a_0}{2 \cdot 3 \cdot 5 \cdot 6} \quad a_9 = -\frac{a_6}{8 \cdot 9} = -\frac{a_0}{2 \cdot 3 \cdot 5 \cdot 6 \cdot 8 \cdot 9}$$

$$a_{12} = -\frac{a_9}{11 \cdot 12} = \frac{a_0}{2 \cdot 3 \cdot 5 \cdot 6 \cdot 8 \cdot 9 \cdot 11 \cdot 12} \text{ and so forth, so the formula is}$$

$$a_{3k} = \frac{(-1)^k a_0}{2 \cdot 3 \cdot 5 \cdot 6 \cdot 8 \cdot 9 \cdot \cdots \cdot (3k-1) \cdot 3k}.$$

Similarly, we can see that

$$a_4 = -\frac{a_1}{3 \cdot 4} \qquad a_7 = -\frac{a_4}{6 \cdot 7} = \frac{a_1}{3 \cdot 4 \cdot 6 \cdot 7}$$

$$a_{10} = -\frac{a_7}{9 \cdot 10} = -\frac{a_1}{3 \cdot 4 \cdot 6 \cdot 7 \cdot 9 \cdot 10}$$

$$a_{13} = -\frac{a_{10}}{12 \cdot 13} = \frac{a_1}{3 \cdot 4 \cdot 6 \cdot 7 \cdot 9 \cdot 10 \cdot 12 \cdot 13},$$

and so forth, so the recurrence relation is

$$a_{3k+1} = \frac{(-1)^k a_1}{3 \cdot 4 \cdot 6 \cdot 7 \cdot 9 \cdot 10 \cdot \cdots \cdot (3k) \cdot (3k+1)}.$$

Putting all the pieces together, we get

$$y(x) = a_0 \left[1 - \frac{x^3}{2 \cdot 3} + \frac{x^6}{2 \cdot 3 \cdot 5 \cdot 6} - \frac{x^9}{2 \cdot 3 \cdot 5 \cdot 6 \cdot 8 \cdot 9} + \cdots \right]$$

$$+ a_1 \left[x - \frac{x^4}{3 \cdot 4} + \frac{x^7}{3 \cdot 4 \cdot 6 \cdot 7} - \frac{x^{10}}{3 \cdot 4 \cdot 6 \cdot 7 \cdot 9 \cdot 10} + \cdots \right]$$

$$= y(0) \left[1 - \frac{x^3}{2 \cdot 3} + \frac{x^6}{2 \cdot 3 \cdot 5 \cdot 6} - \frac{x^9}{2 \cdot 3 \cdot 5 \cdot 6 \cdot 8 \cdot 9} + \cdots \right]$$

$$+ y'(0) x \left[1 - \frac{x^3}{3 \cdot 4} + \frac{x^6}{3 \cdot 4 \cdot 6 \cdot 7} - \frac{x^9}{3 \cdot 4 \cdot 6 \cdot 7 \cdot 9 \cdot 10} + \cdots \right]$$

$$= y(0) \cdot Ai(x) + y'(0) x \cdot Bi(x),$$

where the two series (convergent for all values of x) define $Ai(x)$ and $Bi(x)$, the **Airy functions of the first and second kind,** respectively, up to constant multiplicative factors.

With the aid of technology, let's look at the graph of the solution of Airy's equation with initial conditions $y(0) = 0$, $y'(0) = 1$ (Figure D1), which is just the graph of $xBi(x)$.

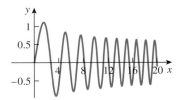

Figure D1
Solution of $y'' + xy = 0$; $y(0) = 0, y'(0) = 1$

Both *Maple* and *Mathematica*, for example, have built-in capabilities to deal with Airy functions numerically and graphically—check out the commands *AiryAi(x)* and *AiryBi(x)* [in *Maple*] or *AiryAi[x]* and *AiryBi[x]* [in *Mathematica*].   ◆

If we want to find a solution near an ordinary point t_0 other than zero, we can use the substitution $u = t - t_0$. This substitution transforms the equation in t to one in the variable u, which we can solve near the ordinary point $u = 0$. When we have solved the equation in u, we can just replace u by $t - t_0$ to return to the original variable.

The method of undetermined coefficients also applies to nonhomogeneous equations and to equations whose coefficients are not polynomials, provided that the function on the right-hand side and the coefficient functions can be expanded in powers of t. When we are trying to solve a nonhomogeneous equation, equating coefficients becomes a little more difficult because some of the coefficients of the solution series $y(t) = \sum_{n=0}^{\infty} a_n t^n$ will include numerical values independent of the two arbitrary constants a_0 and a_1. This part of the general solution y_{GNH} constitutes y_{PNH}. Check this out for yourself by using series to solve the equation $y'' - y = e^x$. (You should recognize your solution as $y = c_1 e^x + c_2 e^{-x} + \frac{1}{2} x e^x$.)

3. REGULAR SINGULAR POINTS: THE METHOD OF FROBENIUS

Some singular points are such that special series methods have been developed to handle situations in which they occur. The point t_0 is a **regular singular point** of $y'' + P(t)y' + Q(t)y = 0$ if t_0 is a singular point and the functions $(t - t_0)P(t)$ and $(t - t_0)^2 Q(t)$ are both analytic at t_0. If t_0 is a singular point that is not regular, it is called an **irregular singular point.**

For example, $t = 1$ is a singular point of the equation $(t^2 - 1)^2 y'' + (t - 1)y' + y = 0$ because $P(t) = \dfrac{t - 1}{(t^2 - 1)^2} = \dfrac{t - 1}{(t + 1)^2 (t - 1)^2}$ and $Q(t) = \dfrac{1}{(t + 1)^2 (t - 1)^2}$ have zero denominators at $t = 1$, so neither $P(t)$ nor $Q(t)$ has a convergent power series expansion in a neighborhood of 1. But if we look at $(t - 1)P(t) = \dfrac{(t - 1)^2}{(t + 1)^2 (t - 1)^2} = \dfrac{1}{(t + 1)^2}$ and $(t - 1)^2 Q(t) = \dfrac{(t - 1)^2}{(t + 1)^2 (t - 1)^2} = \dfrac{1}{(t + 1)^2}$, we see that both $(t - 1)P(t)$ and $(t - 1)^2 Q(t)$ are analytic at $t = 1$, so $t = 1$ is a regular singular point.

Near a regular singular point—say $t = 0$ for convenience—we write equation (D.2) as

$$t^2 y'' + t p(t) y' + q(t) y = 0, \tag{D.3}$$

where $p(t) = tP(t)$ and $q(t) = t^2 Q(t)$. Because $t = 0$ is a regular singular point, p and q are analytic at $t = 0$. The usual power series method will not work, and we use the **method of Frobenius**,[1] which produces at least one solution of the form

$$y(t) = t^r \sum_{n=0}^{\infty} a_n t^n = \sum_{n=0}^{\infty} a_n t^{n+r}, \tag{D.4}$$

where we assume that $a_0 \neq 0$.

---

1. The German mathematician Ferdinand Georg Frobenius (1849–1917) published his method in 1878. It was based on a technique that originated with Euler. (Who else?) Frobenius made many contributions to analysis and especially to algebra.

It is important to note that three of the most popular computer algebra systems (*Maple*, *Mathematica*, and MATLAB) cannot apply the method of Frobenius directly to get power series solutions near regular singular points. You must develop a solution in a step-by-step fashion, using the capabilities of your system to handle power series and recursion relations.

We'll illustrate the method of Frobenius using a famous equation in applied mathematics, one that first arose in an investigation of the motion of a hanging chain and has since appeared in such problems as the analysis of vibrations of a circular membrane and planetary motion.

EXAMPLE

Bessel's equation of order p is $x^2 y'' + xy' + (x^2 - p^2)y = 0$, which is of the form (D.3) and has $x = 0$ as a regular singular point.[2] We'll take the parameter p to be an arbitrary nonnegative real number.

Substituting the type of series given in (D.4) for y, we find that

$$y' = \sum_{n=0}^{\infty} a_n (n + r) x^{n+r-1}$$

and

$$y'' = \sum_{n=0}^{\infty} a_n (n + r)(n + r - 1) x^{n+r-2},$$

so that we have

$$
\begin{aligned}
x^2 y'' + xy' + (x^2 - p^2)y &= \sum_{n=0}^{\infty} a_n (n + r)(n + r - 1) x^{n+r} + \sum_{n=0}^{\infty} a_n (n + r) x^{n+r} \\
&+ \sum_{n=0}^{\infty} a_n x^{n+r+2} - \sum_{n=0}^{\infty} a_n p^2 x^{n+r} \\
&= \sum_{n=0}^{\infty} \{a_n(n + r)(n + r - 1) + a_n(n + r) - a_n p^2\} x^{n+r} \\
&+ \sum_{n=0}^{\infty} a_n x^{n+r+2} \\
&= \sum_{n=0}^{\infty} \{(n + r)^2 - p^2\} a_n x^{n+r} + \sum_{n=0}^{\infty} a_n x^{n+r+2} = 0.
\end{aligned}
$$

Transposing series and making the substitution (actually a shift of subscripts) $n + 2 = n$ on the right-hand side, we get

$$\sum_{n=0}^{\infty} \{(n + r)^2 - p^2\} a_n x^{n+r} = -\sum_{n=0}^{\infty} a_n x^{n+r+2} = -\sum_{n=2}^{\infty} a_{n-2} x^{n+r}.$$

Now we equate coefficients of equal powers. To start, we have

$$n = 0: \quad (r^2 - p^2)a_0 = 0$$
$$n = 1: \quad [(1 + r)^2 - p^2]a_1 = 0.$$

---

2. Among other achievements, the German astronomer Friedrich Wilhelm Bessel (1784–1846) was the first to measure accurately the distance to a fixed star.

Because we have assumed that $a_0 \neq 0$, we must have $r^2 - p^2 = 0$. This last equation is called the **indicial equation**[3] and implies that $r = \pm p$.

Let's assume that $r = p \geq 0$. Then when $n = 1$, the equation $[(1 + r)^2 - p^2]a_1 = 0$ reduces to $(2r + 1)a_1 = 0$, so we can conclude $a_1 = 0$.

For $n \geq 2$, equating coefficients of equal powers of x gives us the recurrence relation

$$\{(n + r)^2 - p^2\}a_n = -a_{n-2}$$

or

$$a_n = -\frac{a_{n-2}}{\{(n + r)^2 - p^2\}} = -\frac{a_{n-2}}{n(n + 2r)}$$

because $r^2 - p^2 = 0$. We can look at a few terms to see the pattern:

$$a_2 = -\frac{a_0}{2(2 + 2r)} = -\frac{a_0}{2^2(1 + r)}$$

$$a_3 = -\frac{a_1}{3(3 + 2r)} = 0 \quad [\text{because } a_1 = 0]$$

$$a_4 = -\frac{a_2}{4(4 + 2r)} = -\frac{\left(\dfrac{-a_0}{2^2(1 + r)}\right)}{2 \cdot 2^2(2 + r)} = \frac{a_0}{2^4 2!(1 + r)(2 + r)}.$$

$$a_5 = -\frac{a_3}{5(5 + 2r)} = 0$$

$$a_6 = -\frac{a_4}{6(6 + 2r)} = -\frac{\left(\dfrac{a_0}{2^4 2!(1 + r)(2 + r)}\right)}{6(6 + 2r)} = -\frac{a_0}{2^6 3!(1 + r)(2 + r)(3 + r)}.$$

We can see, for example, that $a_k = 0$ for k odd.

Letting $n = 2k$ and remembering that we're assuming $r = p$, we can express the even coefficients in the form

$$a_{2k} = \frac{(-1)^k a_0}{2^{2k}k!(r + 1)(r + 2) \cdot \cdots \cdot (r + k)}$$

$$= \frac{(-1)^k a_0}{2^{2k}k!(p + 1)(p + 2) \cdot \cdots \cdot (p + k)}.$$

In working with Bessel's equation, it is common practice to make things neater by taking $a_0 = \dfrac{1}{2^p p!}$,[4] so that

$$a_{2k} = \frac{(-1)^k}{2^{2k + p}k!(p + k)!}.$$

---

3. In general, for the method of Frobenius, the indicial equation has the form $r(r - 1) + rp_0 + q_0 = 0$, where p_0 and q_0 are the constant terms of the series expansions of $p(t)$ and $q(t)$ in equation (D.3).

4. Actually, $a_0 = \dfrac{1}{2^p \Gamma(p + 1)}$, where Γ denotes Euler's *gamma function* (see Appendix A.6).

The final result is the **Bessel function of order p of the first kind, $J_p(x)$:**

$$y(x) = J_p(x) = \sum_{n=0}^{\infty} \frac{(-1)^n x^{2n+p}}{2^{2n+p} n!(p+n)!} = \sum_{n=0}^{\infty} \frac{(-1)^n (\frac{x}{2})^{2n+p}}{n!(p+n)!}$$

$$= \left(\frac{x}{2}\right)^p \sum_{n=0}^{\infty} \frac{(-1)^n (\frac{x}{2})^{2n}}{n!(p+n)!}.$$

It can be shown that this series converges for all real values of x.

Using technology, we can produce a graph of Bessel functions of order p for $p = 0, 1, 2, 3,$ and 4 (Figure D2).

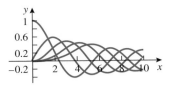

Figure D2

$J_p(x)$ for $p = 0,1,2,3,4; 0 \le x \le 10$

Both *Maple* and *Mathematica*, for example, can deal with Bessel functions of the first kind numerically and graphically via the command *BesselJ(mu, x)* [in *Maple*] or *BesselJ* [*m*, x] [in *Mathematica*]. The parameter *mu* or *m* represents the order that we have called p. It is interesting to note that a CAS could express the solution of the IVP $y'' + xy = 0$; $y(0) = 0, y'(0) = 1$ that we considered in Section 2 of this appendix as

$$y(x) = \frac{2}{9} \frac{3^{5/6}\pi}{\Gamma\left(\frac{2}{3}\right)} \sqrt{x} BesselJ\left(\frac{1}{3}, \frac{2x^{3/2}}{3}\right).$$

◆

We should make several comments about the last example:

1. In our analysis, we have actually assumed that $x > 0$ to avoid the possibility of fractional powers of negative numbers.
2. In the indicial equation $r^2 - p^2 = 0$, we have assumed that $r = p$, a nonnegative number. If r is in fact a nonnegative *integer*, then the Frobenius series is an ordinary power series with first term $a_0 x^n$. For applications, the choices $p = 0$ and $p = 1$ occur the most often.
3. All our efforts have produced just one solution of Bessel's equation for a fixed value of p. It can be shown that when $2p$ is not a positive integer,

$$J_{-p}(x) = \left(\frac{2}{x}\right)^p \sum_{n=0}^{\infty} \frac{(-1)^n (\frac{x}{2})^{2n}}{n!(-p+n)!}$$

defines a second, linearly independent solution of Bessel's equation. When p is an integer, it can be shown that $J_p(x) = (-1)^p J_{-p}(x)$, so the two solutions are dependent.

4. If p is not an integer, the general solution of Bessel's equation has the form

$$y(x) = k_1 J_p(x) + k_2 J_{-p}(x)$$

for arbitrary constants k_1 and k_2.

5. The function

$$Y_p(x) = \frac{(\cos p\pi) J_p(x) - J_{-p}(x)}{\sin p\pi}$$

is the standard **Bessel function of the second kind.** Then

$$y(x) = c_1 J_p(x) + c_2 Y_p(x)$$

is the general solution of Bessel's equation in all cases, whether p is an integer or not. Both *Maple* and *Mathematica* have commands—*BesselY(mu, x)* and *BesselY[m, x]*, respectively—that enable users to explore Bessel functions of the second kind numerically and graphically.

There are many treatments of the properties and applications of Bessel functions. An accessible source of information is the text *Differential Equations with Applications and Historical Notes, 2nd ed.*, by George F. Simmons (New York: McGraw-Hill, 1991).

4. THE POINT AT INFINITY

There are situations in which we want to determine the behavior of solutions of the equation

$$y'' + P(t)y' + Q(t)y = 0$$

for large values of the independent variable t—the behavior "in the neighborhood of infinity." The way to deal with this problem is to use the substitution $t = \dfrac{1}{u}$ and investigate the resulting equation near $u = 0$. This substitution converts a problem in large values of t to one in small values of u. Once the "u-problem" is solved near $u = 0$, we make the substitution $t = \dfrac{1}{u}$ in the u-solution to get the solution near the t point of infinity.

Let $u = \dfrac{1}{t}$. Then, by the Chain Rule,

$$y' = \frac{dy}{dt} = \frac{dy}{du} \cdot \frac{du}{dt} = \frac{dy}{du}\left(-\frac{1}{t^2}\right) = -u^2 \cdot \frac{dy}{du}$$

and

$$y'' = \frac{d}{dt}\left(\frac{dy}{dt}\right) = \frac{d}{du}\left(\frac{dy}{dt}\right) \cdot \frac{du}{dt} = \left(-u^2\frac{d^2y}{du^2} - 2u\frac{dy}{du}\right)(-u^2).$$

Let's use this transformation method to solve an equation for large values of the independent variable.

EXAMPLE

Find the general solution of the equation

$$4t^3\frac{d^2y}{dt^2} + 6t^2\frac{dy}{dt} + y = 0$$

for large values of t.

First we write the equation in the standard form

$$\frac{d^2y}{dt^2} + \frac{3}{2t}\frac{dy}{dt} + \frac{1}{4t^3}y = 0.$$

Making the substitution $u = \dfrac{1}{t}$ and using the calculations for y' and y'' given above, we transform our equation into

$$\left(-u^2\frac{d^2y}{du^2} - 2u\frac{dy}{du}\right)(-u^2) + \frac{3u}{2}\left(-u^2\frac{dy}{du}\right) + \frac{u^3}{4}y = 0$$

or

$$4u\frac{d^2y}{du^2} + 2\frac{dy}{du} + y = 0,$$

which has $u = 0$ as a regular singular point.

If we use the Frobenius method, we find the general solution

$$\Psi(u) = c_1\sum_{n=0}^{\infty}\frac{(-1)^n u^n}{(2n)!} + c_2\sum_{n=0}^{\infty}\frac{(-1)^n u^{n+\frac{1}{2}}}{(2n+1)!}.$$

Substituting $u = \dfrac{1}{t}$, we get the solution

$$y(t) = c_1\sum_{n=0}^{\infty}\frac{(-1)^n}{(2n)!}\left(\frac{1}{t}\right)^n + c_2\sum_{n=0}^{\infty}\frac{(-1)^n}{(2n+1)!}\left(\frac{1}{t}\right)^{n+\frac{1}{2}}$$

$$= c_1\cos\left(\frac{1}{\sqrt{t}}\right) + c_2\sin\left(\frac{1}{\sqrt{t}}\right).$$

(See Appendix A.3.)   ◆

5.   SOME ADDITIONAL SPECIAL DIFFERENTIAL EQUATIONS

There are many famous functions, such as Airy and Bessel functions, that arise as power series solutions of second-order differential equations. These functions form a particular class of what are usually called *special functions*.

Among these important second-order equations that have been significant in solving problems in applied mathematics, science, and engineering are the following, which you are invited to tackle using the methods of this appendix.

Chebyshev's equation: $(1 - x^2)y'' - xy' + p^2y = 0$, where p is a constant. (When p is a nonnegative integer, the solution is an nth-degree polynomial.)

Gauss's hypergeometric equation:

$$x(1 - x)y'' + [c - (a + b + 1)x]y' - aby = 0,$$

where a, b, and c are constants.

Hermite's equation: $y'' - 2xy' + 2py = 0$, where p is a constant.

Laguerre's equation: $xy'' + (1 - x)y' + py = 0$, where p is a constant.

Legendre's equation: $(1 - x^2)y'' - 2xy' + \left[k(k + 1) - \dfrac{m^2}{1 - x^2} \right]y = 0$, where m and k are constants, $k > 0$.

Answers/Hints to Odd-Numbered Exercises

EXERCISES 1.1

1. The independent variable is x, the dependent variable is y; first-order, linear

3. The independent variable is unknown, the dependent variable is x; second-order, nonlinear because of the exponent $-x$

5. The independent variable is x, the dependent variable is y; first-order, nonlinear because the equation contains a second-degree term in y, as well as products of y' with itself and with y

7. The independent variable is x, the dependent variable is y; fourth-order, linear

9. The independent variable is x, the dependent variable is y; first-order, nonlinear because of the exponent y'

11. $a = 1$

EXERCISES 1.2

13. a. $(y')^2$ would have to equal -1, which is impossible for a real-valued function.

  b. The absolute value of a function is nonnegative. The only way for the sum of two nonnegative functions to be the zero function is for each summand itself to be zero. Therefore, $y(x) \equiv 0$ is the only solution.

15. If $x \geq c$ or $x \leq -c$, then $x^2 \geq c^2$, so $c^2 - x^2 \leq 0$. If $c^2 - x^2 < 0$, the two given functions are not real-valued. If $c^2 - x^2 = 0$, then each of the two functions is equal to the zero function, which is not a solution of the differential equation.

17. $y(t) = t^3 + \frac{1}{2}t^2 + 1$

19. $y(x) = e^{x/2} + e^{-x/2}$

23. The length of the runway must be $\frac{5}{6}$ mile (five-sixths of a mile).

25. $y(x) = \dfrac{1}{EI}\left\{\left(\dfrac{-W}{24L}\right)x^2(x - L)^2\right\} = -\left(\dfrac{W}{24EIL}\right)x^2(x - L)^2$

29. One possibility is $(xy - 1)y' + y^2 = 0$.

31. $y(x) = \frac{1}{2}(\sin x - \cos x)$

35. b. $x(t) = (1 - 3t)e^{3t}$, $y(t) = -9te^{3t}$

37. a. $V_1 = V_0 e^{-ct}$

  c. The number of infected cells goes to zero as $t \to \infty$.

EXERCISES 2.1

1. $y = \dfrac{A}{2} + \dfrac{C}{x^2}$, where C is arbitrary

3. $y = (t - 2)^3 = t^3 - 6t^2 + 12t - 8;$ $y \equiv 0$ is a singular solution

5. $y = 2 - 3\cos x$

7. $\dfrac{y^2}{2} + y + \ln|y - 1| = -\dfrac{1}{x} + C;$ $y \equiv 1$ is a singular solution

9. $z = -\dfrac{\ln(C - 10^x)}{\ln 10};$ the solution is defined only for $10^x < C$ (or $x < \log_{10} C$)

11. $y = -\dfrac{x^2}{2} + C$ or $y = Ce^{-x}$

13. $x + 2y - 2\ln|x + 2y + 2| = x + C;$ $y = -(x + 2)/2$ is a singular solution.

15. $\arctan\left(\dfrac{y}{x}\right) - \tfrac{1}{2}\ln\left(\dfrac{x^2 + y^2}{x^2}\right) - \ln|x| - C = 0$

17. $y = x\sqrt{2\ln|x| + C}$ and $y = -x\sqrt{2\ln|x| + C}$

21. a. $x(t) = \dfrac{1}{2 - t}$

b. The interval I can be as large as $(-\infty, 2)$. The interval I includes $t = 1$ but cannot include the point $t = 2$, at which $x(t)$ is not defined.

c.

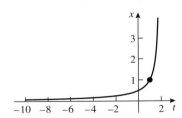

d. $x \equiv 0$

23. $t = 60$

25. $m = (e^x - e^{-x})/2 = \sinh(x)$, the *hyperbolic sine* of x

27. $t = \dfrac{1}{a}\left(L + \dfrac{bL^{1-n}}{1 - n}\right)$

29. a. $C = C(t) = 14e^{\frac{-t}{6}}$

c. 8.06 hours

d. 5.08 hours

e.

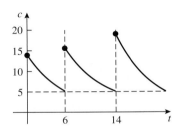

31. a. $P = \dfrac{Ce^t}{1 + Ce^t};$ $P \equiv 1$ is a singular solution

b. $P(t) \to 1$ as $t \to \infty.$

c. $P(t) \to 1$ as $t \to \infty.$

EXERCISES 2.2

1. $y = 2x - 1 + Ce^{-2x}$

3. $x = \dfrac{t^2}{2} - \dfrac{1}{2} + Ce^{-t^2}$

5. $y = \dfrac{t^3}{6} - \dfrac{t^2}{5} + \dfrac{C}{t^3}$

7. $y = x \sin x + Cx$

9. $x = e^t(\ln|t| + t^2/2 + C)$

11. $y(x) = \dfrac{e^x + ab - e^a}{x}$

13. For $m \neq -a$, we have $y = \dfrac{e^{mx}}{a + m} + Ce^{-ax}$. If $m = -a$, then $y = xe^{-ax} + Ce^{-ax} = (x + C)e^{-ax}$. *Note*: A CAS that can solve ODEs may miss the need for an analysis of two cases.

15. $x(t) = \left(\dfrac{t}{t + 1}\right)(t + \ln|t| - 1)$

17. $y = -\ln(x + Cx^2)$

19. $y = \dfrac{t^4}{t^6 + C}$; $y \equiv 0$ is a singular solution

21. $y = \dfrac{\pm 1}{\sqrt{x + \frac{1}{2} + Ce^{2x}}}$; $y \equiv 0$ is a singular solution

23. a. $W(t) = \left(\dfrac{\alpha}{\beta} + Ce^{-\frac{\beta t}{3}}\right)^3$

  b. $W_\infty = \left(\dfrac{\alpha}{\beta}\right)^3$

  c. $W(t) = W_\infty(1 - e^{-\beta t/3})^3$

  d.

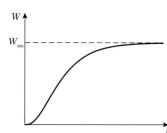

25. a. $I(t) = \dfrac{E}{R} - \dfrac{E}{R}e^{-\left(\frac{R}{L}\right)t} = \dfrac{E}{R}\left(1 - e^{-\left(\frac{R}{L}\right)t}\right)$

  b. $\lim\limits_{t\to\infty} I(t) = \dfrac{E}{R}$

  c. $t = \dfrac{L}{R}\ln 2$

  d. $I(t) \equiv \dfrac{E}{R}$

27. $Q(t) = \dfrac{E_0 C[\sin(\omega t) - \omega RC \cos(\omega t)]}{1 + (RC\omega)^2} + \dfrac{\omega E_0 RC^2}{1 + (RC\omega)^2}e^{-\frac{t}{RC}} = \dfrac{E_0 C}{1 + (RC\omega)^2}\{\sin(\omega t) - \omega RC \cos(\omega t) + \omega RCe^{-\frac{t}{RC}}\}$

29. a. If $S(T) = S_T$, we can write the solution as

$$S(t) = \begin{cases} \dfrac{r\overline{A}}{\left(\dfrac{r\overline{A}}{M} + \lambda\right)} + \left(S_0 - \dfrac{r\overline{A}}{\left(\dfrac{r\overline{A}}{M} + \lambda\right)}\right)e^{-\left(\frac{r\overline{A}}{M} + \lambda\right)t} & \text{for } 0 < t < T \\[2em] S_T e^{-\lambda(t-T)} & \text{for } t \geq T \end{cases}$$

b. Choosing $\overline{A} = 1000$, $r = 10$, $\lambda = 0.1$, $S_0 = 20000$, $S_T = 36000$, $M = 60000$, and $T = 10$, we have the following graph:

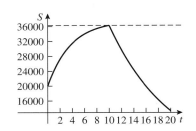

31. The concentration of potassium is 9.375 grams per gallon.

33. a. The tank is half full 100 seconds after the valve and drain are opened. At this time, the chlorine is a 0.75% solution.

    b. The final concentration is 0.875%.

35. a. The size of the agency staff is 4000, $33\frac{1}{3}$% of whom are female.

    b. 50%

EXERCISES 2.3

1.

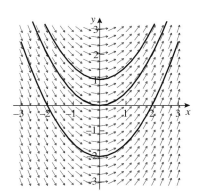

3.

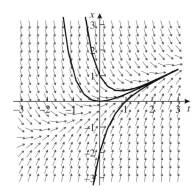

5.

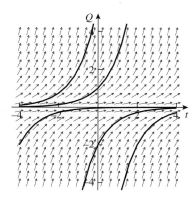

7. Look carefully at the slope field. The *t*-axis, corresponding to $r = 0$, divides the solution curves into two families of (almost) semicircles.

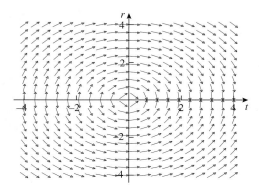

9.

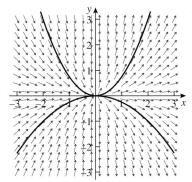

11. Choose the ranges for your variables carefully. There are vertical asymptotes lurking in the bushes.

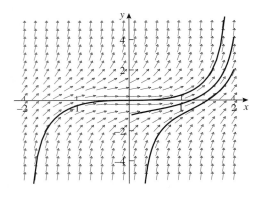

13.

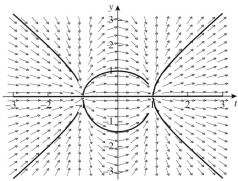

15. Look at the slope field very carefully, noting where the arrows change directions. Depending on the initial conditions specified, your graphing calculator or CAS may produce strange solution curves.

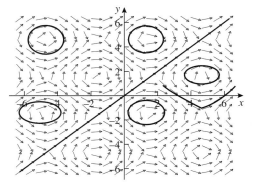

17. a.

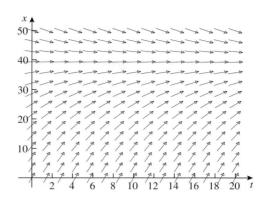

b. $x(t)$ seems to be approaching 40 as $t \to \infty$.

19. $y = -\{(1 + C)/(1 - C)\}t$ for $C \neq 1$—a one-parameter family of straight lines through the origin. For $C = 1$, we get the y-axis as the isocline.

23. *Hint*: equations (a) and (b) are nonautonomous, whereas equation (c) is autonomous.

25. a. Field 3
    b. Field 1
    c. Field 2

27.

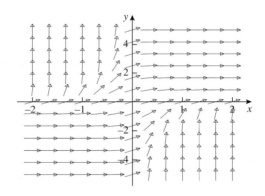

EXERCISES 2.4

1.

3.

5.

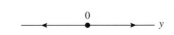

7.

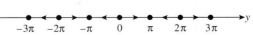

9.

11. a.

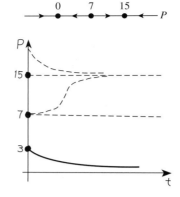

b. $P(t) \to 0$ as $t \to \infty$

13. a.

b. $x(t) \to 40$

15.

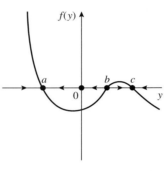

17. a.

b. $x(t) \to \sqrt{\dfrac{a}{b}}$

c. $x(t)$ stays at zero.

d. $x(t) \to \sqrt{\dfrac{a}{b}}$

EXERCISES 2.5

Confirm the answers for Exercises 1–12 by looking at slope fields or phase portraits.

1. $y = 0$ and $y = 1$ are both nodes.

3. $y = 0$ is a source.

5. $x = -a/b$ is a sink and $x = 0$ is a source.

7. $y = -2$ is a source and $y = 5$ is a sink.

9. $x = 0$ is a sink.

11. $y = -1$ is a sink and $y = 0$ is a source.

13. a. $Q \equiv Q^*$

b. The solution is a sink and therefore is stable.

15. a.

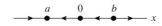

b. $x \equiv a$ is a sink; $x \equiv b$ is a source.

17. There is no such equation.

EXERCISES 2.6

1. a.

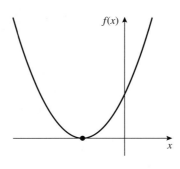

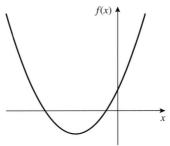

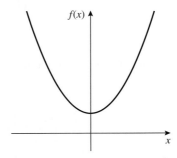

b. $c = -2$ and $c = 2$ are bifurcation points.

c.

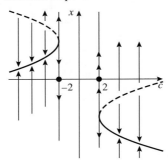

3. a.

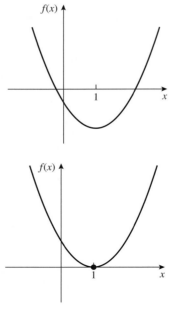

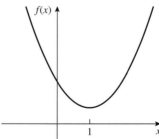

b. $c = 1$ is the only bifurcation point.

c.

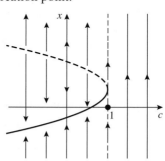

5.

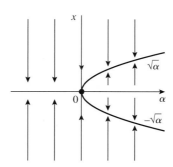

7. c.

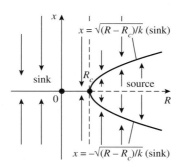

EXERCISES 2.7

1. For example, take any rectangle centered at $(0, 3)$ that avoids the t-axis ($x = 0$).
3. There is no such rectangle.
5. There is no such rectangle.
7. π. See Example 2.7.1.
9. $x(t) = \left(\dfrac{t}{3} + \sqrt[3]{x_0} \right)^3$
11. **a.** $2\sqrt{|y|} - 2k \ln(\sqrt{|y|} + k) = t + C$
    b. *any* initial values (t_0, y_0)
    c. If $k < 0$, the equation has a unique solution for *any* initial condition. When $k = 0$, we have no unique solution for the IVP with initial condition $y(0) = 0$.
13. Yes.
17. **a.** Yes.
19. It does not follow that the reaction cannot take place; nor does the fact that the equation has a solution guarantee that the reaction *does* take place.

EXERCISES 3.1

1.

t_k	y_k
0	1
0.25	0.75
0.50	0.625
0.75	0.589844
1.00	0.643490

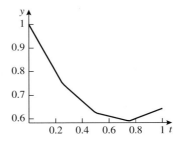

3.

t_k	y_k
1	2
1.50	3.359141
2.00	4.266010
2.50	5.065065
3.00	5.807155

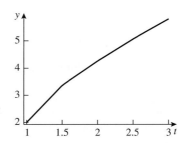

5. $y(\pi/2) \approx 1.148841$. The actual value is 1, so the absolute error is approximately 0.148841.

7. $y(1) \approx 1.385561$

9. a. $P(1) \approx 1.330624 = 1,330,624$ people
   b. $P(0) \approx 1.285 = 1,285,000$ people

11. $V(0) \approx 166.39$ meters per second

13.

t_k	y_k
0	1.000000
0.1	1.500000
0.2	2.190000
0.3	3.146000
0.4	4.474400
0.5	6.324160
0.6	8.903824
0.7	12.505354
0.8	17.537496
0.9	24.572494
1.0	34.411492

The given IVP (involving a linear equation) has the solution $y(t) = \frac{19}{16}e^{4t} + \frac{1}{4}t - \frac{3}{16}$. Therefore, $y(1/2) = 8.712004$, which gives us an absolute error of $|8.712004 - 6.324160| = 2.387844$. Then $y(1) = 64.897803$, giving us an absolute error of $|64.897803 - 34.411492| = 30.486311$. The solution curve rises so steeply that the tangent-line approximations can't keep up.

15. With $h = 0.5$, we get $x(2) \approx 2.746746$, with absolute error about 0.253254.
With, $h = 0.25$, we get $x(2) \approx 2.870814$, with absolute error about 0.129186.

17. All IVPs of the form $\dfrac{dy}{dx} = C$, where C is a constant, $y(x_0) = y_0$.

19. a. $y(t) = [(1 - \alpha)t]^{\frac{1}{1-\alpha}}$
   c. Euler's method succeeds.

21. a. $\dfrac{2500}{2501}\cos x + \dfrac{50}{2501}\sin x - \dfrac{2500}{2501}e^{-50x}; \quad y(0.2) = 0.9836011240\ldots$
   b. $y(0.2) \approx 1.7466146068$. The absolute error is 0.7630134828.
   c. $y(0.2) \approx 1.1761983279$. The absolute error is 0.1925972039.
   d. $y(0.2) \approx 1.8623800769$. The absolute error is 0.8787789529.
   e.

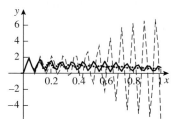

EXERCISES 3.3

1.

	True Value	Euler's Method	Absolute Error	Improved Euler Method	Absolute Error
$h = 0.1$	5.93977	5.69513	0.24464	5.93266	0.00711
$h = 0.05$	5.93977	5.81260	0.12717	5.93791	0.00186
$h = 0.025$	5.93977	5.87490	0.06487	5.93930	0.00047

3. a. $x(t) = -t - 1 + 2e^t$

b. $x(1) \approx 3.42816$

c. The absolute error for $t = 1$ is approximately $|3.43656 - 3.428161| = 0.0084$.

5. $V(0) \approx 166.27517$ m/s.

EXERCISES 3.4

1.

	True Value	Euler's Method	Improved Euler Method	RK4 Method
$h = 0.1$	2.7182818	2.5937425	2.7140808	2.7182797
$h = 0.05$	2.7182818	2.6532977	2.7171911	2.7182817
$h = 0.025$	2.7182818	2.6850638	2.7180039	2.7182818

3. $y(1) = e \approx 2.71828181139414093$

5. a. $x(t) = \dfrac{2}{t^2 + C}$

b. $x(1) \approx 0.99999999727228860$. The actual value of $x(1)$ is 1.

7. a.

t	$V(t)$
5	100.163
10	104.984
15	105.045
16	105.046
17	105.046
18	105.046
19	105.046
20	105.046

The terminal velocity is approximately 105.046 feet per second.

b.

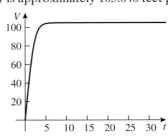

EXERCISES 4.1

1. $y(t) = (c_1 + c_2 t)e^{2t}$

3. $x(t) = e^t(c_1 \cos t + c_2 \sin t)$

5. $x(t) = c_1 + c_2 e^{-2t}$

7. $y(t) = c_1 \cos 2t + c_2 \sin 2t$

9. $r(t) = e^{2t}(c_1 \cos 4t + c_2 \sin 4t)$

15. a. $I(t) = -\frac{3}{2}e^{-50t} + \frac{3}{2}e^{-10t}$

   b.

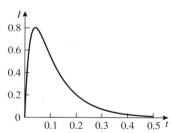

   c. $I_{max} \approx 0.802$

   d. $t = \dfrac{\ln 5}{40}$

17. $x(t) = -e^{2t} + 2e^t$

19. $y(t) = \frac{1}{4}e^{2t-\pi}\sin 4t$

21. $x(t) = 3\cos 12t + \frac{5}{6}\sin 12t$

23. a. $x(t) = -\frac{1}{30}e^{-2t}(11\sqrt{3}\sin(2\sqrt{3}t) + 3\cos(2\sqrt{3}t))$

   b.

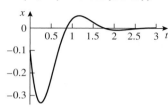

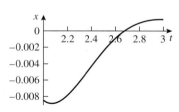

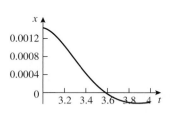

   c. The greatest distance is approximately 33 cm.

EXERCISES 4.2

1. $y(t) = \frac{1}{5}e^{4t} + c_1 e^{-t} + c_2 e^{3t}$

3. $x(t) = e^t(c_1 \sin t + c_2 \cos t + 1) + (-\frac{2}{5}t - \frac{14}{25}) \sin t + (\frac{1}{5}t + \frac{2}{25}) \cos t$

5. $x(t) = c_1 - c_2 e^{-t} - 2 \sin t - 2 \cos t$

7. $y(x) = c_1 \cos(2x) + c_2 \sin(2x) - x \cos(2x) + \sin x \cos x \cos(2x) - \sin(2x) \cos^2 x$
$\qquad + \sin(2x) \ln|\cos x|$

9. $r(t) = c_1 \sin t + c_2 \cos t + (\ln|\sin t|) \sin t - t \cos t$

13. $I(t) = \dfrac{1640}{323}\sqrt{19}e^{-50t} \sin(50\sqrt{19}t) - \dfrac{40}{17}e^{-50t} \cos(50\sqrt{19}t) - \dfrac{160}{17} \sin (100t)$
$\qquad + \dfrac{40}{17} \cos(100t)$

15. The general solution is

$$y(x) = e^{-x/10}(c_1 \cos(\tfrac{3}{10}\sqrt{11}x) + c_2 \sin(\tfrac{3}{10}\sqrt{11}x)) - \frac{25(1 - \omega^2) \sin(\omega x)}{25 - 49\omega^2 + 25\omega^4}$$

$$- \frac{5 \cos(\omega x)\omega}{25 - 49\omega^2 + 25\omega^4}$$

17. $y(x) = \frac{3}{5}xe^{4x} - \frac{3}{25}e^{4x} + \frac{3}{25}e^{-x}$

EXERCISES 4.3

1. $y(t) = c_1 e^{2t} + c_2 e^{-2t} + c_3 e^{3t} + c_4 e^{-3t}$

3. $y(t) = c_1 + c_2 \cos t + c_3 \sin t + t(c_4 \cos t + c_5 \sin t)$

5. $y(t) = c_1 + c_2 e^{-2t} + (c_3 + c_4 t)e^t$

7. $y(t) = c_1 e^{10t} + e^t(c_2 \cos t + c_3 \sin t)$

9. $y(t) = (c_1 + c_2 t + c_3 t^2)e^t + (c_4 + c_5 t)e^{2t} + c_6 e^{3t} + c_7 e^{4t}$

11. $y(t) = \left(\dfrac{t}{4} - \dfrac{9}{16}\right)e^{-t} + \dfrac{9}{16}e^{t/3}$

13. $y(t) = -2 + e^t + \cos t$

15. $y(x) = e^{3x}(c_1 \cos 2x + c_2 \sin 2x) + \frac{3}{5} \cos 2x - \frac{4}{5} \sin 2x$

17. $y(x) = c_1 e^{-3x} + c_2 e^{-x} + c_3 e^{-2x} + x - 3$

19. $y(x) = \frac{3}{7}xe^x + 1$

EXERCISES 4.4

1. The system is $\{dx_1/dt = x_2, dx_2/dt = 1 + x_1\}$.

3. The system is $\{x_1' = x_2, x_2' = 1 - 3x_2 - 2x_1; \quad x_1(0) = 1, x_2(0) = 0\}$.

5. The nonautonomous system is $\{w_1' = w_2, w_2' = w_3, w_3' = w_4, w_4' = 6\sin(4t) + 2w_4 - 5w_3 - 3w_2 + 8w_1\}$. To get an *autonomous* system, replace t by w_5 and add the equation $w_5' = 1$.

7. The system is $\{y_1' = y_2, y_2' = (5\ln x - 4y_1 + 3xy_2)/x^2\}$. To get an autonomous system, let $y_3 = x$, so that $y_3' = 1$. Then replace x by y_3 wherever x occurs.

9. The system is $\{dx_1/dt = x_2, dx_2/dt = -x_1, dy_1/dt = y_2, dy_2/dt = y_1\}$.

11. The system is $\{dy_1/dt = y_2, dy_2/dt = -(g/s_0)y_1\}$.

13. The system is $\{y_1' = y_2, y_2' = y_3, y_3' = \cos(y_1) - y_2\}$.

15. The system is $\{u_1' = u_2, u_2' = u_3 u_2 + u_3^2 u_1, u_3' = 1; \quad u_1(0) = 1, u_2(0) = 2, u_3(0) = 0\}$.

17. One such system is $\left\{ \dfrac{dx_1}{dt} = x_2, \dfrac{dx_2}{dt} = \dfrac{(x_1 - y_2^2)}{2}, \dfrac{dy_1}{dt} = y_2, \dfrac{dy_2}{dt} = \dfrac{(4t + y_1)}{x_1} \right\}.$

19. Using the Chain Rule, we see, for example, that $\dfrac{dx}{dt} = \dfrac{dx}{dw}\dfrac{dw}{dt} = \dfrac{1}{t}\dfrac{dx}{dw}$, so

$t\dfrac{dx}{dt} = -3x + 4y$ becomes $\dfrac{dx}{dw} = -3x + 4y.$

EXERCISES 4.5

1. a. The system is $\{x_1' = x_2, x_2' = -x_2;\ \ x_1(0) = 1, x_2(0) = 2\}.$
   b. The graph of the solution in the phase plane is

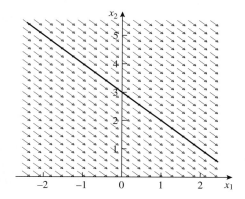

   c. The graphs of the solutions x_1 and x_2 relative to the t-axis are

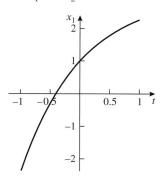

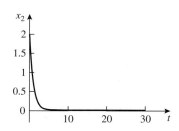

3. a. The system is $\{\dot{y}_1 = y_2, \dot{y}_2 = -y_1; \quad y_1(0) = 2, y_2(0) = 0\}$.

b. The graph of the solution in the phase plane is

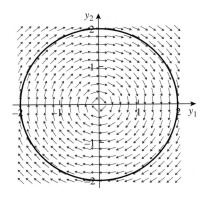

c. The graphs of the solutions y_1 and y_2 with respect to the t-axis are

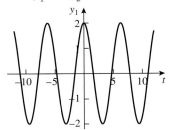

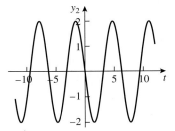

5. a. The system is $\{\dot{x}_1 = x_2, \dot{x}_2 = x_2; \quad x_1(0) = 1 = x_2(0)\}$.

b. The graph of the solution in the phase plane is

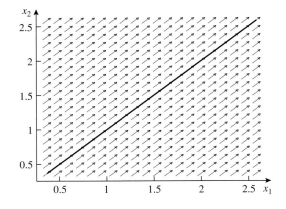

c. The graphs of the solutions x_1 and x_2 relative to the t-axis are

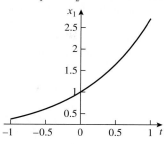

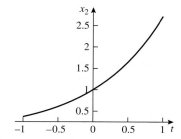

7. The homogeneous equation has the implicit solution

$$\arctan\left(\frac{y}{x}\right) + \tfrac{1}{2}\ln\left(\frac{x^2 + y^2}{x^2}\right) + \ln|x| - C = 0$$

9. a. The system is $\{\dot{x}_1 = x_2,\ \dot{x}_2 = -20x_2 - 64x_1;\quad x_1(0) = 1/3,\ x_2(0) = 0\}$.
  b. The graph of the solution in the phase plane, for $0 \leq t \leq 2$, is

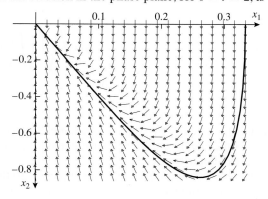

c. The graph of $x(t)\ (= x_1(t)\,)$ relative to the t-axis is

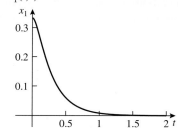

    d. The mass approaches its equilibrium position but doesn't quite reach it because of the large damping force. In particular, the mass doesn't overshoot its equilibrium position.

11. a. The system is $\{\dot{x}_1 = x_2, \dot{x}_2 = 16\cos 8t - 64x_1; \quad x_1(0) = 0, x_2(0) = 0\}$.
    b. The graph of the solution in the phase plane for $0 \le t \le 5$ is

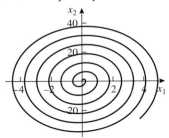

    c. The graph of $x(t)$ $(= x_1(t))$ with respect to t is

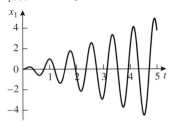

    d. The half-lines serve as an "envelope" for the solution curve:

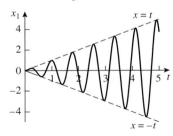

13. a. The system is equivalent (for example) to the single equation $Q'' + 8Q' + 15Q = 0$, which could represent a spring-mass system with spring constant 15 and a coefficient of friction of 8.
    b. The system is equivalent (for example) to the single equation $\ddot{x} - 6\dot{x} + 10x = 0$, which could *not* represent a spring-mass system, because the coefficient -6 would imply a resistance (friction force) acting in the same direction as the motion—an impossibility in a spring-mass system.

15. The system is $\{x_1' = x_2, x_2' = -x_1^3 + x_1 - x_2\}$, with equilibrium points $(0, 0)$, $(-1, 0)$, and $(1, 0)$.

EXERCISES 4.6

3. *Hint*: How many conditions are given?

5. b. The graph of $(x_1(t), y_1(t))$ in the x-y phase plane is

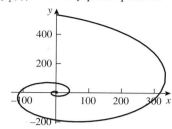

The graph of $(x_2(t), y_2(t))$ in the x-y phase plane is

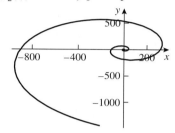

c. Both graphs represent the same spiral. This does not contradict the uniqueness part of our theorem because they are never in the same place at the same time—that is, for any particular value of t, say t^*, we have $(x_1(t^*), y_1(t^*)) \neq (x_2(t^*), y_2(t^*))$.

7. Using the initial points $(x(0), y(0)) = (1, 2), (-1, 2), (-1, -2),$ and $(1, -2)$, for example, we get the trajectories

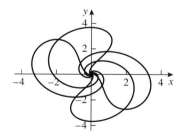

This situation does not contradict the Existence and Uniqueness Theorem.

EXERCISES 4.7

1. a. $x_{k+1} = x_k + \dfrac{h}{2}\{f(t_k, x_k, y_k) + f(t_{k+1}, x_k + hf(t_k, x_k, y_k), y_k + hg(t_k, x_k, y_k)\}$

$y_{k+1} = y_k + \dfrac{h}{2}\{g(t_k, x_k, y_k) + g(t_{k+1}, x_k + hf(t_k, x_k, y_k), y_k + hg(t_k, x_k, y_k)\}$

b. $x(0.5) \approx 1.1273, y(0.5) \approx 0.5202$

c. For $x(0.5)$, the absolute error is approximately 0.0003, and for $y(0.5)$, the absolute error is approximately 0.0009.

3. a. The system is $\{u_1' = u_2, u_2' = 2x + 2u_1 - u_2; \quad u_1(0) = 1, u_2(0) = 1\}$.

   b. $u_1(0.5) \approx 1.8774$ and $u_2(0.5) \approx 4.1711$; $u_1(1.0) \approx 5.5515$ and $u_2(1.0) \approx 13.3031$

   c. $u_1(0.5) \approx 2.1784$ and $u_2(0.5) \approx 4.7536$; $u_1(1.0) \approx 6.7731$ and $u_2(1.0) \approx 14.7205$

5. a. $(x(t), y(t), z(t)) \approx (0, 5, 0)$ for all the values of t specified. The particle doesn't seem to be moving.

   b. x, y, and z seem to be increasing without bound as t grows larger, with the values of x, y, and z approaching each other.

7. $t \approx 3.72$

9. a. $\dfrac{dS}{dt} + \dfrac{dI}{dt} + \dfrac{dR}{dt} = \dfrac{d}{dt}(S + I + R) = 0$. This means that the total population does not change.

   b. (1)

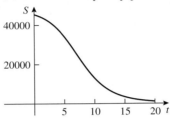

S-I-R model—susceptible population

       (2)

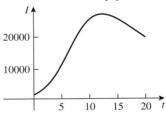

S-I-R model—infected population

       (3)

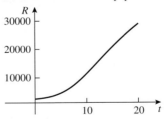

S-I-R model—recovered population

   c.

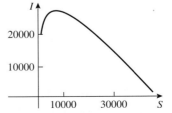

S-I-R model—infected vs. susceptible

S-I-R model—recovered vs. susceptible

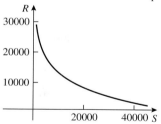

S-I-R model—recovered vs. infected

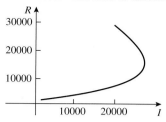

d. All values are rounded to the nearest whole number. The values show the steady increase in the number of people who have recovered, the decreasing number of susceptible people, and the fact that the number of infected people probably peaks between days 10 and 15.

t	S	I	R
1	44255	3062	2682
2	42649	4405	2947
3	40460	6217	3323
10	13044	25547	11408
15	3447	25638	20915
16	2681	24609	22710
17	2108	23464	24428

e. $t \approx 161$ if you round down; but $t \approx 171$ if you round I to the nearest integer.

11. a.

t	$x(t)$	$y(t)$
0.01	0.4492	−0.0158
0.02	0.4468	−0.0113
0.03	0.4432	−0.0068
0.04	0.4385	−0.0024
0.05	0.4330	0.0019
0.06	0.4266	0.0062
0.07	0.4196	0.0105
0.08	0.4120	0.0146
0.09	0.4039	0.0187
0.10	0.3952	0.0227

The direction of the solution curve is counterclockwise.

b.

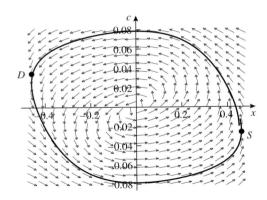

c. $t \approx 1.1$

d. Diastole: $(x, c) \approx (-0.46, 0.02)$ when $t \approx 0.52$; Systole: $(x, c) \approx (0.46, -0.02)$ when $t \approx 1.05$.

EXERCISES 5.1

1. a. $\begin{bmatrix} 3 & 4 \\ -1 & -2 \end{bmatrix} \begin{bmatrix} x \\ y \end{bmatrix} = \begin{bmatrix} -7 \\ 5 \end{bmatrix}$

b. $\begin{bmatrix} \pi & -3 \\ 5 & 2 \end{bmatrix} \begin{bmatrix} a \\ b \end{bmatrix} = \begin{bmatrix} 4 \\ -3 \end{bmatrix}$

c. $\begin{bmatrix} 1 & -1 & 1 \\ -1 & 2 & -3 \\ 2 & -3 & 5 \end{bmatrix} \begin{bmatrix} x \\ y \\ z \end{bmatrix} = \begin{bmatrix} 7 \\ 9 \\ 11 \end{bmatrix}$

3. $\dot{X} = \begin{bmatrix} \dot{x} \\ \dot{y} \end{bmatrix} = \begin{bmatrix} 1 & -1 \\ -4 & 1 \end{bmatrix} \begin{bmatrix} x \\ y \end{bmatrix}$

5. $\dot{X} = \begin{bmatrix} \dot{x} \\ \dot{y} \end{bmatrix} = \begin{bmatrix} 1 & 0 \\ 0 & 1 \end{bmatrix} \begin{bmatrix} x \\ y \end{bmatrix}$

7. a. Let $y_1 = y$, $y_2 = y'$. Then the system is $\{y_1' = y_2, y_2' = 3y_2 - 2y_1\}$, which can be written as

$$\begin{bmatrix} y_1' \\ y_2' \end{bmatrix} = \begin{bmatrix} 0 & 1 \\ -2 & 3 \end{bmatrix} \begin{bmatrix} y_1 \\ y_2 \end{bmatrix}$$

b. Let $y_1 = y$, $y_2 = y'$. Then the system is $\{y_1' = y_2, y_2' = \frac{1}{5}y_1 - \frac{3}{5}y_2\}$, which can be written as

$$\begin{bmatrix} y_1' \\ y_2' \end{bmatrix} = \begin{bmatrix} 0 & 1 \\ \frac{1}{5} & -\frac{3}{5} \end{bmatrix} \begin{bmatrix} y_1 \\ y_2 \end{bmatrix}$$

9. *Hint:* $A(t)B(t) = \begin{bmatrix} a_{11}(t)b_{11}(t) + a_{12}(t)b_{21}(t) \\ a_{21}(t)b_{11}(t) + a_{22}(t)b_{21}(t) \end{bmatrix}$. Now just apply the Product Rule to each entry and separate the matrix into the sum of two matrix products.

11. $x = 1$, $y = 0$

EXERCISES 5.2

1. a. 17

b. 0

c. $6t^4 + 4 \sin t$

d. $\cos^2 \theta + \sin^2 \theta = 1$

3. a. $\begin{bmatrix} \dot{x} \\ \dot{y} \end{bmatrix} = \begin{bmatrix} 2 & 1 \\ 3 & 4 \end{bmatrix} \begin{bmatrix} x \\ y \end{bmatrix}$

   b. $\lambda^2 - 6\lambda + 5 = 0$

   c. $\lambda_1 = 5, \lambda_2 = 1$

   d. $\lambda_1 \leftrightarrow V_1 = x \begin{bmatrix} 1 \\ 3 \end{bmatrix}, \lambda_2 \leftrightarrow V_2 = x \begin{bmatrix} 1 \\ -1 \end{bmatrix}$

5. a. $\begin{bmatrix} \dot{x} \\ \dot{y} \end{bmatrix} = \begin{bmatrix} -4 & 2 \\ 2 & -1 \end{bmatrix} \begin{bmatrix} x \\ y \end{bmatrix}$

   b. $\lambda^2 + 5\lambda = 0$

   c. $\lambda_1 = 0, \lambda_2 = -5$

   d. $\lambda_1 \leftrightarrow V_1 = x \begin{bmatrix} 1 \\ 2 \end{bmatrix}, \lambda_2 \leftrightarrow V_2 = x \begin{bmatrix} 2 \\ -1 \end{bmatrix}$

7. a. $\begin{bmatrix} \dot{x} \\ \dot{y} \end{bmatrix} = \begin{bmatrix} -6 & 4 \\ -3 & 1 \end{bmatrix} \begin{bmatrix} x \\ y \end{bmatrix}$

   b. $\lambda^2 + 5\lambda + 6 = 0$

   c. $\lambda_1 = -2, \lambda_2 = -3$

   d. $\lambda_1 \leftrightarrow V_1 = x \begin{bmatrix} 1 \\ 1 \end{bmatrix}, \lambda_2 \leftrightarrow V_2 = x \begin{bmatrix} 4 \\ 3 \end{bmatrix}$

9. a. $\dfrac{8C_1 e^{2t} - C_2 e^{-t}}{2C_1 e^{2t} - C_2 e^{-t}}$

11.

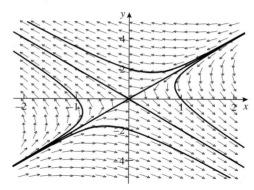

The representative eigenvectors are $\begin{bmatrix} 1 \\ -2 \end{bmatrix}$ and $\begin{bmatrix} 1 \\ 2 \end{bmatrix}$.

13. a. $c_1(x) = \alpha_1 \left(\dfrac{c_0 - C_0}{\alpha_1 - \alpha_2} \right) e^{(\alpha_2 - \alpha_1)x} + \dfrac{\alpha_1 C_0 - \alpha_2 c_0}{\alpha_1 - \alpha_2}$

   $c_2(x) = \alpha_2 \left(\dfrac{c_0 - C_0}{\alpha_1 - \alpha_2} \right) e^{(\alpha_2 - \alpha_1)x} + \dfrac{\alpha_1 C_0 - \alpha_2 c_0}{\alpha_1 - \alpha_2}$

   b. For example, you could solve $\dfrac{d^2 c_1}{dx^2} + (\alpha_1 - \alpha_2) \dfrac{dc_1}{dx} = 0.$

EXERCISES 5.3

1. a. $\lambda_1 = 3, V_1 = \begin{bmatrix} 1 \\ 0 \end{bmatrix}$; $\lambda_2 = 2, V_2 = \begin{bmatrix} 0 \\ 1 \end{bmatrix}$.

   b.

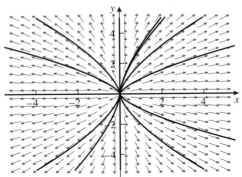

3. a. $\lambda_1 = 1, V_1 = \begin{bmatrix} 1 \\ -4 \end{bmatrix}$; $\lambda_2 = -2, V_2 = \begin{bmatrix} 1 \\ -1 \end{bmatrix}$.

   b.

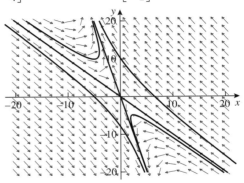

5. a. $\lambda_1 = 2, V_1 = \begin{bmatrix} 5 \\ 1 \end{bmatrix}$; $\lambda_2 = -4, V_2 = \begin{bmatrix} 1 \\ -1 \end{bmatrix}$.

   b.

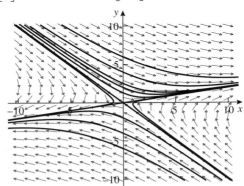

7. a. $\lambda_1 = \dfrac{-5 + \sqrt{17}}{2}, V_1 = \begin{bmatrix} \frac{-1 + \sqrt{17}}{8} \\ 1 \end{bmatrix}; \quad \lambda_2 = \dfrac{-5 - \sqrt{17}}{2}, V_2 = \begin{bmatrix} \frac{-1 - \sqrt{17}}{8} \\ 1 \end{bmatrix}.$

b.

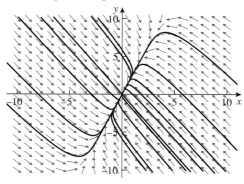

9. a. $\lambda_1 = \sqrt{5}, V_1 = \begin{bmatrix} 2 - \sqrt{5} \\ 1 \end{bmatrix}; \quad \lambda_2 = -\sqrt{5}, V_2 = \begin{bmatrix} 2 + \sqrt{5} \\ 1 \end{bmatrix}.$

b.

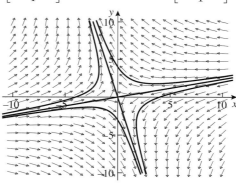

11. a. $\lambda_1 = 0, \lambda_2 = -2.$

b. $V_1 = \begin{bmatrix} 3 \\ 4 \end{bmatrix}, V_2 = \begin{bmatrix} 1 \\ 2 \end{bmatrix}.$

c.

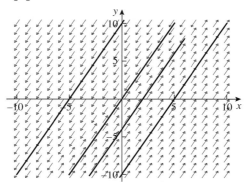

Every point of the line $y = \frac{4}{3}x$ is an equilibrium point. The origin is a *sink*, and every other point is a *node*.

d. $X(t) = c_1 \begin{bmatrix} 3 \\ 4 \end{bmatrix} + c_2 e^{-2t} \begin{bmatrix} 1 \\ 2 \end{bmatrix}.$

13. *Hint:* Construct the characteristic polynomial. Then remember that the coefficient of λ is the negative of the trace of the matrix of coefficients A and that the constant term is the determinant of A.

b. For example, we could have

$$\dot{x} = x + 2y$$
$$\dot{y} = 3x + 2y$$

15. a. $x(t) = \left(\dfrac{x_0 + y_0}{V_1 + V_2}\right)V_1 + \left(\dfrac{x_0 V_2 - y_0 V_1}{V_1 + V_2}\right)e^{-P\left(\frac{1}{V_1} + \frac{1}{V_2}\right)t}$

$y(t) = \left(\dfrac{x_0 + y_0}{V_1 + V_2}\right)V_2 - \left(\dfrac{x_0 V_2 - y_0 V_1}{V_1 + V_2}\right)e^{-P\left(\frac{1}{V_1} + \frac{1}{V_2}\right)t}$

b. $\lim\limits_{t \to \infty} x(t) = \left(\dfrac{x_0 + y_0}{V_1 + V_2}\right)V_1, \lim\limits_{t \to \infty} y(t) = \left(\dfrac{x_0 + y_0}{V_1 + V_2}\right)V_2$

c. $\lim\limits_{t \to \infty} [x(t) + y(t)] = \left(\dfrac{x_0 + y_0}{V_1 + V_2}\right)V_1 + \left(\dfrac{x_0 + y_0}{V_1 + V_2}\right)V_2 = x_0 + y_0$

After a long time, the total quantity of solution on both sides of the membrane still equals the total quantity present originally—that is, the total at $t = 0$. Nothing is gained; nothing is lost.

d. The chemical moves across the membrane from the side with a higher concentration to the side with a lower concentration.

EXERCISES 5.4

1. a. $\lambda_1 = 3 = \lambda_2; V_1 = \begin{bmatrix} 1 \\ 0 \end{bmatrix}$ and $V_2 = \begin{bmatrix} 0 \\ 1 \end{bmatrix}$ are linearly independent eigenvectors.

b.

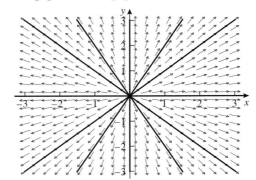

3. a. $\lambda_1 = 3 = \lambda_2; V_1 = \begin{bmatrix} 1 \\ 1 \end{bmatrix}$ is the only linearly independent eigenvector.

b.

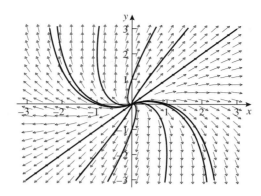

5. a. $\lambda_1 = -1 = \lambda_2$; $V_1 = \begin{bmatrix} 1 \\ 1 \end{bmatrix}$ is the only linearly independent eigenvector.

b.

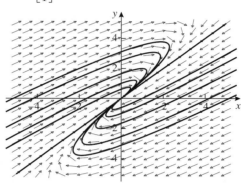

7. For example,

$$\dot{x} = -2x$$
$$\dot{y} = -2y$$

EXERCISES 5.5

1. a. $\lambda_1 = -1 + 2i$, $V_1 = \begin{bmatrix} i \\ 1 \end{bmatrix}$; $\lambda_2 = -1 - 2i$, $V_2 = \begin{bmatrix} -i \\ 1 \end{bmatrix}$.

b.

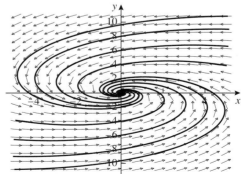

3. a. $\lambda_1 = -0.5 + i, V_1 = \begin{bmatrix} -1 \\ i \end{bmatrix}; \lambda_2 = -0.5 - i, V_2 = \begin{bmatrix} -1 \\ -i \end{bmatrix}.$

b.

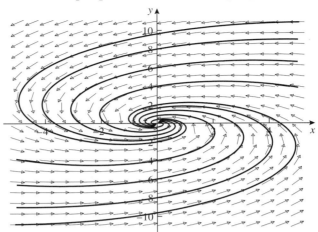

5. a. $\lambda_1 = \frac{1}{2} + \frac{\sqrt{3}}{2}i, V_1 = \begin{bmatrix} 1 \\ -\frac{3}{2} + \frac{\sqrt{3}}{2}i \end{bmatrix}; \lambda_2 = \frac{1}{2} - \frac{\sqrt{3}}{2}i, V_2 = \begin{bmatrix} 1 \\ -\frac{3}{2} - \frac{\sqrt{3}}{2}i \end{bmatrix}.$

b.

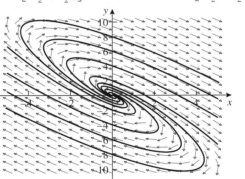

7. a. $\lambda_1 = -6 + i, V_1 = \begin{bmatrix} 1 \\ 1 + i \end{bmatrix}; \lambda_2 = -6 - i, V_2 = \begin{bmatrix} 1 \\ 1 - i \end{bmatrix}.$

b.

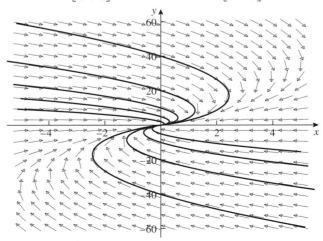

9. **i.**

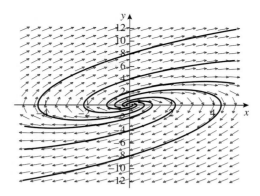

ii.

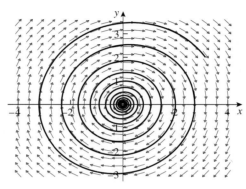

iii.

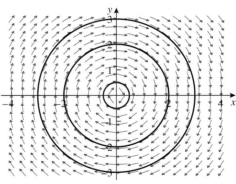

iv.

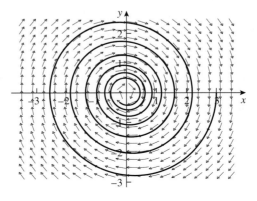

v.

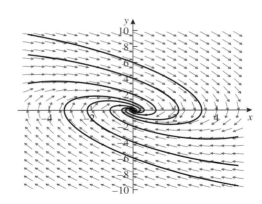

b. Yes: $\beta = 0$.

c. The eigenvalues are $\lambda_1 = \dfrac{-\beta + \sqrt{\beta^2 - 4}}{2}$ and $\lambda_2 = \dfrac{-\beta - \sqrt{\beta^2 - 4}}{2}$.

d. When $\beta = -1$, $\lambda_{1,2} = 0.5 \pm \frac{\sqrt{3}}{2}i$, so the origin is a *spiral source*. When $\beta = -0.1$, the eigenvalues are $\lambda_{1,2} = 0.05 \pm 0.9987i$, so the origin is a *spiral source*. The value $\beta = 0$ gives us eigenvalues $\lambda_{1,2} = \pm i$, and the origin is a *center*. When $\beta = 0.1$, the eigenvalues are $\lambda_{1,2} = -0.05 \pm 0.9987i$, so the origin is a *spiral sink*. The value $\beta = 1$ gives us eigenvalues $\lambda_{1,2} = -0.5 \pm \frac{\sqrt{3}}{2}i$, and the origin is a *spiral sink*.

EXERCISES 5.6

1. $x(t) = \frac{1}{2}(e^t - \cos t - \sin t)$, $y(t) = t - 1 + 2e^{-t}$

3. $x(t) = te^t - t^2 - 2 - \frac{1}{2}e^t + c_1e^t + c_2e^{-t}$
$y(t) = -\frac{3}{2}e^t + te^t - 2t + c_1e^t - c_2e^{-t}$

5. $x(t) = 3e^{5t} + c_1e^t + c_2e^{4t}$, $y(t) = e^{5t} - c_1e^t + \frac{1}{2}c_2e^{4t}$

7. $x(t) = te^{2t} + c_1e^{3t} + c_2e^{2t}$, $y(t) = 2e^{2t} - 2te^{2t} - c_1e^{3t} - 2c_2e^{2t}$

9. $x(t) = -4e^{3t} - e^{-t} + c_1e^{2t} + c_2e^{4t}$
$y(t) = -2e^{3t} - 2e^{-t} + c_1e^{2t} + \frac{1}{3}c_2e^{4t}$

11. $x(t) = 2e^t \cos t - e^t \sin t + c_1e^t + c_2e^{3t}$
$y(t) = 3e^t \cos t + e^t \sin t + c_1e^t - c_2e^{3t}$

17. c. $V_1 = \begin{bmatrix} 1 \\ \frac{-1 + \sqrt{57}}{14} \end{bmatrix}$, corresponding to λ_1; $V_2 = \begin{bmatrix} 1 \\ \frac{-1 - \sqrt{57}}{14} \end{bmatrix}$, corresponding to λ_2.

d. $X_{GH}(t) = c_1 e^{(\sqrt{57}-9)t/2} \begin{bmatrix} 1 \\ \frac{-1 + \sqrt{57}}{14} \end{bmatrix} + c_2 e^{-(\sqrt{57}+9)t/2} \begin{bmatrix} 1 \\ \frac{-1 - \sqrt{57}}{14} \end{bmatrix}$

$= \begin{bmatrix} c_1 e^{(\sqrt{57}-9)t/2} + c_2 e^{-(\sqrt{57}+9)t/2} \\ \left(\frac{-1 + \sqrt{57}}{14}\right) c_1 e^{(\sqrt{57}-9)t/2} - \left(\frac{1 + \sqrt{57}}{14}\right) c_2 e^{-(\sqrt{57}+9)t/2} \end{bmatrix}$

e. For example, try $x(t) \equiv 2$ and $y(t) \equiv 3$.

f. $X_{GNH} = \begin{bmatrix} c_1 e^{(\sqrt{57}-9)t/2} + c_2 e^{-(\sqrt{57}+9)t/2} \\ \left(\frac{-1 + \sqrt{57}}{14}\right) c_1 e^{(\sqrt{57}-9)t/2} - \left(\frac{1 + \sqrt{57}}{14}\right) c_2 e^{-(\sqrt{57}+9)t/2} \end{bmatrix} + \begin{bmatrix} 2 \\ 3 \end{bmatrix}$

$= \begin{bmatrix} c_1 e^{(\sqrt{57}-9)t/2} + c_2 e^{-(\sqrt{57}+9)t/2} + 2 \\ \left(\frac{-1 + \sqrt{57}}{14}\right) c_1 e^{(\sqrt{57}-9)t/2} - \left(\frac{1 + \sqrt{57}}{14}\right) c_2 e^{-(\sqrt{57}+9)t/2} + 3 \end{bmatrix}$

As $t \to \infty$, $X_{GNH} \to \begin{bmatrix} 2 \\ 3 \end{bmatrix}$ because all the exponential terms have negative exponents.

19. **a.** $x(t) = \dfrac{I}{k_1}(1 - e^{-k_1 t})$, $y(t) = \dfrac{I}{k_2}\left[1 + \dfrac{k_2}{k_1 - k_2}e^{-k_1 t} - \dfrac{k_1}{k_1 - k_2}e^{-k_2 t}\right]$

b. $\displaystyle\lim_{t \to \infty} x(t) = \dfrac{I}{k_1}$ and $\displaystyle\lim_{t \to \infty} y(t) = \dfrac{I}{k_2}$.

c.

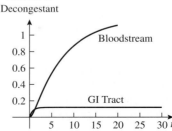

Amount of Decongestant

d.

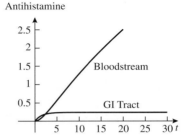

Amount of Antihistamine

EXERCISES 5.7

1. **a.** $\dot{X} = \begin{bmatrix} 1 & -1 & 1 \\ 1 & 1 & -1 \\ 2 & -1 & 0 \end{bmatrix}\begin{bmatrix} x \\ y \\ z \end{bmatrix}$.

b. $\lambda_1 = -1, V_1 = \begin{bmatrix} 1 \\ -3 \\ -5 \end{bmatrix}$; $\lambda_2 = 1, V_2 = \begin{bmatrix} 1 \\ 1 \\ 1 \end{bmatrix}$; $\lambda_3 = 2, V_3 = \begin{bmatrix} 1 \\ 0 \\ 1 \end{bmatrix}$.

c. $X(t) = \begin{bmatrix} c_1 e^{-t} + c_2 e^{t} + c_3 e^{2t} \\ -3c_1 e^{-t} + c_2 e^{t} \\ -5c_1 e^{-t} + c_2 e^{t} + c_3 e^{2t} \end{bmatrix}$.

3. **a.** $\dot{X} = \begin{bmatrix} 3 & -1 & 1 \\ 1 & 1 & 1 \\ 4 & -1 & 4 \end{bmatrix}\begin{bmatrix} x \\ y \\ z \end{bmatrix}$.

b. $\lambda_1 = 1, V_1 = \begin{bmatrix} 1 \\ 1 \\ -1 \end{bmatrix}$; $\lambda_2 = 2, V_2 = \begin{bmatrix} 1 \\ -2 \\ -3 \end{bmatrix}$; $\lambda_3 = 5, V_3 = \begin{bmatrix} 1 \\ 1 \\ 3 \end{bmatrix}$.

c. $X(t) = \begin{bmatrix} c_1 e^{t} + c_2 e^{2t} + c_3 e^{5t} \\ c_1 e^{t} - 2c_2 e^{2t} + c_3 e^{5t} \\ -c_1 e^{t} - 3c_2 e^{2t} + 3c_3 e^{5t} \end{bmatrix}$.

5.

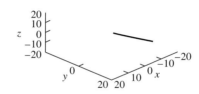

7. a. $x(t) = \frac{2}{9}e^{-t} + \frac{2}{3}e^{-2t} + \frac{1}{9}e^{2t}$, $y(t) = \frac{2}{9}e^{-t} - \frac{2}{3}e^{-2t} + \frac{1}{9}e^{2t}$,
   $z(t) = -\frac{2}{9}e^{-t} + \frac{2}{9}e^{2t}$

b. $x(0.5) \approx 0.6820688660$, $y(0.5) \approx 0.1915629444$, $z(0.5) \approx 0.4692780374$.

c. *Improved Euler method*, with $h = 0.01$: $x(0.5) \approx 0.6820667761$, $y(0.5) \approx 0.1915276431$, $z(0.5) \approx 0.4692372355$.
   RKF45: $x(0.5) \approx 0.6820688625$, $y(0.5) \approx 0.1915629444$, $z(0.5) \approx 0.4692780338$.

9. Denoting the volumes of tanks A, B, and C by $A(t)$, $B(t)$, and $C(t)$, respectively, the system is

$$\dot{X}(t) = \begin{bmatrix} -0.02 & 0.02 & 0 \\ 0.02 & -0.04 & 0.02 \\ 0 & 0.02 & -0.02 \end{bmatrix} \begin{bmatrix} A(t) \\ B(t) \\ C(t) \end{bmatrix},$$

with solution

$$A(t) = 11000e^{-0.02t} + (11000/3)e^{-0.06t} + 25000/3$$
$$B(t) = -(22000/3)e^{-0.06t} + 25000/3$$
$$C(t) = -11000e^{-0.02t} + (11000/3)e^{-0.06t} + 25000/3.$$

11. a. Let $x_1(t)$ and $x_2(t)$ denote the amount of compound A in tanks I and II, respectively. Similarly, define $y_1(t)$, $y_2(t)$, $z_1(t)$, and $z_2(t)$ for compounds B and C in tanks I and II. Then the system is

$$\dot{X}(t) = \begin{bmatrix} -0.1 & 0.02 & 0 & 0 & 0 & 0 \\ 0.1 & -0.14 & 0 & 0 & 0 & 0 \\ 0 & 0 & -0.1 & 0.02 & 0 & 0 \\ 0 & 0 & 0.1 & -0.14 & 0 & 0 \\ 0 & 0 & 0 & 0 & -0.1 & 0.02 \\ 0 & 0 & 0 & 0 & 0.1 & -0.14 \end{bmatrix} \begin{bmatrix} x_1(t) \\ x_2(t) \\ y_1(t) \\ y_2(t) \\ z_1(t) \\ z_2(t) \end{bmatrix} + \begin{bmatrix} 4 \\ 2 \\ 0 \\ 0 \\ 0 \\ 0 \end{bmatrix}$$

with $x_1(0) = 0$, $x_2(0) = 0$, $y_1(0) = 50$, $y_2(0) = 0$, $z_1(0) = 0$, and $z_2(0) = 50$.

b. $x_1(t) = 50 - 4.59e^{-0.169t} - 45.41e^{-0.071t}$
   $x_2(t) = 50 + 15.82e^{-0.169t} - 65.82e^{-0.071t}$
   $y_1(t) = 14.79e^{-0.169t} + 35.21e^{-0.071t}$, $y_2(t) = -51.03e^{-0.169t} + 51.03e^{-0.071t}$
   $z_1(t) = -10.21e^{-0.169t} + 10.21e^{-0.071t}$, $z_2(t) = 35.21e^{-0.169t} + 14.79e^{-0.071t}$

c.

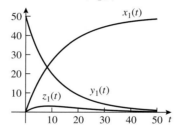

d.

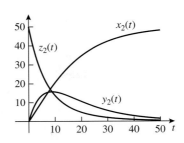

15. a.–b. $x(t) = t^2 + c_1$, $y(t) = t^3 + 3c_1 t + t^2 + c_2$
$z(t) = \frac{5}{3}t^3 + c_1 t + t^4 + 6c_1 t^2 + 4c_2 t + \frac{1}{2}t^2 + c_3$

EXERCISES 6.1

1. $2/s^3$

3. $s/(s^2 + a^2)$

5. $1/(s - a)(s - b)$

7. $(s + e^{-2s} - e^{-s})/s^2$

9. $\mathcal{L}[y(t)] = 1/(s - 1)$

11. $\mathcal{L}[y(t)] = (s + 5)/(s^2 + 4s + 4)$

13. $\mathcal{L}[y(x)] = 2(s^2 + s - 1)/(s^3(s - 1)^3)$

15. $\sqrt{\dfrac{\pi}{s}}$

EXERCISES 6.2

3. $\dfrac{1}{2}\{1 - e^{-t}(\cos t + \sin t)\}$

9. $s/(s^2 + 1)^2$

11. $y(t) = 2t^2 - 6t + e^{-2t} - 8e^{-t} + 7$

13. $y(x) = x^2 + 4x + 4 + x^2 e^x - 4e^x$

15. $Q(t) = \begin{cases} e^{-t}\sin t & \text{for } 0 \le t < \pi \\ \frac{2}{5}\cos t - \frac{1}{5}\sin t - \frac{1}{5}e^{(-t+\pi)}\{2\cos t + \sin t\} + e^{-t}\sin t & \text{for } t \ge \pi \end{cases}$

17. $f(t) = 4t + \frac{2}{3}t^3$

19. $x(t) = 2 - e^{-t}$

21. $x(t) = \frac{1}{10}t^5 e^{2t} + \frac{1}{4}t^4 e^{2t} = \frac{1}{2}t^4 e^{2t}\left(\dfrac{t}{5} + \dfrac{1}{2}\right) = \frac{1}{20}t^4 e^{2t}(2t + 5)$

EXERCISES 6.3

1. a.

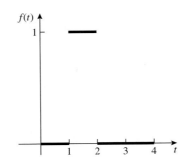

b. $f(t) = U(t - 1) - U(t - 2)$

3. a.

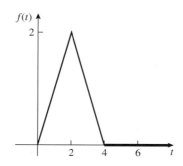

b. $f(t) = tU(t) + (4 - 2t)\,U(t - 2) + (t - 4)\,U(t - 4)$

7. $\mathcal{L}[f(t)] = (e^{-s} - e^{-2s})/s$

9. $\mathcal{L}[f(t)] = (1 - 2e^{-2s} + e^{-4s})/s^2$

11. a. $P(t) = \begin{cases} Ae^{kt} + \dfrac{h}{k}(1 - e^{kt}), & 0 \le t \le 30 \\[2mm] \dfrac{h}{k}(e^{-k(30-t)} - e^{kt}) + Ae^{kt}, & t > 30 \end{cases}$

b. $A = \dfrac{h}{k}(e^{330k} - e^{360k})/(1 - e^{360k})$

13. $y(t) = \begin{cases} \frac{1081}{1049}e^{-5t/4} - \frac{32}{1049}\cos(8t) + \frac{5}{1049}\sin(8t), & t \le 2 \\[1mm] e^{(-5t/4 + 5/2)}\left(\frac{5}{1049}\sin(16) - \frac{32}{1049}\cos(16)\right) + \frac{1081}{1049}e^{-5t/4}, & t > 2 \end{cases}$

15. $y(t) = \frac{5}{2} - \frac{1}{2}e^{-t/2}(5\cos(\frac{\sqrt{15}}{6}t) + \sqrt{15}\sin(\frac{\sqrt{15}}{6}t))$ for $0 \le t < 5$

$y(t) = \frac{1}{2}e^{(-t+5)/2}(5\cos(\frac{\sqrt{15}}{6}(t-5)) + \sqrt{15}\sin(\frac{\sqrt{15}}{6}(t-5)))$
$\quad - \frac{1}{2}e^{-t/2}(5\cos(\frac{\sqrt{15}}{6}t) + \sqrt{15}\sin(\frac{\sqrt{15}}{6}t))$ for $t \ge 5$

17. $g(t) = \begin{cases} t & \text{for } 0 \le t < 1 \\ t - 1 & \text{for } 1 \le t < 2 \end{cases}$ and $g(t) = 0$ for $t \ge 2$.

$y(t) = \begin{cases} t - 1 + e^{-t} & \text{for } 0 \le t \le 1 \\ t - 2 + e^{-t} + e^{(1-t)} & \text{for } 1 \le t \le 2 \\ e^{-t} + e^{1-t} & \text{for } t > 2 \end{cases}$

19. a.

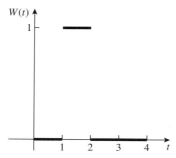

b. $y(t) \equiv 0$

c. $y(t) = \begin{cases} \frac{1}{2}e^{2(1-t)} + \frac{1}{2} - e^{1-t} & \text{for } 1 \le t \le 2 \\ \frac{1}{2}e^{2(1-t)} - e^{1-t} + e^{2-t} - \frac{1}{2}e^{2(2-t)} & \text{for } t > 2 \end{cases}$ and $y(t) = 0$ for $0 \le t < 1$

d.

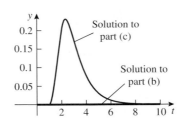

EXERCISES 6.4

1. $y(t) = (t - a)U(t - a) = \begin{cases} 0 & \text{for } t \leq a \\ t - a & \text{for } t \geq a \end{cases}$

3. $y(t) = \begin{cases} 0 & \text{for } t \leq 5 \\ \frac{2}{15}\sqrt{15}e^{(5-t)/4} \sin(\frac{\sqrt{15}}{4}(t - 5)) & \text{for } t > 5 \end{cases}$

5. $y(t) = \begin{cases} 0 & \text{for } t < 1 \\ \frac{1}{10}e^{3(1-t)} \sin(10(t - 1)) & \text{for } 1 \leq t \leq 7 \\ \frac{1}{10}e^{3(1-t)} \sin(10(t - 1)) - \frac{1}{10}e^{3(7-t)} \sin(10(t - 7)) & \text{for } t > 7 \end{cases}$

7. $y(t) = \begin{cases} \sin t & \text{for } t \leq \frac{3}{2}\pi \\ \sin t + 4\cos t & \text{for } t > \frac{3}{2}\pi \end{cases}$

9. $y(x) = \frac{W}{6EI}\left\{ \frac{3L}{2}x^2 - x^3 + \left(x - \frac{L}{2}\right)^3 U\left(x - \frac{L}{2}\right) \right\}$

$= \begin{cases} \frac{W}{EI}\left(\frac{L}{4}x^2 - \frac{1}{6}x^3\right) & \text{for } 0 \leq x < \frac{L}{2} \\ \frac{WL^2}{4EI}\left(\frac{1}{2}x - \frac{L}{12}\right) & \text{for } \frac{L}{2} \leq x \leq L \end{cases}$

13. b. $y(t) = \begin{cases} 0 & \text{for } t \leq 1 \\ (t - 1)e^{1-t} & \text{for } t > 1 \end{cases}$

EXERCISES 6.5

1. $\mathcal{L}[y(t)] = \frac{3s + 11}{(s + 4)(s + 2)}, y(t) = \frac{1}{2}e^{-4t} + \frac{5}{2}e^{-2t}$

3. $x(t) = 5e^{7t} - 5e^{6t}, y(t) = -5e^{7t} + 6e^{6t}$

5. $x(t) = \frac{7}{5}e^{-4t} - \frac{2}{5}e^{t}, y(t) = \frac{7}{5}e^{-4t} - \frac{7}{5}e^{t}$

7. $x(t) = 2 - 3e^{-t} - te^{-t} + e^{-2t}$
$y(t) = -3 + 5e^{-t} + 2te^{-t} - 2e^{-2t}$

9. $x(t) = e^{t/2}, y(t) = e^{t/2}, z(t) = e^{t/2}$

11. $x(t) = 2t + 2e^{t} - \frac{10}{3}e^{t/2} + \frac{4}{3}e^{-t}$
$y(t) = \frac{1}{2}t^2 - t + 1 + 2e^{t} - \frac{5}{3}e^{t/2} - \frac{4}{3}e^{-t}$

13. $x(t) = \frac{1}{20}(11 \cos(2t) + 9 \cos(\sqrt{2}t)), \theta(t) = \frac{1}{20}(11 \cos(2t) - 9 \cos(\sqrt{2} + t))$

15. $I_1(t) = \frac{11}{4} - \frac{1}{20}e^{-6t} - \frac{27}{10}e^{-t}, I_2(t) = \frac{3}{4} + \frac{3}{20}e^{-6t} - \frac{9}{10}e^{-t}$

EXERCISES 6.6

1. $x(t)$ goes to 0 as t gets larger.

3. $x(t)$ oscillates with decreasing amplitude as t gets larger.

5. $x(t)$ oscillates with decreasing amplitude as t gets larger.

7. $x(t)$ oscillates as t grows larger.

9. a. $\mathcal{L}[x(t)] = \dfrac{40s^2 + 94s + 19}{(10s + 1)(s^2 + 2s + 2)}$

   b. The poles are $s = -\frac{1}{10}, -1 + i$, and $-1 - i$.

   c. The solution $x(t)$ goes to 0 as t gets larger.

11. a. $\mathcal{L}[x(t)] = \dfrac{2s^3 + 7s^2 + 4s + 7}{2s^4 + 7s^3 + 5s^2 + 7s + 3}$

   b. The poles are $s = -\frac{1}{2}, -3, i$, and $-i$.

   c. As t grows larger, the solution $x(t)$ becomes oscillatory. (There are two transient terms.)

15. a. The transfer function is $1/(s^2 - s - 6) = 1/[(s - 3)(s + 2)]$.

   b. The impulse response function is $-\frac{1}{5}e^{-2t} + \frac{1}{5}e^{3t}$.

   c. $y(t) = 2e^{3t} - e^{-2t} + \frac{1}{5}e^{3t}\displaystyle\int_0^t e^{-3u}g(u)\,du - \frac{1}{5}e^{-2t}\displaystyle\int_0^t e^{2u}g(u)\,du$

17. a. The transfer function is given by $1/(a_1s + a_0)$.

EXERCISES 7.1

1. $(0, 0)$ and $(\frac{1}{2}, 1)$

3. $(0, 0), (-1, 1)$, and $(1, 1)$

5. $(0, 0)$ and $(-1, -1)$

7. $(0, 0), (0, \frac{3}{2}), (1, 0)$, and $(-1, 2)$

9. $(0, 2k\pi), k = 0, \pm 1, \pm 2, \ldots ; (2, (2k + 1)\pi)), k = 0, \pm 1, \pm 2, \ldots .$ (Graph the two equations in x and y on the same set of axes.)

11. $(0, 0), \left(0, -\dfrac{k_2 - G_2N_0}{a_2G_2}\right)$, and $\left(-\dfrac{k_1 - G_1N_0}{a_1G_1}, 0\right)$

13. a. $\left\{\dfrac{dx_1}{dt} = x_2, \dfrac{dx_2}{dt} = -\dfrac{g}{L}\sin(x_1)\right\}$

   b. $(2k\pi, 0), k = 0, \pm 1, \pm 2, \ldots$

EXERCISES 7.2

1. d. If $a < 0$, then $\dot{r} < 0$, which implies that a trajectory will spiral into $(0, 0)$, so the origin is a *stable spiral point* (a sink). If $a = 0$, then $\dot{r} = 0$, so r is a constant and the origin is a *stable center*. If $a > 0$, then the origin is an *unstable spiral point* (a source). In the language of Section 2.6, the parameter value $a = 0$ is a *bifurcation point*.

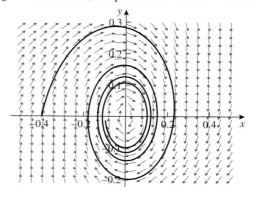

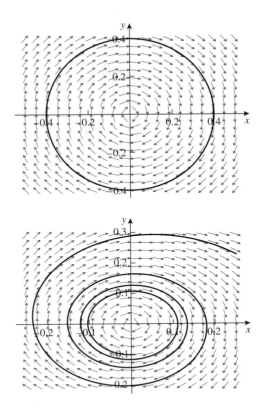

3. c. The origin is a spiral source.
5. c. The origin is a stable node.
7. c. The origin is a saddle point.
9. c. The eigenvalues of the almost linear system are pure imaginary numbers, so the origin is either a center or a spiral point of the nonlinear system. The phase portrait of the nonlinear system indicates that the origin is a center.
11. c. The origin is a spiral sink.
13. a.

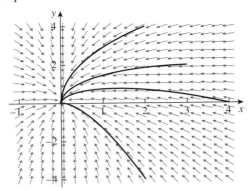

The origin is a sink. The boat eventually winds up at $(0, 0)$.

b.

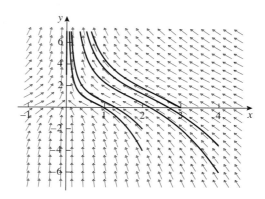

The origin is a source. The boat moves farther and farther away from $(0, 0)$ as time passes.

EXERCISES 7.3

1. $(0, 0)$ and $(\frac{1}{4}, \frac{3}{2})$

3. $(0, 0)$ and $(\frac{3}{4}, \frac{1}{4})$

7. a. $(0, 0)$ and $(100, 100)$

    b.

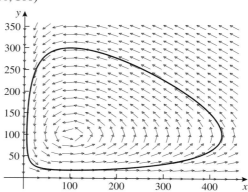

    c. Min $x(t) \approx 6$, Max $x(t) \approx 425$; Min $y(t) \approx 18$, Max $y(t) \approx 300$.

    d. $x - 100 \ln x + 2y - 200 \ln y + 901.3 = 0$

9. a. $\dfrac{du}{dv} = \left(-\dfrac{b^2 c}{a d^2}\right) \cdot \dfrac{v}{u}$

    d. $u(t) = c_1 \cos(\sqrt{ac}\, t) + c_2 \sin(\sqrt{ac}\, t),$

        $v(t) = -\dfrac{d\sqrt{ac}}{bc}(-c_1 \sin(\sqrt{ac}\, t) + c_2 \cos(\sqrt{ac}\, t))$

13. a. Values of $C > 2$ give the top or bottom trajectories, depending on which sign you give the square root in the solution formula.

    b. $C = 2$

15. a. $\{\dot{x}_1 = x_2, \dot{x}_2 = -kx_2 - \sin(x_1)\}$

b.

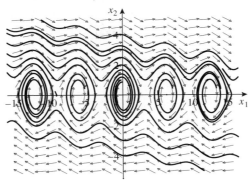

c.

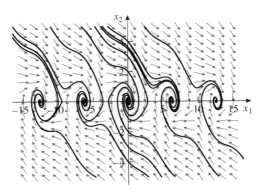

d. The phase portrait in part (b) shows that the pendulum makes a number of revolutions (dependent on the initial velocity imparted to the pendulum) and then settles into a decaying oscillation around the equilibrium point $\theta = 2k\pi$. In the phase portrait for part (d), the larger coefficient of friction leads to fewer revolutions before the decaying oscillation.

EXERCISES 7.4

1. a. When $-1 < x < 1$, we have $x^2 < 1$, so $x^2 - 1 < 0$. Because ε is a positive parameter, we see that $\varepsilon(x^2 - 1) < 0$ for $-1 < x < 1$.

b. When $|x| > 1$ (that is, when $x < -1$ or $x > 1$), we have $x^2 > 1$, so $x^2 - 1 > 0$. Because ε is a positive parameter, we see that $\varepsilon(x^2 - 1) > 0$ for $|x| > 1$.

3. a.

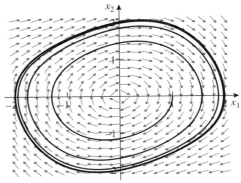

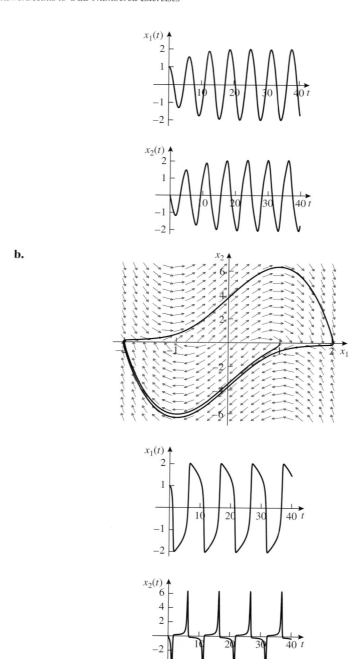

b.

c. Each trajectory indicates the existence of a stable limit cycle. However, the shapes of the trajectories and the limit cycles change as ε changes. Similarly, $x_1(t)$ and $x_2(t)$ are periodic but not trigonometric; and when ε changes from $\frac{1}{4}$ to 4, $x_1(t)$ changes to a flatter shape, whereas $x_2(t)$ develops spikes.

5. a. $r(t) = 1/\sqrt{1 + Ce^{2t}}$

   b. $\theta(t) = t + K$

   c. $x(t) = \dfrac{\cos(t + K)}{\sqrt{1 + Ce^{2t}}}, \ y(t) = \dfrac{\sin(t + K)}{\sqrt{1 + Ce^{2t}}}$

   d.

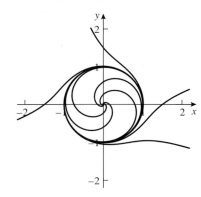

7. b. The stable limit cycle is the unit circle, $x^2 + y^2 = 1$.

9. a. $(0, 0)$ and $(-2, -1)$

   b.

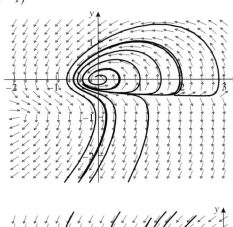

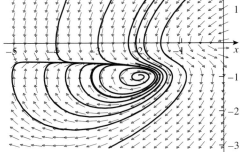

   c. There seems to be an unstable limit cycle around $(0, 0)$ and a stable limit cycle around $(-2, -1)$.

11. $\dfrac{\partial f}{\partial x} + \dfrac{\partial g}{\partial y} = (1 + 2y + 3x^2) + (-2y + x^2) = 1 + 4x^2 \geq 1 > 0$

13. $\dfrac{\partial f}{\partial x} + \dfrac{\partial g}{\partial y} = (-2 - \sin y) + (-3x^2y^2) < 0$

15. $\dfrac{\partial f}{\partial x} + \dfrac{\partial g}{\partial y} = (12 + 2xy - 3x^2) + (14 - 2xy - 3y^2)$

$\qquad\qquad = 26 - 3(x^2 + y^2) \geq 26 - 3(8) = 2 > 0$

INDEX